Advanced Turbulent Combustion Physics and Applications

Explore a thorough and up to date overview of the current knowledge, developments, and outstanding challenges in turbulent combustion and application. The balance among various renewable and combustion technologies is surveyed, and numerical and experimental tools are discussed along with recent advances. The book covers combustion of gaseous, liquid, and solid fuels and subsonic and supersonic flows. This detailed insight into the turbulence–combustion coupling with turbulence and other physical aspects, shared by a number of the world's leading experts in the field, makes this an excellent reference for graduate students, researchers, and practitioners in the field.

N. Swaminathan is Professor of Mechanical Engineering at the University of Cambridge and Director of Studies and Fellow at Robinson College, Cambridge. He has been a fellow of The Combustion Institute since 2018, Member of the Engineering and Physical Sciences Research Council College of Peer Reviewers, holds visiting professorships in many overseas universities, and consults to a number of industries in the transport and energy sectors.

X.-S. Bai is Professor of Fluid Mechanics at Lund University, Lund, Sweden. He has been a fellow of The Combustion Institute since 2018.

N. E. L. Haugen is a senior research scientist at SINTEF Energy Research, Trondheim, Norway.

C. Fureby is Professor of Heat Transfer at Lund University, Lund, Sweden. He is an associate fellow of the American Institute of Aeronautics and Astronautics, and a member of The Combustion Institute.

G. Brethouwer is a senior researcher at FLOW, Department of Engineering Mechanics, KTH Royal Institute of Technology, Stockholm, Sweden.

Advanced Turbulent Combustion Physics and Applications

Edited by

N. SWAMINATHAN
University of Cambridge

X.-S. BAI
Lund University, Sweden

N. E. L. HAUGEN
SINTEF Energy Research

C. FUREBY
Lund University, Sweden

G. BRETHOUWER
KTH Royal Institute of Technology, Stockholm

CAMBRIDGE UNIVERSITY PRESS

CAMBRIDGE
UNIVERSITY PRESS

University Printing House, Cambridge CB2 8BS, United Kingdom

One Liberty Plaza, 20th Floor, New York, NY 10006, USA

477 Williamstown Road, Port Melbourne, VIC 3207, Australiav

314–321, 3rd Floor, Plot 3, Splendor Forum, Jasola District Centre, New Delhi – 110025, India

103 Penang Road, #05–06/07, Visioncrest Commercial, Singapore 238467

Cambridge University Press is part of the University of Cambridge.

It furthers the University's mission by disseminating knowledge in the pursuit of education, learning, and research at the highest international levels of excellence.

www.cambridge.org
Information on this title: www.cambridge.org/9781108497961
DOI: 10.1017/9781108671422

© Cambridge University Press 2022

This publication is in copyright. Subject to statutory exception and to the provisions of relevant collective licensing agreements, no reproduction of any part may take place without the written permission of Cambridge University Press.

First published 2022

Printed in the United Kingdom by TJ Books Limited, Padstow Cornwall

A catalogue record for this publication is available from the British Library.

Library of Congress Cataloging-in-Publication Data
Names: Workshop on Physics of Turbulent Combustion (2016 : Stockholm, Sweden) | Swaminathan, Nedunchezhian, editor. | Nordisk institut for teoretisk fysik, sponsoring body.
Title: Recent advances in turbulent combustion / edited by N. Swaminathan, X.-S. Bai, N.E.L. Haugen, C. Fureby and G. Brethouwer.
Description: New York : Cambridge University Press, 2021. | Report of the month-long Workshop on Physics of Turbulent Combustion, held September 2016 at the Nordic Institute for Theoretical Physics (Nordita) in Sockholm, sponored by Nordita and the Combustion Institute. | Includes bibliographical references and index.
Identifiers: LCCN 2020041947 (print) | LCCN 2020041948 (ebook) | ISBN 9781108497961 (hardback) | ISBN 9781108671422 (epub)
Subjects: LCSH: Combustion engineering–Congresses. | Turbulence–Congresses. | Combustion–Mathematical models–Congresses.
Classification: LCC TJ254.5 .W68 2021 (print) | LCC TJ254.5 (ebook) | DDC 621.402/3–dc23
  LC record available at https://lccn.loc.gov/2020041947
  LC ebook record available at https://lccn.loc.gov/2020041948

ISBN 978-1-108-49796-1 Hardback

Cambridge University Press has no responsibility for the persistence or accuracy of URLs for external or third-party internet websites referred to in this publication and does not guarantee that any content on such websites is, or will remain, accurate or appropriate.

Contents

List of Contributors	*page* xi
Preface	xiii

1 Introduction   1
N. Swaminathan, X.-S. Bai, G. Brethouwer, and N. E. L. Haugen

1.1	Role of Combustion Technology	6
1.2	Setting the Stage	8
1.3	Governing Equations	10
1.4	Aerothermodynamic and Constitutive Relations	11
1.5	Chemical Kinetic Relations	13
1.6	Direct Numerical Simulation	14
1.7	Large-Eddy Simulation	15
	1.7.1 LES Equations	15
	1.7.2 SGS Closures	17
	1.7.3 Challenges for LES	18
1.8	RANS Approach	19
1.9	Aims and Objectives	21
	References	22

2 Turbulent Flame Structure and Dynamics: Combustion Regimes   25
V. Sabelnikov, A. Lipatnikov, X.-S. Bai, and N. Swaminathan

2.1	Historical and Physical Perspective of Turbulent Combustion	26
	V. Sabelnikov, A. Lipatnikov, X.-S. Bai, and N. Swaminathan	
	2.1.1 Regime Classification for Non-premixed Flames	26
	2.1.2 Regime Classification of Turbulent Premixed Flames	28
	2.1.3 Shetinkov's Microvolume Regime of Turbulent Premixed Combustion	32
	2.1.4 Propagation of Thin Reaction Zone in Intense Turbulence	40
2.2	Direct Numerical Simulation Perspective	44
	H. Im, J. H. Chen, and A. Aspden	
	2.2.1 DNS of Canonical Configurations	45
	2.2.2 DNS of Complex Flames	51

	2.3	Experimental Perspective and Challenges	54
		B. Zhou, J. H. Frank, B. Coriton, Z. Li, M. Alden, and X.-S. Bai	
		2.3.1 Simultaneous Multiscalar Visualization of Turbulent Premixed Flames	55
		2.3.2 Advances in Dimensionalities High-Speed Diagnostics	65
		2.3.3 Concluding Remarks	83
		References	84
3	**Premixed Combustion Modeling**	100
	C. Dopazo, N. Swaminathan, L. Cifuentes, and X.-S. Bai	
	3.1	Introduction	100
	3.2	Phenomenological Models	103
		3.2.1 Eddy Breakup Model	103
		3.2.2 Thickened Flame Model	104
		3.2.3 Linear Eddy Model	105
	3.3	Geometrical Models	106
		3.3.1 FSD Approach	106
		3.3.2 G-Equation Model	111
	3.4	Statistical Models	115
		3.4.1 Transported PDF	115
		3.4.2 Bray–Moss–Libby Model	118
		3.4.3 Conditional Moment Closure Approach	121
		3.4.4 Tabulated Chemistry Approach	125
	3.5	Improvements for the Flamelets: *FlaRe* Approach	126
	3.6	Turbulent Mixing in Premixed Combustion Revisited: Some Fundamental Considerations	133
		3.6.1 Characteristic Mixing Times	133
		3.6.2 Length Scales: Spectral Analysis of a Dynamically Passive Scalar Field Undergoing a Linear Chemical Reaction	138
		3.6.3 Further Relationships Relevant to Modeling	144
		3.6.4 Remarks on the G-Equation	145
	3.7	Summary	146
		References	150
4	**Non-premixed and Partially Premixed Combustion Modeling**	162
	C. Fureby and X.-S. Bai	
	4.1	Introduction and Background	162
	4.2	Regime Diagrams for Turbulent Non-premixed and Partially Premixed Combustion	163
	4.3	Theoretical Models of Non-premixed and Partially Premixed Combustion	166
		4.3.1 Premixed Combustion	166

		4.3.2	Non-premixed Combustion	167
		4.3.3	Partially Premixed and Stratified Combustion	169
	4.4	Modeling of Non-premixed and Partially Premixed Combustion		171
		4.4.1	Flamelet Models	171
		4.4.2	Thickened Flame Model	173
		4.4.3	Localized Time Scales Combustion Models	174
		4.4.4	Transported PDF LES Combustion Models Based on Eulerian Stochastic Fields Models	176
		4.4.5	Conditional Moment Combustion Model	177
		4.4.6	Linear Eddy Combustion Model	179
	4.5	Discussion		181
	4.6	Case Studies		182
		4.6.1	The KAUST Diffusion Flame Burner	182
		4.6.2	Stratified Flames: The Lund Low-Swirl Burner	185
		4.6.3	Non-Premixed and Partially Premixed Flames: The Turchemi Model Combustor	187
		4.6.4	Non-Premixed And Partially Premixed Flames: Prediction of Thermoacoustically Unstable Flames	190
	4.7	Concluding Remarks		192
	References			193
5	**Chemical Kinetics**			200
	E. J. K. Nilsson, C. Fureby, and A. Aspden			
	5.1	Introduction		200
	5.2	Combustion Chemistry		202
		5.2.1	Chemical Kinetics	203
		5.2.2	Global Combustion Characteristics	207
	5.3	Chemical Kinetic Mechanisms		212
		5.3.1	Complete or Detailed Mechanisms	213
		5.3.2	Skeletal Mechanisms	214
		5.3.3	Reduced Mechanisms	215
		5.3.4	Global Mechanisms	217
		5.3.5	Mathematical Stiffness	218
	5.4	Mechanisms for Natural Gas, Heavy Liquid Fuels, and Ethanol		218
		5.4.1	Natural Gas and/or Methane	218
		5.4.2	Heavy Liquid Fuels	220
		5.4.3	Alcohols	220
	5.5	Incorporating Chemistry in Combustion Modeling		222
		5.5.1	Turbulence–Chemistry Interactions in LES	222
		5.5.2	The Turchemi Combustor: Methane	222
		5.5.3	The Volvo Bluff Body Combustor: Propane	224
		5.5.4	A Supersonic Cavity Stabilized Combustor: Ethylene (and Hydrogen)	229

	5.6 Future Directions	232
	References	233
6	**MILD Combustion**	240
	Y. Minamoto, N. A. K. Doan, and N. Swaminathan	
	6.1 Introduction	240
	6.2 Definitions of MILD Combustion	244
	6.3 Investigations Using Zero- and One-Dimensional Model Reactors	247
	6.3.1 Well-Stirred Reactor	247
	6.3.2 Counterflow Flame	248
	6.3.3 Plug Flow Reactor	250
	6.3.4 Insights from Laminar Calculations	251
	6.4 Past Experimental Explorations	251
	6.4.1 Typical Configurations to Achieve MILD Combustion	251
	6.4.2 Structure and Identification of Reaction Zones	254
	6.5 DNS of MILD Combustion	257
	6.5.1 Reaction Zone Shape and Structure	257
	6.5.2 Are There Flames?	263
	6.5.3 Comments on Markers	265
	6.6 Modeling of MILD Combustion	266
	6.6.1 RANS Calculations	267
	6.6.2 LES	270
	6.7 Potential Applications and Future Outlook	271
	References	273
7	**Supersonic Combustion**	281
	A. Mura and V. Sabelnikov	
	7.1 Introduction and Background	281
	7.2 Supersonic Reactive Flows: Governing Equations	284
	7.3 Steady Premixed Combustion Waves in High-Speed Flows	290
	7.3.1 Situations with Released Energy Below the Critical Value	291
	7.3.2 Situations with Released Energy Above the Critical Value	297
	7.3.3 Fast Compressible Flames and Their Transition to Detonation	297
	7.4 Influence of Wall Friction and Heat Transfer	298
	7.5 Thermal Choking in Constant Cross Section Area Channel	301
	7.6 Stability Analysis in the Vicinity of Thermal Choking Conditions	303
	7.6.1 Constant Cross Section with Wall Friction and Heat Exchange	303
	7.6.2 Variable Cross Section: Divergent Channel	306
	7.7 Turbulent Mixing in Compressible Flows	307
	7.8 Turbulent Combustion in High-Speed Flows	314
	7.9 Current Challenges and Future Research Needs	319
	References	320

8 Liquid Fuel Combustion — 328
Z. Bouali, A. Mura, and J. Reveillon

- 8.1 Two-Phase Flow Topology and Spray Statistics — 330
- 8.2 Mathematical Framework and Description of Two-Phase Flows — 336
- 8.3 Modeling Issues Relevant to Evaporation and Combustion — 340
  - 8.3.1 Description of the Gaseous Phase — 340
  - 8.3.2 Description of the Liquid Phase — 341
  - 8.3.3 Boundary Conditions — 341
  - 8.3.4 Eulerian/Lagrangian Couplings — 349
- 8.4 Two-Phase Flow Turbulent Combustion Regimes and Diagrams — 351
  - 8.4.1 Genesis of Two-Phase Flow Combustion Diagrams — 351
  - 8.4.2 Spray Flame Structures — 352
  - 8.4.3 Combustion Diagrams — 354
- 8.5 Conclusion — 363
- References — 363

9 Solid Fuel Combustion — 367
N. E. L. Haugen, K. Umeki, M. Liberman, I. Rogachevskii, and F. Picano

- 9.1 Introduction — 367
  - 9.1.1 Governing Equations — 367
  - 9.1.2 Thermal Conversion Reactions — 368
- 9.2 The Effect of Turbulence on the Heterogeneous Conversion of Powders — 370
  - 9.2.1 Velocity-Induced Mass Transfer Increase — 371
  - 9.2.2 Cluster-Induced Mass Transfer Decrease — 372
- 9.3 Radiation-Induced Mechanism of Unconfined Dust Explosions — 374
  - 9.3.1 Effect of Turbulent Clustering of Dust Particles on Radiative Heat Transfer — 375
  - 9.3.2 Radiation-Induced Secondary Explosions — 378
- 9.4 Intraparticle Transport Phenomena in Solid Fuel Combustion — 380
  - 9.4.1 Time-scale Analyses — 380
  - 9.4.2 Resolved Particle Models — 382
  - 9.4.3 Effect of Thermal Conduction on the Devolatilization of Biomass — 383
  - 9.4.4 Simplified Models and Application in Burner Simulation — 385
- 9.5 Turbulent Transport of a Dispersed Phase with Implications for Combustion — 386
  - 9.5.1 Dispersed Solid Phase in Turbulent Flows: Coupling Mechanisms — 386
  - 9.5.2 Preferential Transport of Solid Dispersed Phases — 388
- 9.6 Final Remarks and Perspectives — 391
- References — 391

10 Challenges in Practical Combustion — 396

10.1 Stationary Gas Turbine Combustion Challenges — 396
D. Lörstad

- 10.1.1 Combustor Design Process — 399
- 10.1.2 Future Simulation Efficiency — 402
- 10.1.3 Challenges for Stationary Gas Turbines — 405
- 10.1.4 Challenges for Combustion Prediction — 406

10.2 Aero-engine Combustor Design Methods: Approach and Challenges — 407
M. Zedda

- 10.2.1 The Role of Low-Order Methods in Combustor Design — 410
- 10.2.2 The Role of High-Order Methods in Combustor Design — 411
- 10.2.3 CFD for System Aerodynamics — 411
- 10.2.4 CFD for Fuel Injector Design — 413
- 10.2.5 CFD for Temperature Traverse — 415
- 10.2.6 CFD for Emissions Ranking — 419
- 10.2.7 CFD and Finite-Element Analyses of Metal Temperature — 423
- 10.2.8 Low- and High-Order Methods for Thermoacoustics — 426
- 10.2.9 Relight and Extinction Methods — 430
- 10.2.10 Fuel Coking Methods — 431
- 10.2.11 The Role of Spray Modeling — 433
- 10.2.12 Trends in Aero-engine Combustor CFD — 435

10.3 Internal Combustion Engines — 436
D. Norling and X.-S. Bai

- 10.3.1 Combustion Concepts in ICE — 437
- 10.3.2 Diesel Engine Combustion Chamber Design — 439
- 10.3.3 The Combustion Process — 441
- 10.3.4 Gas Motion and Turbulence in the Combustion Chamber — 444
- 10.3.5 Challenges in Diesel Combustion Measurements — 449
- 10.3.6 Concluding Remarks — 451

References — 453

11 Closing Remarks — 460
X.-S. Bai, N. E. L. Haugen, C. Fureby, G. Brethouwer, and N. Swaminathan

Index — 464

Contributors

M. Alden
Lund University, Lund, Sweden

A. Aspden
Newcastle University, Newcastle upon Tyne, UK

X.-S. Bai
Lund University, Lund, Sweden

Z. Bouali
ISAE-ENSMA, Poitier, France

G. Brethouwer
KTH Royal Institute of Technology, Stockholm, Sweden

J. H. Chen
Sandia National Laboratory, Livermore, CA, USA

L. Cifuentes
Universität Duisburg-Essen, Duisburg, Germany

B. Coriton
Sandia National Laboratory, Livermore, CA, USA

N. A. K. Doan
Engineering Department, Cambridge University, Cambridge, UK

C. Dopazo
University of Zaragoza, Zaragoza, Spain

J. H. Frank
Sandia National Laboratory, Livermore, CA, USA

C. Fureby
Lund University, Lund, Sweden

N. E. L. Haugen
SINTEF Energy Research, Trondheim, Norway

H. Im
KAUST, Thuwal, Saudi Arabia

Z. Li
Lund University, Lund, Sweden

M. Liberman
NORDITA, Stockholm, Sweden

A. Lipatnikov
Chalmers University of Technology, Gothenburg, Sweden

D. Lörstad
Siemens, Finspang, Sweden

Y. Minamoto
Tokyo Institute of Technology, Tokyo, Japan

A. Mura
ISAE-ENSMA, Poitier, France

E. J. K. Nilsson
Lund University, Lund, Sweden

D. Norling
Scania, Sweden

F. Picano
University of Padova, Padova, Italy

J. Reveillon
CORIA, Rouen, France

I. Rogachevskii
Ben Gurion University of the Negev, Beersheba, Israel

V. Sabelnikov
ONERA, Palaiseau, France

N. Swaminathan
Engineering Department, Cambridge University, Cambridge, UK

K. Umeki
Luleå University of Technology, Luleå, Sweden

M. Zedda
Aerothermal Methods, Rolls-Royce, London, UK

B. Zhou
Sandia National Laboratory, Livermore, CA, USA

Preface

Turbulent combustion is ubiquitous in practical systems used for generating power for land, sea, and air transports and also for electricity production. This socioeconomically important topic is multidisciplinary and rich in physics. Although the use of renewable and sustainable energy sources is gaining popularity, the importance of turbulent combustion cannot be underestimated, especially in view of the current outcry on energy demand, environmental impacts, and climate change. This is because more than 90% of the world's total primary energy supply comes from fossil or biofeul combustion. Renewable technologies are being developed to reduce this combustion share, which is to be welcomed, but the rate of replacing the fossil fuel technology is nowhere near what one would like to have. The replacement of combustion by renewable and sustainable sources and their technologies can be accelerated but it is important to realize that the energy required to produce renewable energy systems is likely to come from fossil fuels. One must therefore consider the environmental impacts of activities related to supplying materials required to produce these systems. Hence, an accelerated growth of renewable technologies can leave its own footprint on the environment and one must carefully consider the period needed to offset the effects of this footprint. This means that improving combustion technologies will help enormously in mitigating global warming and climate change, at least for the next several decades to come, and also in finding avenues for carbon-neutral or carbon-free energy technologies.

In view of all of these considerations, the topic of turbulent combustion becomes important, and major advances in our understanding of this topic have been achieved only in the past 60 years or so, which is not long compared to hundreds of years spent on other areas of physics and chemistry. The advent of computing technologies and laser diagnostics has helped us to further our understanding of the subtleties of this complex topic and to find avenues for "greener" (highly efficient with ultralow emissions) combustion along with the scientific and technological challenges involved. The broader objective of this volume is to bring together the recent scientific and technological advances in turbulent combustion and to identify further challenges involved in achieving "greener" combustion. We have tried to keep the discussions on a physical basis while drawing attention to the detail required.

The introduction given in Chapter 1 tries to identify the role of combustion technologies with a view to promoting renewable technologies. This is followed by a brief introduction to the standard conservation principles and equations required to study turbulent combustion. The various perspectives obtained using theoretical, direct numerical simulation, and laser diagnostic tools in the past are discussed in Chapter 2. The modeling of premixed and non-premixed turbulent combustion of gaseous fuels is discussed in Chapters 3 and 4. The chemical kinetics of combustion is presented in Chapter 5. The various physical aspects of combustion in vitiated flow are discussed in Chapter 6. The intricacies of supersonic combustion, which is a potential candidate for high-speed long-haul flights, are explained in Chapter 7. The various physical aspects involved in the combustion of liquid and solid fuels are presented in Chapters 8 and 9 respectively. The practical challenges in gas turbines and internal combustion engines are discussed in Chapter 10 and the final chapter provides a concluding summary. Ample references are given in each chapter for further reading.

This work is based on various discussions that took place during a month-long *Turbulent Combustion Physics* workshop held in the late summer of 2016 at NORDITA and KTH, Stockholm, Sweden. The funding from NORDITA is gratefully acknowledged and we would also like to note that almost all of the contributors to this volume were participants in the workshop and we thank them for their contributions and dedication in the preparation of this volume.

1 Introduction

N. Swaminathan, X.-S. Bai, G. Brethouwer, and N. E. L. Haugen

Combustion is a socioeconomically important topic in view of its role in civilization for many millenia, with the energy from its reactions used to produce bronze, iron, pottery, and so forth; cook food; produce mechanical power for transportation vehicles; and generate electricity. It is still the case even now, since more than 90% of the world's total primary energy supply (TPES) is met through combustion of fossil and biofuels and waste [1]. Fossil fuels include coal, oil, and natural gas. Shale gas also belongs to this category although its contribution is currently very small. Figure 1.1 shows the share of various fuels for TPES in terms of Mtoe (million tons of oil equivalent). There is 47 GJ of energy per ton of oil, which translates to about 647 EJ of energy for the year 2016 and about 287 EJ for 1973. There is an about 125% increase over this period of 43 years, suggesting a nearly 2.9% increase per year. This is in line with an estimate from the National Academies of Sciences, Engineering, and Medicine of an increase in the global energy consumption of an increase by about 40% in the next two decades [2]. In reality, this rate of energy consumption may increase further because of the recent surge in the "energy-hungry" technologies associated with consumer electronics such as smartphones, smart televisions, and so forth, and also the upcoming technologies such as Internet of things (IoTs), 5G networks required to meet the communication speeds needed for autonomous vehicles such as self-driving cars, and so forth. On the one hand, we all like to have these technologies without asking whether they are needed or not. On the other hand, we decry and worry about the environmental impacts resulting from anthropogenic sources leading to crises such as global warming affecting the weather pattern, agriculture and life on the planet. These two, the energy-hungry modern technologies and mitigation of global warming, are at opposite ends. Bringing them together is a grand challenge, and carefully considered, fully analyzed, and completely evaluated solutions are needed. Otherwise, these technologies may accelerate global warming further. The reports from the Intergovernmental Panel on Climate Change (IPCC) estimate that the global temperature will rise in the next 100 years [3, 4]. If the greenhouse gas (GHG) emissions are low (RCP 2.6[1]), then

---

[1] Representative Concentration Pathways. These are proposed by IPCC to have a common basis to compare the climate models used across the world. RCP 2.6 suggests that the GHG emissions in gigatons of carbon (GtC) will be close to zero in the year 2100 and the CO_2 concentration in the atmosphere will be about 400 ppm. Under this scenario, the temperature rise because of the radiative loading on the planet ranges from 0.3 to 1.7°C. For the RCP 8.5 scenario, the CO_2 emission is about 30 GtC, giving about 950 ppm of CO_2 concentration in the atmosphere in the year 2100.

the temperature rise will be about 0.3 to 1.7°C and if the GHG emissions are high (RCP 8.5) then the temperature rise could be about 2.6 to 4.8°C. These estimates are relative to 1986–2005 data.

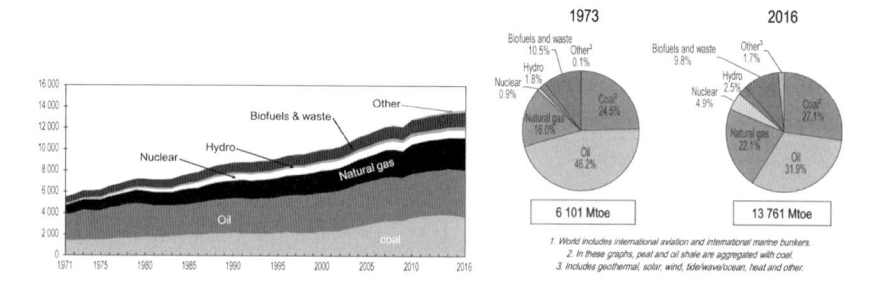

Figure 1.1 World total primary energy supply in Mtoe, million tons of oil equivalent, by fuel type. Adapted from [1], © IEA, 2018.

To reduce the rate of GHG emission and thereby the rate of global warming, energy production using renewable and sustainable sources is gaining popularity and has become widespread since the past decade. The renewable sources include hydro, solar, wind, and tidal. Including nuclear power among renewables is debatable. There are arguments to include it in the list of renewables for the following two reasons: (1) uranium deposits could provide energy for as long as the relationship between Earth and the sun is expected to last (about 5 billion years) [5] and (2) it does not emit GHGs while satisfying the increasing demand for energy in the world [6, 7]. The arguments to exclude it from the list of renewable energies are based on safety issues and the concept of clean energy. Figure 1.1 shows the nuclear energy had an about 5% share of the world TPES in 2016 whereas the contribution of renewables, excluding hydropower and biofuels, was only 1.7%. However, there is a substantial increase from 0.1% in 1973 for the renewables because of the advancement in the associated technologies in the past couple of decades. Both rooftop and commercial solar photovoltaic (PV) systems have become popular but the capital cost projections reported in [8] (see their figure 4.1), about £1000 per kilowatt for 2019, seems to be about half of the actual cost in the year. Also, the efficiency of the solar panel is only about 22%, which is improving with time.

If one considers the levelized cost of electricity (LCOE)[2] from utility-scale power generation using renewable technologies in the period from 2010 to 2017 then it becomes clear that these technologies are becoming competitive with the traditional power technologies using fossil fuels, with a cost ranging from about 0.05 to 0.18 USD per kilowatt-hour [9]. This comparison is shown in Fig. 1.2 for various renewable technologies. Biomass, geothermal, hydro, and onshore wind technologies are becoming highly competitive with prices ranging from 0.05 to 0.07 USD per kilowatt-hour. Both PV and concentrating solar power (CSP) technologies are becoming less expensive

---

[2] This is projected life cycle cost and is calculated as the ratio of costs for (investment + operation + maintenance + fuel) to the electricity energy generated.

and their LCOE is projected to range between 0.04 and 0.06 USD per kilowatt-hour for the year 2022 [9]. It is also possible that the LCOE for renewables will undercut that of fossil fuel-fired technologies. These trends and projections are encouraging, giving us hope to mitigate the global warming issues in due course. Nevertheless, one must recognize that there is no "free lunch" – because there are economical and environmental implications and costs associated with the renewable technologies as well. The typical *payback* period that is quoted for these technologies, for example 5 to 7 years for solar PV, is based purely on the initial investment cost for the system but one needs to consider the payback period for the environmental impact. For example, one can ask how long does it take to offset the GHG emissions produced in creating and constructing the renewable systems?

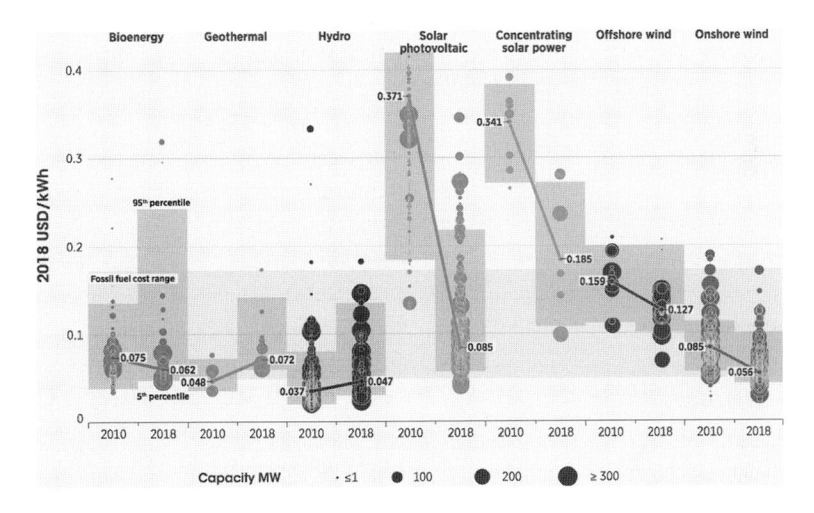

Figure 1.2 Global levelized cost of electricity for utility-scale renewable power technologies for the period from 2010 to 2018. Reproduced from [9], © IRENA, 2018. The diameter of the circle represents the size of the plant and the global averaged cost is shown by the line.

Careful considerations of these implications bring up two questions [10]: (1) Is there sufficient energy production, required as per the TPES data in Fig. 1.1, using renewable technologies to meet the total world primary energy demand? (2) What are the environmental impacts in terms of kilograms of CO_2 equivalents per megawatt-hour of electricity produced using the renewable technologies? The data shown in Fig. 1.3 provide some answers to the first question. The second question is somewhat controversial and needs a careful and honest full life cycle analysis/assessment (LCA) for each alternative technology.

The projections shown in Fig. 1.3 suggest that there will be about 1 TW of electricity power produced using solar and wind technologies in the year 2020. Using a conservative estimate of about 12 hours per day of operation gives about 4380 TWh of energy, which is equivalent to 376.8 Mtoe. The data collected from 2010 to 2018 reported in [11] suggest that there will be about 1.5 TW of electricity produced using other renewable technologies, which gives 3 Mtoe if one takes 24 hours (again a conservative estimate) of operation per day. The data in Fig. 1.1 and the projected 3% rise yield TPES required in the year 2020 of about 15,412.32 Mtoe. These estimates show that the supply of renewables is only about 2.5% of the world TPES. Hence, the answer to our first question is clear.

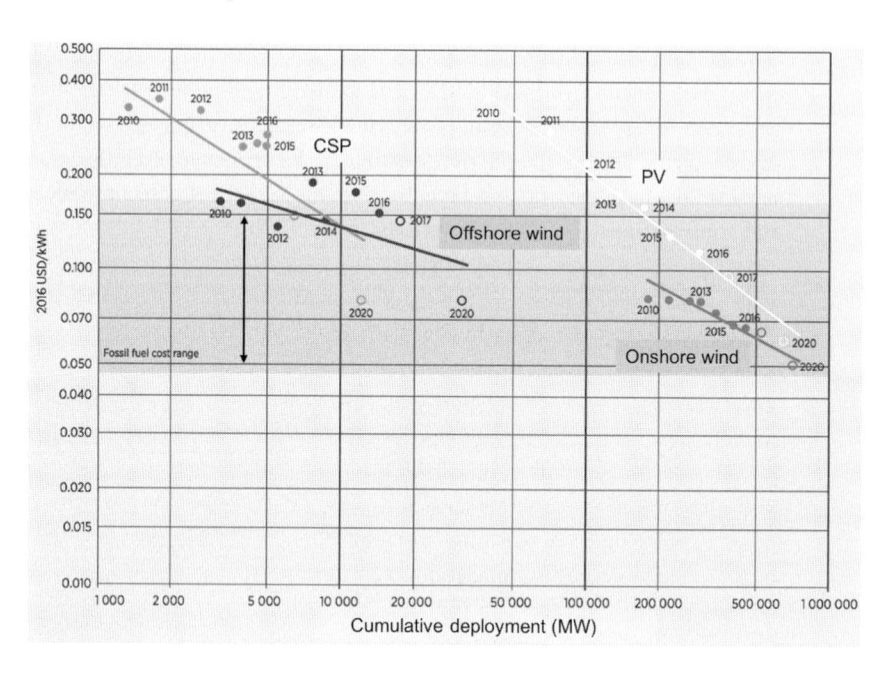

Figure 1.3 Variation of global averaged LCOE for solar, CSP and PV, and wind, onshore and offshore, technologies. Reproduced from [9], © IRENA, 2018.

To find a meaningful answer to the second question, one needs a careful LCA for all renewable technologies. This assessment must include the energy spent and GHGs emitted in *upstream*, operation, and *downstream* stages of the technology [12]. The upstream stage includes raw material extraction, manufacturing of construction materials, components of the power plant, and construction of the power plant. This stage includes the fuel cycle – extraction of fuel materials, their processing or conversion, and their delivery to the power plant side. The contribution from the fuel cycle stage is zero for most of the renewable technologies except for biofuels, biomass, and biowaste technologies. The operation stage includes combustion (for conventional technologies), plant operation, and maintenance. The downstream stage involves dismantling and decommissioning of the plant and disposal and recycling of the equipment and components. This type of assessment can help to determine energy and environmental burdens from *cradle to grave* and facilitate more consistent comparisons of energy technologies for making investment and policy decisions [12]. However, it must be recognized that the LCA can be quite difficult because of the scatter in the required data available in the literature. The scatter arises from the assumptions and approximations invoked, method of analysis employed, and so forth. Hence, it is a common practice to employ a procedure called harmonisation to reduce the variability in the required data (GHG emissions or energy spent) and details can be found in [13, 14]. The estimates of GHG emissions in grams of CO_2 equivalent per kilowatt-hour of electricity energy produced using both renewable and nonrenewable technologies. There are some differences in the data from these two studies but there is a general agreement in the trend. The maximum values reported in [15] seem to correspond to the median values reported in [12], shown in Fig. 1.4b and 1.4a respectively. Switching from coal to natural gas reduces the GHG emissions by nearly 50%, which is well known today. The emission levels of waste treatment and biomass technologies are comparable to those of natural gas because these technologies involve combustion in one form or another. However, the emission level shown in Fig. 1.4a for the biopower from the IPCC report [12] is substantially different (lower) compared to the estimate in [15] shown in Fig. 1.4b. A similar observation is also made for PV systems between these two studies. The data in Fig. 1.4b show that the life cycle GHG emissions for PV systems about 60% of those for natural gas technology. It seems that there is still a great deal of variation and disparity in these types of data coming from various studies and thus one needs to be cautious in drawing conclusions. One can hope that these differences are likely to decline with time as more careful investigations in the future will provide further data. Nevertheless, one can see that renewable technologies can help to mitigate global warming. However, the current trend and projected capacities of these technologies are inadequate to meet the world total primary energy demand (shown by the TPES data in Fig. 1.1), as discussed earlier. It is probable that fossil fuel–derived energy is required for constructing renewable technologies. So, a utopian solution would be to reduce energy usage by every individual, which would require social engineering and drastic energy policies but these are impractical for political and personal reasons.

A heavily accelerated introduction of renewables to replace fossil-fuel technologies may sound like a plausible solution but the GHG emissions from the upstream stage of LCA and various activities (e.g., mining to meet the huge demand for resources, the required energy coming from fossil fuel sources) associated with this stage may cause irreversible changes to the planet. Alternatively, a gradual (an appropriate question to ask here would be at what rate?) shift toward renewables while improving the technologies for nonrenewable may be a more pragmatic way forward. Combustion-based technologies are likely to play a central role for applications such as transports requiring high energy densities but the form and type of combustion are likely to change for future applications.

Electric vehicles (EVs) are projected to have zero emission but this depends on where the boundary for the analysis is drawn. The LCA of this technology shows that one has to drive an EV of Nissan LEAF size for about 35,000 km (about 3 years) to offset the CO_2 emissions coming from just the battery pack production part in the upstream stage of the LCA. It is important to note that the GHG emissions coming from the production of a vehicle or the electricity required to charge the battery pack for 35,000 km of driving are not included. Also, this amount of CO_2 is equivalent to producing 6400 liters of gasoline and the amount of energy spent to produce the

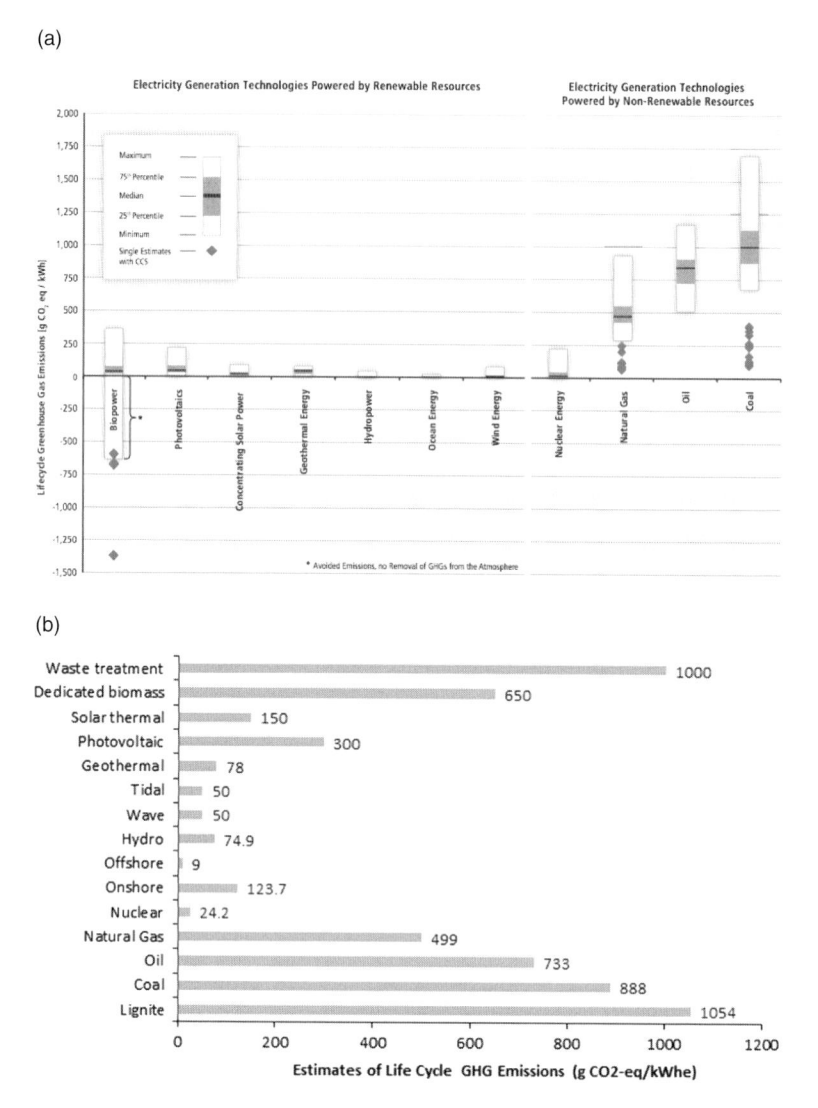

Figure 1.4 Estimates of GHG emissions in g CO_2-eq/kWh for various electricity generation methods. (a) A broad range with median reported in [12]. ©Cambridge University Press, 2011. (b) Maximum levels as reported in [15]. Reproduced with permission, © Elsevier, 2014.

battery pack is equivalent to 8500 liters of gasoline delivered to the tank. The details of this analysis, known as *cave to case* (CtC), can be found in [16].

1.1 Role of Combustion Technology

Global warming is truly a global problem and hence a globally agreed upon solution is going to be more effective than countrywise measures. Thus, concerted efforts across various nations (more aptly continents) are required so that GHG emissions across

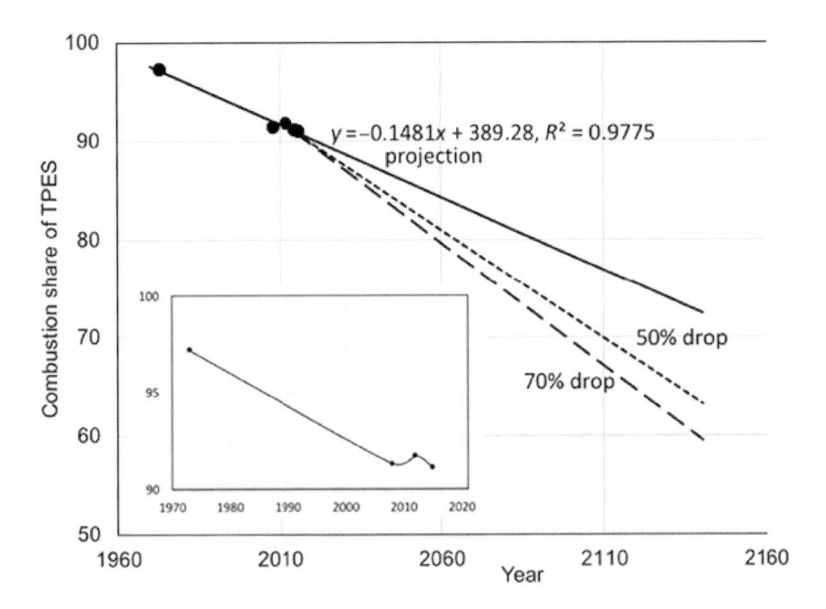

Figure 1.5 Combustion share of world TPES and its future projections.

the planet are reduced. While a complete shift toward renewable energies sounds attractive, it does not seem to be plausible in a short timescale set by various governments across the globe. These accelerative shifts, although desired in view of global warming mitigation, may likely accelerate global warming because the additional energy required to increase the TPES share of renewables (related to manufacturing and construction of these systems) has to come from nonrenewables. Thus, one needs a balanced approach to meet the ever increasing energy demands without aggravating global warming. Combustion technologies play an important role in this respect, which is clear from the results in Fig. 1.5 showing future projections of the potential combustion share of world TPES under three different scenarios [17]. The inset is the actual data from the International Energy Agency [1] showing a gradual decrease in the share of combustion. A small rise in 2012 is attributable to the increase in coal combustion in some countries around the world in that year. If one naively projects these data by assuming that the progress in renewable technologies to replace combustion to meet the energy demand is steady and organic following the current trends then the combustion share is likely to be more than 70% even by the year 2110 (the solid line). The slope of this curve is related to the progress and advancement of alternative energy technologies. If one keeps an optimistic view for these technologies progressing at an about 50% faster pace compared to the current trend then the combustion share falls just below 70% by 2110. This share decreases to 67% by the year 2110 even if one assumes optimistically that the alternative technologies progress at a 70% faster pace. It seems that a radical paradigm shift is required for a significant reduction of the combustion share. Whether this is practical or not is an open question. A pragmatic approach to mitigate the impact of combustion on the environment is to seek alternative combustion concepts and technologies that can significantly reduce GHG emissions and can act as retrofits to the existing fossil and biofuel systems involving combustion. Fuel-lean and MILD (moderate, intense or low dilution) combustion concepts emerge as potential solutions, since they have potentials to deliver both low emissions and high efficiency. However, using them for practical applications brings their own challenges, as pointed out in the book edited by Swaminathan and Bray [18].

1.2 Setting the Stage

The discussion in the previous section identified that combustion science and technology will play an important role in meeting our future energy demand. The question is how can one minimize the impacts of combustion (of not only fossil fuels but also other fuels such as biomasses, biofuels, solid waste, etc.) on the environment, since combustion will emit oxides of carbon, hydrogen, nitrogen, and other elements present in the fuel. Many of these oxides are known to be hazardous and detrimental to the environment and also to life. For example, the aldehydes coming from biofuel combustion are toxic to both life and the environment. Hence, the challenging question in front of us is, *how can we control and minimize the formation of these oxides and chemicals in combustion processes?* Also, how do we use hydrocarbon fuels in the most efficient and environmentally responsible way to meet our ever-growing demand for energy? These questions will remain even if one moves away from the carbon economy completely or into a carbon-neutral economy in view of sustainability. Advancing combustion science is imperative to find answers to these challenging questions and it is not quite right to say that the intricacies in this field are well understood. If that is the case then we should be able to build "green combustion systems" (with high efficiency and low impact on the environment) and one knows that we are quite far from this!

Turbulent combustion in many practical systems is complicated for a number of reasons such as complex geometries, multiple phases involved and energy exchange among these phases, operating conditions involving elevated temperature and pressure, and also radiative heat exchange between hot gases and the walls of the system. The reactivity and chemical pathways at elevated temperature and pressure conditions can yield unexpected behaviors compared to normal temperature and pressure conditions as noted in [19]. For example, the laminar flame speed can increase with pressure for elevated temperature while it is expected to decrease with pressure based on studies conducted at normal temperature. The chemical pathways and their attributes and significance can change at elevated temperature and pressure, leading to unexpected behavior. Hence one needs to be cautious in extending our knowledge and wisdom to new operating conditions.

Radiative heat transfer is dealt with elaborately in many textbooks on heat transfer. So, this topic is discussed only briefly when and where required in this book and other complexities are covered to good extent. Liquid fuels have to be atomized and vaporized first whereas solid fuels undergo pyrolysis, producing combustible gases. Therefore, combustion typically occurs in the gas phase in almost all practical systems. The final phase of solid fuel combustion involves a heterogeneous reaction on the char surface because of oxygen transported to the surface through diffusion processes. Nevertheless, the fraction of heat release from the char oxidation is relatively small compared to that from the gas phase combustion. Also, the minerals in the solid fuels end up as ash in various forms. Handling all these complexities with scientific rigor is very challenging and typically some empirical approximations are made. A similar approach of employing empirical relations is also used for descriptions of liquid spray and atomization processes.

Turbulent combustion of gases is classically investigated by separating turbulence and combustion by assuming that the relevant combustion timescales are shorter than those for turbulence. This led to the development of the *flamelet* modeling approach, which has been discussed elaborately in many textbooks, for example, [20, 21]. Although the aforementioned separation is artificial, this approach is shown to work well for many practical applications. However, alternative theories are required to predict pollutants and their formation, since their timescales are not small enough to separate them from those for turbulence. So, many approaches such as transported probability density function and conditional moment closure were proposed in past studies and these approaches are being developed further to improve their performances. The flamelet and nonflamelet approaches have been developed for both non-premixed and premixed combustion, which are two classical limits studied extensively. Combustion in modern practical systems such as gas turbine engines and direct injection internal combustion engines belongs to neither of these two limits but falls within partially premixed combustion. This mode is more complex compared to the two classical modes because it involves a wide range of equivalence ratios and the fuel–air mixing and combustion proceed together. These aspects create further complexities, creating a close coupling among turbulence, mixing, and combustion. Thus, one needs to be cautious and careful while extending the learnings from the two classical modes of combustion to the partially premixed combustion. Also, combustion with vitiated air having low oxygen is employed for stationary gas turbines to reduce nitrogen oxides (NOx) emissions. Recent investigations using direct numerical simulation (DNS) methodology showed that the combustion with vitiated air has its own distinctive features while sharing some attributes of classical combustion modes. The propulsion devices for high-speed air transport applications involve combustion in a supersonic stream and it is likely that supersonic combustion may share some of the distinctive features of vitiated air combustion.

Three numerical approaches, namely DNS, large eddy simulation (LES), and Reynolds-averaged Navier–Stokes (RANS) calculation, are used to study turbulent combustion. These approaches involve different levels of detail, approximations, and modeling. The complete set of conservation equations is solved with no models using high-order numerical schemes in the DNS approach and further details are discussed in Section 1.6. In the LES approach discussed in Section 1.7, the contributions and effects of small scales, which are known as subgrid scales, are filtered out and modeled. The combustion and associated chemical reactions are typically small-scale phenomena and thus they need to be modeled. Appropriately averaged conservation equations are solved in the RANS approach discussed briefly in Section 1.8. These averaged equations need quite a large number of models and approximations, which are discussed elaborately in many past works; for example, see the books edited by Libby and Williams [22, 23] and the works in [18]. Thus, we aim to provide a discussion in view of LES of turbulent combustion, with related modeling, and physical insights. The importance of laser diagnostics cannot be understated in providing valuable information to improve our physical understanding of turbulent combustion and reliable statistics for model validations and developments. It is imperative that there be closer interactions among experimentalists, modelers, and theoreticians working in the field of combustion and related areas to find answers to the challenging questions such as those identified in the preceding text and to recognize the role of combustion in meeting the ever-increasing energy demand with minimal environmental impact.

1.3 Governing Equations

In this section, governing equations for combustion of gaseous fuels are discussed. These equations require additional source terms or interphase boundary conditions when combustion of liquid and solid fuels is considered. Details of the source terms or interphase boundary conditions in liquid and solid fuel combustion will be discussed in Chapters 8 and 9 respectively.

Combustion of gaseous fuels involves multiple constituents, say N_s different species. Some of these species will be consumed and some will be produced in chemical reactions. The turbulent combustion process is governed by three basic conservation laws: the conservation of mass, momentum, and energy. Transport equations for the species mass fraction, momentum, and energy can be derived from these basic conservation laws, as discussed in the text that follows.

Since different species in the system have different molecular weights, different species generally have different macroscopic velocities. This velocity is different from the microscopic velocity of each individual molecule. The macroscopic velocity of species α ($\alpha = 1, 2, 3, \ldots, N_s$) is an averaged quantity of many molecules and it is not directly related to the temperature as is the microscopic (molecule) velocity.

Let us write the velocity of species α as $\mathbf{u}_\alpha$. One may define a mass-weighted average velocity for the mixture as $\mathbf{u}$, viz.,

$$\mathbf{u} = \sum_{\alpha=1}^{N_s} Y_\alpha \mathbf{u}_\alpha, \tag{1.1}$$

where Y_α is the mass fraction of species α. The difference between velocity of species α and the mass-weighted average velocity of the mixture is known as the diffusion velocity of species α,

$$\mathbf{U}_\alpha = \mathbf{u}_\alpha - \mathbf{u}. \tag{1.2}$$

By definition,

$$\sum_{\alpha=1}^{N_s} Y_\alpha \mathbf{U}_\alpha = \sum_{\alpha=1}^{N_s} Y_\alpha \mathbf{u}_\alpha - \mathbf{u} \sum_{\alpha=1}^{N_s} Y_\alpha = 0, \tag{1.3}$$

since $\sum_{\alpha=1}^{N_s} Y_\alpha = 1$. By considering the mass conservation of species α, the conservation equation for Y_α can be derived,

$$\frac{\partial \rho Y_\alpha}{\partial t} + \nabla \cdot (\rho \mathbf{u} Y_\alpha) = \nabla \cdot \left(-\rho Y_\alpha \mathbf{U}_\alpha\right) + \rho \dot{\omega}_\alpha, \tag{1.4}$$

where $\rho \dot{\omega}_\alpha$ is the net formation rate of species α per unit volume. Summation of Eq. (1.4) over all the species gives the well-known continuity equation,

$$\frac{\partial \rho}{\partial t} + \nabla \cdot (\rho \mathbf{u}) = 0, \tag{1.5}$$

after using Eq. (1.3) and recognizing that $\sum_{\alpha=1}^{N_s} \dot{\omega}_\alpha = 0$ to conserve the mass.

The momentum conservation, also known as the Navier–Stokes, equation is

$$\frac{\partial \rho \mathbf{u}}{\partial t} + \nabla \cdot (\rho \mathbf{u}\mathbf{u}) = -\nabla p + \nabla \cdot \boldsymbol{\tau}, \tag{1.6}$$

where p is the pressure of the gas mixture and $\boldsymbol{\tau}$ is the viscous stress tensor. The conservation equation of energy can be written in terms of specific enthalpy of the mixture, h,

$$\frac{\partial \rho h}{\partial t} + \nabla \cdot (\rho \mathbf{u} h) = \frac{Dp}{Dt} - \nabla \cdot \mathbf{q} + \boldsymbol{\tau} : \nabla \mathbf{u} + Q_r, \tag{1.7}$$

where $\mathbf{q}$ is the heat flux vector and $D/Dt = \partial/\partial t + \mathbf{u} \cdot \nabla$ is the substantial or material derivative. The terms on the right-hand side (RHS), from left to right, of Eq. (1.7) denote the change of enthalpy due to compression or expansion, the enthalpy flux due to the diffusion of heat and mass, viscous dissipation, and thermal radiation including external heat addition, respectively. The specific enthalpy of the mixture (per unit mass) is defined as the mass-weighted average of specific enthalpies of all species,

$$h = \sum_{\alpha=1}^{N_s} Y_\alpha h_\alpha, \tag{1.8}$$

where h_α is the specific enthalpy of species α. The effects of body force, such as buoyancy, are neglected in Eqs. (1.6) and (1.7).

1.4 Aerothermodynamic and Constitutive Relations

To close the system of governing equations (1.4) to (1.7) certain aerothermodynamic relations are required. Generally, the gas mixture is taken to be an ideal gas and its equation of state relating the pressure, temperature, density, and species mass fractions is

$$p = \rho R_u T \sum_{\alpha=1}^{N_s} \frac{Y_\alpha}{W_\alpha}, \quad (1.9)$$

where W_α is molecular weight of species α, $R_u = 8.314$ kJ kmol^{-1}K^{-1} is the universal gas constant, and T is the mixture temperature. The relationship between the temperature and specific enthalpy is given by the calorific equation of state, written as

$$h_\alpha = h_{f,\alpha}^0 + \int_{T^0}^{T} c_{p,\alpha}(\theta) \, d\theta, \quad (1.10)$$

where $h_{f,\alpha}^0$ is the enthalpy of formation at standard reference temperature T^0. The constant pressure specific heat capacity of species α is $c_{p,\alpha}$ and it depends only on temperature for an ideal gas. The specific heat capacity of the gas mixture is defined as

$$c_p = \sum_{\alpha=1}^{N_s} Y_\alpha c_{p,\alpha}. \quad (1.11)$$

The species diffusion velocity, stress tensor, and heat flux vector in Eqs. (1.4), (1.6), and (1.7) are given by [24–27]

$$\mathbf{U}_\alpha = \frac{1}{X_\alpha} \sum_{j=1}^{N_s} \frac{Y_j}{X_j} \mathcal{D}_{\alpha j} \mathbf{d}_j - \frac{\mathcal{D}_\alpha^T}{\rho Y_\alpha} \nabla (\ln T), \quad (1.12)$$

$$\boldsymbol{\tau} = \mu \left[\nabla \mathbf{u} + (\nabla \mathbf{u})^T \right] - \left(\frac{2}{3} \mu - \kappa \right) \mathbf{I} (\nabla \cdot \mathbf{u}), \quad (1.13)$$

$$\mathbf{q} = -\lambda \nabla T + \rho \sum_{\alpha=1}^{N_s} Y_\alpha \mathbf{U}_\alpha h_\alpha - \sum_{\alpha=1}^{N_s} \frac{p}{\rho Y_\alpha} \mathcal{D}_\alpha^T \mathbf{d}_i, \quad (1.14)$$

respectively. The symbol $\mathbf{I}$ denotes a unit tensor and μ is the dynamic viscosity of the gas mixture. The bulk viscosity, κ, is typically negligible for low-speed reacting flows. The mole fraction of species α is X_α, $\mathcal{D}_{\alpha j}$ is the multicomponent diffusion coefficient, $\mathcal{D}_\alpha^T$ is the thermal diffusion coefficient for species α, and λ is the thermal conductivity for the mixture. The thermal diffusion flux $\mathbf{d}_\alpha$ is given by [24–26],

$$\mathbf{d}_\alpha = \nabla X_\alpha + (X_\alpha - Y_\alpha) \nabla (\ln p). \quad (1.15)$$

The first term on the right-hand side of Eq. (1.15) is due to concentration gradients and the second term is due to pressure gradients. It is clear that $\sum_{\alpha=1}^{N_s} \mathbf{d}_\alpha = 0$.

A mixture averaged diffusion coefficient is often used to simplify Eq. (1.12),

$$\mathbf{U}_\alpha = -\frac{\mathcal{D}_\alpha^m}{X_\alpha} \mathbf{d}_\alpha - \frac{\mathcal{D}_\alpha^T}{\rho Y_\alpha} \nabla (\ln T), \quad (1.16)$$

where the mixture averaged diffusion coefficient $\mathcal{D}_\alpha^m$ is given as

$$\mathcal{D}_\alpha^m = \frac{1 - X_\alpha}{\sum_{j \neq \alpha} X_\alpha / \mathcal{D}_{\alpha j}}. \tag{1.17}$$

The binary diffusion coefficient matrix is symmetric ($\mathcal{D}_{\alpha j} = \mathcal{D}_{j \alpha}$), and the diagonal elements are zero ($\mathcal{D}_{\alpha \alpha} = 0$). In general, using Eq. (1.12) or (1.16) can violate the mass conservation given in Eq. (1.3), i.e., $\sum_{\alpha=1}^{N_s} Y_\alpha \neq 1$. This can be overcome by using a correction velocity as suggested by Coffee and Heimerl [28]. Another attractive and simpler approach is to compute the mass fraction of the species in excess by subtracting the sum of the remaining mass fractions from unity [29]. The latter approach is commonly used in reacting flow calculations.

1.5 Chemical Kinetic Relations

The net formation rate of species i, required for the species conservation equation given as Eq. (1.4), is obtained using chemical kinetic relations, known as the law of mass action, for all the elementary reactions involving the species. For example, consider the elementary reaction

$$\text{H} + \text{O}_2 \rightleftharpoons \text{O} + \text{OH}.$$

The law of mass action states that

$$\frac{dC_\text{H}}{dt} = \frac{dC_{\text{O}2}}{dt} = -\frac{dC_\text{O}}{dt} = -\frac{dC_\text{OH}}{dt} = -k_f C_\text{H} C_{\text{O}2} + k_b C_\text{O} C_\text{OH}, \tag{1.18}$$

where C_i is the molar concentration of species i, which is related to the mass fraction through $C_i = \rho Y_i / W_i$. The specific rate constants for the forward, k_f, and backward, k_b, reactions depend on temperature and this dependence is given by the Arrhenius relation. For example, for the forward reaction

$$k(T) = A T^n \exp\left(\frac{-E_a}{R_u T}\right), \tag{1.19}$$

where A, n, and E_a are empirical constants and E_a is the activation energy. There could be hundreds or thousands of such elementary reactions involved in a combustion process. Hence, a chemical kinetic mechanism may contain N_r elementary reactions and species i may participate in a number of reactions among N_r. These elementary reactions can be described in a general and compact form as

$$\sum_{i=1}^{N_s} v'_{ij} M_i \rightleftharpoons \sum_{i=1}^{N_s} v''_{ij} M_i, \quad j = 1, \ldots, N_r$$

where v'_{ij} is the stoichiometric coefficient for species M_i on the reactant side of reaction j, and v''_{ij} is the species stoichiometric coefficient on the product side of the reaction. The net rate of formation of species i is

$$\frac{dC_i}{dt} = \sum_{j=1}^{N_r}(v''_{ij} - v'_{ij})q_j, \quad \text{where} \quad q_j = k_{f,j}\prod_{i=1}^{N_s} C_i^{v'_{ij}} - k_{b,j}\prod_{i=1}^{N_s} C_i^{v''_{ij}}. \tag{1.20}$$

Further details regarding chemical kinetic mechanisms and chemical reaction rates, including the determination of empirical constants, A, n, and E_a and reaction rates for elementary as well as global reaction mechanisms, are discussed in Chapter 5. The global mechanisms may not depend on the elementary reactions and thus they are empirical.

1.6 Direct Numerical Simulation

The most accurate way of simulating turbulent gas-phase combustion is by solving the conservation equations, Eqs. (1.4)–(1.8), together with the necessary constitutive relations without involving any kinds of turbulence modeling. This is called a direct numerical simulation (DNS). Assuming that initial and boundary conditions are known, and that the mean free path of the molecules in the gas is much smaller than the smallest turbulent scale, such a simulation is referred to by some researchers as a "numerical experiment" due to its independence of models and its corresponding high fidelity. For reactive conditions, it is required, however, to have a model for the chemical reaction rate, $\dot{\omega}_i$ in Eq. (1.4). Hence, it is not strictly true that a reactive DNS is model free.

All relevant spatial and temporal scales must be properly resolved in direct simulations. This implies that for a compressible nonreactive case, the size of the grid cells is determined by the smallest turbulent eddies, which is typically taken to be the Kolmogorov scale, while the time step is dictated by the acoustic Courant–Friedrichs–Lewy (CFL) criterion. Since the Kolmogorov scale can be approximated by $\eta = L/\text{Re}_t^{3/4}$, the size of a grid cell scales as $\Delta x \sim L/\text{Re}_t^{3/4}$, such that the total number of grid points for a three-dimensional homogeneous mesh becomes $N_{\text{grid}} = (L/\Delta x)^3 \sim \text{Re}_t^{9/4}$. This scaling also depends on the numerical schemes used to approximate the spatial derivatives in the conservation equations. The scaling given in the preceding is for spectral methods and if one employs finite difference/volume method then the above scaling must be at least three to four times larger. The time step is linearly dependent on Δx when an explicit time integration method is used, which means that the total computational cost of a compressible nonreactive DNS can be found from the following relation: $C_{\text{CPU}} \sim \text{Re}_t^3$. Both spatial and temporal resolutions requirements will change when chemical reactions for combustion are included. For such cases, the spatial resolution is also limited by the thickness of the flame, which has to be resolved by at least 6–10 grid cells. So, in particular for flamelet regime combustion (see Chapter 2), the spatial resolution is constrained by the flame thickness. Furthermore, since chemical kinetic mechanisms may be rather stiff, the time step may have to be smaller than what is required to satisfy the acoustic CFL number criterion for numerical stability.

Despite the fact that direct simulations of turbulent combustion are computationally very expensive and some modeling is required for the chemical kinetics, it is the most accurate method with high fidelity available for simulating flames. As such, direct simulations are often used to develop and validate models for RANS or LES. Recently, DNS have even been used to investigate some specific (very small) sections of real industrial combustors in order to guide the development of such combustors. Nevertheless, it is clear that for the foreseeable future, DNS tools will not be used to study full-scale industrial application because of the prohibitively large computational cost.

1.7 Large-Eddy Simulation

1.7.1 LES Equations

The basic principle of large-eddy simulation (LES) of turbulent combustion is to solve the low-pass filtered governing equations for the flow, energy, and species mass fraction. This means that the dynamics of the large scales, larger than the filter cutoff wavelength, are resolved in LES whereas only the small, unresolved subgrid scales (SGS) are modeled. The computational costs of LES are thus much lower than that of DNS because the grid can be coarser and the time steps larger while maintaining a similar level of fidelity for the simulations. The filtering, or separation of the scales, is done with a spatial filter, which is applied to the governing equations discussed in Section 1.3. The spatial filter of a quantity varying in space and time, $f = f(\mathbf{x}, t)$, is defined as its convolution with a filter function G according to

$$\overline{f}(\mathbf{x}, t) = \int_\Omega f(\mathbf{x}', t)\, G(\mathbf{x} - \mathbf{x}'; \Delta(\mathbf{x}))\, d\mathbf{x}', \qquad (1.21)$$

defined over the entire flow domain Ω with a characteristic filter width Δ, which, in general, may vary with position. In LES the filtering can be explicit, meaning that the solution or convective terms are explicitly filtered during the computation, or implicit, meaning that the governing equations are solved on a numerical grid that is too coarse to resolve all scales without using an explicit filter. Commonly applied spatial filters in LES such as the top-hat, Gaussian, and various discrete filters have different properties in physical and wavenumber spaces [30–33]. In general, the filtering and spatial differentation in finite difference and finite volume codes does not commute on nonuniform grids and this leads to commutation errors that can be significant [34, 35]. However, it is possible to construct discrete filters with reduced commutation errors [32].

It is common to use Favre-filtering for flows with strong density variations because this minimizes the number of extra SGS terms that appear due to the filtering operation. Favre-filtered variables are defined as

$$\widetilde{f} = \frac{\overline{\rho f}}{\overline{\rho}}. \qquad (1.22)$$

When this Favre-filtering is applied to Eqs. (1.4), (1.5), (1.6), and (1.7) the following set of equations governing the resolved scales in LES of turbulent combustion is obtained:

$$\frac{\partial \overline{\rho} \widetilde{Y_i}}{\partial t} + \nabla \cdot \left(\overline{\rho} \widetilde{\mathbf{u}} \widetilde{Y_i} \right) = \nabla \cdot \left(-\overline{\rho Y_i \mathbf{U}_i} \right) + \overline{\rho \dot{\omega}_i} - \nabla \cdot \overline{\boldsymbol{\psi}}_i^S \tag{1.23}$$

$$\frac{\partial \overline{\rho}}{\partial t} + \nabla \cdot \left(\overline{\rho} \widetilde{\mathbf{u}} \right) = 0 \tag{1.24}$$

$$\frac{\partial \overline{\rho} \widetilde{\mathbf{u}}}{\partial t} + \nabla \cdot \left(\overline{\rho} \widetilde{\mathbf{u}} \widetilde{\mathbf{u}} \right) = -\nabla \overline{p} + \nabla \cdot \overline{\boldsymbol{\tau}} - \nabla \cdot \overline{\boldsymbol{\tau}}^S \tag{1.25}$$

$$\frac{\partial \overline{\rho} \widetilde{h}}{\partial t} + \nabla \cdot \left(\overline{\rho} \widetilde{\mathbf{u}} \widetilde{h} \right) = \frac{\partial \overline{p}}{\partial t} + \widetilde{\mathbf{u}} \cdot \nabla \overline{p} - \nabla \cdot \overline{\mathbf{q}} - \nabla \cdot \overline{\left(\rho \sum_{i=1}^{N_s} Y_i \mathbf{U}_i h_i \right)}$$
$$+ \overline{\boldsymbol{\tau} : \nabla \mathbf{u}} + \overline{Q_r} + \Pi_{dil} - \nabla \cdot \overline{\boldsymbol{\theta}}^S. \tag{1.26}$$

As a result of the filtering operation extra terms appear, namely the SGS species flux $\overline{\boldsymbol{\psi}}_i^S$, SGS stress tensor $\overline{\boldsymbol{\tau}}^S$, SGS enthalpy flux $\overline{\boldsymbol{\theta}}^S$, and SGS pressure-dilation Π_{dil}, which are respectively given by

$$\overline{\boldsymbol{\psi}}_i^S = \overline{\rho} \left(\widetilde{\mathbf{u} Y_i} - \widetilde{\mathbf{u}} \widetilde{Y_i} \right) \tag{1.27}$$

$$\overline{\boldsymbol{\tau}}^S = \overline{\rho} \left(\widetilde{\mathbf{u} \mathbf{u}} - \widetilde{\mathbf{u}} \widetilde{\mathbf{u}} \right) \tag{1.28}$$

$$\overline{\boldsymbol{\theta}}^S = \overline{\rho} \left(\widetilde{\mathbf{u} h} - \widetilde{\mathbf{u}} \widetilde{h} \right) \tag{1.29}$$

$$\Pi_{dil} = \overline{\mathbf{u} \cdot \nabla p} - \widetilde{\mathbf{u}} \cdot \nabla \overline{p}. \tag{1.30}$$

These terms account for the influence of unresolved small scales on the resolved large scales and are unknown, since quantities like $\widetilde{\mathbf{u} Y_i}$ and $\widetilde{\mathbf{u} \mathbf{u}}$ are undetermined. They, therefore, need to be modeled and it is usually done in terms of the known resolved quantities. Closures for the SGS terms in Eqs. (1.27)–(1.29) are discussed in some more detail in the next section. The SGS pressure-dilation in Eq. (1.30) is sometimes less important in compressible flows and therefore neglected [36, 37]. A plausible modeling for this term and its limitation are explored in [38]. Also some other terms, i.e., the first term on the right-hand side of Eq. (1.23), the second term on the right-hand side of Eq. (1.25) and from the third to sixth terms on the right-hand side of Eq. (1.26) are not expressed in terms of resolved quantities and therefore they need to be approximated or modeled. For further discussions about the LES governing equations for compressible flows and the importance and modeling of the different terms we refer the reader to Ref. [31]. However, we note that it is common to neglect fluctuations in viscosity and diffusivity [36, 39].

The term $\overline{\rho \dot{\omega}_i}$ in the species equation, Eq. (1.23), is naturally of major importance for turbulent combustion. In general, it is a highly nonlinear function of temperature, T, and species mass fractions, Y_i. Also, the physical processes represented by this term typically occur at SGS level. It cannot directly or accurately be expressed in terms of the resolved quantities $\widetilde{T}$ and $\widetilde{Y}_i$. Formulating a closure for this unclosed term in LES is in fact one of the main topics in turbulent combustion research. This is discussed in more detail in Chapters 3 and 4, but we also refer to some review papers [39–41] and books [18, 20, 21] on LES of turbulent combustion.

1.7.2 SGS Closures

We now discuss a few common explicit models for the SGS terms given in Eqs. (1.27)–(1.29). The most simple models for the SGS stress in Eq. (1.28) are the eddy viscosity models and one of the most simple and popular of these is the classical Smagorinsky model [42], with an extension in [43] to model the SGS kinetic energy

$$\tau_{ij}^S - \frac{\delta_{ij}}{3}\tau_{kk}^S = -C_s^2\, 2\Delta^2 \overline{\rho} |\widetilde{\mathbf{S}}| \left(\widetilde{S}_{ij} - \frac{\delta_{ij}}{3}\widetilde{S}_{kk} \right)$$

$$= -2\overline{\rho}\, \nu_{\mathrm{SGS}} \left(\widetilde{S}_{ij} - \frac{\delta_{ij}}{3}\widetilde{S}_{kk} \right), \text{ and} \quad (1.31)$$

$$\tau_{kk}^S = 2\, C_I \overline{\rho}\, \Delta^2 |\widetilde{\mathbf{S}}|^2, \quad (1.32)$$

where $\widetilde{S}_{ij} = 0.5\left(\partial \widetilde{u}_i / \partial x_j + \partial \widetilde{u}_j / \partial x_i\right)$ is the resolved strain-rate tensor and $|\widetilde{\mathbf{S}}| = \left(2\,\widetilde{S}_{ij}\widetilde{S}_{ij}\right)^{1/2}$. The SGS eddy viscosity, ν_{SGS}, is defined by Eq. (1.31) and C_s and C_I are model constants. As customary, repeated indices imply summation. The trace of the SGS stress tensor, τ_{kk}^S, appears to be small or negligible in flows with low Mach numbers [37] but probably not so for reacting flows with nonnegligible heat release rate as in many practical systems.

The Smagorinsky models is relatively simple and robust, but it does not work well near walls and in regions with flow transitions, since it can give a nonvanishing contribution. This can partly be remedied by invoking damping functions, but an alternative approach used widely is to employ a dynamical procedure to determine the model constants C_s and C_I, as proposed in [44]. In the dynamical procedure, a second filter of typical width $\hat{\Delta} = 2\Delta$ is applied to the resolved scales to compute the resolved stress near the filter cutoff. Assuming similarity of the stresses near the filter cutoff scale, Δ, this resolved stress can be used to find an expression for C_s and C_I in terms of the resolved velocity gradients; see [30, 31, 37] for further discussions and details. It has been found in past studies that the model constants display large spatial and temporal variations. These variations are reduced by averaging in either time or space.

The advantage of the dynamic procedure is that the model adapts itself to the local flow changes and that the SGS stress naturally approaches zero near solid walls and in laminar regions, as it should. In theory, the dynamic procedure can produce negative values for C_s and thus represent the instantaneous backscatter, transport of energy from the unresolved to the resolved scales, that is known to occur in turbulent flows. However, this can lead to numerical instabilities, and therefore it is a common practice to clip C_s to avoid negative viscosities.

In addition to the SGS stress in Eq. (1.28), the SGS species flux given by Eq. (1.27) and enthalpy flux in Eq. (1.29) need to be modeled. A straightforward approach is to use an eddy diffusivity model, i.e.,

$$\overline{\psi}_i^S = \frac{-\overline{\rho}\,\nu_{SGS}}{Sc_{SGS}} \nabla \widetilde{Y}_i, \quad \text{and} \quad \overline{\theta}^S = \frac{-\overline{\rho}\,\nu_{SGS}}{Pr_{SGS}} \nabla \widetilde{h}, \qquad (1.33)$$

where Sc_{SGS} and Pr_{SGS} are the SGS Schmidt and Prandtl numbers respectively. The SGS viscosity can be computed using, for example, the Smagorinsky model given in Eq. (1.31) with or without dynamic procedure. A common approach is to assume that the SGS Schmidt and Prandtl numbers are constant but it is also possible to use a dynamic procedure to compute these parameters [31, 37, 44].

Besides the Smagorinsky model with or without dynamical procedure and the eddy diffusivity models given in Eq. (1.33), many other models for the SGS stresses and fluxes have been developed and tested in past studies. These are not discussed here; instead we refer the reader to other works, for example [31, 37]. The study reported in [45] gives a recent overview and discussion of SGS stress models although limited to incompressible flows. With the aim to improve the description of the SGS physics, models that do not rely on the eddy viscosity assumption but have nonlinear dependence on the resolved velocity gradient have been developed in the past.

The SGS terms in Eqs. (1.27)–(1.29) can be modeled as discussed earlier but one should note that the errors introduced by the numerical scheme can be substantial, especially for scales near the filter cutoff. The numerical scheme and modeling are thus in general intrinsically coupled. An alternative approach is therefore to use tailored numerical algorithms or schemes that let the numerical scheme take care of the influence of the unresolved scales on the resolved ones. This approach is called implicit LES [46]. Deconvolution methods, whereby the unfiltered solution is reconstructed by using an inverse filter operation, can also be viewed as an implicit LES approach [47, 48].

1.7.3 Challenges for LES

As discussed in the previous subsection, the SGS closures are predominantly based on the gradient flux hypothesis and it is well known that in reacting flows there are processes that defy this hypothesis. Hence, modeling counter-gradient subgrid scalar fluxes are still an outstanding issue. However, many LES calculations using the gradient hypothesis show a good agrement between the computed and measured statistics, suggesting that the gradient flux models may be sufficient and that the counter-gradient scalar fluxes observed in turbulent premixed flames caused by the heat release effects are large-scale phenomena that are resolved and captured in a careful LES. Another challenge from a practical perspective for LES is on the near-wall flow characteristics. It is quite well known that practical LES cannot recover the law of the wall and some special numerical treatments are required. The details can be found in [49, 50]. Recovering the law of the wall becomes important when the heat and momentum fluxes through the walls (of the combustor, for example) need to be evaluated as design variables.

It is generally observed that the numerical grids used for LES of reacting flows resolve instantaneous flame structure to some extent, which is acceptable for atmospheric pressure. Practical applications involve invariably high pressures and complex geometries. The flame thickness approximately scales as $\delta \sim p^{-1/2}$ [51] and thus resolving the instantaneous flame structure would lead to impractical grid cell counts. Furthermore, the geometry details need to be captured to get the relevant flow features correctly. Thus, the common practice of using grids having cell sizes of the order of the instantaneous (laminar) flame thickness is not attractive for practical LES. Consequently, SGS combustion models have to be robust and accurate in representing the relevant physical processes. Also, physical and mathematical consistencies among the various SGS closures must be maintained. Recent advances in these direction are discussed in later chapters; see specifically Chapters 3 and 4. It is probably useful to design or select a grid resolving most of the kinetic energy in the flow and let the SGS combustion closure handle the turbulence–chemistry interactions and their intricacies for LES of reacting flows in practical systems. The guidance suggested by Pope [30], which is $\mathcal{K} = k_{sgs}/(k_{res} + k_{sgs}) \leq 0.2$, where k_{sgs} and k_{res} are subgrid scale and resolved kinetic energies respectively, may be used. It is to be noted that this condition can be evaluated only after completing a preliminary LES of nonreacting flow in a given geometry. Alternative measures to evaluate LES grid requirement have also been suggested in past studies. However, the parameter $\mathcal{K}$ is quite practical and useful, and thus it is recommended.

1.8 RANS Approach

In the two previous sections, we have seen how DNS resolves all the scales of the flow, while LES resolves the largest scales and models the smaller ones. In the so called "Reynolds averaged Navier–Stokes" (RANS) approach, all turbulent and other fluctuating scales are modeled, which means that only the mean flow and mean species mass fractions are computed. This is done by realizing that any quantity f can be split into mean and fluctuating components as

$$f = \widetilde{f} + f''; \tag{1.34}$$

where Favre averaging is used as explained in the previous section; see Eq. (1.22). The averaging can be done over time or space or ensembles depending on the flow conditions. However, these three averages converge to the same value according to the central limit theorem, if the sample size is sufficiently large.

Using the preceding decomposition in the conservation equations (1.4)–(1.7) and then Favre averaging the resulting equations, one recovers Eqs. (1.23)–(1.26), which are the same governing equations as for the LES approach. The unclosed terms $\overline{\psi}_i^S$,

$\overline{\tau}^S$, $\overline{\theta}^S$, and Π_{dil} in the RANS equations, however, have a different meaning and are typically written as follows:

$$\overline{\psi}_i^S = \overline{\rho}\widetilde{\mathbf{u}''Y_k''} \tag{1.35}$$

$$\overline{\tau}^S = \overline{\rho}\widetilde{\mathbf{u}''\mathbf{u}''} \tag{1.36}$$

$$\overline{\theta}^S = \overline{\rho}\widetilde{\mathbf{u}''h''} \tag{1.37}$$

$$\Pi_{\text{dil}} = \overline{\mathbf{u}\cdot\nabla p} - \widetilde{\mathbf{u}}\cdot\nabla\overline{p}. \tag{1.38}$$

The difference is that these unclosed terms represent the effects of turbulent fluctuations at all scales on the evolution of mean quantities. Models for the unclosed turbulent Reynolds stress in Eq. (1.36), turbulent species and enthalpy fluxes in Eqs. (1.35) and (1.37), and pressure-dilation in Eq. (1.38) have been developed over several decades. The momentum equation can, for example, be closed by using the k-ε [52] or k-ω [53] models, which relate the Reynolds stresses to the mean flow stresses through the eddy viscosity. Similarly, the turbulent flux terms, Eqs. (1.35) and (1.37), can be approximated by eddy diffusivity models. However, the presence of counter-gradient fluxes needs to be considered carefully, although their influences may become weaker at flow Reynolds number observed typically in practical systems.

A major unclosed term in RANS simulations of turbulent combustion is of course the reaction term, the third term on the right-hand side of Eq. (1.23). Modeling this term is more challenging than in LES, since even less information about spatial and temporal distribution of the species, heat, and their correlations are available in the RANS approach. In LES, at least the large-scale turbulent transport of heat and mixing of reacting species is resolved, but in the RANS approach these have to be modeled. Many models for the unclosed terms, depending on the combustion regime, exist. This is thoroughly covered in a number of good textbooks and we will therefore not go into more detail on this here.

It is clear that among the three computational fluid dynamics (CFD) approaches (DNS, LES, and RANS) discussed in this chapter, the RANS approach is the least computationally expensive. For this reason, the RANS approach has been the preferred choice for many applications. The first RANS CFD simulations were performed several decades ago, and it is still the most commonly used approach for industrial applications. LES, however, is gaining popularity and has become feasible for larger industrial application [39]. This is because of the recent advances in the computational power and numerical algorithms. Also, LES is more suited to capture highly transient events such as local extinction, flame lift-off, and thermoacoustic oscillations. The main reasons for choosing LES instead of RANS for simulations of turbulent reacting flows are that

1. The largest turbulent eddies are solved and captured (do not rely on modeling). This means that more of the turbulent intermittency is recovered. RANS calculations average out intermittency.

2. The subgrid modeling is only for small scale, and hence isotropic, turbulence. This means that the subgrid modeling developed typically by assuming isotropy is more applicable.

1.9 Aims and Objectives

The broad aim here is to review the recent progress made in the field of turbulent combustion and its application along with fundamental concepts required for the review. The role of combustion technology in the next many decades, probably well into the next century also, cannot be underestimated, as discussed earlier in this chapter. Hence, it is hoped that reviewing the recent advances in the field will provide a consolidated overview and thereby help us to design new combustion concepts and systems that would enable us to meet the two conflicting demands: (1) ever increasing energy requirement in the world and (2) mitigation of global warming. We hope that the discussions in the following chapters, which have a physical basis, would help readers to understand the complexities and challenges involved in turbulent combustion and probably motivate them to embark on an exploration to seek a neat solution for the two aforementioned conflicting demands.

The fundamental conservation equations and constitutive relations governing turbulent reacting flows are reviewed and discussed in earlier sections of this chapter. As discussed in many places of this book, turbulent combustion involves a wide range of physical and chemical time and length scales. Hence, studying turbulent reacting flows is challenging, but excellent progress has been made in the past 60 years or so, which is not long compared to the time spent for other fields such as physics and chemistry (of the order of a couple of hundred years). Sound theoretical approaches were developed and used to investigate the turbulent flame structure only in the 1970s although Damköhler's classical analysis using scaling arguments was done in the 1940s. A historical perspective of the turbulent flame structure and dynamics is provided in Section 2.1. The insights gathered using high-fidelity numerical simulations and sophisticated laser diagnostics are presented and discussed in the rest of Chapter 2. The two classical, premixed and non-premixed, modes of turbulent combustion and the recent advances in their modeling are discussed in Chapters 3 and 4 respectively. The subgrid modeling for partially premixed combustion is also discussed in Chapter 4. The chemical kinetics aspects are elaborated in Chapter 5.

Turbulent combustion (either premixed or non-premixed) with vitiated air is also known as MILD (moderate, intense, or low dilution) combustion, if the oxygen level is less than or equal to 5% on a molar basis in the vitiated air with temperature typically larger than the autoignition temperature of a given fuel. This combustion concept has the potential to achieve high overall efficiency with very low emission level and thus it can serve as a strong candidate to cater for the two conflicting questions (on energy demand and environmental impact) identified earlier in this chapter. However, achieving this combustion condition safely and robustly over a wide range of operating conditions is a challenge, as it does not involve just classical flames or autoignition but a combination of these two phenomena, which is determined by the operating and thermo-physical and -chemical conditions. Thus, modeling turbulent MILD combustion poses challenges that are discussed in Chapter 6. Combustion in high-speed flows has received renewed interest because of recent interest in building intercontinental high-speed civil transport. Supersonic combustion is known to have spotty characteristics that are similar to those of MILD combustion [17]. The fundamentals of and recent advances in supersonic combustion are discussed in Chapter 7.

Most of the practical fuels are liquids and thus they have to be atomized, vaporized and mixed with air before combustion can begin. These aspects bring their own scientific and technological challenges. These are discussed in Chapter 8. In terms of renewables, combustion of biomass, a solid fuel, is growing. The combustion of solid fuels brings its own fundamental challenges because it involves both homogeneous and heterogenous phase combustion. The heterogenous part involves pyrolysis, char oxidation, and ash formation, which are quite complicated in terms of chemical and physical aspects. However, some simplified approaches were developed in past studies. These are reviewed and discussed in Chapter 9. The challenges involved in practical applications such as power generation, aero propulsion, and road transport are presented and discussed in Chapter 10. The closing remarks are made in the final chapter of this book.

References

[1] IEA, "Key world energy statistics," International Energy Agency, Paris, Tech. Rep., 2018.

[2] "How we use energy," http://needtoknow.nas.edu/energy/energy-use/. Accessed January 9, 2019.

[3] "Future climate changes, risks and impacts," https://ar5-syr.ipcc.ch/topic_futurechanges.php. Accessed: January 9, 2019.

[4] K. Hayhoe et al., "Climate models, scenarios, and projections," in *Climate Science Special Report: Fourth National Climate Assessment, Volume I*, D. J. Wuebbles, D. W. Fahey, K. A. Hibbard, D. J. Dokken, B. C. Stewart, and T. K. Maycock, Eds. Washington, DC: U.S. Global Change Research Program, 2017, pp. 133–160.

[5] B. L. Cohen, "Breeder reactors: A renewable energy source," *Am. J. Phys.*, **51**, 75–76, 1983.

[6] R. Vasques, "Breeder reactors: A renewable energy source," *Energy Res. J.*, **5**, 33–34, 2014.

[7] P. Moore, "Nuclear re-think," *IAEA Bull*, **48**, 56–58, 2006.

[8] M. Winskel et al. "Decarbonising the UK energy system: Accelerated development of low carbon energy supply technologies," UK Energy Research Centre, Edinburgh University, Tech. Rep., 2009.

[9] IRENA, "Renewable power generation costs in 2018," International Renewable Energy Agency, Abu Dhabi, Tech. Rep., 2019.

[10] N. Swaminathan, "The role of combustion technologies in 21st and 22nd centuries," unpublished data, 2019.

[11] IRENA, "Renewable capacity statistics 2019," International Renewable Energy Agency, Abu Dhabi, Tech. Rep., 2019.

[12] J. Sathaye, O. Lucon, A. Rahman, J. Christensen, F. Denton, J. Fujino, G. Heath, S. Kadner, M. Mirza, H. Rudnick, A. Schlaepfer, and A. Shmakin, "Renewable energy in the context of sustainable development," in *IPCC Special Report on Renewable Energy Sources and Climate Change Mitigation*, O. Edenhofer, R. Pichs-Madruga, Y. Sokona, K. Seyboth, P. Matschoss, S. Kadner, T. Zwickel, P. Eickemeier, G. Hansen, S. Schlömer, and C. von Stechow, Eds. Cambridge University Press, Cambridge, United Kingdom and New York, NY, USA, 2011, 707–789.

[13] N. Santero and J. Hendry, "Harmonization of LCA methodologies for the metal and mining industry," *Int. J. Life Cycle Assess.*, **21**, 1543–1553, 2016.

[14] Q. Tu, M. Eckelman, and J. Zimmerman, "Meta-analysis and harmonization of life cycle assessment studies for algae biofuels," *Environ. Sci. Technol.*, **51**, 9419–9432, 2017.

[15] N. Y. Amponash, M. Troldborg, B. Kington, I. Aalders, and R. L. Hough, "Greenhouse gas emissions from renewable energy sources: A review of lifecycle considerations," *Renew. Sust. Energy Rev.*, **39**, 461–475, 2014.

[16] J. A. A. Alvarez, "Cave-to-case analysis of batteries for electric vehicles," MPhil thesis, University of Cambridge, 2019.

[17] N. Swaminathan, "Physical insights on MILD combustion from DNS," *Front. Mech. Eng. Therm. Mass Trans.*, **5**, Article 59, 2019.

[18] N. Swaminathan and K. N. C. Bray, *Turbulent Premixed Flames*. Cambridge: Cambridge University Press, 2011.

[19] G. Ghiasi, I. Ahmed, and N. Swaminathan, "Gasoline flame behavior at elevated temperature and pressure," *Fuel*, **238**, 248–256, 2019.

[20] N. Peters, *Turbulent Combustion*. Cambridge: Cambridge University Press, 2000.

[21] T. Poinsot and D. Veynante, *Theoretical and Numerical Combustion*. Philadelphia, PA, USA: R. T. Edwards, 2005.

[22] P. A. Libby and F. A. Williams (Eds.), *Turbulent Reacting Flows*. New York: Springer-Verlag, 1980.

[23] P. A. Libby and F. A. Williams (Eds.), *Turbulent Reacting Flows*. New York: Academic Press, 1994.

[24] J. O. Hirschfelder, C. F. Curtiss, and B. R. Bird, *Molecular Theory of Gases and Liquids*. New York: Wiley, 1965.

[25] R. B. Bird, W. E. Stewart, and E. N. Lightfoot, *Transport Phenomena*. New York: Wiley, 1960.

[26] V. Giovangigli, *Multicomponent Flow Modeling*. Boston, MA: Birkhauser, 1999.

[27] J. H. Chen, et al., "Terascale direct numerical simulations of turbulent combustion using S3D," *Comput. Sci. & Discovery*, **2**, p. 015001, 2009.

[28] T. P. Coffee and J. M. Heimerl, "Transport algorithms for premixed, laminar steady-state flames," *Combust. Flame*, **43**, 273–289, 1981.

[29] R. J. Kee, G. Dixon-Lewis, J. Warnatz, M. E. Coltrin, and J. A. Miller, *A Fortran computer code package for the evaluation of gas-phase multicomponent transport properties*. Sandia National Laboratories Report SAND86-8246 13, 80401-1887, 1986.

[30] S. B. Pope, *Turbulent Flows*. Cambridge: Cambridge University Press, 2000.

[31] E. Garnier, N. Adams, and P. Sagaut, *Large Eddy Simulation for Compressible Flows*. Springer Science+Business Media B.V. 2009.

[32] O. Vasilyev, T. Lund, and P. Moin, "A general class of commutative filters for LES in complex geometries," *J. Comput. Phys.*, **146**, 82–104, 1998.

[33] C. Bogey and C. Bailly, "A family of low dispersive and low dissipative explicit schemes for flow and noise computations," *J. Comput. Phys.*, **194**, 194–214, 2004.

[34] S. Ghosal and P. Moin, "The basic equations for the large eddy simulation of turbulent flows in complex geometry," *J. Comput. Phys.*, **118**, 24–37, 1995.

[35] F. van der Bos and B. Geurts, "Commutator errors in the filtering approach to large-eddy simulation," *Phys. Fluids*, **17**, p. 035108, 2005.

[36] U. Piomelli, "Large-eddy simulation: achievements and challenges," *Prog. Aero. Sci.*, **35**, 335–362, 1999.

[37] M. Martin, U. Piomelli, and G. Candler, "Subgrid-scale models for compressible large-eddy simulations," *Theoret. Comput. Fluid Dynamics*, **13**, 361–376, 2000.

[38] I. Langella, N. Swaminathan, Y. Gao, and N. Chakraborty, "Large eddy simulation of premixed combustion: Sensitivity to subgrid scale velocity modeling," *Combust. Sci. Technol.*, **189**, 43–78, 2017.

[39] L. Gicquel, G. Staffelbach, and T. Poinsot, "Large eddy simulations of gaseous flames in gas turbine chambers," *Prog. Energy Combust. Sci.*, **38**, 782–817, 2012.

[40] H. Pitsch, "Large-eddy simulation of turbulent combustion," *Ann. Rev. Fluid Mech.*, **38**, 453–482, 2006.

[41] C. Rutland, "Large-eddy simulations for internal combustion engines: A review," *Int. J. Engine Res.*, **12**, 421–451, 2011.

[42] J. Smagorinsky, "General circulation experiments with the primitive equations. I. The basic experiment," *Monthly Weather Review*, **91**, 99–164, 1963.

[43] A. Yoshizawa, "Statistical theory for compressible turbulent shear flows, with the application to subgrid modeling," *Phys. Fluids*, **29**, 2152–2164, 1986.

[44] P. Moin, K. Squires, W. Cabot, and S. Lee, "A dynamic subgrid-scale model for compressible turbulence and scalar transport," *Phys. Fluids*, **3**, 2746–2757, 1991.

[45] M. Silvis, R. Remmerswaal, and R. Verstappen, "Physical consistency of subgrid-scale models for large-eddy simulation of incompressible turbulent flows," *Phys. Fluids*, **29**, p. 015105, 2017.

[46] F. Grinstein, L. Margolin, and W. Rider, *Implicit Large Eddy Simulation*. Cambridge: Cambridge University Press, 2007.

[47] S. Stolz, N. Adams, and L. Kleiser, "The approximate deconvolution model for large-eddy simulations of compressible flows and its application to shock-turbulent-boundary-layer interaction," *Phys. Fluids*, **13**, p. 2985, 2001.

[48] S. Hickel, C. Egerer, and J. Larsson, "Subgrid-scale modeling for implicit large eddy simulation of compressible flows and shock-turbulence interaction," *Phys. Fluids*, **26**, p. 106101, 2014.

[49] N. V. Nikitin, F. Nicoud, B. Wasistho, K. D. Squires, and P. R. Spalart, "An approach to wall modeling in large-eddy simulations," *Phys. Fluids*, **12**, 1629–1632, 2000.

[50] J. G. Brasseur and T. Wei, "Designing large-eddy simulation of the turbulent boundary layer to capture law-of-the-wall scaling," *Phys. Fluids*, **22**, 021 303–1–21, 2010.

[51] S. R. Turns, *An Introduction to Combustion: Concepts and Applications*, 2nd ed., Singapore: McGraw-Hill International, 2006.

[52] B. Launder and D. Spalding, "The numerical computation of turbulent flows," *Comput. Methods App. Mech. Eng.*, **3**, 269–289, 1974.

[53] D. Wilcox, *Turbulence Modeling for CFD*, DCW Industries, 1993, **93**.

2 Turbulent Flame Structure and Dynamics
Combustion Regimes

V. Sabelnikov, A. Lipatnikov, X.-S. Bai, and N. Swaminathan

Turbulent combustion can occur in different modes depending on the arrangement for fuel–air mixing, e.g., premixed, non-premixed and partially premixed combustion. Premixed combustion occurs when the fuel is mixed homogeneously with oxidizer, whereas in non-premixed combustion the fuel and oxidizer react as they mix in the reaction zone as they are supplied through different streams feeding the reaction zone. When the chemical reaction timescale is shorter than the fuel–air mixing timescale the non-premixed combustion becomes mixing controlled and it is referred to as diffusion combustion. If the reaction is slower than the mixing then the oxidizer will leak into the fuel side and as a result the temperature in the reaction zone will be lowered. Subsequently the chemical reactions can quench locally, leading to partial premixing of the fuel and oxidizer in the system and partially premixed combustion ensues. One example of this partially premixed combustion is the lifted non-premixed jet flame, where local extinction occurs in the proximity of the burner rim, leading to some amount of fuel mixing with oxidizer and vice versa at the base of this flame.

In premixed combustion, chemical reactions sustain for mixture temperature above a critical value, sometimes referred to as the crossover temperature, T_c [1], above which the chain branching reactions are faster than the chain termination reactions, leading to self-sustained combustion (see Chapter 5 for definition of these reactions). This condition can be achieved through external heat addition, for example using a spark, which heats up the local mixture to a temperature well above T_c. Alternatively, if the reactant mixture is preheated but below T_c, the mixture may auto-ignite after a certain induction time (ignition delay time), as in compression ignition engines. A shock wave may heat up the reactant behind it, leading to ignition of the mixture in high-speed flows with shock waves. The reaction zone of premixed combustion will propagate toward the unreacted fuel–air mixture by continuous heat/mass diffusion from the high-temperature reaction zone to heat up the fuel–air mixture to T_c, and this is referred to as premixed flame propagation or deflagration wave propagation; see Fig. 3.1. In some premixed combustion processes the unburned mixture auto-ignites successively at different spatial locations during which heat diffusion from the reaction zone to the igniting reactant mixture plays a minor role, and this combustion process is referred to as ignition wave propagation, which is the main process in homogeneous charge compression ignition (HCCI) engines.[1] Premixed combustion

---

[1] The recently announced Mazda Skyactiv X engine is based on the HCCI combustion concept.

can also be driven by the shock wave propagation. If the shock is strong enough, the fuel–oxidant mixture can react immediately behind the shock and the reaction zone propagates following the shock. If the lag between the propagations of the shock wave and the flame behind it is small then this process of premixed combustion is referred to as detonation.

The turbulence–chemistry interaction is a vital process in turbulent flames. This highly nonlinear interaction is quite challenging to study and much progress has been made in the past using direct numerical simulations and laser diagnostics, which are discussed in Sections 2.2 and 2.3 respectively. The propagation and structure of the turbulent flames are different under different turbulence conditions. The various regimes of turbulent combustion and associated hypotheses were proposed as early as 1940s by various researchers and these are reviewed in the next section. As one shall see there, although many of these regimes and hypotheses are well understood using experiments and numerical simulations there are still many debatable views regarding the regime classification that need to be reconciled in future investigations.

2.1 Historical and Physical Perspective of Turbulent Combustion

V. Sabelnikov, A. Lipatnikov, X.-S. Bai, and N. Swaminathan

2.1.1 Regime Classification for Non-premixed Flames

The Damköhler number, Da, defined as the ratio of mixing time scale to chemical reaction timescale, is useful to characterize turbulent non-premixed flames. When Da $\gg$ 1, the chemical reactions are sufficiently fast such that no fuel could leak through the reaction zone to the oxidizer side and vice versa. This implies that the reaction zone is around the stoichiometric fuel/air ratio. The structures of turbulent non-premixed flames can be described using mixture fraction, Z, which defines the extent of mixing of the materials originating from the fuel stream and those originating from the oxidizer stream. The reaction zone locates around the stoichiometric mixture fraction, Z_{st}. In the limiting case of Da $\to \infty$, the fuel and oxidizer cannot coexist and Burke and Schumann modeled this limiting case using a one-step irreversible reaction [2]. In this model, the fuel and oxidizer are consumed fully in the reaction zone with a negligible thickness and hence this structure of diffusion flame is commonly referred to as the Burke–Schumann flame sheet.

For turbulent non-premixed flames in practical applications Da is large but finite and thus chemical reactions occur in a thin diffusive-reactive layer [3–7]. From the fuel-rich and fuel-lean sides of the diffusive-reactive layer the fuel and oxidizer are transported to the reaction layer by diffusion and convection. These diffusive-convective layers are of the thickness of the mixing layer, on the same order of magnitude as the integral length scale of turbulence, Λ. The thickness of the diffusive-reactive layer may be denoted as δ_f, whose value depends on the Da value. In general $\Lambda \gg \delta_f$. If the smallest turbulent eddy (Kolmogorov eddies) of a length scale η_k is larger than δ_f then the diffusive-reactive layer in a turbulent non-premixed flame is essentially the same as that of a stretched laminar diffusion flame. Hence, the combustion in turbulent flows with $\eta_k \geq \delta_f$ may therefore be classified as laminar flamelet combustion regime. One may further argue that if the reactive-diffusive layer is thicker than the mixing layer, i.e., $\Lambda < \delta_f$, the diffusion process of the reactive-diffusive layer is by virtue of turbulent eddy motion. This regime of combustion may be referred to as a thickened flame regime [8]. In-between the laminar flamelet and the thickened flame regimes, i.e., $\eta_k < \delta_f < \Lambda$, the combustion may be referred to as the perturbed flamelet regime, since the smallest turbulent eddy can enter into the diffusive-reactive layer and disrupt it [8].

The thickness of the reactive-diffusive layer is not a well defined quantity for laminar diffusion flame, since it varies with Da, or more explicitly with the scalar dissipation rate, χ, defined as $\chi = 2\mathcal{D}(\nabla Z \cdot \nabla Z)$, where $\mathcal{D}$ is the molecular mass diffusivity of Z [9]. The inverse of χ at the stoichiometric location, $1/\chi_{st}$, represents the mixing time scale in the diffusive-reactive layer. If one defines a chemical time scale, τ_c, using the fuel consumption rate then from $\text{Da} = (\chi_{st}\tau_c)^{-1}$, one has $\text{Da} \to \infty$ in the limit of $\chi_{st} \to 0$ and combustion will move toward equilibrium condition having a timescale of $\tau_{c,eq}$. As the rate of mixing or χ increases the reactions will be slower compared to the equilibrium condition, and thus Da would decrease. When χ_{st} is larger than a critical value χ_q, the mixing rate becomes too high for the chemical reactions to sustain and the reactants start to flow across the reaction layer, leading to quenching of the flame [10]. The value of χ_q defines the maximum mixing rate that could be sustained by a flame and this condition gives $\tau_c = \tau_{c,q}$ and $\text{Da} = \text{Da}_q$. Hence, the laminar diffusion flame structure depends solely on χ_{st}. In turbulent non-premixed flames the smallest mixing timescale may be taken as the Kolmogorov time, τ_k. The laminar flamelet combustion regime exists when the condition $\tau_k \geq \tau_{c,q}$ is met or alternatively $\tau_c/\tau_k \leq \tau_c/\tau_{c,q} < 1$. Note that τ_c/τ_k is the Karlovitz number, Ka, frequently used in turbulent premixed flames. Sometimes the strain rate at the location of Z_{st} is used, instead of χ_{st}, to define the mixing time scale, since the strain rate and scalar dissipation rate are related to each other, which was shown using a counterflow diffusion flame in earlier studies [5, 8].

Peters estimated the thickness of the diffusive-reactive layer by assuming the layer thickness in the mixture fraction space is of the order of $\Delta Z \sim 2Z_{st}$ [11]. He classified turbulent non-premixed flames into three regimes, namely, separated flamelets, connected flame zones, and flame extinction. The separated flamelets regime is defined as $\chi_{st} < \chi_q$ and $\sigma_{Z,st} > \Delta Z$, where $\sigma_{Z,st}$ is the standard deviation of the mixture fraction fluctuations at the stoichiometry location. The connected flame zones regime is seen when the conditions $\chi_{st} < \chi_q$ and $\sigma_Z < \Delta Z$ are met. The third regime with flame extinction exists for $\chi_{st} > \chi_q$, which is obvious.

The diffusive-reactive layer thickness in the physical space can also be estimated using the reaction zone thickness in the mixture fraction space and the scalar dissipation rate [12]. The mixture fraction space thickness, ΔZ, is typically defined to be the thickness over which the fuel consumption (or heat release) rate drops to 5% of its maximum value, occurring near the Z_{st} location. The physical-space thickness can then be defined as $\ell_r = \Delta Z/|\nabla Z|_{st}$, where the magnitude of the mixture fraction gradient at the stoichiometry is given by $\sqrt{\chi_{st}/2\mathcal{D}}$. Thus, the laminar flamelet combustion is valid if $\ell_r < \eta_k$, as Bilger suggested [12]. Further details on the regime classification for non-premixed combustion and their validities are discussed in Chapter 4.

2.1.2 Regime Classification of Turbulent Premixed Flames

The classification of turbulent premixed combustion regime can be traced back to the work of Damköhler in 1940s on turbulent Bunsen flames of propane–oxygen mixtures [13], in which he discussed two types of turbulence causing completely different effects on premixed flames. The first one is coarse-body turbulence, with a larger mixing length compared to the thickness of the laminar combustion zone, in which the flame front is "roughened and ultimately scattered" and "as a result, the surface of the flame is enlarged." The second type of turbulence is the "extremely fine-body turbulence with very short mixing length compared to laminar flame zone thickness." Under this condition, "there is no roughening of flame surface," but the "transport processes in it are greater," Damköhler suggested that the increase of turbulent flame speed, S_T, over the laminar flame speed, S_L, for the two turbulent conditions is due to different reasons, i.e., "enlarged flame surface" and "enhanced turbulent mixing in the flame."

Significant progress in turbulent premixed combustion regime classification has been made by improving our understanding of turbulence and laminar premixed flames structure. The large-scale energy containing turbulence has a length scale of Λ, a velocity scale of u', and a timescale $\tau_\Lambda = \Lambda/u'$. The reference length and velocity scales for flames in premixed combustion are the unstrained laminar flame thickness, δ_L, and its burning velocity, S_L. These quantities are intrinsic properties of the fuel–oxidizer mixture and they are not affected by the flow speed or turbulence intensity. This is in contrast to the reference scales for the non-premixed flames discussed in the previous section. Using the aforementioned flame reference scales, a timescale for the flame (or chemical time scale) can be obtained conveniently as $\tau_c = \delta_L/S_L$, which is proportional to the inverse of the overall heat release rate or fuel-consumption rate in the planar unstretched laminar flame. These specifications of reference scales essentially identify that Damköhler's "coarse-body" regime has $\Lambda \gg \delta_L$ whereas the "fine-body" regime has $\Lambda \ll \delta_L$.

Laminar premixed flame theory (see Section 3.1) gives

$$S_L \sim \left(\frac{\alpha}{\tau_c}\right)^{1/2}; \quad \delta_L \sim (\alpha \tau_c)^{1/2}; \quad S_L \delta_L \sim \alpha,$$

where α is the thermal diffusivity. The Reynolds number based on large-scale turbulence is thus

$$\mathrm{Re}_t = \frac{u'\Lambda}{\nu} \sim \frac{u'\Lambda}{\alpha} \sim \frac{u'\Lambda}{S_L \delta_L}. \qquad (2.1)$$

Here ν is the fluid kinematic viscosity, which is of the same order of α since the Prandtl number, $\mathrm{Pr} = \nu/\alpha$, for most reactant mixtures is of the order of unity.

The Reynolds number for Kolmogorov eddies is of the order of unity, $\mathrm{Re}_k = u_k \eta_k / \nu \sim 1$, since viscous dissipation is dominant for scales smaller than η_k. Thus, $u_k \eta_k \sim S_L \delta_L$. The ratio of τ_c to τ_k is known as the Karlovitz number, Ka, which is related to the ratio of flame thickness to the Kolmogorov length scale,

$$\mathrm{Ka} = \frac{\tau_c}{\tau_k} = \frac{\delta_L}{S_L} \frac{u_k}{\eta_k} \sim \frac{\delta_L^2}{S_L \delta_L} \frac{u_k \eta_k}{\eta_k^2} \sim \frac{\delta_L^2}{\eta_k^2} \frac{u_k \eta_k}{S_L \delta_L} \sim \left(\frac{\delta_L}{\eta_k}\right)^2$$
$$= \left(\frac{u'}{S_L}\right)^{3/2} \left(\frac{\Lambda}{\delta_L}\right)^{-1/2}. \tag{2.2}$$

It is unclear when this particular definition of Ka was introduced as a tribute to Bela Karlovitz, who introduced the concept of *flame stretch* [14]. The flame stretch is the percentage change in the flame area, A, per unit time, that is, $\gamma = (dA/dt)/A$. The nondimensional stretch factor introduced by Klimov [15] and Williams [16] is $\varphi = \gamma \delta_L / S_L$ which yields $\varphi \sim (\delta_L / \eta_k)^2$ for $|\gamma| \sim (u'/\lambda)$, where λ is the Taylor microscale.[2] This nondimensional stretch factor is the same as Ka in Eq. (2.2).[3] Bradley and his co-workers [19] introduced Karlovitz stretch factor for turbulent premixed flames, defined as $K = \tau_c / \tau_\lambda$ using the timescale for turbulent eddies of Taylor microscale, λ, and thus $\tau_\lambda = \lambda / u'$. This specific timescale was used because it was thought that the turbulent viscous dissipation might influence the flame propagation phenomena [20, 21]. The appropriateness of using the Taylor microscale and a related Reynolds number $\mathrm{Re}_\lambda = u' \lambda / \nu$ instead of Re_t given in Eq. (2.1) was critiqued by Williams [22]. Furthermore, one could simply see that $K = \varphi$ using the relationships given earlier. We shall adopt the commonly used definition given in Eq. (2.2) for our discussion here. Also, it is worth noting that this Ka can be interpreted as the inverse of a Damköhler number based on the Kolmogrov timescale, $\mathrm{Da}_k = \tau_k / \tau_c$ [23].

If $\mathrm{Ka} \leq 1$ then $\delta_L \leq \eta_k$, which indicates that the heat and mass transfer in the reactive-diffusive layers of the flame is essentially governed by molecular diffusion. The condition of $\mathrm{Ka} \leq 1$ defines a regime of turbulent premixed combustion in which turbulent eddies are usually taken to affect the flame wrinkling (roughening as stated by Damköhler), increasing the surface area of the flame front, while the structure of the local reactive-diffusive layer is similar to that of laminar flame. This regime is known as the laminar flamelet regime [24, 25], which can be subdivided into *wrinkled flamelets* when $u'/S_L < 1$ and *corrugated flamelets* for $u'/S_L > 1$ [11]. The condition $\delta_L = \eta_k$ corresponding to $\varphi = 1$ for the laminar flamelets to exist is referred to as the Klimov–Williams criterion [15, 16]. This condition also corresponds to $\mathrm{Da}_k \geq 1$ or $\mathrm{Ka} \leq 1$. Figure 2.1 shows these limits in the combustion regime diagram as plotted by Peters [26]. This diagram is a variant of various propositions made in the 1980s [24, 25, 27–30]. Figure 2.1 also depicts lines of constant Re_t and $\mathrm{Da} = 1$ along with $\mathrm{Ka} = 1$ and 100 lines to mark the thin reaction zones regime proposed by Peters [26].

---

[2] $|\gamma| \sim \mathrm{Re}_t^{1/2} \tau_\Lambda^{-1}$

[3] The term Karlovitz number or stretch factor has been introduced in the books by Lewis and von Elbe [17] and Strehlow [18], published as early as in the 1960s while discussing laminar spherical flame propagation during ignition and extinction of laminar flames in shear flows.

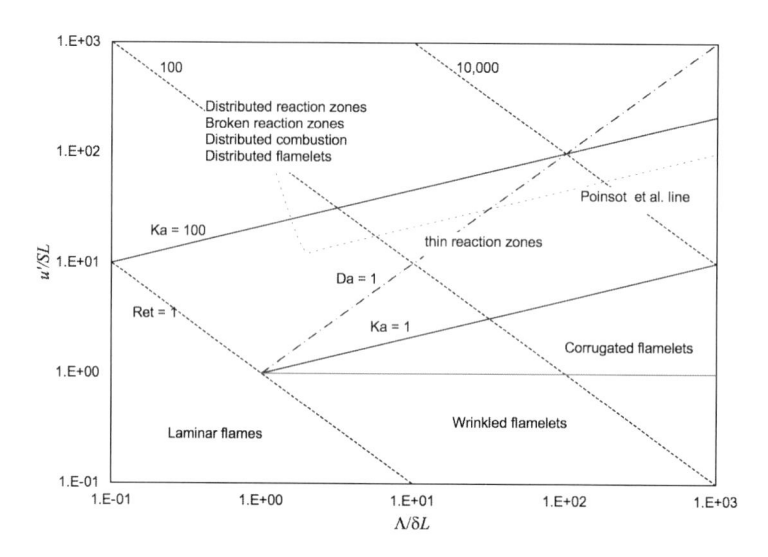

Figure 2.1 Turbulent premixed combustion regime diagram depicting various regimes and their limits.

The combustion regime diagram can also be plotted in Re_t–Da space, as preferred by some researchers, since the diagram so plotted can be used for both premixed and non-premixed combustion. Figure 2.2 shows the various regimes using the turbulence Reynolds number, Re_t, and Damköhler number, Da. The $u'/S_L = 1$ line is plotted to distinguish the wrinkled and corrugated flamelet regimes of turbulent premixed combustion. Since Da is very small in the distributed combustion regime the terminology of distributed flamelets can become contentious but one must be aware that the Da used here is not defined based on the local competition between turbulence and chemical reaction. If one defines the Damköhler number using the local turbulence and chemical timescales then it will surely be larger than unity, where substantial heat is being released in conventional (or classical) combustion situations. Hence, the terminology of *distributed flamelets* may be acceptable and there is ample experimental and numerical evidence for this.

The regime diagram shown in Fig. 2.2 can also be used for non-premixed combustion if one defines the Damköhler number using the mixing timescale, scalar dissipation rate at stoichiometric mixture fraction, that is, $Da = (\chi_{st}\tau_c)^{-1}$ as discussed in the previous subsection. The Karlovitz number can be expressed as an inverse of a Damköhler number based on Kolmogorov timescale as discussed earlier and hence $Da = Da_k Re_t^{1/2}$. The non-premixed combustion occurs in thin sheets for large Da and this sheet is continuous when the turbulence level is low when $(u'\ell_r/\nu) < 1$. This condition is denoted using the $(u'/S_L) = 1$ line in Fig. 2.2. This sheet may be multiply connected at large turbulence level and the flamelet combustion may be taken to prevail in the region bounded by the $Da_k \geq 0.01$ line. Local extinction of non-premixed reaction zones may occur in broken reaction zones regime marked in the figure. Further discussions are given in Chapter 4. Also, it is worth noting here that these regime diagrams characterize the influence of turbulence on combustion but not the influence of combustion on turbulence.

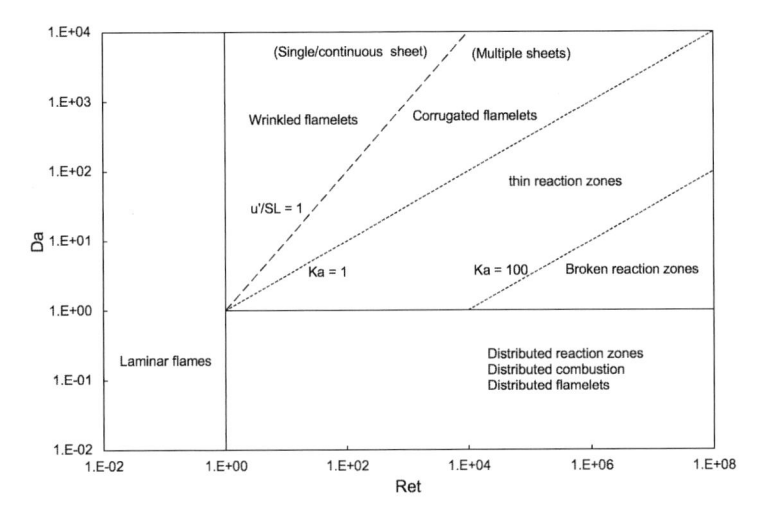

Figure 2.2 Turbulent combustion regime diagram plotted using Re_t and Da parameters.

Knowledge of the structure and propagation of laminar flamelet regimes has been important for the development of models for turbulence–chemistry interaction (TCI) in premixed flames. Various TCI models based on the flamelet concepts are discussed in Chapter 3. It has been shown that models based on the flamelet concepts can indeed be valid for Ka > 1. Numerical simulations [31, 32] and experimental study [33] of flame–vortex interaction suggested that the flame structure remains similar to the laminar flamelet at Ka much larger than unity. These flame–vortex interaction studies considered a single vortex interacting with a premixed flame propagating through the vortex or vice versa, suggesting that the interaction of turbulence and flame might be seen as a superposition of such flame–vortex interactions, which could be at odds with the reality that the turbulence is nonlinear.

More recent laser-induced fluorescence (LIF) imaging of two-dimensional distribution of CH or HCO in turbulent methane–air jet flames at Ka as high as 100 indicated that these radical species are in thin zones of thickness similar to that in laminar flames [34, 35]. This supports the proposition of Peters [11] to divide the regime of Ka > 1 into the *thin reaction zones* regime with 1 < Ka < Ka_c, and a high turbulent intensity flame regime of Ka > Ka_c. The line marked "Poinsot et al. line" in Fig. 2.1 was obtained by investigating the interaction of an initially laminar flame with a vortex pair of a given size and intensity in a few two-dimensional (2D) direct numerical simulations [31]. The upward turn of this line near $u'/S_L \sim 10$ is caused by local quenching induced by fluid dynamic strain. It is also worth noting that this line is parallel to lines of constant Ka values as seen in Fig. 2.1. The flamlet combustion is suggested to be valid in the region bounded by this line rather than the Klimov–Williams line. However, recent direct numerical simulations (see Section 2.2) and experiments [36] show that the flamelet combustion may extend well above Ka = 100 line and also large eddy simulations [37–40] with flamelet-based subgrid scale (SGS) closure offered support for this. Indeed, the upper limit for this regime is still an open question.

In the thin reaction zones regime, the structure of the reaction zone is similar to that in the flamelet regime, whereas the preheat zone is significantly broader than that in the laminar flamelets. The broadening of the preheat zone is by the virtue of heat and mass transfer by turbulent eddies. This implies that the propagation of the local flame is by the mechanism of turbulent eddy diffusion and reaction. Similar to the flamelet regime, the thin reaction zones are wrinkled by the turbulence eddies, which increases the volumetric heat release owing to the increased *flame surface density* (FSD, Σ, defined as flame surface area per unit volume), although locally the reaction rates are of the same order of those in laminar flames.

At high turbulent intensities, that is, $Ka > Ka_c$, all layers of premixed flames are expected to be affected by turbulence. The preheat zone may be broadened and the reaction zones may be distorted and disrupted. This regime of flame is sometimes referred to as the *broken reaction zones regime* [11] or *distributed reaction zones regime*, since it is hypothesized that reactions in this flame regime are locally extinguished or homogeneously distributed. This regime is also called as *distributed combustion regime* some times. Recent direct numerical simulation (DNS) of high Karlovitz number flames [41] has confirmed that indeed local flame extinction of reaction rates can occur at sufficiently high Ka conditions; however, local reaction zones can also be compressed, leading to enhanced reaction rates locally. In fact, homogeneously distributed reactions are seldom found in DNS or experiments (readers are referred to Section 2.2 for a full discussion). The flamelet structures are shown to prevail at high Ka in DNS studies [42]. Indeed, flamelets-based models are also shown to work well for the so-called distributed reaction zones regime combustion [37] and thus this combustion regime may be called a *distributed flamelets regime*. There is also experimental evidence showing flamelet-like structures even in very intense turbulence characterized by $Ka \gg 1$ and $Da \ll 1$. Hence, further exploration is required to understand the upper limit for the regime of flamelet combustion. However, laser diagnostics of turbulent premixed combustion of preheated reactants did not show thin flamelet structure but the reactions are distributed over a broad region, which is indicative of volumetric combustion [43]. Hence, the regime diagram shown in Fig. 2.1 is indeed incomplete and further parameters influencing the turbulence–chemistry interactions and the competing roles of turbulence and chemistry should be included. For example, the chemical timescale is strongly influenced by reactant preheating and dilution using combustion products, which could lead to other modes of combustion without flame propagation. This microvolume combustion is discussed next.

2.1.3 Shetinkov's Microvolume Regime of Turbulent Premixed Combustion

In the discussion that follows, we would like to draw attention to two eventual regimes of highly turbulent[4] premixed burning, which have yet been beyond the mainstream

---

[4] In the rest of this section, words "highly turbulent," "strong turbulence," or "intense turbulence" refer to flows characterized by a large turbulent Reynolds number $Re_t \gg 1$, a low Damköhler number $Da \ll 1$, and a high Karlovitz number $Ka \gg 1$.

discussions in the combustion literature. One regime was pioneered by Shetinkov[5] [44–46] a long time ago. As early as 1955 [46], he introduced a microvolume regime of premixed burning in strong turbulence by hypothesizing that combustion might occur in multiple small-scale, practically homogeneous, pockets/patches/blobs (i.e., spatial regions with low gradients of temperature and species concentrations) of different scales, with the local mixture composition and temperature of a single patch randomly varying within the flame brush. In spite of publication of those ground-breaking ideas in *Proceedings of the Seventh Symposium (International) on Combustion* [44] and a great importance of those ideas for development of advanced engines, the work by Shetinkov was almost forgotten by the turbulent combustion community, probably because flamelet approaches to modeling turbulent premixed burning dominated in the literature over several decades. However, the rapidly growing interest in investigating and applying lean premixed combustion under extreme (Da $\ll 1$ and Ka $\gg 1$) conditions makes the pioneering ideas by Shetinkov highly relevant. Accordingly, these ideas are discussed first.

In the second part, another eventual regime of chemical reactions in highly turbulent medium is discussed. More specifically, propagation of a thin reaction zone in intense turbulence is numerically and theoretically explored under conditions that are well outside the range of conditions for the thin reaction zone regime that was introduced by Peters [11].

2.1.3.1 Physical Reasoning

E. S. Shetinkov (1907–1976) was a Russian engineer and scientist. In 1936, he started his professional activity in the field of ramjet technology in a team led by S. P. Korolev, who later became the chief designer of the space program in USSR. As described in detail elsewhere [47], Shetinkov invented and patented a scramjet (supersonic combustion ramjet) engine in 1957. When compared to a conventional ramjet combustor, a hypersonic flow at the inlet of his scramjet engine is weakly decelerated. Consequently, the flow is supersonic and has an elevated temperature at the scramjet combustor entrance, whereas the flow is subsonic at the entrance of a ramjet combustor [48]. When working on such a scramjet combustor in the 1950s, Shetinkov realized that combustion in very intense turbulence might mainly be controlled by volumetric ignition/autoignition in an ensemble of reactant or reactant–product patches, rather than by propagation of a flame front. In other words, under conditions typical for the scramjet combustor entrance, $\tau_{ign,u}(\phi_u, T_u) < \tau_L(\phi_u, T_u)$, with both the ignition time $\tau_{ign,u}$ and the laminar flame residence time, $\tau_L = \delta_L/S_L$, depending on the local equivalence ratio, ϕ_u, and temperature, T_u, of reactants. Accordingly, the ignition time $\tau_{ign,u}$ is a fundamental characteristic of such a combustion regime and an ignition-delay Damköhler number $\text{Da}_{ign,u} = \tau_\Lambda/\tau_{ign,u}$ naturally appears via an analogy with a common Damköhler number $\text{Da} = \tau_\Lambda/\tau_L$. Moreover, because the ignition time $\tau_{ign,u}$ depends strongly on the local temperature and mixture composition, appearance of patches characterized by a short ignition delay is highly sensitive to small-scale

---

[5] Sometimes, his Russian name is written as Shchetinkov or Schetinkov in English-language literature.

mixing (micromixing) of reactants and products. Accordingly, one more characteristic timescale, i.e., a micromixing timescale τ_{mix}, and one more Damköhler number, $Da_{mix} = \tau_{mix}/\tau_L$, are required to properly describe this combustion regime. The micromixing time is a function of turbulent timescale, τ_Λ, and turbulent Reynolds number, Re_t, i.e., $\tau_{mix} = \tau_\Lambda\, g\,(Re_t)$, where $g\,(Re_t) \ll 1$ is a nondimensional function. If one assumes that $\tau_{mix} \propto \tau_k$, i.e., the micromixing timescale is determined by the Kolmogorov timescale, $\tau_k \propto \tau_\Lambda Re_t^{-1/2}$, then $g\,(Re_t) \propto Re_t^{-1/2}$. In the following discussion of strong turbulence, we shall assume that $Re_t \gg 1$ and the micromixing Damköhler number, Da_{mix}, is much less than unity, i.e., $Da_{mix} \ll 1$.

To recognize the novelty of the paradigm of microvolume burning, developed by Shetinkov by considering reactants at elevated temperatures, it is worth remembering that the two classical regimes of premixed turbulent combustion, i.e., (1) the wrinkled laminar flame regime (known also as flamelet regime), which was introduced by Damköhler [13] and Shchelkin [49] and is commonly associated with $Da \gg 1$ and $Ka \ll 1$, and (2) the distributed reaction regime, which was pioneered by Damköhler [13] and is commonly associated with $Da \ll 1$ and $Ka \gg 1$, address the case of chemically frozen reactants [11, 50, 51]. In other words, the two classical regimes are relevant when $\tau_{ign,u} \gg \tau_L$. Accordingly, the chemical timescale, τ_L, and the corresponding Damköhler number, Da, are used in the classical regime diagrams of premixed turbulent combustion, which are discussed in detail elsewhere [11, 50, 51] and reviewed briefly in Section 2.1.2.

Shetinkov did not restrict himself to the aforementioned relatively simple reasoning but also argued that volumetric ignition/autoignition in reactant–product patches could occur in highly turbulent premixed flames even without reactant preheating, i.e., even if $\tau_{ign,u} \gg \tau_L$ or $Da_{ign,u} \ll Da$. As will be discussed later, Shetinkov demonstrated that a substantial probability of volumetric ignition/autoignition in reactant patches could result solely from the action of turbulence on the mixture provided that the turbulence is sufficiently strong. In other words, Shetinkov hypothesized that the limit of $Da_{ign,u} \to 0$ was singular in strong turbulence! This pioneering idea, which appears to be new even today, was presented as early as in 1958 [44]. This idea led Shetinkov to introduce a new (not only in 1958, but also in 2018) regime and develop a new concept of highly turbulent premixed combustion, i.e., the microvolume combustion regime/concept.

Let us expose the main reasoning and results by Shetinkov [44]. It should be noted that the concept of microvolume combustion was presented by Shetinkov with great clarity. It is difficult, if not impossible, to do better. Therefore, to give readers an opportunity to follow Shetinkov's ideas as closely as possible, but not to deform his style and implied meaning, we shall cite the key points of his analysis from [44].

Shetinkov began the paper with the following statement, which is still valid.

Despite the considerable amount of research work that has been carried out up to the present there is still no clear understanding of the nature of turbulent burning, owing to the exceptional complexity of the phenomenon. For this reason, it would be more correct to speak not of the theory but of the conception or model of turbulent flame. The construction of a proper model is one of the most important present day tasks. The wrinkled laminar flame model has been widely recognized. The model was first formulated by Damköhler and later developed by Schelkin and a number of other research workers. It is based on the assumption that laminar flame which has initially originated cannot be extinguished. The flame is mechanically split into parts or wrinkled by turbulent fluctuations, attended with a proportional increase in the total flame surface area. It is, however, known from experiments that the free flame globules of limited dimensions are extinguished under the effect of strong turbulence, and further propagation of the flame ceases,"

e.g., see Ref. [52] cited by Shetinkov. Subsequently, to resolve the problem, Shetinkov hypothesized the appearance of small-scale patches in which ignition/autoignition occurred and developed following the scenario of creation of such patches by strong turbulence.

The first step consists of local breakup of the quasi-laminar flame by strong turbulent eddies. To argue a significant likelihood of such local breakup events, Shetinkov referred to papers by Karlovitz et al. [14] and by Kovasznay [53] and has written that "theoretical analysis of the laminar flame in a stream with a gradient of velocity points to possible extinction of flame under such conditions" [14] and that the study by Kovasznay [53] "likewise acknowledges the break-up of laminar flame at sufficiently high turbulent velocity."

The second step consists of a rapid (when compared to the laminar flame time, τ_L) micromixing of cold reactants and hot combustion products. The characteristic timescale of this step is $\tau_{mix} \ll \tau_L$. During this step, multiple small-scale inherently homogeneous patches of different scales are randomly distributed by turbulent eddies over the entire turbulent flame brush, with the composition of each single patch being also a random characteristic of the micromixing process. Here, patch composition (1) means not only mass fractions of major species and radicals, but also temperature and (2) corresponds to a randomly selected volume fraction V_p of the products in the reactant–product mixture within the patch. Due to rapid mixing, gradients of temperature and mass fractions of various species are assumed to be low in each patch. Accordingly, each patch is considered to be a homogeneous reactor. In particular, Shetinkov writes that[6]

it is quite likely that in a turbulent brush the laminar flame need not necessarily appear as the boundary between individual moles consisting of combustion products and fresh mixture. If this is so, the mutual diffusion of the combustion products and of the fresh mixture, occurring on the boundaries of and inside individual moles, acquires great importance: the microdiffusion proceeds at a sufficiently rapid rate and the laminar flame has no time to start. Instead, a homogeneous chemical reaction may begin in the mixture of the combustion products, the fuel and the oxidant, which has been formed inside the mole.

The key point of Shetinkov's approach is in using the local ignition time of the lth patch, $\tau_{ign,l}(\phi_l, T_l, V_{p,l})$, consisting of both cold reactants and hot products rather than the local ignition time for the cold reactants, $\tau_{ign,u}(\phi_u, T_u) \gg \tau_L$ (the latter ignition time was used earlier when considering preheated reactants at the scramjet combustor entrance). Contrary to $\tau_{ign,u}(\phi_u, T_u)$, the local ignition time $\tau_{ign,l}(\phi_l, T_l, V_{p,l})$ is not

---

[6] Shetinkov's terms "turbulent mole" and "microdiffusion" correspond to the terms "patch" and "micromixing" with $\tau_{mix} \ll \tau_L$ respectively, used in the present section.

known a priori and it depends on the local patch composition and consequently is controlled by the whole process of (1) the flame break-up, i.e., appearance of holes on the flame surface, and (2) subsequent mixing of the cold reactants and the hot products by turbulent eddies. Due to $T_l > T_u$ and the presence of radicals in lth patch, there could be a substantial probability for $\tau_{\text{ign},l}(\phi_l, T_l, V_{p,l}) < \tau_L$ in spite of the dilution of reactants with products.

As stressed by Shetinkov [44], the microvolume regime is a limiting case, i.e., in this regime, the amount of fuel consumed by propagating wrinkled quasi-laminar flame is assumed to be negligible when compared to the amount of fuel consumed due to ignition/autoignition in the patches. In a general case, both propagation of the flame (the so-called surface mechanism of premixed turbulent combustion) and the ignition/autoignition in the patches (the microvolume mechanism) could simultaneously occur in a turbulent flame brush. Accordingly, Shetinkov [44] has written that

in general, it may be assumed that both surface and homogeneous volume burning may take place in a turbulent brush. The development of a method for calculating such a complex model appears, however, quite difficult. At the present stage of development of the combustion theory, the model should be considerably simplified so that it might be defined quantitatively. Simplification may be effected in two directions: (1) we may disregard the homogeneous reaction and consider that the basic mass of fuel burns out according to the surface mechanism – this is the way chosen by most investigators; (2) the model of the phenomenon may be simplified in another direction, namely, by disregarding the amount of fuel consumed according to the surface mechanism and considering that the bulk of fuel burns out in volume reactions occurring inside individual moles. This paper deals with the second method. For the sake of brevity, such a simplified model is to be subsequently termed the model of micro-volume burning.

It should be stressed that fluctuations of species concentrations and temperature are not negligible for the microvolume combustion regime, because compositions of different patches differ from one another and from the mean mixture composition. Consequently, the mean chemical source terms cannot be calculated using the mean temperature and species concentrations. Therefore, the microvolume combustion regime differs fundamentally from the distributed combustion regime hypothesized by Damköhler [13], because fluctuations of temperature and species concentrations are disregarded when calculating the mean chemical source terms in the latter regime.

We conclude the summary of Shetinkov's concept of microvolume combustion of premixed gas in intense turbulence with the following comment. The microvolume combustion regime is relevant solely to flames with complex chemistry, different molecular-transfer coefficients, or heat losses. Otherwise (i.e., if a flame is adiabatic, all species have the same molecular diffusivities equal to the molecular heat diffusivity of the mixture, and combustion chemistry is reduced to a single reaction), (1) the state of the mixture is fully characterized with a single scalar variable, e.g., the combustion progress variable $c = (T - T_u)/(T_b - T_u)$, and (2) local "holes" in an instantaneous reaction surface cannot appear even in a very strong turbulence (however, a confined flame can be extinguished globally under the influence of the strong turbulence). Accordingly, the surface is continuous even if it is a multiply connected surface.

Let us prove these statements following [54, 55]. In the simple case considered, the heat release rate $\dot{\omega}_T$ depends solely on c with $\dot{\omega}_T(0) = \dot{\omega}_T(1) = 0$ and the dependence $\dot{\omega}_T(c)$ being continuous and peaks at some value of the progress variable, e.g., at $0 < c_m < 1$. Let us (1) consider a statistically planar turbulent premixed flame propagating from right to left and (2) select an arbitrary path starting in reactants ($c = 0$) at $x = -\infty$ and ending in combustion products ($c = 1$) at $x = \infty$. Obviously, a continuous function $c(x,t)$, attains any value $0 \leq c \leq 1$ at least once when following this path and, in particular, the value of c_m. At a point where $c(x,t) = c_m$, the heat release rate $\dot{\omega}_T(c_m)$ reaches its maximal possible value, which is equal to the peak value of $\dot{\omega}_T(c)$ in the unperturbed laminar flame. Since an arbitrary path is selected, the foregoing reasoning proves that there are no "holes" in the reaction surface and the patches do not appear in the simplest case considered.

2.1.3.2 Numerical Simulation of the Microvolume Combustion Regime: Lattice Automata

Due to the complexity of the microvolume regime of turbulent premixed combustion, it is not evident how to apply available turbulence models for simulating such a regime. As a first step to resolving the problem, Shetinkov [44] developed a simplified 2D Monte Carlo method for numerically modeling the microvolume combustion regime. The following two simplifications were invoked for that purpose. First, the influence of combustion-induced thermal expansion on the flow was neglected (the same assumption was invoked in many classical [13, 49] and more recent, e.g., [11, 50], studies). Second, the mean flow (longitudinal) velocity U was assumed to be much larger than the rms turbulent velocity u', i.e. $U \gg u'$, with the x-axis being parallel to the direction of the mean flow. It may be of interest to note that Shetinkov's method strongly resembles a lattice (cellular) automata approach to studying various phenomena described commonly by Navier–Stokes equations, e.g., reaction–diffusion systems [56] and, in particular, turbulent premixed flames [57]. The lattice (cellular) automata approach offers an alternative to numerically integrating Navier–Stokes equations.

Since, in spite of its originality, the numerical method developed by Shetinkov does not add a new physics to the concept of microvolume combustion in intense turbulence, we will restrict ourselves to a brief summary of the method. The reader interested in details is referred to the original paper [44].

Within the framework of the method, during each time step Δt, the following processes were simulated: (1) turbulent diffusion controlled by the velocity field, (2) micromixing between hot combustion products and cold reactants, and (3) chemical reactions in the patches. To perform such simulations, the 2D flow was divided into a number of equally small (not necessarily square) cells. Each cell contained zero, one, or several patches (*turbulent moles* as described in [44]), with compositions (species mass fractions and temperature) of all patches within any single cell being assumed to be the same and homogeneous at each time step t_n. Between two subsequent time steps t_n and t_{n+1}, evolution of the composition of each cell was divided into two fractional steps and was calculated invoking the following physical models and simplified hypotheses.

During the first fractional step, evolution of the patch composition due to chemical reactions was directly computed and turbulent diffusion was modeled by simulating a random walk of the patches and by assuming that the turbulent velocity field was homogeneous. More specifically, for each patch, (1) an absolute value v of its transverse velocity was randomly chosen from a finite interval of $v_{min} \leq v \leq v_{max}$, (2) one of two possible directions of advection of the patch in the transverse direction was also selected randomly, (3) a length ℓ of the patch advection was calculated as follows: $\ell/\ell_{min} \propto (v/v_{min})^3$ following the Kolmogorov theory of inertial turbulence [58–60], and (iv) duration Δt_d of this fractional step was set equal to ℓ_{min}/v_{min}. During this time interval, different patches moved the same (different) distances $\Delta x = U\Delta t_d = U\ell_{min}/v_{min}$ ($\Delta y = v\Delta t_d = v\ell_{min}/v_{min}$, respectively) in the longitudinal (transverse, respectively) direction, with $\Delta x \gg \Delta y$. Accordingly, patches occupying the same cell at instant t_n could arrive in different cells at instant $t_n + \Delta t_d$.

During the second fractional step of duration $\Delta t_m = \tau_{mix}$ equal to the micromixing time, micromixing between patches that occupied the same cell at instant $t_n + \Delta t_d$ was simulated. For that purpose, different methods might be used, e.g., the compositions of all patches in a single cell after the micromixing might be assumed to be the same and calculated by simply averaging the compositions of all patches that were in the cell at instant $t_n + \Delta t_d$. The reader interested in details is referred to [44, 45]. Thus, after the two steps, i.e., at $t_{n+1} = t_n + \Delta t = t_n + \Delta t_d + \Delta t_m$, each cell that contained at least one "new" (with respect to instant t_n) patch had its new composition.

In Refs. [44, 45], calculations were performed in the case of two-scale turbulent velocity field, i.e., two lengths, $\ell_1 = \ell_{min}$ and $\ell_2 = 8\ell_1$, and two velocity, $v_1 = v_{min}$ and $v_2 = v_{max} = 2\ell_1$, scales were used. The micromixing process was assumed to be very fast when compared to turbulent diffusion and chemical reactions. Consequently, $\Delta t \approx \Delta t_d$. Combustion of a hydrocarbon fuel was addressed invoking a single-step mechanism. Shetinkov [44] performed those simulations to examine the influence of the rms turbulent velocity u', equivalence ratio ϕ, reactant temperature T_u, laminar flame speed S_L, and pressure p on the turbulent burning velocity. He also studied the blowdown of turbulent flames of finite dimensions. As discussed in detail in Refs. [44, 45], the obtained numerical results were in the qualitative agreement with experimental data available at that time.

2.1.3.3 Microvolume Paradigm and the State-of-the-Art of Highly Turbulent Premixed Combustion

In this subsection, we will briefly review and discuss recent data that show the importance of Shetinkov's microvolume concept of highly turbulent premixed combustion for both fundamental and applied research into turbulent flames. Over the past years, there appeared a growing body of experimental [36, 43, 61–68] and DNS [69–81] evidence that seemed to support the microvolume concept. Such evidence was obtained not only in the case of preheated reactants but also in cases with sufficiently low reactant temperature. Based on this evidence, the microvolume combustion regime appears to play a role in various advanced combustion systems such as moderate, intense or low dilution (MILD) combustion [71–79], gas turbine engines (e.g., see the review paper [82] and references therein), scramjet engines [66–68], and so forth.

For instance, Yoshida et al. [43] experimentally investigated turbulent premixed combustion in a stirred reactor, with the reactants being preheated (from 800 to 1200 K) in order to simulate conditions in an adiabatically compressed mixture or in a cylinder of an internal combustion engine. These authors have concluded that their "OH-PLIF images suggested that there are no thin laminar flamelets and that the reacting eddies were distributed throughout the reactor"; see the OH-PLIF image in fig. 1 of [43].

Summarizing experimental results obtained from highly turbulent methane–air flames stabilized using a piloted Bunsen burner under room conditions, Driscoll has concluded that "the small amount of evidence to date indicates that turbulence alone does not create large, uniform reaction zones, but instead produces pockets of distributed reactions that are connected to each other by thin flamelets. This situation has been called partially-distributed" [36] (p. 5). Moreover, he has noted that "partially distributed reactions (PDR) are local blobs of distributed reactions that are connected by thin flamelets. They were found to be associated with either merging or breaking of the flamelets" [36] (p. 1). This type of combustion is also called as distributed flamelets regime [37]. Furthermore, based on the same experiments performed at the University of Michigan, he proposed that the "Borghi regime diagram needs third (and fourth) axes" to take into account residence time, mixing time, and ignition time (in the case of preheated reactants) [43, 61–83]. The same group has also written about "new autoignition physics if reactants are preheated" [84]. The aforementioned ideas of "partially-distributed" combustion and "new autoignition physics if reactants are preheated" appear to be basically close to Shetinkov's concept of a microvolume combustion regime and to Shetinkov's analysis of the combustion mechanism in a scramjet combustor, respectively. It may be of interest to note that the interplay between autoignition and flame propagation mechanisms was theoretically studied by considering a single irreversible reaction in the one-dimensional case by Zaidel and Zeldovich [85] more than 55 years ago. A thorough discussion of that problem and of theoretical results obtained by Zaidel and Zeldovich [85] can be found in Refs. [86, 87].

It is also worth noting that, for large hydrocarbon molecules, the autoignition process is mainly controlled by the low-temperature chemistry of fuel oxidation. As recently claimed by Ju [82], "turbulent combustion inadvanced engines is highly governed by the low temperature chemistry and transitions between ignition and flame propagation. The existence of low temperature chemistry and the increase of ignition Damköhler number will significantly modify the turbulent flame regimes and the regime diagram." In particular, the appearance of patches containing a mixture of fresh reactants and hot combustion products and subsequent interplay between the autoignition and flame propagation mechanisms seems to be of great importance for MILD combustion. For instance, "there are possibilities for the fuel to be partly mixed, but not completely, with the products and air for a short mixing time; this thereby creates spatially and temporally non-uniform mixtures of exhaust and fresh gas pockets. The size of these pockets is random and determined by the turbulence conditions" [72] (p. 10), and "thin reaction zones propagating at S_L solely are no longer predominant in MILD combustion due to the reaction zone interactions and autoignition producing distributed reaction zones" and "MILD combustion consists of volumetric reaction zones which spread out over a large portion of the domain" [72] (p.126). This physical picture is not influenced by the presence of spatiotemporal variation of mixture fraction [88]. These statements agree very well with Shetinkov's pioneering ideas about highly turbulent combustion of preheated reactants.

To conclude this section, it is worth noting that MILD combustion is commonly simulated using Perfectly Stirred Reactor (PSR) and Partially Stirred Reactor (PaSR) approaches [71–79, 89], which are discussed in detail elsewhere [71, 72, 79]. Both the PSR and PaSR approaches (see Section 3.2) place the focus on "volumetric reaction zones which spread out over a large portion of the domain having a representative residence time" [72]. Both the PSR and PaSR approaches are basically similar to the Eddy Dissipation Concept (EDC) [90] containing two key input parameters: (1) residence time τ^* in fine structures (pockets) and (2) volume fraction of the fine structures. As far as the residence time τ^* is concerned, it may be associated with the micromixing time τ_{mix} discussed earlier. The EDC assumes that $\tau^* \approx \tau_k$. The choice of τ^* for simulations of MILD combustion is discussed elsewhere [71–79, 89]. As far as the volume fraction is concerned, DNS of MILD combustion [72–76] showed that the characteristic length of the fine structures is of the order of the laminar flame thickness, δ_L, i.e., their volume is of the order of δ_L^3, whereas the EDC assumes that the characteristic length is of the order of the Kolmogorov microscale η_k, i.e., the volume is on the order η_k^3. Accordingly, it would be of interest to assess an Extended PaSR (EPaSR) model [91] for MILD combustion. In the EPaSR model, (1) the characteristic length of the fine structures depends not only on both η_k and δ_L, but also on τ_k and S_L and (2) the residence time τ^* is equal to the geometrical mean of the Kolmogorov time τ_k and either a timescale associated with the subgrid velocity stretch in large eddy simulations (LES) or the eddy turnover time τ_Λ in Reynolds averaged Navier–Stokes (RANS) simulations.

2.1.4 Propagation of Thin Reaction Zone in Intense Turbulence

In this subsection, another eventual regime for reaction wave propagation in a highly turbulent medium is addressed. It may be of interest to note that, contrary to Shetinkov's concept of microvolume burning, the physical scenario discussed in the following does not require complex chemistry.

Over the past years, the growing body of experimental [61, 92–94] and DNS [95–99] data obtained from highly turbulent premixed flames indicated that, contrary to the widely accepted paradigm of distributed burning, thickening of reaction zones was statistically moderate even in very intense turbulence characterized by Ka $\gg$ 1 and Da $\ll$ 1. These data put this classical paradigm into question and call for better exploration of the governing physical mechanisms of highly turbulent premixed combustion. However, this task is very difficult for a number of reasons. In particular, owing to extreme complexity and high costs of the aforementioned experiments or simulations, each (especially, numerical) study cited earlier has yet dealt with a few cases only.

A significantly wider range of highly turbulent conditions can be explored in DNS of propagation of a dynamically passive single-reaction wave in intense constant-density turbulence not affected by the wave. While such a problem is extremely simplified when compared to premixed turbulent combustion (and Shetinkov's concept of microvolume burning clearly shows this), the problem is an important building block for various models of turbulent flames. Moreover, the problem is directly relevant to discussion of various combustion regimes, because the classical combustion-regime diagrams [11, 24, 25, 30] and, in particular, the concept of distributed burning in intense turbulence were introduced by neglecting variable-density, preferential diffusion, and complex-chemistry effects. Accordingly, propagation of a dynamically passive single-reaction wave in intense constant-density turbulence was recently investigated [100, 101] by analyzing a big subset (23 cases characterized by $0.01 \leq \text{Da} < 1$) of DNS data, extracted from a larger data set of 45 cases [102–104].

The results obtained in [100, 101] indicate that (1) thickening of reaction zones of the waves is statistically weakly pronounced even at Da as low as 0.01, in qualitative agreement with the aforementioned experimental [61, 92–94] and DNS [95–99] data obtained from flames; (2) an increase in burning rate by turbulence is mainly controlled by an increase in the area of the reaction-zone surface, in qualitative agreement with other recent DNS data [105] obtained from highly turbulent flames; but (3) a ratio of turbulent burning velocity S_T to the laminar flame speed S_L scales as $\text{Re}_t^{1/2}$, in line with experimental [106, 107] and DNS [108] data. It is worth stressing that the same scaling of $S_T \propto S_L \text{Re}_t^{1/2}$ was predicted by Damköhler [13] and is commonly associated with the distributed reaction zones, whereas reaction zones were thin under conditions of the discussed DNS [100–104].

To reconcile the aforementioned findings (1) and (2) and the experimental [61, 92–94] and DNS [95–99, 105] data with the aforementioned finding (3) and experimental [106, 107] and DNS [108] data, an alternative (to distributed reactions) regime of propagation of a thin reaction zone in intense turbulence was studied in Refs.[100, 101] and the scaling of $S_T \propto S_L \text{Re}_t^{1/2}$ was theoretically predicted for that regime. These recent findings are briefly summarized in the text that follows.

Let us consider a statistically 1D planar reaction wave that propagates from right to left along the x-axis in 3D homogeneous isotropic intense ($\text{Re}_t \gg 1$, $\text{Da} \ll 1$, and $\text{Ka} \gg 1$) turbulence, but does not affect it, i.e., the density and viscosity are assumed to be constant. To explore a physical scenario opposite to the widely accepted paradigm of distributed reactions, the reaction rate $\dot{\omega}$ is assumed to depend on $c(\boldsymbol{x},t)$ in a highly nonlinear manner, i.e., $\dot{\omega}(c)$ vanishes outside a very thin reaction zone, whose thickness δ_r is much less than the Kolmogorov length scale η_k. Accordingly, $\delta_r \ll \delta_L$ by virtue of $\text{Ka} \gg 1$. When considering length scales $\ell \gg \delta_L \gg \eta_k \gg \delta_r$, the thin reaction zone may be reduced to a reaction sheet, where the following constraints hold [86]:

$$c(\boldsymbol{x},t)|_r = 1, \quad |\boldsymbol{n} \cdot \nabla c|_r = |\nabla c|_r = \frac{S_L}{\mathcal{D}} = \frac{1}{\delta_L}, \qquad (2.3)$$

i.e., the reaction progress variable is continuous, but its gradient drops from $1/\delta_L$ at the reaction sheet to zero on either sides. The mass diffusivity of c is denoted using $\mathcal{D}$, which is equal to the thermal diffusivity if c is defined using temperature or under conditions with Pr = 1. The subscript r designates the reaction sheet, $\boldsymbol{n} = -\nabla c/|\nabla c|$ is the unit vector normal to an isosurface of $c(\boldsymbol{x},t)$, and n is spatial distance counted from the reaction sheet along the $\boldsymbol{n}$-direction. Equation (2.3) warrants that the reactant flux $\mathcal{D}|\nabla c|_r$ toward the reaction sheet is equal to the rate S_L of the reactant consumption per unit sheet area, but the speed $S_{d|r} = \mathcal{D}\left(\nabla^2 c/|\nabla c|\right)_r$ of self-propagation of the sheet in an inhomogeneous flow can significantly differ from S_L [109].

Under the preceding assumptions, the structure of a reaction wave in a turbulent flow is modeled by a standard transport equation

$$\frac{\partial c}{\partial t} + \boldsymbol{u}\cdot\nabla c = \nabla\cdot(\mathcal{D}\nabla c), \qquad (2.4)$$

provided that Eq. (2.4) holds at the reaction sheet and the boundary condition of $c(-\infty,y,z,t)=0$ is set far ahead of the sheet. Thus, while ∇c at the reaction sheet is determined solely by the molecular diffusion and reaction [see Eq. (2.3)] the transport Eq. (2.4) does not involve the reaction rate. Accordingly, turbulence affects the gradient ∇c at a finite distance from the reaction sheet. When the distance is increased, the influence of the boundary conditions given by Eq. (2.3) on the fields $c(\boldsymbol{x},t)$ and ∇c is likely to be reduced and, consequently, the relative magnitude of the effect of the turbulence on these fields is likely to be increased. Therefore, it is tempting to hypothesize that, outside a transition layer of a thickness n_r, i.e. at $n > n_r$, the turbulence overwhelms the influence of the reaction and the local concentration gradients are controlled solely by inert turbulent advection and molecular transport under considered conditions of $\text{Re}_t \gg 1$, $\text{Da} \ll 1$, and $\text{Ka} \gg 1$.

This hypothesis can be supported by the following order-of-magnitude estimates. First, let us compare the magnitude $|\nabla c|_t$ of the concentration gradient outside the transition layer with the magnitude $1/\delta_L$ of the concentration gradient at the reaction sheet. The order of the former magnitude may be estimated based on the widely accepted view that, in an inert flow, the mean scalar dissipation rate $\bar{N} = \overline{\mathcal{D}(\nabla c)^2} \propto (1/\tau_\Lambda)$ is independent of Re_t [55, 110]. Accordingly, $|\nabla c|_t \propto (\mathcal{D}\tau_\Lambda)^{-1/2}$ and

$$\frac{|\nabla c|_t}{|\nabla c|_r} \propto \text{Da}^{-1/2} \gg 1, \qquad (2.5)$$

provided that the molecular diffusivity $\mathcal{D}$ and kinematic viscosity ν are of the same order, i.e., the Schmidt number $\text{Sc} = O(1)$.

Second, the transition-layer thickness n_r can be estimated by expanding $c(n)$ to Taylor series in the vicinity of the reaction sheet, i.e.,

$$c(n) \approx 1 - \left|\frac{\partial c}{\partial n}\right|_r + \frac{1}{2}\left(\frac{\partial^2 c}{\partial n^2}\right)_r n^2. \qquad (2.6)$$

Since the expansion coefficient in the linear term is controlled solely by the reaction [see Eq. (2.3)], let us estimate the thickness n_r by equating the linear and quadratic terms in Eq. (2.6). Then, we have to estimate $(\partial^2 c/\partial n^2)_r$. This could be done by considering the simplest relevant problem, i.e., a stationary 1D planar laminar reaction wave stabilized in a 2D flow ($u = -\gamma x, v = \gamma y$), with the velocity gradient being on the order of the inverse Kolmogorov timescale τ_k^{-1}. In such a case, $c(-\infty) = 0$, Eq. (2.3) holds, Eq. (2.4) reads

$$-\frac{x}{\tau_k}\frac{dc}{dx} = \mathcal{D}\frac{d^2c}{dx^2}, \tag{2.7}$$

and $n = x_r - x$. As shown elsewhere [54], integration of Eq. (2.7) results in $x_r = \eta_k \{\ln(\text{Ka}/2\pi)\}^{1/2}$ if $\text{Sc} = \mathcal{O}(1)$ and $\tau_k \ll \delta_L^2/\mathcal{D}$, i.e. $\text{Ka} \gg 1$.

Consequently, Eqs. (2.3), (2.6), and (2.7) yield

$$n_r = 2\eta_k \left\{\ln\left(\frac{\text{Ka}}{2\pi}\right)\right\}^{-1/2} \ll \eta_k. \tag{2.8}$$

Moreover, at the boundary of the transition layer, the difference between unity and the boundary value c^* of the reaction progress variable is less than $\epsilon = 2\{\text{Ka}\ln(\text{Ka}/2\pi)\}^{-1/2} \ll 1$ [100].

The aforementioned order-of-magnitude estimates support the following scenario. If $\text{Re}_t \gg 1$ and $\text{Da} \ll 1$, the reaction significantly affects the $c(\mathbf{x},t)$-field in a narrow layer ($c^* < c < 1$ with $c^* > 1 - \epsilon$) in the vicinity of the reaction sheet, with the thickness of this layer being less than η_k. Since the distance between the reaction sheet and the inert isosurface $c(\mathbf{x},t) = c^* < 1$ is so small, we may assume that the two surfaces move in a close correlation with one another and, hence, their areas are roughly equal, i.e., $A_r \approx A_{c^*}$. The latter area can be estimated as follows $A_{c^*} \approx A_c$ invoking knowledge on the area A_c of an isoscalar surface in the case of inert turbulent mixing.

Finally, using the following scaling [55],

$$A_c \propto A_0 \text{Re}_t^{1/2} \tag{2.9}$$

for the latter area, we arrive at the well-known Damköhler's expression for turbulent consumption velocity:

$$\frac{S_T}{S_L} = \frac{A_r}{A_0} \propto \frac{A_{c^*}}{A_0} \propto \frac{A_c}{A_0} \propto \text{Re}_t^{1/2}. \tag{2.10}$$

Here, A_c and A_0 are areas of instantaneous $c(\mathbf{x},t) = C$, and mean, $\bar{c}(\mathbf{x},t) = C$, iso-surfaces, respectively. Equation (2.10) agrees well with (1) the recent DNS data [100, 104] simulated for a dynamically passive, single-reaction wave; (2) DNS data computed for thermonuclear explosion in Type Ia supernovae [108] and highly turbulent premixed flames [111]; and (3) experimental data obtained from chemically reacting aqueous solutions [106] and premixed turbulent flames characterized by a low Da [107]. Moreover, the preceding analysis is also consistent with experimental [61, 92–94] and DNS [95–99] data that indicate statistically moderate thickening of reaction zones in highly turbulent premixed flames. Furthermore, the preceding analysis is also consistent with recent DNS data [105], which show that a ratio of S_T/S_L is controlled by flame surface area at large u'/S_L, high Ka, and low Da.

While Eq. (2.10) coincides with the classical scaling by Damköhler [13], the two results were obtained for different regimes, i.e., distributed reactions in Ref. [13], but infinitely thin reaction zones in the preceding analysis. The governing physical mechanisms of the influence of turbulent eddies on S_T are different in the two regimes. These are intensification of mixing within broad reaction zones in the former (classical) regime and an increase in the area of the reaction sheet in the latter (alternative) regime. A common feature of the two approaches consists in highlighting turbulent mixing.

The considered study [100, 101] implies that, even at a low Da, the reaction zone may be thin and an increase in the consumption velocity by turbulence may be controlled by an increase in the area of the reaction surface, with the latter increase being well described by the theory of inert turbulent mixing. The summarized results [100, 101] offer an opportunity to reconcile experimental [106, 107] and DNS [108, 111] data that validate the classical Damköhler expression, i.e., $S_T/S_L \propto \mathrm{Re}_t^{1/2}$, with the latest experimental [61, 92–94] and DNS [95–99] data that show thin heat-release zones in highly turbulent premixed flames. It is worth noting that a similar scaling is reported not only by Damköhler [13], but also by other researchers based on different reasoning, e.g. see Refs. [26, 112–116]. The discussed results [100, 101] do not disprove the concept of distributed burning, but indicate that eventual transition to distributed reactions requires Damköhler and Karlovitz numbers significantly lower and higher, respectively, than the commonly accepted boundaries of Da $=$ 1 and Ka $=$ 100, repectively.

In summary, both the microvolume combustion concept put forward by Shetinkov [44–46] about 60 years ago and the recent study [100, 101] imply that the governing physical mechanisms of highly turbulent combustion are still poorly recognized and combustion regimes alternative to the classical paradigm of distributed reactions are worth exploring. Perhaps the microvolume combustion regime can help us to reduce the impact of combustion on the environment and MILD combustion, discussed in Chapter 6, is an example for this.

Acknowledgments

V. A. S. gratefully acknowledges the financial support by ONERA and by the grant of the Ministry of Education and Science of the Russian Federation (Contract No. 14.G39.31.0001 of 13.02.2017). A. N. L. gratefully acknowledges the financial support by the Combustion Engine Research Center (CERC).

2.2 Direct Numerical Simulation Perspective

H. Im, J. H. Chen, and A. Aspden

This section attempts to provide an overview of the current state of the art in direct numerical simulation (DNS) to serve as a *numerical experiment* for turbulent combustion modeling development. Here the word "direct" is taken in the context of conventional turbulent flow simulations within the continuum limit, in that no closure models are used to describe the Reynolds or subgrid stress terms. As such, the premise of DNS is to provide accurate solutions to the conservation equations for the reacting flows with full temporal and spatial resolutions. Within the DNS applications, however, the complexities in the number of reactive scalars and the description of the chemical reaction pathways vary, from single-step chemistry with one-step irreversible reaction to detailed chemistry description with hundreds of species and thousands of elementary reactions incorporating elementary diffusive transport formulas. Considering that the physically correct turbulent kinetic energy spectrum is possible only in three-dimensional (3D) flow fields, the present discussion covers 3D simulations only.

There are significant restrictions in attempting DNS of turbulent premixed flames, in terms of the range of length and timescales that can be reached. In particular, three length scale ratios are of interest. First, $R = l_F/\Delta x$ is the number of computational cells across the flame thickness, l_F, commonly defined by the thermal thickness based on the temperature profile. Second, $\Lambda = l/l_F$ is the number of flames across the integral length scale, and finally $\Omega = L/l$ is the number of integral length scales across the domain. The product of these three ratios gives the number of computational cells required in each direction of the domain $N = L/\Delta x = \Omega \Lambda R$, from which the computational cost can be estimated. For example, taking the flame thickness $l_F \approx 100$ μm, which is resolved by 10 computational cells, $l \approx 1$ cm, and $L \approx 10$ cm quickly leads to the conclusion that even a small combustion system requires about 10,000 cells in each direction, or a trillion cells in 3D simulations. The number will increase even more with increasing pressure as the flame thickness decreases, implying that DNS of any canonical combustion system will remain infeasible for the foreseeable future.

The largest combustion DNS to date typically have a few billion cells (two or three orders of magnitude smaller than the preceding example), and these kinds of simulations are out of reach for all but those with the largest computational allocations; even then, only one or two simulations can be run at a cost of tens of millions of CPU hours. As a result, DNS of laboratory scale burners are either restricted to relatively small cases (e.g., the Lund burner [34, 117]), or with a limited number of cells across the flame (e.g., the low-swirl burner [118, 119]). For fundamental studies of turbulence–flame interaction, an alternative canonical configuration that has been used extensively in the past decade is a rectangular "flame in a box" in which a statistically planar premixed flame propagates toward an incoming homogeneous isotropic turbulence that is either decaying or forced [42, 120–123]. In the following, DNS of canonical configurations will be discussed first in Section 2.2.1, followed by DNS of complex flames in Section 2.2.2.

2.2.1 DNS of Canonical Configurations

The flame-in-a-box approach is a commonly adopted configuration to study the turbulence-flame interaction in an unbiased way by imposing homogeneous and isotropic turbulent flows at the inflow boundary. As such, the flame front is planar on statistical average, eliminating any potential skewness in turbulent statistics associated with mean flame curvature or flow shear. Periodic boundary conditions are used in both transverse directions to avoid complexities of flame–wall interaction. For the turbulent inflows, a nonreacting turbulent velocity field generated from a prescribed kinetic energy spectrum or by a cold flow DNS in a box is fed into the inflow boundary. The level of turbulence intensity inherently decays over the streamwise direction, as in experimental turbulent channel flow [124]. To sustain the level of turbulence intensity over a longer distance, various ways of forcing strategies have been developed in the spectral space [42, 120] or in the physical space [122, 125, 126].

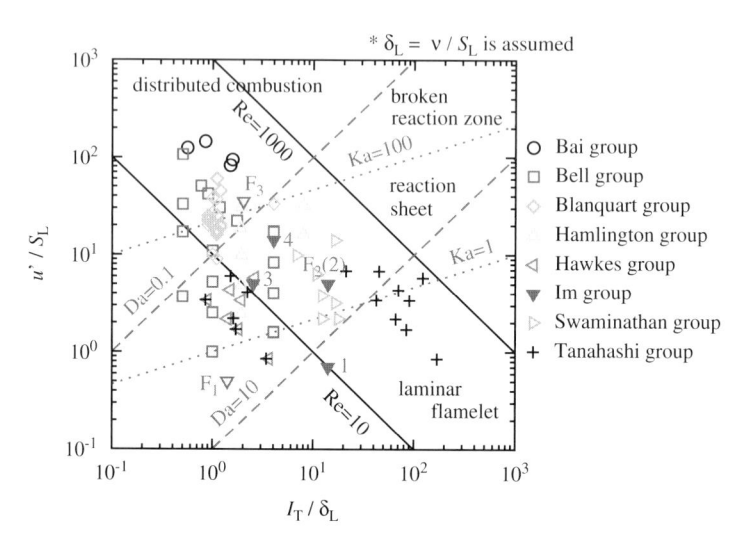

Figure 2.3 Compilation of parametric conditions of DNS cases in recent studies. The Zeldovich flame thickness, $\delta_L = \nu/S_L$, was used for normalization.

Figure 2.3 shows a compilation of DNS cases in recent publications [97, 121–123, 127–143] shown on the Borghi diagram. Isolines of the key nondimensional numbers, such as Re, Da, and Ka, are overlaid to show the relevant combustion regimes. Note that most of the conditions are overestimated based on the initial conditions of the simulations and the effective level of turbulence is expected to be somewhat lower than reported. In particular, the estimation of Ka depends on the definition of the flame thickness. Considering that the actual thermal thickness accounting for the temperature-dependent properties is usually larger than the nominal Zeldovich thickness typically by several factors, in many papers the reported Ka tends to be overestimated.

The simplicity of the configuration accounts for its prevalence in the literature, but there are a number of limitations that should be taken into account. For example, turbulence decay in the streamwise direction leads to a fundamental instability; as the flame progresses toward the inflow, it experiences higher turbulence levels, leading to an increased turbulent flame speed. Conversely, if the flame burns slower than the inflow speed, it is pushed further downstream, where it experiences lower turbulence, leading to a reduced turbulent flame speed.

The density jump across the reaction zone makes the flame front vulnerable to the Darrieus–Landau instability. In addition, the density gradient combined with an accelerating mean flow may also induce Rayleigh–Taylor instabilities depending on the sign of the acceleration. Since a simulation usually starts as a flat flame, these instabilities developed over a long initial transient period (until the statistically steady condition is reached) may lead to a highly complex and a wide extent of the turbulent flame brush in the streamwise direction, thus demanding a substantial domain length. To suppress these large-scale phenomena that require significantly higher computational costs to capture, a high aspect ratio domain (longer in the streamwise) with the domain size in the periodic dimension limited to a few to fewer than 10 integral scales is commonly adopted.

2.2.1.1 Small-Eddy Simulation

The flame-in-a-box DNS with maintained turbulence has earned the moniker of "small eddy simulation" (SES), implying that this approach is to capture the turbulence–flame interactions at the smallest scales, while the large-scale are implicitly filtered out by the small (periodic) domain size. While the large-scale effects may be mimicked by a momentum source term such as a "super-grid model," e.g. a mass source term representing the compression stroke of an internal combustion engine, the small box DNS inherently omits the dynamical behavior associated with turbulent eddies and flame scales that are larger than the domain size.

The classical description of turbulence is a spectrum of eddies from the energy-containing large scales responsible for the supply of kinetic energy through the inertial subrange to the dissipation subrange, all the way down to the Kolmogorov eddies where the turbulent kinetic energy is dissipated into heat by molecular diffusion. If the level of turbulence is sufficient for the Komogorov scale to be comparable to the flame scale or smaller, the ensemble of these small eddies interacting with the local flame structure provides a universal statistics of the turbulent combustion characteristics. According to the prevailing theory, the turbulent combustion regimes are identified based on the turbulent Karlovitz number, $Ka = \tau_f/\tau_K$, defined as the ratio of the flame time to the Kolmogorov time; as the laminar flamelet regime if $Ka < 1$, and the reaction sheet regime if $1 < Ka < 100$, and the distributed combustion regime if $Ka > 100$ [11]. Note that in real turbulent combustion systems, the dilatation associated with heat release may significantly modify the turbulence structure in the direction of laminarization, such that the flame structure truly exhibiting the distributed combustion regime has been much more difficult to realize in laboratories [61], but recent simulations using large-scale turbulence forcing was able to reveal such conditions [123, 144] to provide fundamental insights into high-Ka flames.

Based on the preceding arguments, SES should be regarded as a canonical model simulation to unravel the universal characteristics of turbulence–flame interaction without the complication by the system-specific nonidealities of real-world burner configurations. If one believes that the interaction of the turbulent eddies and flames at comparable physical scales is the essential feature that determines the turbulent combustion dynamics, the flame-in-a-box SES database serves as a valuable reference for a priori validations of combustion closure models.

2.2.1.2 Characteristics of Flame Structure and Propagation in the Flamelet Regime

Due to the limitations in the computational cost, earlier DNS cases mainly reproduced the flames in the laminar flamelet or reaction sheet regimes, at Ka of $O(10)$ or less, either with simple one-step chemistry or simple fuels with detailed mechanisms. At these conditions, turbulent eddies at all scales are virtually comparable or larger than the flame thickness, and thus their interactions are manifested as the wrinkling of laminar flames whose internal structure across the normal direction remain nearly intact. The database was subsequently analyzed to investigate the statistical behavior of typical flamelet observables, such as curvature and tangential strain rate, on the local velocities of the flame surface commonly defined by an isosurface of a chosen scalar variable.

Turbulent flame *thickness* relative to the reference laminar flame thickness is one of the key quantities of interest. To this end, there are two different thicknesses of interest. First, under the condition that the elementary flame segments retain either the laminar flamelet or reaction sheet regimes, the turbulent eddies interfering with the molecular transport within the preheat zone are expected to modify the flame thickness. On the statistical average integrated over the entire laminar flame length, one can determine whether turbulence has a net thickening or thinning effect. Whether the net effect is thinning or thickening also depends on whether the integrated stretch on the entire surface is net positive or negative. Aspden et al. [127] analyzed the DNS data for different fuel/air mixtures and reported the "thickening factor" defined as the ratio of the mean local flame thickness to the reference laminar flame thickness. The results are found to be overall thickening as Ka increases, while the level of thickening depends strongly on the Lewis number of the mixture; high-Le flames (e.g., n-dodecane) are more susceptible to mixing in the preheat zone and become thicker, while low-Le flames (e.g., hydrogen) actually become thinner with stretch. These behaviors correspondingly translated to a decrease and increase in the global flame speed, respectively. This finding contrasts another DNS study of turbulent Bunsen flame [95] with methane–air mixtures, in which the turbulence showed a net thickening effect and then levels off as Ka increases in the thin reaction sheet regime. Whether one can draw a universal conclusion on the prevailing effect of turbulence on flame thickness, including the sensitivity to the preferential diffusion effect, awaits further investigation.

Another relevant turbulent flame thickness of practical interest is the thickness of the statistical average profile of a reactive scalar (e.g., temperature or fuel concentration). In the flame-in-a-box configuration, this quantity is easily computed by the streamwise temperature profile conditionally averaged at each cross section, i.e., $\langle T | x \rangle$, where x is the streamwise coordinate. This quantity is practically relevant as it literally represents the "flame brush" thickness, which also serves as a measure of the extent of flame corrugation and hence the flame speed enhancement. Sankaran et al. [95] reported the comparison of the average temperature profile for the turbulent Bunsen flames at different Ka conditions along the flame normal direction. It was found that the main effect of Ka was to thicken the preheat layer by way of elevating the upstream temperature, while the temperature gradient in the main preheat zone remain close to each other. In contrast, Im et al. [123] found in the flame-in-a-box DNS that the flame brush thickness actually becomes thinner at significantly higher Ka conditions.

The implication is that the variation in the flame brush thickness at higher Ka depends strongly on the specific configuration, as well as on the dimension of the entire flame scales that determine the extent of the large-scale flame wrinkles. As demonstrated by Poludnenko et al. [145], even at the same Ka and integral scales, an increased domain size of the flame-in-a-box DNS leads to a much larger extent of flame corrugations leading to a much faster turbulent flame speed. The possibility of turbulent energy backscatter by the flame into the flow field has also been postulated [136, 146–148]. This brings an attention to the role of intrinsic flame instabilities, especially the Darrieus–Landau mechanism, which is more pronounced at larger scales. Creta and coworkers [149, 150] conducted moderately turbulent Bunsen flames at different scales to demonstrate that the large-scale corrugation, mainly arising from the Darrieus–Landau instability, can lead to a significant differences in the flame topology and subsequently the flame brush thickness. This is an important practical problem to be further investigated, but is difficult to be addressed by SES with a limited dimension. This may also explain the large discrepancies in the attempt to reproduce the experimental S_T/S_L vs. u'/S_L curves by using DNS with limited domain size, as S_T inherently depends strongly on the large-scale phenomena.

Lewis number effects has long been investigated in laminar flame theory, and how it translates to turbulent flames has also been examined in a number of SES studies [122, 129, 131] by comparing the detailed structure against that of the stretched laminar flame. For simulations with detailed chemistry, a global Lewis number based on the deficient species is used as a representative parameter. Consistent with the laminar flame theory, a low Le combined with a positive curvature results in an increase in reaction rates and a decrease in local chemical timescale, thereby resisting turbulent mixing and giving rise to the leading points phenomenon [121, 128, 130]. Conversely, for high *Le*, positive curvature leads to a decrease in reaction rates and an increase in local chemical timescale, thereby allowing turbulence a greater opportunity to mix partially reacting species in the preheat region [131, 151]. As Ka increases, the Le effect is found to gradually vanish and the flame structure degenerates to that of the reference unity Lewis number flame, as the transport in the preheat zone becomes dominated by turbulent mixing.

The overall effect on global turbulent flame speed is that S_T is much higher for low *Le* at the same Da and Ka [131]. Note that the turbulent premixed flame parameter space is multidimensional, so attempts to correlate S_T/S_L with just one parameter, u'/S_L, may be a too simplistic hypothesis for unified scaling laws. Turbulence–flame interactions at the flame scale will most likely depend on many other parameters, such as Ka, Le, and Re. Larger scale effects such as the Darrieus–Landau instability would require DNS at much larger dimensions, and is thus beyond the current capability of SES.

2.2.1.3 High-Ka Flames and Distributed Combustion

As Ka increases further by two orders of magnitude, a substantial range in the energetic spectrum extends down below the flame thickness scale, such that turbulent eddies interact with scales internal to the flame. At this condition, turbulence is sufficiently strong that turbulent mixing overrides the molecular diffusion as a means to drive flame propagation. Damköhler [13] identified a limiting behavior now referred to as the "small-scale turbulence" regime [11], and suggested scaling laws analogous to laminar flames with molecular diffusion dominated by turbulent diffusion. The term "distributed reaction" can be traced back to Summerfield et al. [152], who referred to modification of reaction rates and transport by turbulence. Pope [153] also considered distributed combustion, in which "reaction is distributed more uniformly in space and is not necessarily accompanied by steep spatial concentration gradients." Williams [24] discusses the case in which turbulent mixing is rapid compared with chemistry, thereby causing combustion to occur "with heat release occurring more or less homogeneously throughout the turbulent flame brush and with local fluctuations in temperature and composition being small." Bray [154] also refers to the distributed reaction regime where chemical times are long compared with the largest timescales of the turbulent flow. Langella et al. [37] suggested that the distributed combustion could result from the spatiotemporal intermittency of flamelets distributed over a large region.

There has been increasing experimental interest in realizing such combustion conditions, e.g., Refs. [34, 61, 155]. As discussed earlier, however, in a typical jet flame configuration turbulence level inherently decays in the streamwise direction, while combustion and heat release lowers the Reynolds number significantly, making it extremely difficult to maintain the anticipated high level of Ka at the flame location. It appears that achieving a truly distributed flame in the laboratory remains elusive, and whether such conditions are practically relevant remains questionable. Nevertheless, there are several strategies that may be explored. For example, increase in the flame timescale can be achieved at a leaner and diluted condition, as would occur in MILD combustion. In fact, a DNS study [156] reported a wide variety of different flame topologies for methane–air premixed flames at MILD conditions even at relatively low turbulence intensity. It is unclear whether equivalence ratio is significant, while differential diffusion among fuel–oxidizer species at the same Ka may shift the preheat zone profile, making it either closer to or further away from the distributed combustion regime. In any case, SES combined with forced turbulence can serve as a numerical experiment to reveal the fundamental flame structures and topologies.

Computationally, distributed burning was first reported in the context of Type Ia supernova in [120]. It was found that the advection–diffusion-reaction system found in the internal working of a Type Ia supernova is essentially a high Lewis number stoichiometric premixed flame, along with Reynolds numbers that can far exceed 10^{10}. The fuel in this case was simply carbon-12 (there was no oxidizer), and the reaction was nuclear fusion forming magnesium-24, hence stoichiometric and premixed. An Arrhenius expression was used for the reaction rate, and flame propagation was driven by optically thick radiation, described mathematically by a diffusion term (with negligible species diffusion), hence a high Lewis number premixed flame. At a Karlovitz number of 266, qualitatively different flame behavior was observed. In particular, the reaction zone was no longer a thin interface between fuel and products, but more spatially distributed, and the temperature field resembled a turbulent mixing zone, reminiscent of mixing in a Rayleigh–Taylor instability (see figures 5 and 6 in [120]).

Attempts to realize distributed combustion in typical turbulent flames have been made by SES for hydrogen, methane, and propane flames [42, 121, 123, 157–161], and for *n*-heptane flames [96, 122, 132, 133, 151], where inflow turbulence was either unforced or forced. In these studies, turbulence was able to influence the preheat zone, but was less able to interact with the reaction zone sufficiently to produce a distributed flame. Aspden [121] observed that despite at a slightly higher Karlovitz number compared with that in the supernova flame, the hydrogen flames only presented transitionally distributed behavior. Poludnenko and Oran [42] concluded that subsonic turbulence would not be able to produce a distributed flame in stoichiometric hydrogen, and later showed that these flames could detonate [162]. The Lund group studied both hydrogen ($Le < 1$) and methane ($Le \approx 1$) flames [41, 128] with forced turbulence, in which a notable observation from these high Karlovitz number studies was that global extinction was not observed, as there is no possibility of cold air entrainment in this configuration.

More recently, SES attempts were made to push Ka up to an order of thousands [144] in order to show the possibility of a truly distributed combustion regime. The fundamental difference at high Karlovitz numbers is that turbulent mixing dominates over molecular diffusion, resulting in smaller scalar gradients [121], such that the normalized probability density functions (pdfs) of scalar gradients tended toward an exponential distribution – a characteristic of passive scalar mixing. A detailed investigation of strain rate, reaction, and diffusion balance within high-Ka flames [144] also reported that at an extreme Ka (> 3000), the energy budget balance is dominated by the normal strain rate and diffusion while the reaction term becomes insignificant. Subsequent implications of these findings in the general combustion closure models in RANS/LES applications need to be further investigated.

2.2.2 DNS of Complex Flames

DNS in complex flow configurations builds on the physical understanding provided by turbulence-in-a-box DNS of premixed flame interactions with homogeneous isotropic turbulence conditions. Forced turbulence advected into the computational domain through the upstream boundary or through additional source terms applied to the governing Navier–Stokes equations throughout the computational domain provides valuable insights into turbulence–chemistry interactions in the flamelet, thin reaction zones, and distributed reaction zones regimes in the Borghi premixed flame diagram. External forcing, although it inherently changes the fundamental nature of turbulence-flame interactions not realizable in physical experiments, provides a way to sustain the intensity of the turbulence, as it interacts with a premixed flame before the kinetic energy is dissipated by viscosity. DNS is limited by the dynamic range of scales that can be simulated due to computational cost even with high performance computing. Hence, forcing is often applied in DNS-in-a-box, for example, to attain very high Karlovitz conditions in the distributed reaction zones regime, since kinetic energy dissipation is exacerbated at the small integral turbulence scales and high-intensity velocity fluctuations of most DNS. Nevertheless, even with these constraints, DNS provides unique insights into causal relationships between unsteady stretch rate and Lewis number effects coupled with the intrinsic chemical scales for hydrogen-rich and hydrocarbon fuels of practical interest used to assess theoretical assumptions.

Recent computer science and high-performance computing advances in supercomputing architectures has afforded increases in computational complexity in DNS of turbulent combustion [163, 164]. A number of computational advances have occurred, making it feasible for DNS of turbulent combustion with complex chemistry to be performed effectively on current petascale (10^{18} floating point operations per second) and eventual exascale (10^{21} floating point operations per second) hardware to appear in the near future between 2021 and 2023. These advances include dynamic asynchronous programming systems adapted for heterogeneous machine architectures with accelerators; domain-specific compilers for fast chemistry evaluation [165] on accelerators; and numerical algorithmic advances, especially adaptive mesh refinement [166, 167] and its efficient implementation on accelerators. In the United States, the Exascale Computing Project (ECP) [168] has supported about two dozen application projects in addition to developments in system software. The ECP combustion project is developing open source software for low-Mach and compressible DNS of turbulent reacting flows with multiphysics (spray, radiation, and soot) and complex geometry encountered in practical devices. These and other similar computational combustion projects worldwide will enable DNS to be performed in canonical geometries, broadening the accessible parameter regimes, (i.e., higher Reynolds and Karlovitz numbers), and in more realistic complex flow configurations representative of those encountered in engines for power generation and transportation.

Engines for propulsion and power generation are characterized by rapid mixing and the requirement for flame stabilization in adverse turbulence environments that often include high mean shear, swirl, and flow separation. Examples include behind bluff bodies, swirl flames, and backward facing steps [98, 169–171]; above flame-holder cavities [172]; and sudden expansion (e.g., downstream of cross-sectional area changes in a rectangular duct or tube [173–175], characterized by flow separation and recirculation of vitiated gases. These products of combustion provide back support to assist in anchoring a flame that has to withstand the oncoming intense shear generated turbulence. The region of influence of the back-support provided by hot vitiated gases is measured by the recirculation zone length, and a scaling law involving the length scales with the surface-averaged root mean square of the turbulent velocity fluctuations normalized by the bulk mean velocity at the base of the bluff body, the heat release rate parameter and by turbulence–chemistry interactions was recently derived [176]. The turbulent burning rate depends on the evolution of the Karlovitz and turbulent Reynolds numbers that embody the competition between micromixing and chemical timescales. Both quantities can evolve locally in a complex flow depending upon the local development of the flame brush and its physical overlap with sources of turbulence generation, e.g., mean shear zones or wall boundary layers. If the flame brush resides in a region of mean shear at a given physical location, the flame may be subjected to a continual source of turbulent generation that may extend across the flame brush from the reactants to the products, unlike for isotropic turbulence, where heat release rate and viscosity nominally result in strong dissipation of turbulence across a premixed flame. Recently, piloted laboratory-scale jet flame experiments have been performed, pushing well into extreme turbulence conditions in the distributed reaction zone regime, characterized by high Reynolds, low Damkohler, and high Karlovitz numbers [61, 93, 177, 178]. These experiments all incorporate a pilot that provides a flame stabilization mechanism in the near field and also shields the flame from heat losses and dilution by entrained ambient gases downstream in the jet. The products of combustion provided by the surrounding pilot serve a similar function as a recirculation zone by providing back-support to the flame. Another interesting experimental configuration is a turbulent premixed counterflow flame in which a turbulent stream of fresh reactants is opposed to a stream of hot combustion products. Under certain nonadiabatic and equivalence ratio conditions of the product stream, the product gas stratification provide back-support to the flame [179, 180]. This configuration is compact and provides a systematic way to evaluate the effects of product stratification on flame structure, localized quenching, and flame stabilization.

DNS of a stratified premixed turbulent ultra-lean methane–air jet flame in the same experimental configuration as Zhou et al. [34] showed significant back-support provided by the near-stoichiometric pilot surrounding the central jet of lean reactants at an equivalence ratio of 0.4. The DNS attained a maximum Karlovitz number of 1400 and a jet Reynolds number of 15,764. The back-supported flame resulting from the gradient in equivalence ratio in the flame normal direction had a significant effect on the local flame structure. In particular, the flame was found to be thinner, the burning rate was higher, and the radical pool was enhanced in regions of strong mixture stratification. These regions correlated with flames with high tangential strain and low curvature. The mean flame structure was affected by the strong shear that was predicted reasonably well by unity Lewis number laminar stratified premixed flames at the local (i.e., as a function of axial position) mean strain rate and thermochemical conditions of the reactant and products of the turbulent premixed flame. This is expected because of the high turbulent diffusivity at these conditions. The mean shear rate was quantified by the ratio of the tangential strain rate imposed by the persistent mean flow to the total tangential strain rate that the flame experiences due to both mean shear and turbulent fluctuations [181]. This ratio is large in the near field and decreases to about 0.1–0.2 downstream. These results are consistent with previous laminar simulations of opposed reactants-to-product stratified flames [182–185] and DNS of turbulent stratified slot jet flames [186]. Another feature of fast jet flames is the existence of spatial separation of the primary and secondary reaction zones at different axial positions in a jet. While fuel oxidation reactions in the primary reaction zone occur upstream in the jet due to transport of OH radicals and enthalpy from the pilot, there is insufficient residence time for the secondary CO and H_2 oxidation reactions to complete, and hence while production of CO and H_2 occurs from fuel oxidation, consumption of these stable intermediates is locally quenched, leading to a net production of stable intermediates. Turbulent diffusion and advection transport these stable intermediates and enthalpy where they accumulate downstream in the jet, leading to enhanced secondary reaction zones behind the primary reaction zone. Downstream, the fuel concentration in the reactants becomes negligible, consistent with the observed separation of combustion stages. The primary source of heat release downstream is from CO oxidation and reactions involving HO_2, H_2, and OH rather than from fuel consumption reactions which are dominant upstream. The downstream spatial evolution of the piloted jet flame structure and local burning rate, along with the role of turbulent advection and diffusion, result in partial quenching of the oxidation layer in the near field and enhanced CO and H_2 oxidation in the far field. These aspects are not captured by DNS of turbulent planar premixed flames in isotropic turbulence.

In addition to DNS and experiments performed of high-Karlovitz flames, scaling laws were developed [187] that show the relation between dilution by burnt gases and the burning rate. They showed that the energy transported to the reaction zone by burnt gases can offset fast mixing rates induced by intense turbulence, as demonstrated experimentally [93, 178] and by DNS of piloted jet flames [188]. The validity of the scaling law was demonstrated in [187] by examining the response of the quenching Karlovitz number as a function of the level of dilution by vitiated gases. In particular, the micromixing rate was varied in a multistream canonical turbulent and reacting stochastic micromixing simulation for a liquid n-decane–air mixture calibrated against practical aero-gas turbine combustion chamber conditions.

To summarize, experiments, DNS, stochastic simulation, and scaling laws all confirm that extremely high Karlovitz combustion is feasible with the assistance of a vitiated mixture. The vitiated mixture can come from different sources in practical systems and in laboratory-scale experiments including coflowing and counterflowing pilots and flow recirculation induced hydrodynamically or by a physical structure such as a bluff body, cavity, or backward facing step, as long as micromixing of the vitiated gases with the reaction zones provides adequate back support to the flame.

2.3 Experimental Perspective and Challenges

B. Zhou, J. H. Frank, B. Coriton, Z. Li, M. Alden, and X.-S. Bai

Turbulent combustion involves a large number of species and elementary reactions and is inherently three-dimensional and time dependent. The structure and propagation of turbulent flames are dictated by turbulence and chemistry interaction over a wide range of length and timescales. Study of turbulent combustion therefore requires multidimensional and high-speed measurement capabilities for various scalars and velocity fields. Industrial combustion systems typically employ nongaseous fuels and operate in a confined environment at elevated pressures, which necessitates adapting diagnostic capabilities to such challenging environments. Advancing diagnostic dimensionality and capabilities is key to addressing the complex multidimensional and multiphysics problems involved in turbulent combustion. Laser-based diagnostics is probably the most suitable tool that carries unique potentials to fulfill such demanding diagnostic requirements for turbulent combustion studies. A great deal has been learned about combustion using laser-based diagnostics to provide nonintrusive single-point, line, and multidimensional imaging measurements. Imaging measurements of scalars provide insights into the flame structure, curvature, orientation, connectivity, and gradients in turbulent flames. Measurements with high repetition rates reveal the spatiotemporal evolution of turbulent flames. However, there is a trade-off among the spatial dimensions, repetition rates of the measurements, the number of scalars that are measured simultaneously, and situations to which diagnostic techniques can be applied.

A comprehensive review of multidimensional and high-speed imaging diagnostic techniques for combustion research is beyond the scope of this chapter, and overviews are available in the literature [189–193]. In this chapter, we present a limited selection of recent developments in imaging diagnostics for studying turbulent combustion based on research contributions from our laboratories at Lund University and Sandia National Laboratories. Section 2.3.1 addresses the capabilities in simultaneous imaging of various key scalars that are used to reveal different local thermochemical states of turbulent premixed flames. Section 2.3.2 highlights recent advances in diagnostic techniques for measuring key quantities in two or three dimensions over a wider range of conditions with repetition rates up to 100 kHz. Examples are given for improved understanding of turbulent combustion as learned from these advanced laser-based measurements. Future directions of experimental turbulent combustion research are discussed at the end of this chapter.

2.3.1 Simultaneous Multiscalar Visualization of Turbulent Premixed Flames

As discussed in Section 2.1, turbulent premixed flames can be categorized into different regimes, that is, laminar flamelet regime where Karlovitz number Ka $\leq$ 1, thin reaction zone regime where 1 < Ka $\leq$ 100, and distributed reaction zone (and/or the broken reaction zone regime) where Ka $\geq$ 100. Current knowledge on the structures of turbulent premixed flames has been developed based on theoretical reasoning and different hypotheses are available. Experimental studies on the structures of turbulent premixed flames have attracted the attention of recent research, but great challenges in the experiments remain to be resolved.

2.3.1.1 Theoretical Hypotheses on the Structure of Turbulent Premixed Flames

The structure of turbulent premixed flames in the laminar flamelet regime has been hypothesized to be similar to that of laminar flames [194]. Theoretical analysis has shown that the structure of laminar premixed flames can be characterized by multiple layers [1, 10]: a thin preheat zone where the temperature is low (such that no significant reactions taking place) and the mixture is heated up owing to heat transfer from the downstream reaction zones where heat is released. The reaction zone can be further divided into different sublayers, in which different species can be found and different elementary reactions are taking place. Downstream the reaction zones is the burned gas zone, which is also referred to as the post-flame zone where chemical reactions are in equilibrium. The reactive-diffusive thin layer, the flame, propagates toward the unburned fuel–oxidizer mixture owing to the molecular diffusion–reaction mechanism.

It has been argued that in the *laminar flamelet regime* the length scale of the diffusion-reaction layer (δ_f) is on the order of, or smaller than, the viscous dissipation scale, known as the Kolmogorov length (η). The effect of turbulence on the flame structure is through wrinkling the flame, which has two consequences: (1) the wrinkling of the flame substantially increases the area of reaction zones, enabling the flames to be more compact in the combustion chamber; and (2) turbulence affects the flame straining and stretching, which subsequently affects the local laminar flame speed. It is clear that a desirable measurement method should be able to resolve the thin molecular diffusion–reaction layer, and the overall 3D flame topology, in particular the flame curvature and strain rate.

The structure of turbulent premixed flames in the *thin-reaction zones regime* is hypothesized to bear a similarity to that in the laminar flamelet regime: the reaction zones are essentially thin since the thickness of the reaction zones are on the order of, or smaller than, the Kolmogorov length η; however, the preheat zone is significantly broadened. The broadening of the preheat zone is by virtue of turbulence eddy transfer of heat and mass. This implies that the propagation of the local flame is by the mechanism of turbulent eddy diffusion and reaction. Similar to the laminar flamelet regime, the thin reaction zones are wrinkled by the turbulence eddies, which increases the volumetric heat release owing to the increased *flame surface density* (FSD, Σ), although locally the reaction rates are on the same order as those of laminar flames. Measurements of the structure of turbulent premixed flames in the thin-reaction zone regime should consider resolving the fine turbulent scales in the preheat zone and the distribution of key scalars in the reaction zone and preheat zone.

At high turbulent intensities (e.g., Ka > 100) it is expected that the preheat zone is broadened and the reaction zones are also distorted. This regime of flame is therefore sometimes referred to as the *broken reaction zones regime* [11] or *distributed reaction zones regime*, since it is hypothesized that reactions in this flame regime are locally extinguished or homogeneously distributed. Recent DNS of high Karlovitz number flames [41, 121] has confirmed that indeed local flame extinction of reaction zone can occur at high-Ka conditions; however, local reaction zones can also be compressed, and local reaction rates can be enhanced. In fact, homogeneously distributed reactions are seldom found in DNS or experiments (readers are referred to Section 2.2 for more discussion).

To understand the structures of turbulent premixed flames in different regimes, it is desirable to visualize the distribution of different species in the reaction–diffusion layers. Since the flames are highly dynamic, simultaneous multiscalar and velocity measurements are advantageous in characterizing the dynamic evolution of the flame structures.

2.3.1.2 PLIF Visualization of Turbulent Premixed Flames

Experimental visualization of flame structure through spatial precise imaging of combustion intermediate species using laser-based techniques was introduced in the earlier 1980s [195–197]. Over the past decades, the tremendous progress in laser diagnostics has provided invaluable results in supporting the understanding of turbulent combustion through high-quality instantaneous spatially resolved measurement [198]. Planar laser-induced fluorescence (PLIF) is one of the most powerful tools in mapping the instantaneous distribution of highly reactive species important in combustion process. A typical optical setup for PLIF measurement is shown in Fig. 2.4a, where the beam of a tunable pulsed laser is shaped into a sheet around 0.1 mm thick and a few cm tall to luminate the selected part of a flame. A snapshot chemiluminescence image of a turbulent jet flame is shown in Fig. 2.4b, where the upper part of the flame is blurred due to the spatiotemporal-integrated projection. By tuning the laser to a resonant absorption line of a selected species and collecting the induced fluorescence signal with an intensified CCD camera, instantaneous (during around tens of nanosecond time period of the typical laser pulse and camera gate) distribution of the selected radicals can be mapped. A typical single-shot-based image of OH radicals from the thin cross section of the luminated part of the flame is shown in Fig. 2.4c. In obtaining the OH PLIF image, the $A^2\Sigma^+ - X^2\Pi$ (1-0) transition was excited with a pulse laser tunable at around 283 nm, and fluorescence at 310 nm from the transitions $A^2\Sigma^+ - X^2\Pi$ (0-0) and (1,1) was collected. Some details of turbulent flame structure can be illustrated in the OH PLIF image. In such a setup, a spatial resolution of tens of micrometers in-plane (due to the camera imaging resolution) and about 100 micrometers out-of-plane (due to the laser sheet thickness) can be achieved.

Most of the reactive radicals in laminar premixed flames locate in a thin layer of about 1 mm thickness, where most of the oxidation and heat release reactions are hosted. A photo of a piloted laminar Bunsen flame fueled with premixed stoichiometric methane–air mixture is shown to the left in Fig. 2.5, where a thin bright reaction

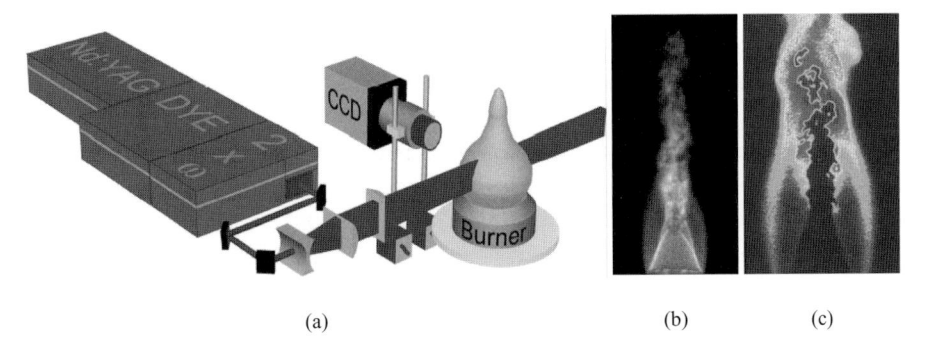

Figure 2.4 (a) A schematic setup of planar laser-induced fluorescence (PLIF) system with a typical laser sheet of a few centimeters tall and 0.1 mm thick. (b) Chemiluminescence image of a typical jet turbulent methane–air flame. (c) A single-shot PLIF image showing the instantaneous OH radical distribution in the flame.

zone is visible as a cone. Using a similar setup as shown in Fig. 2.4a, 2D distribution of temperature from planar Rayleigh scattering and PLIF of selected radical species can be obtained and demonstrated to the right in Fig. 2.5. Technical details and strategy in obtaining high quality single-shot-based 2D temperature and species distributions will be discussed later.

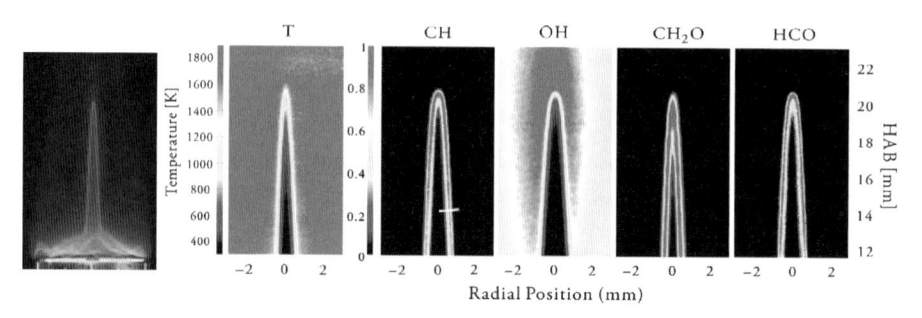

Figure 2.5 Photo of a laminar piloted laminar Bunsen flame fueled with premixed stoichiometric methane–air mixture and a planar laser measurement of temperature (Rayleigh scattering) and species (laser-induced fluorescence) [34].

The results of planar measurements of laminar flame illustrate the basic structure of flame front characterized as a thin layer. Figure 2.6 shows profiles of temperature and the selected radicals across the flame front, as indicated by the bar in the CH PLIF imaging in Fig. 2.5. The experimental measured scalar profiles agree well with the results from numerical simulations using CHEMKIN software. These scalars can be used to characterize the preheat zone and reaction zone of the flame. Following the theory of laminar premixed flames [1, 10], in the preheat zone the mixture is heated up to the crossover temperature (T_c) of chain branching reactions (e.g., H + O_2 = O + OH) and terminating reactions (e.g., H + O_2 + M = HO_2 + M) [10]. For the flame condition shown in Fig. 2.6, $T_c \approx 1000$ K. It is seen that the PLIF intensity of reactive radicals, e.g., CH, OH, and HCO, starts to be significant when the temperature is above T_c. The temperature increases monotonically from that of the unburned reactant to T_c over a layer of thickness of 0.5 mm, which defines the thickness of the preheat zone. Following the preheat zone, reactive radicals, e.g., CH and HCO, can be found in a thin layer of about 0.2 mm thick. This layer corresponds to the inner layer where fuel is oxidized to CO. CH_2O is formed in the inner layer and oxidized in the inner layer toward the high-temperature side. There is no reaction that consumes CH_2O in the low-temperature side of the inner layer. As a result, CH_2O diffuses into the preheat zone together with other combustion products, e.g., CO_2. Thus, CH_2O PLIF signal can be used as a marker of preheat zone in turbulent premixed flames [199].

OH radicals are seen in the CH/HCO layer (inner layer) and survive over a thick layer downstream of the inner layer where the oxidation of CO and H_2 takes place. OH PLIF signal can also be found in the postflame hot gas. Thus, OH marks the beginning of the reaction zone but not the end of reaction zone. An ideal marker for the oxidation layer would be CO or H_2. A single photo-based technique for the PLIF imaging of these species is not yet available.

When investigating the structure of laminar premixed methane–air flames, Paul et al. [200] found that the HCO profile has a good overlap with the heat relate profile. This finding suggested the possibility of marking the heat release rate distribution with PLIF imaging of HCO. Due to the limited quality of achievable HCO PLIF technique at that time, as an alternative, spatial overlaps of simultaneously recorded PLIF signal distribution of OH and CH_2O were suggested to represent the heat release rate mapping [201]. Figure 2.6 shows that the overlap region of OH and CH_2O PLIF signals coincides well with the HCO PLIF region. Due to the relatively high signal-noise-ratio experimentally achievable in PLIF of OH and CH_2O, simultaneous PLIF measurements of OH and CH_2O have been performed and the product of $[OH] \times [CH_2O]$ has been used in turbulent counterflow flames [202] and turbulent jet flames [93, 203]. There are, however, difficulties in achieving high spatial resolution and high signal-to-noise ratio for the PLIF signal product, $[OH] \times [CH_2O]$, to represent heat release rate in turbulent premixed flames. This is due to the fact that the dynamic range of the PLIF signal is limited and the overlap region of OH and CH_2O has a lower concentration for both species, as shown in Fig. 2.6. Overlap of PLIF signals obtained from two independent systems will also reduce the spatial resolution. It is therefore preferred to perform PLIF of a single species to represent the reaction zone of the flame.

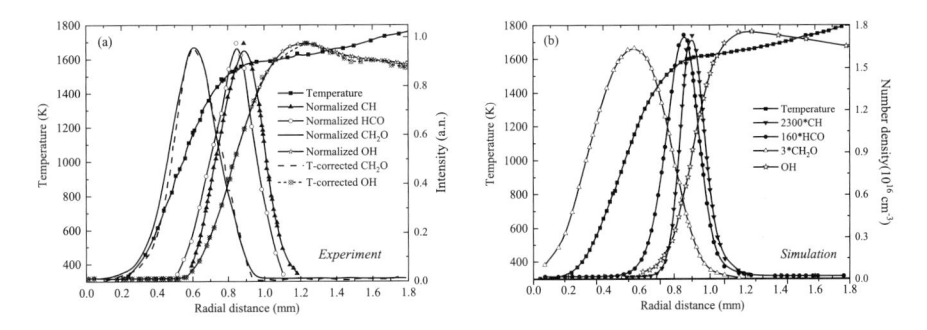

Figure 2.6 Comparison of temperature and species profiles from experimental data (a) across the flame front as indicated by the bar shown in Fig. 2.5 and (b) CHEMKIN simulation [34].

A comprehensive survey of strategies of LIF detection of various species has been given by Smyth and Crosley (chapter 2 of *Applied Combustion Diagnostics*) [216]. Table 2.1 lists a number of key combustion intermediate species for flame structure visualization, with suggested excitation/detection strategies and relevant remarks. Formaldehyde can be excited with a 355 nm laser pulse from the third harmonic of the commonly available Nd:YAG lasers in most laser combustion diagnostic laboratories. A comprehensive investigation of CH_2O PLIF measurement using 355 nm for excitation has been reported by Brackmann et al. [205, 217]. Li et al. discussed the possibility of using CH_2O PLIF to detect the broadening of the preheat zone in piloted methane/air turbulent jet flames [199]. A schematic structure of the jet flame is shown in Fig. 2.7, where the instantaneous distribution of CH_2O PLIF signal clearly indicated

Table 2.1 Selected combustion intermediate species for flame structure visualization

Species	Excitation	Fluorescence	Refs.	Remarks
OH	$X^2\Pi \to A^2\Sigma^+$ (0-1) @283 nm	$A^2\Sigma^+ \to X^2\Pi$ (0-0), (1,1) @310 nm	[204]	Narrow band excitation, off-resonant detection, well-established for OH PLIF
CH$_2$O	$X^1A_1 \to A^2A_1 4_0^1$ @355 nm	$A^2A_1 \to X^1A_1$ @450–650 nm	[205, 206]	High-power laser excitation, off-resonant broadband detection, well established for PLIF
CH	$X^2\Pi \to A^2\Delta$ @431 nm	$A^2\Delta \to X^2\Pi$ @431 nm	[207, 208]	On-resonant detection, easily saturated and moderate fluorescence yield for PLIF detection
	$X^2\Pi \to B^2\Sigma^-$ @387 nm	$A^2\Delta \to X^2\Pi$ @431 nm	[209, 210]	Off-resonant detection, high-pulse-energy laser for high signal level, very sensitive PLIF visualization demonstrated in flames with low CH concentrations
	$X^2\Pi \to C^2\Sigma^+$ @313 nm	$C^2\Sigma^+ \to X^2\Pi$ @313 nm	[211]	On-resonant detection (can be improved using narrow filters), low-pulse-energy laser is preferred for excitation, while high spontaneous fluorescence yield enables good PLIF detection in flames with relatively high CH concentrations
HCO	$X \to B$ @257 nm	$B \to X$ @280–450 nm	[212, 213]	Interference-free imaging measurement available for fuel-lean flames
CH$_3$	$(X^2A_2'')\text{CH}_3 + 213\text{ nm} \to (B^2A_1')\text{CH}_3$ $\to (X^2\Pi)\text{CH} + \text{H}_2$, $(X^2\Pi)\text{CH} + 427\text{ nm} \to (A^2\Delta)\text{CH}$ $\to (X^2\Pi)\text{CH} + \text{fluores. @ 431 nm}$		[214, 215]	High-quality imaging can be obtained, especially in fuel-lean flames, at least 10 times better in signal to background ratio than CH PLIF

the broadening of the CH$_2$O region, whereas the CH PLIF signal remained in a thin layer. A broadened CH$_2$O layer was recently reported under different turbulent flame conditions [34, 35, 177, 203, 218].

As shown in Fig. 2.6, CH radicals are seen in the inner layer of the reaction zone. The detection of CH in combustion diagnostics has attracted the attention of numerous researchers, and the summary shown in Table 2.1 represents part of the works reported in the literature. A substantial improvement of the CH PLIF detectability was reported by Li et al. by using an Alexandrite laser [206] to excite the $X - B$ (0,0) transition at

387 nm, and detection of the resonant $B - X$ (0,1), $A - X$ (1,1), and (0,0) transition at 431 nm. The high pulse energy and longer pulse duration provide a much better excitation and the off-resonant detection provides the possibility of effective spurious scattering reduction [209]. An example of a single-shot CH PLIF image is shown in Fig. 2.7 [199].

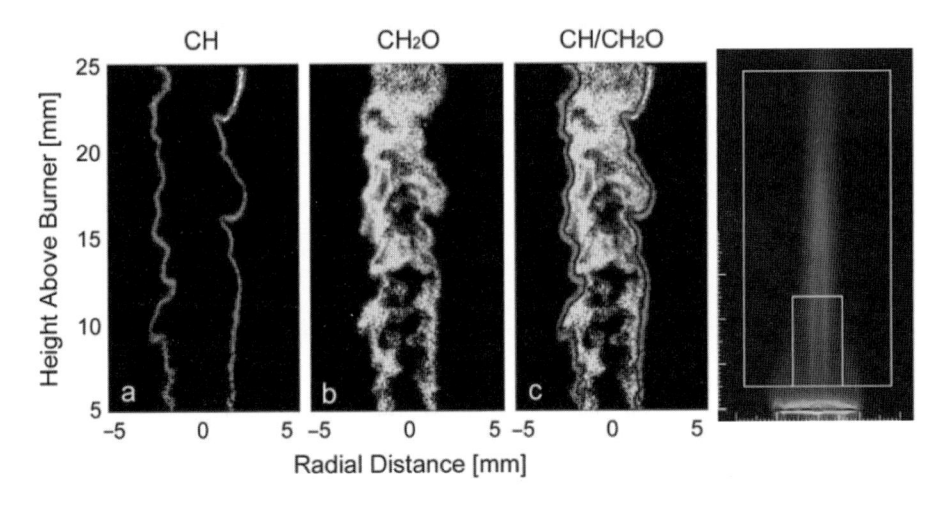

Figure 2.7 Chemiluminescence photo and PLIF images of a stoichiometric methane–air turbulent jet flame at a jet exit bulk flow speed of 60 m/s [199].

Employing the $C-X$ transition at 310 nm, CH PLIF detection has been achieved with high-speed lasers [211]. One important advantage is that the 313 nm laser can be obtained with the recently available high-speed diode-pumped solid-state (DPSS) laser or burst lasers [219]. When applying CH PLIF, one has to note that the CH concentration changes dramatically with local stoichiometry. As predicted for laminar flames, the mole fraction varies from a few ppm to a few ppb when the equivalence ratio is varied from 1.2 to 0.7. An experimental investigation of the detectability of CH PLIF using the $B-X$ transition has been performed [209]. Thus, a highly sensitive CH detection is needed to provide enough dynamic range when a turbulent flame is to be investigated.

It has been debated for decades whether there exists a distributed reaction zones combustion regimes as the turbulent intensity is increased such that the Kolmogorov scale eddies are smaller enough to penetrate into the inner layer of the reaction zone to cause broadening of the entire reaction zone. Driscoll [220] (p. 128) stated that "a documented distributed reaction zones is defined as a set of measured images of the heat release region or images of radicals such as CH that are spread out over distances that are many times larger than the thickness of a laminar flame." This indicates the importance of CH PLIF visualization in determining instantaneous flame structures. To compare the structural change of the flame at different turbulent intensities, simultaneous multispecies measurements are preferred. A schematic setup for a

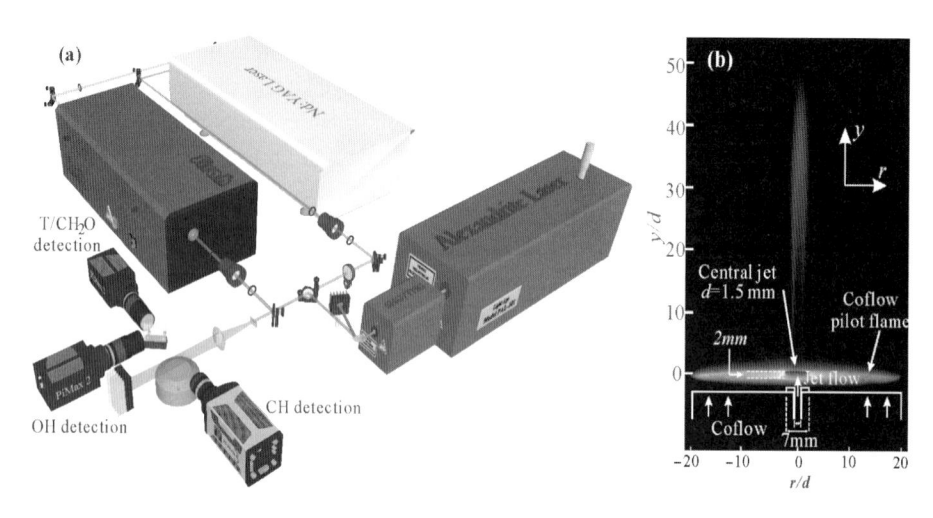

Figure 2.8 (a) Experimental setup for simultaneous multiscalar imaging measurements and (b) a photograph of a piloted jet flame with a schematic plot of the burner employed [93].

simultaneous three-scalar planar detection system is shown in Fig. 2.8. A simultaneous PLIF detection of CH, CH_2O, and OH was performed in a piloted jet flame as shown in Fig. 2.8, and the obtained distributions of CH, CH_2O, and OH radicals are shown in Fig. 2.9. In the left frames of Fig. 2.9, one can see that in the case of 110 m/s of the jet exit bulk flow speed, the preheat zone represented by the formaldehyde distribution is broadly distributed, while the CH layer is still thinner than half a millimeter; to the right of Fig. 2.9, one can see that in the case of 418 m/s of the jet speed, the CH layer is significantly broadened by the turbulent eddies.

Although the CH layer is in general thin for the flame with a jet speed of 110 m/s, careful inspection of different instantaneous distributions of the multiscalar PLIF signals has shown that local broadening of the CH layer can be found in, e.g., the zoom-in region shown Fig. 2.10. The broadened CH region is likely formed due to rapid roll-up of the large pocket of hot gas with OH radicals, which oxidized CH_2O but not CH. The roll-up hot pocket has a scale of about 1 mm, where CH has a higher intensity and the CH_2O and OH signal has lower intensity. This is an example of information that simultaneous imaging of multiscalars can provide for the understanding of flame–turbulence interaction in highly turbulent flames.

The same multiscalar PLIF technique has been applied to study a swirl-stabilized premixed methane–air flame and the results are shown in Fig. 2.11. Broadened CH layers were observed in the recirculation zone downstream of the thin reaction zone where the flame anchored. The fast mixing of entrained fresh fuel and the hot products in the recirculation zone enabled a favorite mixture composition (below the lean flammability limit) and high temperature for the reactions to continue, therefore generating CH [222]. This result further confirmed the existence of a broadened reaction zone in highly turbulent combustion, or alternatively under very fuel-lean conditions in the swirl-stabilized flames.

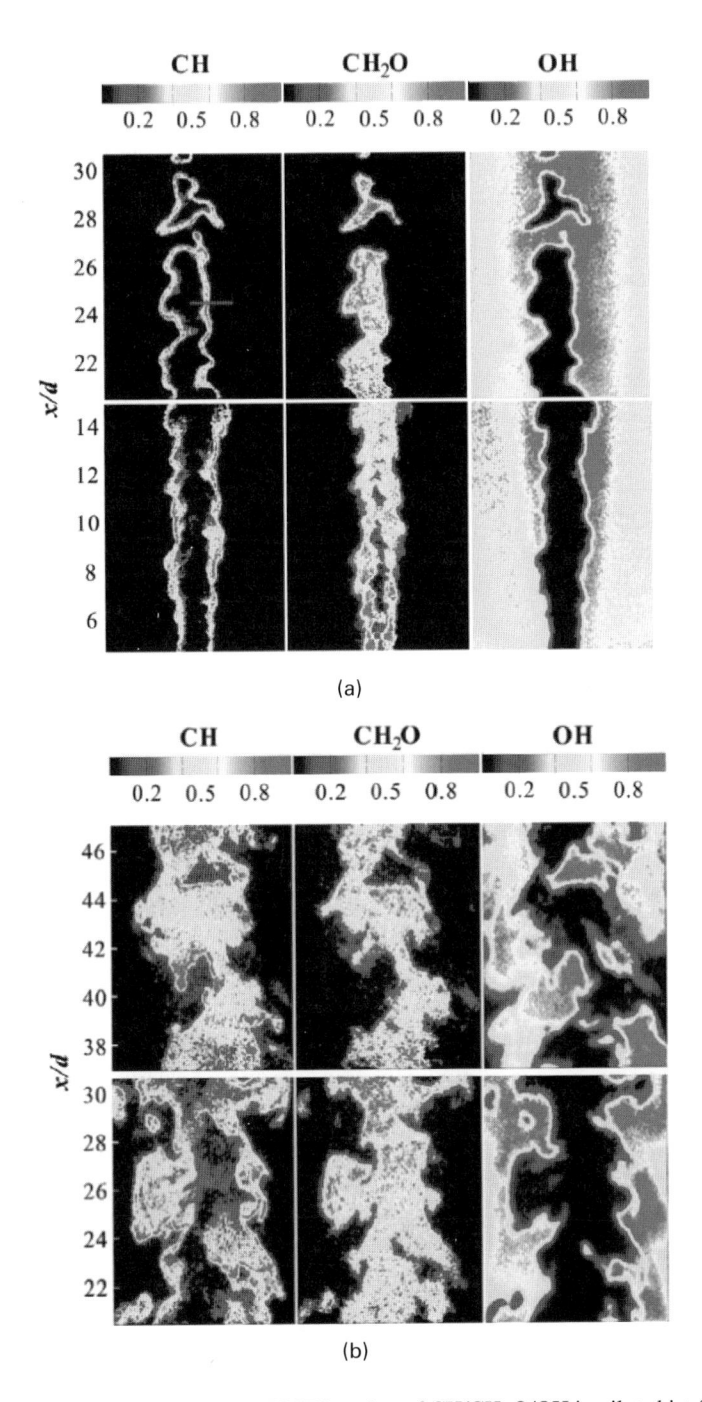

Figure 2.9 Simultaneous PLIF imaging of CH/CH$_2$O/OH in piloted jet flames (as shown in Fig. 2.8) of a jet exit bulk flow speed of 110 m/s (top) and 418 m/s (bottom) [34].

2.3.1.3 Further Comments on PLIF Imaging of Flame Structures

One limitation of using CH PLIF for flame front visualization is the low CH concentration, especially for lean premixed flames. Furthermore, CH radicals, as a minor intermediate species, are not included in many reduced kinetic models that are used in numerical simulations. In this regard, HCO PLIF was developed to better represent the flame front [213]. Zhou et al. [212] later proved the viability of HCO PLIF for flame front visualization at lean methane–air premixed flames. Application of HCO PLIF to study of high-turbulence premixed flames can be found in Zhou et al. [93].

Methyl radical, CH_3, has a relatively higher concentration in most premixed hydrocarbon flame and is a main combustion intermediate species in most reduced kinetics. Thus, it is a suitable marker for the fuel-consumption layer (the inner layer) of methane–air flames [10]. Li et al. [215] reported a sensitive planar detection of CH_3 employing a two-step method, namely photo-fragmentation–planar fluorescence imaging. This method has been proved to have a potential to visualize the instantaneous flame reaction zone structure in the otherwise most challenging lean premixed flames.

One important scalar characterizing the flames is temperature. Temperature can be used to define the reaction progress variable; as such, a 2D temperature field in a turbulent flame can give insightful information about the flame–turbulence interaction. For instantaneous 2D temperature imaging, Rayleigh scattering is still one of the best options, especially for a clean environment where spurious elastic scattering from particles and surface can be negligible [223, 224]. For temperature mapping in a particle-laden environment or near a surface, filtered Rayleigh scattering thermometry has been proven useful by deliberately filtering out the elastic scattering interference [225]. However, Rayleigh scattering measurements in turbulent reacting flows require prior knowledge of local mixtures that could consist of species that have significantly different Rayleigh cross sections, leading to potential uncertainties of

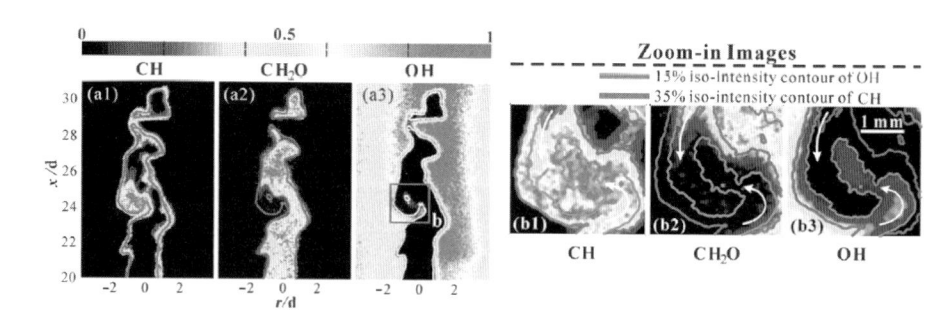

Figure 2.10 (a1–a3) Overall flame structure for a piloted jet flame at the jet exit bulk flow velocity of 110 m/s illustrated by simultaneous $CH/CH_2O/OH$ images, featuring overall flamelet-like structure. An OH isosurface (10% of maximum intensity) is superimposed on the image of each scalar for comparison. Detailed flame structures of a local CH broadening identified are further illustrated in the zoom-in images of (b1–b3) together with isosurfaces of the measured scalars to facilitate comparison. (For interpretation of the references to color in this figure, the reader is referred to the web version of this article [93]).

the measurement. Alternatively, temperature mapping using LIF signals from tracer species [226, 227] or laser-induced phosphorescence from thermographic phosphor tracer particles [228] has also been successfully demonstrated.

Finally, one other important nature of turbulent flames is their high unsteadiness. In order to capture the temporal evolution of the turbulent flame structure, a PLIF system at high repetition rates is needed. Recent advances in high repetition rate diagnostics includes pulsed solid state–pumped dye lasers capable of rates exceeding 50 kHz and burst-mode lasers exceeding 100 kHz [229]. A burst laser system can provide temporally resolved scalar imaging in turbulent reacting flows and shed light on the temporal evolution of the flow and scalar structures [230]. In the following section we will focus on the recent development of the state-of-the-art flame diagnostics at high repetition rates and multidimensions.

2.3.2 Advances in Dimensionalities High-Speed Diagnostics

2.3.2.1 Tomographic Particle Image Velocimetry Methodology

Velocity field measurements are critical for understanding the feedback between turbulence and combustion. In turbulent flames, particle image velocimetry (PIV) is commonly employed for velocity field measurements, which until recently have been limited to planar and stereo PIV (SPIV). Development of tomographic particle image velocimetry (TPIV) and its application to combustion research provides unprecedented opportunities for accessing key fluid dynamic quantities. To fully capture the turbulent flow field requires volumetric measurements of all three velocity components using either multiplane SPIV or tomographic particle image velocimetry. Pseudo-3D velocity field measurements in flames have been performed using high-repetition rate SPIV by invoking Taylor's hypothesis [231]. Tomographic particle image velocimetry is an advanced PIV technique that provides instantaneous measurements of all three velocity components within a probe volume [232–234]. The results enable the determination of the complete velocity gradient tensor, from which key quantities for understanding the turbulent flow field can be derived, such as vorticity, divergence, and the strain rate tensor. The application of TPIV to turbulent combustion studies is relatively recent [235–237]. A brief overview of TPIV is presented in this section along with a method for coupling simulations and experiments to evaluate TPIV measurement uncertainties.

For TPIV measurements, the laser scattering from particles seeded into the flow is recorded on multiple cameras with different viewing angles. Figure 2.12a shows an example of a combined TPIV and OH-LIF imaging configuration that uses two TPIV cameras on each side of the laser beam and a dual-head high-repetition rate, diode-pumped solid-state laser that illuminates a slice of the flow with typical probe volume dimensions on the order of $20 \times 15 \times 1$–3 mm^3. Alternatively, the TPIV cameras can be configured with all cameras on one side of the laser beam in a fan geometry or with cameras having views from above or below the plane of the optical table. The camera acquisition is synchronized with the laser pulses in a frame straddling mode. For example, 10 kHz velocity measurements are obtained by operating each laser head

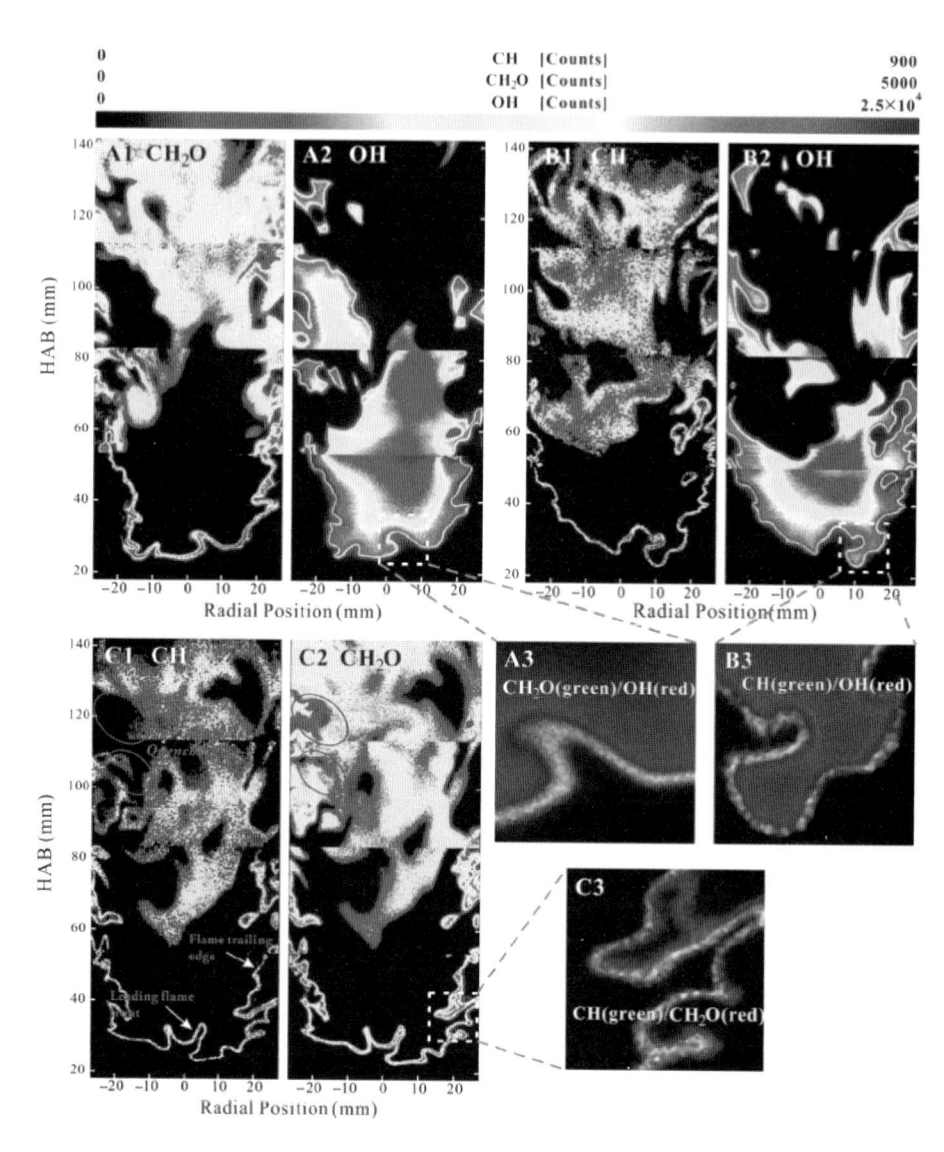

Figure 2.11 Examples of simultaneous PLIF measurements in a swirl flame, where (A1–A2) CH$_2$O/OH, (B1–B2) CH/OH, and (C1–C2) CH/CH$_2$O, and local structures as marked are illustrated in (A3–C3) [222].

at a repetition rate of 10 kHz and the cameras at a framing rate of 20 kHz. The time interval (Δt) between every pulse pair from the two laser heads is set to accommodate the velocity magnitudes of interest, which results in particle displacements of approximately one fourth of the interrogation window size used for cross-correlation calculation. Figure 2.12b1 shows an example of two sets of sequential images of particles recorded from a laser pulse pair on each camera in a lifted turbulent jet flame. The 3D particle field is then reconstructed at each time step using a tomography algorithm such as multiplicative algebraic reconstruction tomography

(MART) [232], as shown in Fig. 2.12b2. The particle displacements in sequential frames are computed using iterative volume cross-correlations in each interrogation region with 50% or 75% overlap of interrogation regions. Strategies for interrogation window size selection to optimize velocity field measurement are elaborated in [238]. For the experimental configuration illustrated here, the final size of the TPIV interrogation region is $24 \times 24 \times 24$ voxels^3, which corresponds to approximately $400 \times 400 \times 400$ μm^3. The final step is removing and replacing spurious vectors and reducing the measurement noise with a spatial filter, which is important for calculating velocity gradients. The result is a volumetric measurement of all three components of the velocity field as exemplified in Fig. 2.12b3. The accuracy of TPIV data evaluation is continuously improving with the development of new analysis methods, such as the Shake-The-Box (STB) scheme, which reduces the adverse effects of reconstruction artefacts for a certain range of particle seeding densities [239]. STB is based on a Lagrangian tracking methodology that iteratively compares subsequent particle images with the predicted particle distribution based on the temporal coherence from high-speed TPIV measurements. Another important element of accurate TPIV measurements is careful registration of multiple camera views, which is performed using a sequence of images of a reference target that is translated across the probe volume.

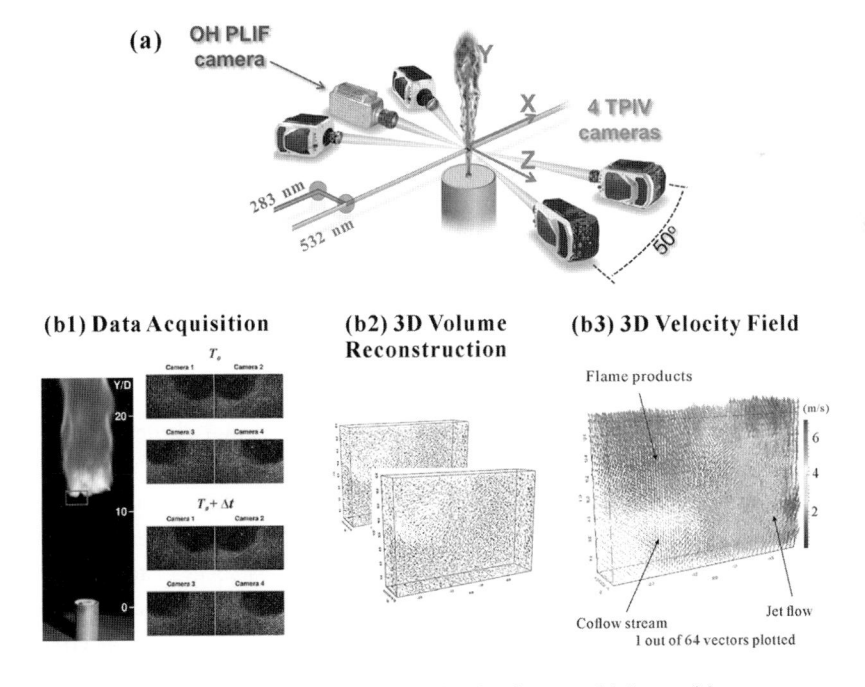

Figure 2.12 (a) Experimental configuration for simultaneous high-repetition rate tomographic PIV and OH-LIF imaging. (b1) Sample particle image pairs recorded from four camera views in a lifted turbulent jet flame. (b2) Reconstructed 3D particle distribution at sequential times using MART algorithm. (b3) 3D velocity field computed using iterative volume cross-correlations. Adapted from Ref. [236].

Quantification of TPIV uncertainties in turbulent flames is particularly challenging because the 3D image reconstruction process involves solving an inherently underdetermined problem, which can result in reconstruction artefacts. The impact of these artefacts and other measurement uncertainties depends on locally fluctuating parameters, such as particle seed density, velocity gradients, and turbulence length scales. Numerical simulations of actual flow geometries from experiments are advantageous for evaluating measurement uncertainties that would otherwise be difficult or impossible to estimate. As part of this effort, the coupling of large eddy simulations (LES) and experimental measurements is considered for parametric studies of measurement uncertainties under a range of different conditions. This approach is illustrated using LES calculations of the Yale turbulent counterflow burner [240] to generate synthetic experimental data for studying uncertainties in TPIV measurements. The LES is performed using the RAPTOR code framework developed by Oefelein [241, 242]. The domain of the simulation is identical to the geometry of the actual counterflow burner, including the internal flow of the nozzles. The geometric details of the turbulence generating plate, which is housed within the nozzle, are fully resolved in the calculations, and boundary conditions precisely match the experimental operating conditions without the use of tuning parameters. The LES grid contains approximately 10 million cells for each nozzle with a grid spacing of 450 μm, which is slightly larger than the TPIV interrogation region used in the corresponding experimental configuration. Calculations are performed using a three-stream flamelet model to simulate the effects of heat release on the particle distributions. To emulate the PIV methodology used in the experiment, the LES calculations are seeded with particles using the Lagrangian–Eulerian approach [241, 242]. Figure 2.13a shows an example of the particles emerging from each of the counterflow nozzles with the particles color-coded by their axial velocities.

The TPIV measurements are simulated by constructing projections of synthetic particle images at the same viewing angles as the four-camera TPIV experimental configuration. These synthetic TPIV data are analyzed using the same MART algorithm and cross-correlation methods that are used for calculating the velocities in the experiments. A single-shot reconstructed particle distribution for a $1.5 \times 1.5 \times 1.5$ mm^3 subset of the domain is shown in Fig. 2.13b with the reconstruction in gray (nearly-spherical shaped volumes) and the actual particle positions in black (points). Discrepancies between the positions of corresponding reconstructed and actual particles constitute an error in the measurement. The size and shape of the reconstructed particles depend in part on the number of cameras and their viewing angles. Reconstruction artefacts are identified by instances of reconstructed particles that are not paired with a corresponding actual particle. These artefacts, referred to as ghost particles in the literature, introduce additional uncertainties in the measured velocities, and their number density increases as the seed density increases. Figure 2.13c shows the probability density functions of the error in the axial velocity for simulated TPIV measurements using three different particle seed densities and a probe volume thickness of 2.5 mm, which is representative of the laser sheet thickness in an experiment. The low, medium, and high seed densities correspond to 60, 120, and 180 particles per mm^3, respectively. The PDFs broaden as the seed density increases with the largest degradation occurring from the low to medium seed density. The velocity error for the lowest seed density is approximately 0.27 m/s, corresponding to a displacement of approximately 0.27 voxels. A similar analysis can be used to evaluate the effects of other experimental parameters on uncertainties in the velocity measurements, such as detector noise or the number of cameras and their viewing angles. In general, this approach of generating synthetic experimental data from simulations of turbulent flows provides a powerful tool for systematic studies of measurement uncertainties as well as for developing and testing new data analysis methods.

2.3.2.2 Turbulence–Flame Interactions

The recent application of TPIV to combustion research has enabled fundamental studies of the interactions between combustion heat release and turbulence. One of the major focuses has been on measuring the effects of combustion on the temporal and spatial structure of the strain rate field as well as the coupling between vorticity and strain rate in turbulent flames [234, 236, 243–245]. Experiments using 2D velocity field measurements do not capture the full vorticity vector or strain rate tensor, precluding studies of fundamental processes in turbulence such as vortex stretching. The TPIV technique provides volumetric, three-component velocity measurements that enable determination of all components of the vorticity vector and strain rate tensor as well as the vorticity-strain rate alignment. This section provides examples of improved understanding of turbulence–flame interactions through combined measurements of TPIV and OH-LIF imaging. The instantaneous three-dimensional velocity field is used to calculate the strain rate tensor eigenvalues, s_i, which are defined such that $s_1 \geq s_2 \geq s_3$, where s_1 and s_3 are the most extensive and compressive strain rates, respectively, and s_2 is the intermediate strain rate. The strain rate tensor eigenvalues are also referred to as the principal strain rates, and the axes of the principal strain rates are determined from the strain rate tensor eigenvectors. Simultaneous 2D OH-LIF imaging in the central plane of the TPIV measurement volume provides a marker of the response of the high-temperature reaction zone.

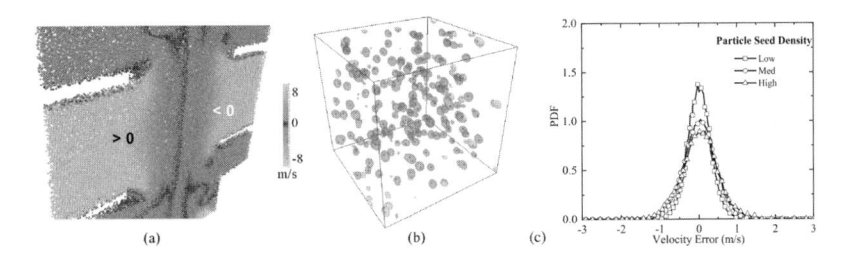

Figure 2.13 Synthetic TPIV experiment using LES solution of a turbulent counterflow seeded with particles. (a) Synthetic particles color-coded by axial velocity. (b) Subset of 3D reconstruction of particle distribution (gray spheres = reconstruction, black points = actual particle positions). (c) PDFs of error in axial velocity as a function of particle seed density.

A. Flame–Strain Rate Interaction

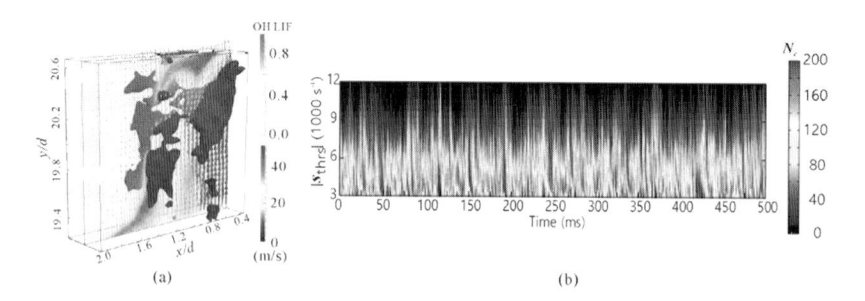

Figure 2.14 (a) Single shot from time sequence of velocity, OH-LIF, and compressive strain rate (3D isosurface = $-15{,}000$ s^{-1}) measurements in a turbulent jet flame. (b) Time evolution of the number of strain rate clusters (singly connected regions of strain rate) present in the TPIV probe volume as a function of strain rate magnitude. Adapted from Refs. [236, 243].

The fluid dynamic strain rate plays a central role in turbulent mixing by enhancing or suppressing scalar gradients, and the turbulence–scalar interaction is strongly dependent on the magnitude of the strain rate tensor eigenvalues and the alignment of the eigenvectors with the scalar gradients [246]. Figure 2.14a shows an example of the structure of the strain rate field in a turbulent partially premixed piloted methane–air jet flame at an axial location of 20 diameters downstream of the jet exit with the jet axis located near the right side of the field of view [236]. The isosurface indicates a compressive strain rate of $-15{,}000$ s^{-1}, which is an order of magnitude larger than the extinction strain rate from a laminar counterflow flame calculation for this fuel–air mixture. The gap in the OH-LIF signal indicates localized extinction that coincides with structures of compressive strain rate. The strain rate field is also highly intermittent, with concentrated regions of large strain rate occurring in bursts, as shown by the time trace in Fig. 2.14b [243]. These bursts are indicated by the narrow spikes in the time series of the number of strain rate clusters, N_c, in the probe volume. Clusters are defined as singly connected groups of voxels for which the strain rate magnitude exceeds a threshold value. These short-lived clusters of excessive strain rate may have a disproportionate effect on the combustion because they are capable of inducing finite-rate chemistry effects including localized extinction.

The fluid dynamic strain rate affects the progress of chemical reactions, which leads to variations in heat release rates and plays a central role in the dynamics of flame extinction and reignition. The heat release in turn introduces dilatation and changes in the density, viscosity, and diffusivity, which can have important effects on the turbulence. For example, the effect of heat release can be manifested in changes of the strain rate eigenvector alignment with respect to the flame front, which plays a critical role in determining the scalar–normal strain rate in the transport equation of the scalar dissipation rate [114]. The flame-front contour and its normal direction (**n**) can be determined from OH-LIF images that are obtained simultaneously with TPIV measurements. Figure 2.15a and b shows examples of single-shot OH-LIF images superimposed with (a) the s_1-eigenvectors and (b) the s_3-eigenvectors in turbulent Bunsen and counterflow flames, respectively. The flame-front contour is identified at the boundary of the region containing an OH-LIF signal, and the angles between the flame-front normal direction and the strain rate eigenvectors are determined as illustrated in the schematic of Fig. 2.15c. PDF plots of the cosine angle, $\cos(\theta_i)$, between the strain rate eigenvectors and the flame-normal at the flame front are shown in Fig. 2.16a1 and b1 for the Bunsen flame and the counterflow flame, respectively. The peaks in the PDFs of s_1 and s_3 at $\cos(\theta_i) = 1$ in the Bunsen and counterflow flames, respectively, indicate that the preferential alignment of the strain rate eigenvectors is quite different in these flames.

The effects of heat release on the local strain rate alignment can be investigated using simultaneous TPIV and OH-LIF imaging measurements to calculate strain rate statistics conditioned on the flame-normal coordinate, i.e., the distance (x_n) from the flame front along its normal direction. The effect of heat release on the alignment angles, θ_i, manifests in the conditional probabilities for the alignment angles less than 45°, $P(\theta_i < 45°)$, as a function of x_n across the flame front, as shown in Figs. 2.16a2 and b2 for the Bunsen flame and the counterflow flame, respectively. In both flames, the heat release has the effect of promoting the alignment with extensive strain along the flame-normal direction and reducing the alignment with the compressive strain. The greatest enhancement of $s_1 - \mathbf{n}$ alignment and reduction of $s_3 - \mathbf{n}$ alignment are observed near the flame front ($x_n = 0$), where the strongest impact of heat release is expected. This implies that the same underlying physical mechanism of dilatation from heat release affects the strain–flame alignment in both flames. However, the preferential $s_3 - \mathbf{n}$ alignment in the counterflow flame remains dominant across the flame front as a consequence of the inherent bulk compressive strain that is imposed by the counterflow burner geometry [247, 251]. In contrast, flames with similar turbulence intensities that are not subject to a bulk strain rate field, such as the Bunsen flame, consistently show preferential alignment of $\mathbf{n}$ with the most extensive eigenvector, s_1, near the flame front due to local dilatation [245, 248–250].

For the purposes of turbulent combustion modeling, it is also of interest to consider strain rate alignment statistics based on the fluctuations in the velocity field by first subtracting the mean velocity field. Corresponding alignment statistics based on the strain rate fields (s'_i) derived from the fluctuating velocity fields are shown in Fig. 2.16a3 and a4 for the Bunsen flame and Fig. 2.16b3 and b4 for the counterflow flame. The difference between the corresponding alignment statistics for the full and fluctuating strain rate fields is highly dependent on the flow configuration. For the counterflow flame, subtraction of the mean bulk strain field leads to a large reduction of the $s'_3 - \mathbf{n}$ alignment and an enhancement of the $s'_1 - \mathbf{n}$ alignment, but it has limited impact on the Bunsen flame due to the lack of strong geometry-imposed bulk strain. The resulting alignment PDFs based on the fluctuating strain rates in Figs. 2.16a3 and b3 show very similar shapes for the two flow configurations. However, the profiles of $P(\theta_i < 45°)$ in Figs. 2.16a4 and b4 remain distinctly different, with the counterflow flame exhibiting a narrower peak in the $s'_1 - \mathbf{n}$ profile and a narrower trough in the $s'_3 - \mathbf{n}$ profile. This result indicates that modeling of the scalar–strain alignment in turbulent flames requires careful consideration of the influence of the bulk flow, even in the fluctuating strain rate field. Readers are referred to Ref. [251] for more details.

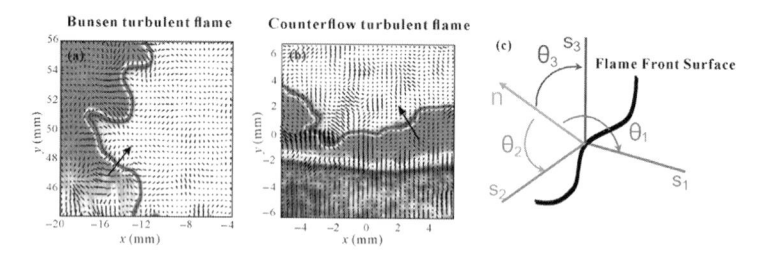

Figure 2.15 Simultaneous OH-LIF images with (a) axes of principal compressive strain s_1-eigenvectors for a turbulent Bunsen flame and (b) axes of principal extensive strain s_3-eigenvectors for a turbulent counterflow flame. Length of strain rate eigenvectors is scaled by their corresponding magnitudes with one out of nine eigenvectors displayed. (c) Schematic of alignment angles between the principal strain-rate axes and the flame front normal direction.

B. Effect of Heat Release on Vorticity–Strain Rate Interaction

Vorticity and strain rate are intrinsically coupled in turbulence, and simultaneous measurements of these quantities contribute to understanding of the production and suppression of vorticity and strain rate in the governing transport equations [252, 253]. In turbulent reacting flows, the effects of chemical reactions on this coupling need to be better understood. Heat release from combustion changes the morphology of the vorticity and strain rate fields as well as the production rates of vorticity and strain. The morphological effects are apparent in comparisons of single-shot measurements [234] of the vorticity and strain rate fields in an isothermal air jet and turbulent partially premixed jet flames with significant localized extinction (C_{LP}) and a stable reaction zone (C) shown in Fig. 2.17. All of these jet flows have a Reynolds number of approximately 13,000, and flame C corresponds to one of the partially premixed piloted methane–air jet flames from the well-established series of target flames in the International Workshop on Measurement and Computation of Turbulent Flames (TNF Workshop) [254]. Structures of elevated vorticity norm (ω) and strain rate norm ($s = \sqrt{(s_1^2 + s_2^2 + s_3^2)}$) are represented by light-gray and dark-gray isosurfaces, respectively. In flame C, elongated overlapping structures of high vorticity and strain rate are predominantly concentrated along the strong shear layer between the jet and the coflow where the flame reaction zone is located. In contrast, the air jet has structures of high vorticity and strain rate that are more fragmented and less tightly overlapped than in flame C. These structures are distributed between the core of the jet and the coflow. The localized extinction in flame C_{LP} results in vorticity and strain rate fields that are a combination of structures observed in the air jet and the fully burning flame C.

A statistical analysis of the orientation of the vorticity, ω, with respect to the eigenvectors of the strain rate tensor (i.e., s_1, s_2, and s_3) reveals the preferential alignment that is induced by the flame. Figure 2.18 shows probability density functions

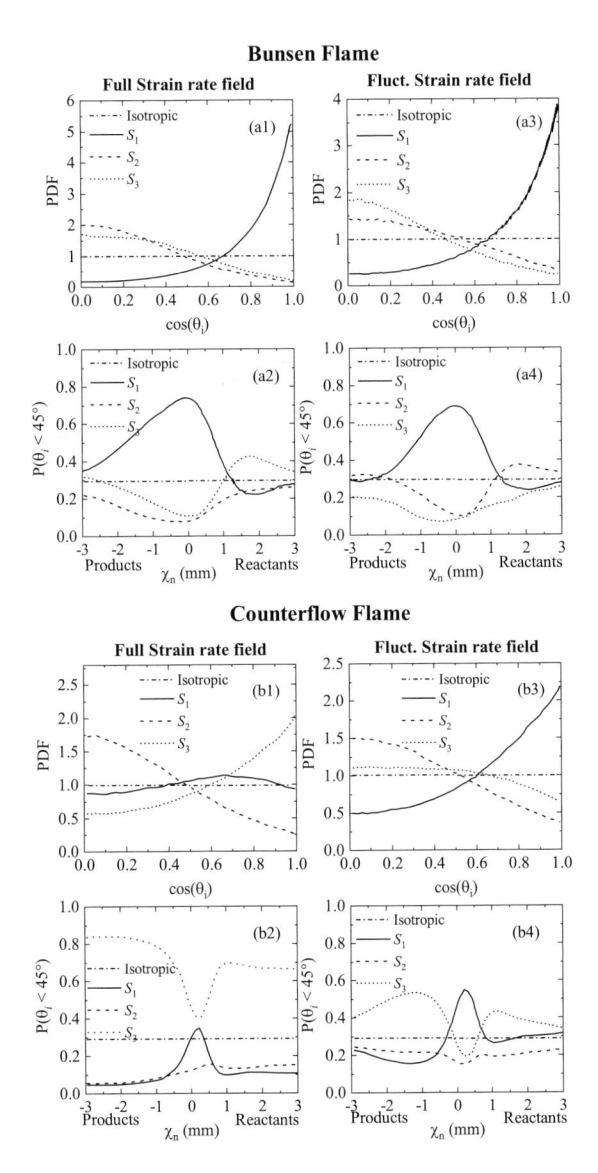

Figure 2.16 PDFs of strain rate eigenvector alignment with the flame-normal at the flame front ($x_n = 0$) for (a1) Bunsen flame and (b1) counterflow flame. Conditional probabilities for $\theta_i < 45°$ as a function of flame-normal coordinate, x_n, across the flame front for (a2) Bunsen flame and (b2) counterflow flame. Black dash-dot line indicates the probability for an isotropic angular distribution. The same alignment statistics on the strain rate fields derived from fluctuating velocity fields are shown in the second column (a3–a4, b3–b4). Adapted from [251].

of $|\cos \alpha_i|$ for the three flow conditions, where α_i is the angle between the vorticity vector and the ith strain rate eigenvector. The plots include both unconditional PDFs and PDFs conditioned on elevated vorticity for $\omega > 12,000 \text{ s}^{-1}$. The preferential alignment of vorticity with the axis of intermediate strain rate, s_2, is evident in the

PDFs of $|\cos\alpha_2|$, which all have peaks at $|\cos\alpha_2| = 1$. The peak of the unconditional PDF of $|\cos\alpha_2|$ for flame C is higher than in the air jet and flame C_{LP}, showing that stable combustion enhances the preferential alignment of ω and s_2. The differences between the alignment statistics of flame C_{LP} and the air jet are moderate, indicating that the localized extinction in flame C_{LP} significantly reduces the overall preferential alignment of vorticity and strain from that of the stable flame. The PDFs of $|\cos\alpha_2|$ conditioned on $\omega > 12,000\ \text{s}^{-1}$ have significantly larger peak values than the unconditional PDFs, indicating that the preferential ω–s_2 alignment is particularly prevalent in regions of large vorticity. With the presence of strong heat release, flame C shows the greatest enhancement of $\omega - s_2$ alignment for large values of vorticity.

Previous studies using numerical simulations of nonreacting turbulent flows have shown that $\omega - s_2$ alignment is enhanced by the strain rate field that is induced by regions of high vorticity [255–257]. The vorticity-induced strain can be so strong that the principal strain axes are switched to favor the $\omega - s_2$ alignment. If the local vorticity-induced strain is subtracted, the $\omega - s_1$ alignment would be expected from the remaining (nonlocal) strain rate field [258], a picture that is consistent with the concept of turbulent flow energy cascade through vortex stretching [259].

The spatial correlation between vorticity and strain rate fields can be different for various types of flames and burner geometries. For example, turbulent premixed counterflow flames have different vorticity and strain distributions than the aforementioned jet flames. Unlike jet flames, the counterflow imposes a bulk compressive strain on the flame, and the turbulence is produced by a turbulence generation plate embedded inside the top nozzle [247]. The effect of heat release in a counterflow can be seen in a comparison of single-shot measurements in a turbulent nonreacting counterflow of N_2-versus-combustion products and a premixed counterflow flame of $CH_4/O_2/N_2$-versus-combustion products in Fig. 2.19. Figure 2.19a shows isosurfaces of vorticity $\omega = 7000\ \text{s}^{-1}$ (ω) as well as extensive and compressive principal strain rates, $s_1 = 7000\ \text{s}^{-1}$ (s_1) and $s_3 = -6000\ \text{s}^{-1}$ (light-gray) and $-1000\ \text{s}^{-1}$ ($s_3 < 0$) measured in the nonreacting counterflow. Nitrogen flows from the upper nozzle, and combustion

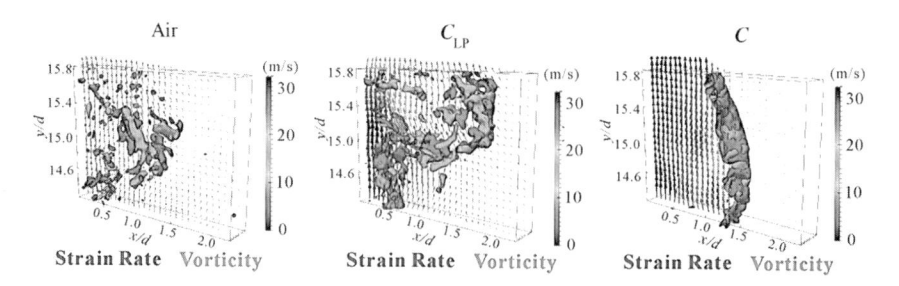

Figure 2.17 Isosurfaces of high strain rate (dark-gray) and vorticity (light-gray) for $s = 2.6\langle s\rangle$ and $\omega = 3.0\langle\omega\rangle$ from single-shot measurements in an air jet, partially premixed flame C_{LP} with significant localized extinction, and flame C without localized extinction. $\langle s\rangle$ and $\langle\omega\rangle$ are the mean strain rate norm and vorticity averaged over the whole probe volume for each condition. Jet axis is near the left edge of the probe volume. Velocity vectors are plotted in the central plane with 1 out of 36 vectors plotted. Adapted from Ref. [234].

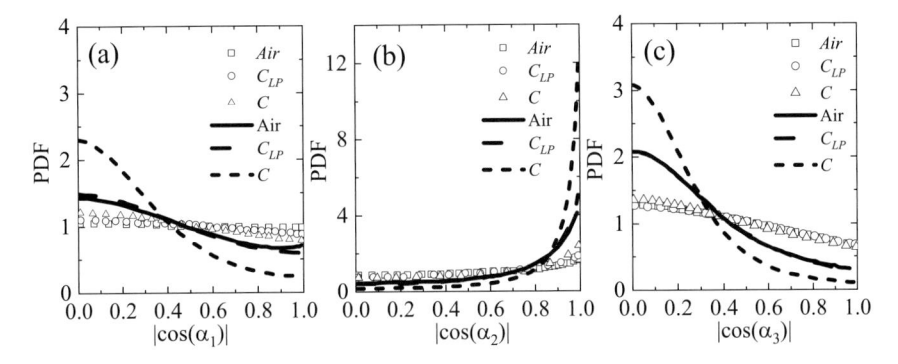

Figure 2.18 Probability density functions of the alignment angle, α_i, between ω and each of the strain rate eigenvectors, s_1, s_2, and s_3. Symbols: unconditional PDFs, lines: conditional PDFs for large vorticity ($\omega > 12,000$ s^{-1}). Adapted from Ref. [234].

products are produced by a pre-burner inside the lower nozzle. An abrupt change in gas density occurs at the interface between the top and bottom flows, which we refer to as the gas mixing layer interface (GMLI). Near the GMLI surface, the velocity vectors are preferentially aligned with the surface tangential direction, indicating that the GMLI surface corresponds to the boundary of the bottom product stream where the flow is stagnated. Highly convoluted and entangled structures of elevated vorticity as well as extensive and compressive strain rates are concentrated along the GMLI. Additional regions of high vorticity and strain rate are observed away from the GMLI. In the counterflow flame shown in Fig. 2.19b, localized regions of high vorticity overlap with strong extensive principal strain rate $s_1 = 7000$ s^{-1} and weak compressive principal strain rate $s_3 = -1000$ s^{-1} along the flame front. Dilatation due to the heat release introduces regions of strong extensive strain that partially counteract the compressive bulk strain imposed by the counterflow configuration, resulting in weaker compressive strain rates near the flame front. In contrast, regions of strong compressive strain $s_3 = -6000$ s^{-1} appear together with high-vorticity regions near the GMLI surface where the flow is compressed the most. Overall, complex morphologies of vorticity and strain rate fields are a result of interactions between turbulence, heat release, density variation, and the bulk flow. The individual roles of these parameters need to be further clarified using the current state-of-the-art experimental measurements.

2.3.2.3 High-Speed Imaging Using a Burst-Mode Laser

Studies of turbulent flame dynamics require imaging measurements at multi-kilohertz rates to capture relevant timescales. Unlike temporally uncorrelated measurements at low repetition rates, images that track the spatiotemporal evolution of transient events enable analysis of data from the Lagrangian or scalar-fixed coordinate with time as an independent dimension. Unique insights can be provided into understanding of transient processes such as fuel injection, turbulent mixing, extinction, ignition, and flashback.

High-speed imaging involves trade-offs between repetition rate, spatial and temporal resolution, and signal-to-noise ratios. High-frame-rate cameras, such as

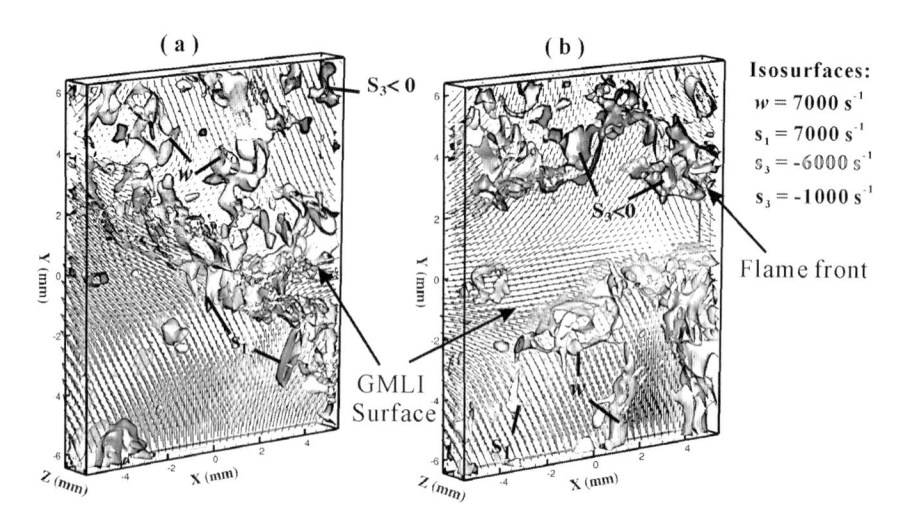

Figure 2.19 Simultaneous OH LIF image with isosurfaces of vorticity $\omega = 7000$ s^{-1} (ω), the most extensive strain rate $s_1 = 7000$ s^{-1} (s_1), as well as the most compressive strain rate $s_3 = -6000$ s^{-1} (light-gray) and -1000 s^{-1} ($S_3 < 0$) for (a) a nonreacting N$_2$-versus-product counterflow and (b) a premixed counterflow turbulent flame. One out of 36 velocity vectors is displayed.

complementary metal–oxide-semiconductor (CMOS) detectors, tend to have significantly higher noise levels than their low-frame-rate counterparts. When using high-frame-rate image intensifiers, consideration should be given to effects such as finite phosphor decay times and nonlinearities from charge depletion of microchannel plates. Laser repetition rates ranging from kilohertz to megahertz are achievable using different configurations, including a cluster of low-repetition rate lasers, burst-mode lasers, and continuously operating diode-pumped solid-state (DPSS) lasers. These configurations provide streams of laser pulses with different durations, repetition rates, and pulse energies. The output power of high-repetition-rate lasers with long streams of pulses is limited by thermal loading issues. Available pulse energies and repetition rates have increased with advances in diode-pumped solid-state lasers. A wide wavelength range can be accessed by a tunable dye laser system with a DPSS pump laser. Examples include joint OH-LIF measurements with planar PIV to study vortex–flame interactions [260], flame propagation [261], local extinction [262], and flashback [263]. However, for scalar imaging measurements at sustained multi-kHz rates, it is much harder to achieve the same quality images as those of conventional 10 Hz measurements due to the significantly lower laser pulse energy available. For many diagnostic applications, high laser pulse energy is needed to ensure sufficient signal detection and imaging quality. Studies of the dynamics of many highly relevant transient events in combustion do not necessarily require sustained high-repetition rate imaging because of the short duration of transient events. Therefore, a burst of laser pulses with a duration matching that of the transient flow is sufficient. Imaging measurements with very short bursts containing fewer than 10 laser pulses have been performed using the combined outputs from a cluster of low-repetition-rate lasers [264–266]. Recent advances in burst-mode lasers deliver a train of pulses from tens of pulses to 1000 pulses in each burst with appealing pulse energies, longer time sequences, as well as pulse repetition rates from kilohertz to megahertz [193]. Various wavelengths can be achieved by an optical parametric oscillator [267] pumped by a burst-mode laser for excitation of different species. Imaging of CH [266], OH [230, 267], CH_2O [230, 268], NO [269], and temperature [270] at multi-kHz frame rates has been demonstrated in various combustion applications.

High-speed cameras are also continually improving with greater framing rates and pixel resolution. The parallel readout architecture of CMOS sensors enables sustained kilohertz framing rates with record lengths determined by on-board camera storage capacity. Currently, commercially available CMOS cameras provide megapixel detection at framing rates exceeding 20 kHz. Higher rates up to a few hundred kilohertz are possible using a subregion of the detector. Because CMOS cameras are not UV sensitive and have relatively long camera gate times, high-speed image intensifiers are commonly implemented with a CMOS camera for detection of low light intensity levels and rejection of background interference through fast electronic gating. Other detector options include framing cameras that record a few image sequences at even higher frame rates [264].

The following sections highlight recent examples of high-speed imaging using a burst-mode diode-pumped Nd:YAG laser developed in our laboratory. Rayleigh scattering imaging of a transient jet flow at 100 kHz is demonstrated under a challenging condition of elevated pressure. A wavelet-based filtering method is developed to correct for the appearance of severe beam-steering artefacts under elevated pressure. Ultimately, the recent development of high-speed three-dimensional scalar imaging combined with TPIV is demonstrated.

A. Beam-Steering Effects for Laser-Based Imaging in High-Pressure Environments

In contrast to path-integrated measurements such as schlieren, shadowgraphs or chemiluminescence imaging [271], laser-based imaging techniques provide spatially and temporally resolved measurements of turbulent flows. However, implementing laser-based imaging techniques at the requisite repetition rates and at engine relevant conditions is particularly challenging. One of the diagnostics challenges at elevated pressures is beam-steering artefacts that complicate the interpretation of laser-based imaging measurements irrespective of the repetition rate. In turbulent flows that have large index-of-refraction gradients, the light rays in the laser beam are refracted, causing artefacts in the recorded image that appear as stripes with either higher or lower laser light intensity along the direction of laser propagation, which severely impedes quantitative data interpretation. For conditions in which the beam steering is modest and the index-of-refraction gradients are well resolved by the imaging system, ray-tracing methods can be used to correct for the effects of beam steering [272]. However, high-pressure turbulent flows have extremely steep index-of-refraction gradients and small turbulence length scales, resulting in severe beam-steering artefacts. At elevated pressures or in large Reynolds number flows, it becomes difficult or even impossible to resolve the fine-scale fluctuations in the index of refraction, which limits the effectiveness of ray-tracing methods. For these important applications, a different approach is required as explained in the text that follows.

The effects of beam-steering artefacts are highlighted by an example of mixture fraction measurements using Rayleigh scattering imaging in a high-pressure transient gas jet [273]. Figure 2.20a shows the experimental configuration for high-speed Rayleigh scattering imaging measurements that were performed in a transient jet of helium injected into a methane bath gas at approximately 8 bar in the pressure vessel. The vessel consists of a six-way stainless steel cube with three flanges containing window ports. Two opposing flanges have narrow slits fitted with fused silica windows with an antireflection coating at 532 nm for transmission of the laser sheet from the second harmonic of the Nd:YAG burst-mode laser. Rayleigh scattering is collected orthogonal to the laser sheet through a circular fused silica window and recorded on an unintensified high-speed CMOS camera that is synchronized with the 100 kHz pulses. The chamber is pressurized with the methane bath gas to the desired pressure. For each injection, the solenoid valve is opened for 30–50 ms, depending on the pressure settings, and the 5 ms burst of laser pulses is triggered during the last 7–10 ms of the gas injection to allow time for flow transients to subside. Prior to each injection, the chamber is flushed to remove the gas mixture from the previous injection.

As an example, the stripe artefacts created by index-of-refraction gradients in a single-shot Rayleigh scattering image of the high-pressure helium jet are shown in the upper image of Fig. 2.20b1. The laser beam propagates from right to left in the image. This image includes corrections for background, average beam profile, throughput of the imaging optics, and camera response using a separate series of Rayleigh scattering measurements in the vessel. This image has also been corrected for variations in the incoming beam profile using a region containing pure methane on the far right. The remaining stripes are generated by the turbulent mixing field of helium and methane, which have disparate indices of refraction. Figure 2.20b2 shows the same Rayleigh scattering image that has been corrected using a wavelet-based filtering algorithm [273]. A comparison of the zoomed-in plots of the corrected and uncorrected images shows that this method is very effective at removing the stripe artefacts while preserving the turbulent mixing data. Figure 2.20c shows a time sequence of the helium mole fraction based on the corrected images. In each image, the helium mole fraction was calculated from the Rayleigh scattering signals using the relative Rayleigh scattering cross sections of helium and methane along with uniform regions of methane on either side of the jet as a reference signal. The imaged region spans 14 mm in the axial direction, centered at $x/D = 21$ and 42 mm in the horizontal direction. The image sequence is a subset of the 500 shots that were recorded during a 5 ms duration burst from the laser. The wavelet method takes advantage of the coherence length scale of the beam-steering artefacts along the direction of the beam propagation, and details of this wavelet-based filtering approach can be found in [273]. This approach enables more quantitative imaging measurements of high-pressure turbulent flows at elevated pressures and other flows with large index-of-refraction variations. The method also is quite robust to noise in the imaging measurements, which is important for applications with weaker Rayleigh scattering signals, enabling studies of the dynamics of turbulent mixing for a range of pressures.

B. Three-Dimensional Scalar Visualization in Reacting Flows

Turbulent flame structures are inherently three dimensional, and the determination of flame surface area, curvature, scalar dissipation, and other quantities involving gradients requires three-dimensional scalar imaging. Furthermore, the interpretation of high-repetition rate planar imaging measurements is complicated by out-of-plane motion. Methods for reducing these ambiguities include simultaneous crossed-plane imaging or three-component velocity field measurements, enabling conditional statistics that exclude instances with significant out-of-plane motion [274, 275]. However, this approach can be inefficient, since a large portion of the data may need to be discarded due to out-of-plane motion. Three-dimensional scalar and velocity measurements are needed to fully address these issues but pose inherent experimental challenges and limitations.

Three-dimensional scalar imaging can be performed using tomography, which involves trade-offs between the number of cameras at different viewing angles and the spatial resolution of the tomographic reconstruction. Because of the spatial continuity of scalar fields, accurate tomographic reconstruction of scalar fields typically requires larger numbers of camera views than that of TPIV. Early demonstration of 3D flame tomographic measurements was based on flame chemiluminescence [276–279]. For better characterization of local thermochemical states of turbulent reacting flows, single-shot tomographic LIF (tomo-LIF) measurements have been demonstrated for combustion relevant species, such as acetone as a fuel tracer [280], OH [281], and CH [282]. Further efforts have pursued time-resolved tomo-LIF. For example, high-speed tomographic OH-LIF imaging has been recently demonstrated in a lifted jet flame using a burst-mode optical parametric oscillator [283]. Despite the rapidly emerging capabilities for extending measurement dimensionalities to better characterize turbulent combustion processes, high-speed tomographic scalar imaging is very demanding in terms of equipment. Multiple high-speed cameras coupled with image intensifiers are typically required, and laser sources with appropriate wavelengths and high peak fluence are also a necessity to achieve homogeneous volumetric illumination for tomo-LIF measurements. As a result, the limited number of views used in the data collection constrains the spatial resolution to considerably less than that available in planar LIF imaging.

Alternatively, quasi-instantaneous 3D scalar measurements can be acquired by rapidly scanning a sheet of laser light from a high-repetition rate laser across the probe volume, provided that the scan time is short compared to flow timescales [284–288]. A 3D scalar distribution is reconstructed from a stack of 2D images. Compared with tomographic measurements, scanning-based techniques are experimentally less expensive, as only one camera is needed, although the camera has to operate at a faster framing rate, which could lead to a compromise in the size of the imaging area. The demand on laser fluence is also greatly reduced for planar imaging,

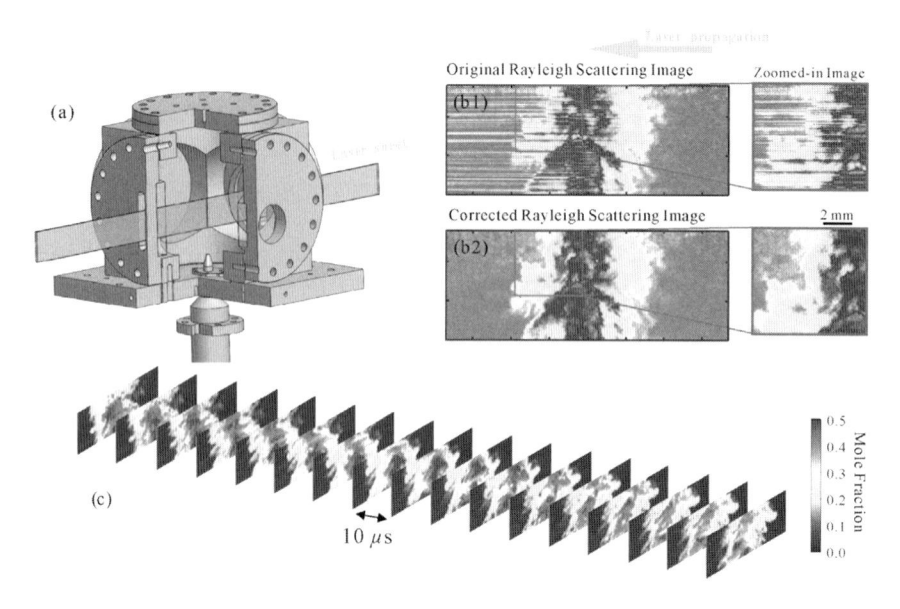

Figure 2.20 (a) Experimental configuration for Rayleigh scattering imaging in a turbulent nonreacting jet at elevated pressure. (b1–b2) Illustration of beam-steering correction applied to a single-shot Rayleigh scattering image. (c) Time sequence of mole fraction of jet gas measured at a repetition rate of 100 kHz using Rayleigh scattering from a burst-mode laser. Adapted from Ref. [273].

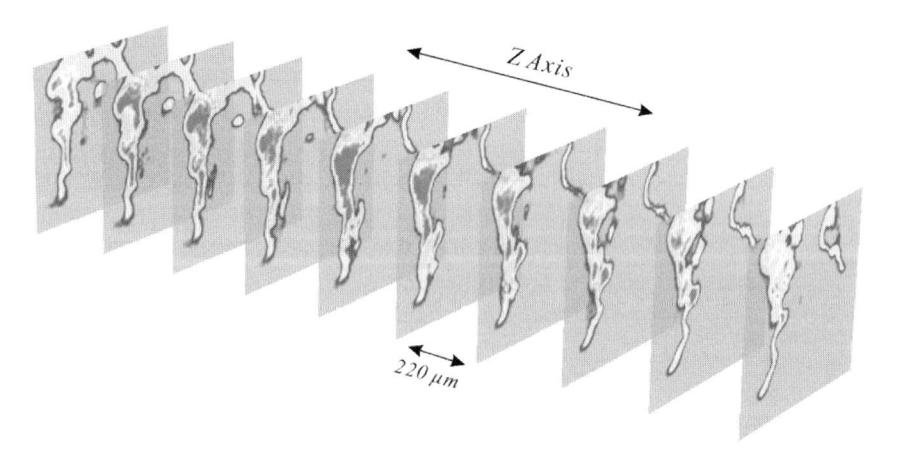

Figure 2.21 A sequence of 10 CH_2O LIF images recorded within a single scan cycle of the probe volume illustrating highly convoluted CH_2O in three dimensions. Each image is acquired from different planes within the lifted jet flame with a 10 μs time between frames and a displacement of 220 μm.

and a better in-plane imaging resolution and a simpler 3D reconstruction algorithm can also be expected. Conventional rapid scanning is performed using mechanically controlled polygonal mirrors [288] or a galvanometric scanner [287, 289]. However, the maximum scan frequency is restricted by the inertia associated with rapid mechanical rotation, and the scanner needs to be optimized for the desired scan frequency. As an alternative to the scanning techniques, the 3D scalar measurements can also be realized using multiple parallel laser sheets slicing the probe volume at several imaging planes simultaneously [290, 291]. The signals from different laser sheets can either be separated by using different wavelengths in the case of elastic scattering [290] or by structured illumination in which each laser sheet is encoded with a periodic spatial modulation in the intensity and enters the probe volume at different angles of incidence [291]. In structured illumination, the signals are collected all at once using one camera with a single exposure. The recorded image is Fourier transformed, and signals from various planes are separated in the frequency domain by their different spatial modulation frequencies or incident angles. Appropriate frequency filters are then employed to select the corresponding signal components in the Fourier domain that are converted back to planar scalar images through an inverse Fourier transform for 3D reconstruction. This technique, however, requires that the camera have sufficient resolution to accommodate the high-frequency spatial modulation, and the recovered images suffer from degraded spatial resolution to certain degrees.

Recent advances in scanning techniques eliminate the need for mechanically moving mirrors by using an acousto-optic deflector (AOD) [292]. A high-frequency acoustic wave that is introduced into the AOD crystal causes a periodic variation in the refractive index such that the AOD acts like an optical Bragg grating. The deflection angle can be rapidly modulated using a voltage-controlled oscillator to vary the acoustic frequency, which enables rapid beam sweeping with greater flexibility and better angle accuracy than a mechanical scanning mirror. We demonstrate the use of acousto-optic scanning for 3D CH_2O-LIF imaging. Figure 2.21 shows a stack of CH_2O-LIF images from 10 adjacent planes that are used to reconstruct the volumetric CH_2O structure near the base of a lifted turbulent jet flame. For this example, the AOD voltage modulation is synchronized with a burst-mode Nd:YAG laser ($\lambda = 355$ nm) operated at 100 kHz and is programmed such that every 10 frames forms a 100 μs scan cycle, and the spacing between each image plane is approximately 200 μm. The CH_2O structure shown in Fig. 2.21 reveals a complex topology that contains wrinkled surfaces and notches extending in all three dimensions. Isolated CH_2O pockets that appear in a single planar image are the result of a highly convoluted 3D structure. These measurements of the volumetric distribution of CH_2O are repeated at a rate of 10 kHz over the burst duration of 5 ms, thereby revealing the temporal evolution of the 3D structure of the flame.

This time resolved 3D CH_2O-LIF imaging capability can be combined with high-speed TPIV to provide volumetric scalar-velocity measurements for studying the dynamics of scalar-turbulence interactions. Figure 2.22 shows a time sequence of the 3D CH_2O topology combined with (a) velocity fluctuation, (b) vorticity magnitude, and (c) strain rate magnitude. From 3D scalar and velocity measurements, 3D curvature and surface density as well as scalar-normal orientation with respect to the vorticity and strain rate eigenvectors can be evaluated. Information from the time domain can be used to characterize turbulent transport/mixing and measure quantities

(a) CH$_2$O LIF - Velocity Fluctuation

(b) CH$_2$O LIF - Vorticity Mag.

(c) CH$_2$O LIF - Strain Rate Mag.

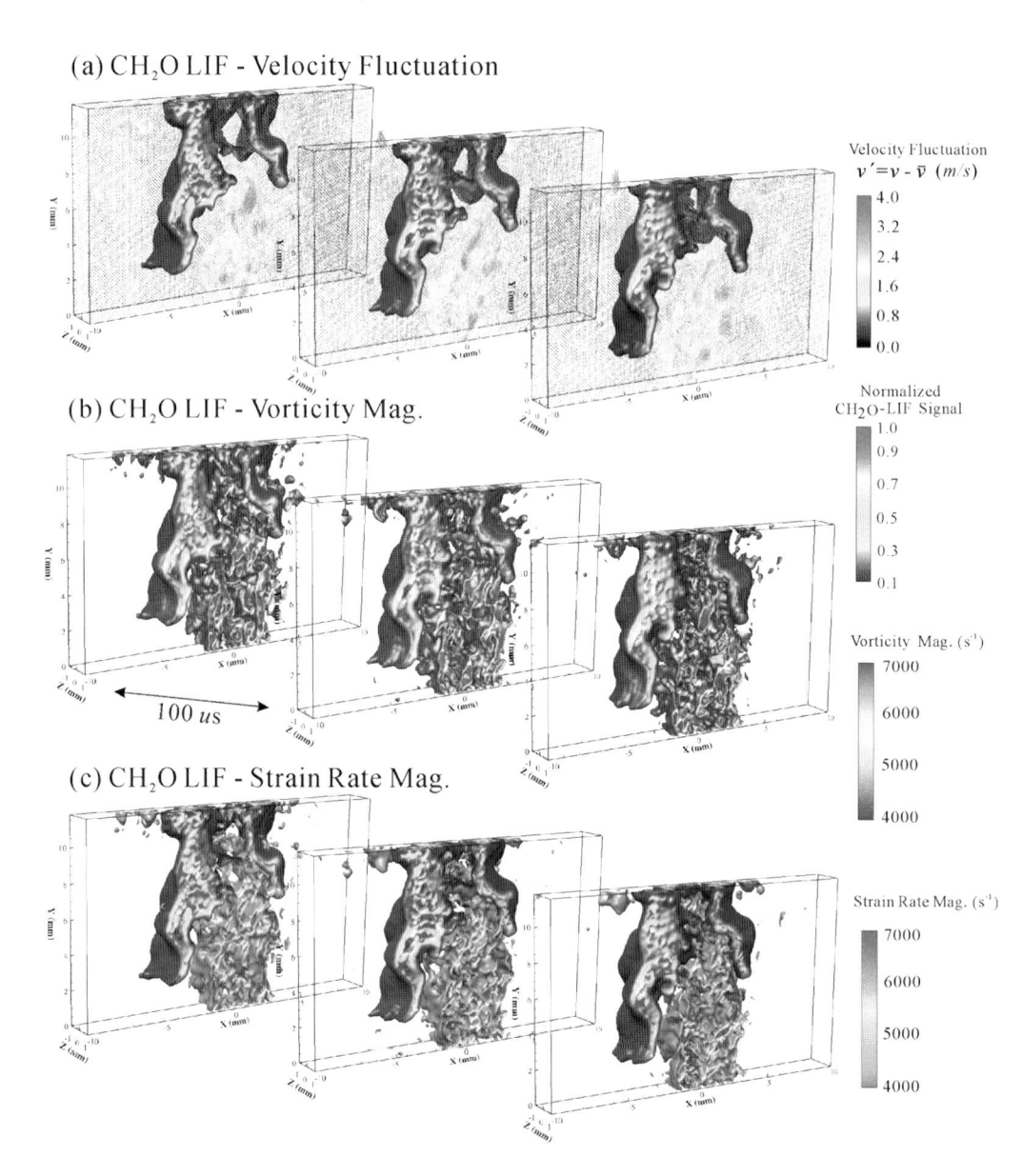

Figure 2.22 Simultaneous 3D CH$_2$O-LIF imaging and TPIV measurements performed at 10 kHz. Time sequence of 3D CH$_2$O topology combined with (a) velocity fluctuation, (b) vorticity magnitude, and (c) strain rate magnitude measured near the base of a lifted jet flame.

such as displacement speed of a scalar iso-surface. Efforts to improve 3D imaging methods with high acquisition rates are ongoing, and methods for data processing and analysis are also under development to accommodate the newly emerged 4D experimental capability for studies of turbulent reacting flows.

2.3.3 Concluding Remarks

Laser-based diagnostic techniques for turbulent combustion research have been continuously advancing along with the development of laser and detector systems. Examples presented in this chapter demonstrate the state-of-the-art diagnostic capabilities in simultaneous multiscalar imaging as well as high-speed multidimensional measurements of scalar and velocity fields. These diagnostic advances enable improvements to our understanding of turbulent combustion, and multidimensional measurements provide direct validation of specific terms in turbulent combustion models. However, the resulting experimental findings can also present challenges to current turbulent combustion models.

Numerical simulations have the advantage of simultaneously resolving a full matrix of scalars in both spatial and temporal domains, which is beyond the capabilities of experimental diagnostics. However, numerical simulations involve empirical models that constrain the fidelity of simulation results to certain degrees. In large eddy simulations, for example, turbulence-chemistry interactions at the subgrid scale are modeled, and the chemical kinetic mechanism is essentially a chemistry model. While direct numerical simulations do not require turbulence models, they are still subject to the uncertainties in chemistry and transport models. In addition, the computational cost of DNS results in additional limitations, such as the domain size, Reynolds number, and simulation time. From this perspective, experimental results provide a benchmark that is free of any models.

An ongoing research need is complementary experimental and numerical studies of turbulent combustion to develop truly predictive modeling capabilities for complex combustion applications. The framework of the TNF workshop [254] has set a great example of cooperative experimental and numerical research on a series of target turbulent flames with progressively greater complexity. Point-based and Reynolds-averaged experimental statistics have been compared extensively with numerical results to evaluate combustion models. With the continuous development of high-speed imaging diagnostic capabilities, there is an open question of how to take full advantage of the spatio-temporal information from experiments and incorporate with numerical results. An example along this path includes the recent work of J. W. Labahn et al. [293] in which multidimensional high-speed measurements have been assimilated into a combustion simulation to investigate transient combustion events of local extinction and evaluate model performance. In addition, the emergence of petascale computational facilitates has increased the capacity of DNS calculations of turbulent flames and expanded the possibilities for comparisons with experimental results to gain fundamental insight of turbulence–flame interactions [294].

Another need for future diagnostic development includes improvements to quantitative multidimensional scalar measurements to facilitate comparisons with numerical results. Depending on the measurement conditions, some laser-based imaging techniques, e.g., planar laser-induced fluorescence, may only provide qualitative measurements of scalars, since signal quantification involves a number of underdetermined parameters such as local mixture composition, temperature, and pressure, which are challenging to measure simultaneously. Furthermore, both fundamental and applied combustion research require advances in diagnostic capabilities for challenging combustion environments. This requires diagnostic techniques at elevated pressures, near interfaces, and in multiphase combustion environments where signal degradation as well as potential cross-talk of laser-induced signals due to the presence of complex nongaseous fuels can be expected.

References

[1] N. Peters and F. A. Williams, "The asymptotic structure of stoichiometric methane/air flames," *Combust. Flame*, **68**, 185–207, 1987.

[2] S. P. Burke and T. E. W. Schumann, "Diffusion flames," *Indust. Eng. Chem.*, **20**, 996, 1928.

[3] P. A. Libby and F. A. Williams, *Turbulent Reacting Flows*, New York: Springer-Verlag, 1980.

[4] P. A. Libby and F. A. Williams, *Turbulent Reacting Flows*, New York: Academic Press, 1994.

[5] N. Peters, "Laminar diffusion flamelet models in non-premixed turbulent combustion," *Prog. Energy Combust. Sci.*, **10**, 319–339, 1984.

[6] R. W. Bilger, "Future progress in turbulent combustion research," *Prog. Energy Combust. Sci.*, **26**, 367–380, 2000.

[7] R. W. Bilger, "Turbulent diffusion flames," *Ann. Rev. Fluid Mech.*, **21**, 101–135, 1989.

[8] R. Borghi, "Turbulent combustion modeling," *Prog. Energy Combust. Sci.*, **14**, 245–292, 1988.

[9] R. W. Bilger, "The structure of diffusion flames," *Combust. Sci. Technol.*, **13**, 155–170, 1976.

[10] K. Seshadri and N. Peters, "The inner structure of methane/air flames," *Combust. Flame*, **81**, 96–118, 1990.

[11] N. Peters, *Turbulent Combustion*. Cambridge, UK: Cambridge University Press, 2000.

[12] R. W. Bilger, "The structure of turbulent non-premixed flames," *Symp. (International) Combust.*, **22**, 475–488, 1988.

[13] G. Damköhler, "Der einfluss der turbulenz auf die flammengeschwindigkeit in gas-gemischen," *Z. Elektrochem. angewandte physikal. Chemie*, **46**, 601–626, 1940.

[14] B. Karlovitz, D. H. Denniston Jr, D. W., Knapschaefer, and F. E. Wells, "Studies on turbulent flames: A. Flame propagation across velocity gradients. B. Turbulence measurement in flames," in *Symp. (International) Combust.*, **4**, no. 1, 1953, 613–620.

[15] A. M. Klimov, "Laminar flame in a turbulent flow," *Zhur. Prikl. Mekh. Tekh. Fiz.*, **3**, 49–58, 1963.

[16] F. A. Williams, "A review of some theoretical considerations of turbulent flame structure," in *Analytical and Numerical Methods for Investigation of Flow Fields with Chemical Reactions, Especially Related to Combustion, AGARD Conference Proceedings*, **164**, 1975, II 1–1 – II 1–25.

[17] B. Lewis and G. von Elbe, *Combustion Flames and Explosions of Gases*. New York: Academic Press, 1961.

[18] R. Strehlow, *Fundamentals of Combustion*. Scranton, PA: International Textbook Co., 1968.

[19] R. G. Abdel-Gayed, K. J. Al-Khishali, and D. Bradley, "Turbulent burning velocities and flame straining in explosions," *Proc. R. Soc. Lond. A*, **391**, 393–414, 1984.

[20] J. Chomiak, "A possible propagation mechanism of turbulent flames at high Reynolds number," *Combust. Flame*, **15**, 319–321, 1970.

[21] G. E. Andrews, D. Bradley, and S. B. Lwakabamba, "Turbulence and turbulent flame propagation – A critical appraisal," *Combust. Flame*, **24**, 285–304, 1975.

[22] F. A. Williams, "Criteria for existence of wrinkled laminar flame structure of turbulent premixed flames," *Combust. Flame*, **26**, 269–270, 1976.

[23] F. A. Williams, "Turbulent combustion," in *The Mathematics of Combustion*, J. D. Buckmaster, Ed. Philadelphia: SIAM, 1985, 97–131.

[24] F. A. Williams, *Combustion Theory*. San Francisco: Benjamin/Cummings, 1985.

[25] R. Borghi, "On the structure and morphology of turbulent premixed flames," in C. Bruno and C. Casci, Eds., *Recent Advances in the Aerospace Sciences*. Berlin, Heidelberg: Springer-Verlag, 1985, pp. 117–138.

[26] N. Peters, "The turbulent burning velocity for large-scale and small-scale turbulence," *J. Fluid Mech.*, **384**, 107–132, 1999.

[27] M. Barrere, "Modèles de combustion turbulente," *Rev. Gen. Therm.*, **148**, 295–308, 1974.

[28] K. N. C. Bray, "Turbulent flows with premixed reactants," in *Turbulent Reacting Flows*, P. A. Libby and F. A. Williams, Eds. Berlin, Heidelberg: Springer-Verlag, 1980, 115–183.

[29] M. Barrère, "Mise au point quelques recherches sur la combustion de la derniere decennie," *J. Chimie Phys.*, **81**, 519–531, 1984.

[30] N. Peters, "Laminar flamelet concepts in turbulent combustion," in *Symp. (International) Combust.*, **21**, no. 1, 1988, 1231–1250.

[31] T. Poinsot, D. Veynante, and S. Candel, "Diagrams of premixed turbulent combustion based on direct simulation," *Proc. Combust. Inst.*, **23**, 613–619, 1990.

[32] C. Meneveau and T. Poinsot, "Stretching and quenching of flamelets in premixed turbulent combustion," *Combust. Flame*, **86**, 311–332, 1991.

[33] W. L. Roberts, J. F. Driscoll, M. C. Drake, and L. P. Goss, "Images of the quenching of a flame by a vortex – to quantify regimes of turbulent combustion," *Combust. Flame*, **94**, 58–69, 1993.

[34] B. Zhou, C. Brackmann, Q. Li, Z. Wang, P. Petersson, Z. Li, M. Aldén, and X.-S. Bai, "Distributed reactions in highly turbulent premixed methane/air flames: Part I. flame structure characterization," *Combust. Flame*, **162**, 2937–2953, 2015.

[35] A. W. Skiba, T. M. Wabel, C. D. Carter, S. D. Hammack, J. E. Temme, T. Lee, and J. F. Driscoll, "Reaction layer visualization: A comparison of two PLIF techniques and advantages of kHz-imaging," *Proc. Combust. Inst.*, **36**, 4593–4601, 2017.

[36] J. F. Driscoll, "Premixed turbulent combustion in high Reynolds number regimes of thickened flamelets and distributed reactions," Dept. Aerospace Engineering, Michigan University, Ann Arbor, Tech. Rep., 2016.

[37] I. Langella, N. Swaminathan, and R. W. Pitz, "Application of unstrained flamelet SGS closure for multi-regime premixed combustion," *Combust. Flame*, **173**, 161–178, 2016.

[38] I. Langella, Z. X. Chen, N. Swaminathan, and S. K. Sadasivuni, "LES of reacting flows in an industrial gas turbine combustor," *J. Prop. Power*, **34**, 1269–1284, 2018.

References

[39] Z. X. Chen, N. Swaminathan, M. Stöhr, and W. Meier, "Interaction between self-excited oscillations and fuel-air mixing in a dual swirl combustor," *Proc. Combust. Inst.*, **37**, 2325–2333, 2019.

[40] J. C. Massey, Z. X. Chen, and N. Swaminathan, "Lean flame root dynamics inside a model gas turbine combustor," *Combust. Sci. Technol.*, **191**, 1019–1042, 2019.

[41] H. Carlsson, R. Yu, and X.-S. Bai, "Flame structure analysis for categorization of lean premixed CH4/air and H2/air flames at high karlovitz numbers: Direct numerical simulation studies," *Proc. Combust. Inst.*, **35**, 1425–1432, 2015.

[42] A. Y. Poludnenko and E. S. Oran, "The interaction of high-speed turbulence with flames: Global properties and internal flame structure," *Combust. Flame*, **157**, 995–1011, 2010.

[43] A. Yoshida, H. Naito, and D. P. Mishra, "Turbulent combustion of preheated natural gas–air mixture," *Fuel*, **87**, 605–611, 2008.

[44] E. S. Shetinkov, "Calculation of flame velocity in turbulent stream," in *Symp. (International) Combustion*, **7**, no. 1, 1958, 583–589.

[45] E. S. Shchetinkov, "Combustion physics of gases," *Nauka*, Moscow, in Russian (1965), (machine translation in English: FTD-HT-23-496-48).

[46] E. S. Shchetinkov, "Theoretical study of the combustion of a homogeneous mixture in a turbulent flow," unpublished report, Moscow, 1955.

[47] V. A. Sabelnikov and V. Penzin, "Scramjet research and development in Russia," *Scramjet Propul.: Prog. Astro. Aero.*, **189**, 2000.

[48] V. A. Sabelnikov and V. V. Vlasenko, "Combustion in supersonic flows and scramjet combustion simulation," in *Modeling and Simulation of Turbulent Combustion*. Springer, 2018, 585–660.

[49] K. I. Shelkin, "On combustion in a turbulent flow," *J. Tech. Phys.*, **13**, 1947.

[50] T. Poinsot and D. Veynante, *Theoretical and Numerical Combustion*. Flourtown, PA: RT Edwards, 2005.

[51] A. Lipatnikov, *Fundamentals of Premixed Turbulent Combustion*. Boca Raton, FL: CRC Press, 2012.

[52] H. L. Olsen, "Incipient flame propagation in a turbulent stream," *J. Jet Propul.*, **25**, 276–283, 1955.

[53] L. S. G. Kovasznay, "Combustion in turbulent flow-a summary of remarks at a special panel meeting," *J. Jet Propul.*, **26**, 480–495, 1956.

[54] V. R. Kuznetsov and V. A. Sabelnikov, "Combustion characteristics of mixed gases in a strongly turbulent flow," *Combust. Explos. Shock Waves*, **13**, 425–435, 1977.

[55] V. R. Kuznetsov and V. A. Sabelnikov, *Turbulence and Combustion*, New York: Hemisphere, 1990.

[56] H. Singh, S. Singh, and B. M. Deb, "Lattice gas automata: A tool for exploring dynamical processes," in *18th Proceedings of the Indian Academy of Sciences-Chemical Sciences*, **106**, no. 2. Berlin, Heidelberg: Springer, 1994, 539–551.

[57] R. Said and R. Borghi, "A simulation with a cellular automaton for turbulent combustion modelling," in *Symp. (Int.) Combust.*, **22**, no. 1, 1989, 569–577.

[58] A. N. Kolmogorov, "The local structure of turbulence in incompressible viscous fluid for very large reynolds numbers," in *Dokl. Akad. Nauk SSSR*, **30**, no. 4, 1941, 299–303.

[59] A. Monin and A. Yaglom, *Statistical Fluid Mechanics, Mechanics of Turbulence*. Cambridge, MA: MIT Press, 1971, **2**.

[60] U. Frisch, *Turbulence: The Legacy of A. N. Kolmogorov*. Cambridge, UK: Cambridge University Press, 1995.

[61] A. W. Skiba, T. M. Wabel, C. D. Carter, S. D. Hammack, J. E. Temme, and J. F. Driscoll, "Premixed flames subjected to extreme levels of turbulence part I: flame structure and a new measured regime diagram," *Combust. Flame*, **189**, 407–432, 2018.

[62] F. Hampp and R. P. Lindstedt, "Quantification of combustion regime transitions in premixed turbulent DME flames," *Combust. Flame*, **182**, 248–268, 2017.

[63] R. Cabra, T. Myhrvold, J.-Y. Chen, R. W. Dibble, A. N. Karpetis, and R. S. Barlow, "Simultaneous laser Raman-Rayleigh-LIF measurements and numerical modeling results of a lifted turbulent H_2/N_2 jet flame in a vitiated coflow," *Proc. Combust. Inst.*, **29**, 1881–1888, 2002.

[64] R. Cabra, J.-Y. Chen, R. W. Dibble, A. N. Karpetis, and R. S. Barlow, "Lifted methane–air jet flames in a vitiated coflow," *Combust. Flame*, **143**, 491–506, 2005.

[65] R. Cabra, "Turbulent jet flames into a vitiated coflow," Ph.D. dissertation, UC Berkeley, 2003.

[66] D. J. Micka, "Combustion stabilization, structure, and spreading in a laboratory dual-mode scramjet combustor," Ph.D. dissertation, University of Michigan, 2010.

[67] D. J. Micka and J. F. Driscoll, "Combustion characteristics of a dual-mode scramjet combustor with cavity flameholder," *Proc. Combust. Inst.*, **32**, 2397–2404, 2009.

[68] D. J. Micka and J. F. Driscoll, "Stratified jet flames in a heated (1390 K) air cross-flow with autoignition," *Combust. Flame*, **159**, 1205–1214, 2012.

[69] Y. Mizobuchi, S. Tachibana, J. Shinio, S. Ogawa, and T. Takeno, "A numerical analysis of the structure of a turbulent hydrogen jet lifted flame," *Proc. Combust. Inst.*, **29**, 2009–2015, 2002.

[70] Y. Mizobuchi, J. Shinjo, S. Ogawa, and T. Takeno, "A numerical study on the formation of diffusion flame islands in a turbulent hydrogen jet lifted flame," *Proc. Combust. Inst.*, **30**, 611–619, 2005.

[71] A. Cavaliere and M. de Joannon, "MILD combustion," *Prog. Energy Combust. Sci.*, **30**, 329–366, 2004.

[72] Y. Minamoto, "Physical aspects and modelling of turbulent MILD combustion," PhD Thesis, Cambridge University, The Department of Engineering, 2014.

[73] Y. Minamoto, N. Swaminathan, R. S. Cant, and T. Leung, "Reaction zones and their structure in MILD combustion," *Combust. Sci. Technol.*, **186**, 1075–1096, 2014.

[74] Y. Minamoto and N. Swaminathan, "Scalar gradient behaviour in MILD combustion," *Combust. Flame*, **161**, 1063–1075, 2014.

[75] Y. Minamoto, N. Swaminathan, R. S. Cant, and T. Leung, "Morphological and statistical features of reaction zones in MILD and premixed combustion," *Combust. Flame*, **161**, 2801–2814, 2014.

[76] Y. Minamoto and N. Swaminathan, "Modelling paradigms for MILD combustion," *Int. J. Adv. Eng. Sci. Appl. Math.*, **6**, 65–75, 2014.

[77] Y. Minamoto and N. Swaminathan, "Subgrid scale modelling for MILD combustion," *Proc. Combust. Inst.*, **35**, 3529–3536, 2015.

[78] Z. X. Chen, V. M. Reddy, S. Ruan, N. A. K. Doan, W. L. Roberts, and N. Swaminathan, "Simulation of MILD combustion using perfectly stirred reactor model," *Proc. Combust. Inst.*, **36**, 4279–4286, 2017.

[79] N. A. K. Doan, N. Swaminathan, and Y. Minamoto, "DNS of MILD combustion with mixture fraction variations," *Combust. Flame*, **189**, 173–189, 2018.

References

[80] C. S. Yoo, R. Sankaran, and J. H. Chen, "Three-dimensional direct numerical simulation of a turbulent lifted hydrogen jet flame in heated coflow: flame stabilization and structure," *J. Fluid Mech.*, **640**, 453–481, 2009.

[81] C. S. Yoo, E. S. Richardson, R. Sankaran, and J. H. Chen, "A DNS study on the stabilization mechanism of a turbulent lifted ethylene jet flame in highly-heated coflow," *Proc. Combust. Inst.*, **33**, 1619–1627, 2011.

[82] Y. Ju, "Recent progress and challenges in fundamental combustion research," *Adv. Mech.*, **44**, 201402, 2014.

[83] J. F. Driscoll, "Premixed turbulent combustion-regimes of thickened and distributed reactions," in *Proceedings of the 9th Mediterranean Combustion Symposium, Rhodes, Greece*, 2015, 7–11.

[84] J. F. Driscoll, A. W. Skiba, T. W. Wabel, and J. E. Temme, "Premixed turbulent combustion in the high Reynolds number regime of thickened flamelets and distributed reactions," in *AFOSR Project FA9550-12-1-0101*, 2015.

[85] R. M. Zaidel and Y. B. Zeldovich, "On possible stationary combustion regimes," *Zh. Prikl. Mekh. Tekh. Fiz.*, **4**, 27–32, 1962.

[86] Y. B. Zeldovich, G. I. Barenblatt, V. B. Librovich, and G. M. Makhviladze, *Mathematical Theory of Combustion and Explosions*. New York: Plenum Press, 1985.

[87] A. G. Merzhanov and B. I. Khaikin, "Theory of combustion waves in homogeneous media," *Prog. Energy Combust. Sci.*, **14**, 1–98, 1988.

[88] N. A. K. Doan, "Physical insights of non-premixed MILD combustion using DNS," PhD Thesis, Cambridge University, The Department of Engineering, 2018.

[89] Z. X. Chen, N. A. K. Doan, X. J. Lv, N. Swaminathan, G. Ceriello, G. Sorrentino, and A. Cavaliere, "Numerical study of a cyclonic combustor under moderate or intense low-oxygen dilution conditions using non-adiabatic tabulated chemistry," *Energy & Fuels*, **32**, 10 256–10 265, 2018.

[90] B. Magnussen, "On the structure of turbulence and a generalized eddy dissipation concept for chemical reaction in turbulent flow," in *19th Aerospace Sciences Meeting*, 1981, 42.

[91] V. Sabelnikov and C. Fureby, "LES combustion modeling for high Re flames using a multi-phase analogy," *Combust. Flame*, **160**, 83–96, 2013.

[92] J. F. Driscoll, "Turbulent premixed combustion: Flamelet structure and its effect on turbulent burning velocities," *Prog. Energy Combust. Sci.*, **34**, 91–134, 2008.

[93] B. Zhou, C. Brackmann, Z. Wang, Z. Li, M. Richter, M. Aldén, and X.-S. Bai, "Thin reaction zone and distributed reaction zone regimes in turbulent premixed methane/air flames: scalar distributions and correlations," *Combust. Flame*, **175**, 220–236, 2017.

[94] B. R. Chowdhury and B. M. Cetegen, "Experimental study of the effects of free stream turbulence on characteristics and flame structure of bluff-body stabilized conical lean premixed flames," *Combust. Flame*, **178**, 311–328, 2017.

[95] R. Sankaran, E. R. Hawkes, C. S. Yoo, and J. H. Chen, "Response of flame thickness and propagation speed under intense turbulence in spatially developing lean premixed methane–air jet flames," *Combust. Flame*, **162**, 3294–3306, 2015.

[96] S. Lapointe, B. Savard, and G. Blanquart, "Differential diffusion effects, distributed burning, and local extinctions in high Karlovitz premixed flames," *Combust. Flame*, **162**, 3341–3355, 2015.

[97] A. J. Aspden, M. S. Day, and J. B. Bell, "Three-dimensional direct numerical simulation of turbulent lean premixed methane combustion with detailed kinetics," *Combust. Flame*, **166**, 266–283, 2016.

[98] S. Chaudhuri, H. Kolla, H. L. Dave, E. R. Hawkes, J. H. Chen, and C. K. Law, "Flame thickness and conditional scalar dissipation rate in a premixed temporal turbulent reacting jet," *Combust. Flame*, **184**, 273–285, 2017.

[99] H. Wang, E. R. Hawkes, J. H. Chen, B. Zhou, Z. Li, and M. Aldén, "Direct numerical simulations of a high Karlovitz number laboratory premixed jet flame–an analysis of flame stretch and flame thickening," *J. Fluid Mech.*, **815**, 511–536, 2017.

[100] V. A. Sabelnikov, R. Yu, and A. N. Lipatnikov, "Thin reaction zones in highly turbulent medium," *Int. J. Heat Mass Tran.*, **128**, 1201–1205, 2019.

[101] V. A. Sabelnikov, R. Yu, and A. N. Lipatnikov, "Thin reaction zones in constant-density turbulent flows at low Damköhler numbers: Theory and simulations," *Phys. Fluids*, **31**, 055 104-1–18, 2019.

[102] R. Yu, X.-S. Bai, and A. N. Lipatnikov, "A direct numerical simulation study of interface propagation in homogeneous turbulence," *J. Fluid Mech.*, **772**, 127–164, 2015.

[103] R. Yu and A. N. Lipatnikov, "Direct numerical simulation study of statistically stationary propagation of a reaction wave in homogeneous turbulence," *Phys. Rev. E*, **95**, 063101, 2017.

[104] R. Yu and A. N. Lipatnikov, "DNS study of dependence of bulk consumption velocity in a constant-density reacting flow on turbulence and mixture characteristics," *Phys. Fluids*, **29**, 065116, 2017.

[105] G. Nivarti and R. S. Cant, "Direct numerical simulation of the bending effect in turbulent premixed flames," *Proc. Combust. Inst.*, **36**, 1903–1910, 2017.

[106] P. D. Ronney, B. D. Haslam, and N. O. Rhys, "Front propagation rates in randomly stirred media," *Phys. Rev. Letters*, **74**, 3804, 1995.

[107] S. S. Shy, C. C. Liu, J. Y. Lin, L. L. Chen, A. N. Lipatnikov, and S. I. Yang, "Correlations of high-pressure lean methane and syngas turbulent burning velocities: Effects of turbulent Reynolds, Damköhler, and Karlovitz numbers," *Proc. Combust. Inst.*, **35**, 1509–1516, 2015.

[108] A. J. Aspden, J. B. Bell, and S. E. Woosley, "Distributed flames in type Ia supernovae," *Astrophys. J.*, **710**, 1654, 2010.

[109] A. M. Klimov, "Laminar flame in a turbulent flow," Foreign Technology Div, Wright-Patterson AFB OH, Tech. Rep., 1988.

[110] B. I. Shraiman and E. D. Siggia, "Scalar turbulence," *Nature*, **405**, 639, 2000.

[111] A. J. Aspden, M. S. Day, and J. B. Bell, "Towards the distributed burning regime in turbulent premixed flames," *J. Fluid Mech.*, **871**, 1-21, 2019.

[112] H. P. Schmidt, P. Habisreuther, and W. Leuckel, "A model for calculating heat release in premixed turbulent flames," *J. Fluid Mech.*, **384**, 107–132, 1998.

[113] H. Eickhoff, "Analysis of the turbulent burning velocity," *Combust. Flame*, **128**, 347–350, 2002.

[114] H. Kolla, J. W. Rogerson, N. Chakraborty, and N. Swaminathan, "Scalar dissipation rate modeling and its validation," *Combust. Sci. Technol.*, **181**, 518–535, 2009.

[115] S. Chaudhuri, V. Akkerman, and C. K. Law, "Spectral formulation of turbulent flame speed with consideration of hydrodynamic instability," *Phys. Rev. E*, **84**, 026322, 2011.

[116] I. Ahmed and N. Swaminathan, "Simulation of spherically expanding turbulent premixed flames," *Combust. Sci. Technol.*, **185**, 1509–1540, 2013.

[117] H. Wang, E. R. Hawkes, and J. H. Chen, "Turbulence-flame interactions in DNS of a laboratory high Karlovitz premixed turbulent jet flame," *Phys. Fluids*, **28**, 095107, 2016.

[118] M. Day, S. Tachibana, J. Bell, M. Lijewski, V. Beckner, and R. K. Cheng, "A combined computational and experimental characterization of lean premixed turbulent low swirl laboratory flames: I. Methane flames," *Combust. Flame*, **159**, 275–290, 2012.

[119] B. Bedat and R. Cheng, "Experimental study of premixed flames in intense isotropic turbulence," *Combust. Flame*, **100**, 485–494, 1995.

[120] A. J. Aspden, J. B. Bell, M. S. Day, S. E. Woosley, and M. Zingale, "Turbulence-flame interactions in Type Ia supernovae," *Astro. J.*, **689**, 1173–1185, 2008.

[121] A. J. Aspden, M. S. Day, and J. B. Bell, "Turbulence–flame interactions in lean premixed hydrogen: transition to the distributed burning regime," *J. Fluid Mech.*, **680**, 287–320, 2011.

[122] B. Savard, B. Bobbitt, and G. Blanquart, "Structure of a high Karlovitz n-C_7H_{16} premixed turbulent flame," *Proc. Combust. Inst.*, **35**, 1377–1384, 2015.

[123] H. G. Im, P. G. Arias, S. Chaudhuri, and H. A. Uranakara, "Direct numerical simulations of statistically stationary turbulent premixed flames," *Combust. Sci. Technol.*, **188**, 1182–1198, 2016.

[124] G. Comte-Bellot and S. Corrsin, "The use of a contraction to improve the isotropy of grid-generated turbulence," *J. Fluid Mech.*, **25**, 657, 1966.

[125] T. Lundgren, "Linear forced isotropic turbulence," Annual Research Briefs. Center for Turbulence Research, Stanford, 461–473, 2003.

[126] M. Klein, N. Chakraborty, and S. Ketterl, "A comparison of strategies for direct numerical simulation of turbulence chemistry interaction in generic planar turbulent premixed flames," *Flow Turbul. Combust.*, 1–17, 2017.

[127] A. J. Aspden, "A numerical study of diffusive effects in turbulent lean premixed hydrogen flames," *Proc. Combust. Inst.*, **36**, 1997–2004, 2017.

[128] H. Carlsson, R. Yu, and X.-S. Bai, "Direct numerical simulation of lean premixed CH_4/air and H_2/air flames at high Karlovitz numbers," *Int. J. Hydrogen Energy*, **39**, 20 216 – 20 232, 2014.

[129] A. J. Aspden, M. S. Day, and J. B. Bell, "Characterization of low Lewis number flames," *Proc. Combust. Inst.*, **33**, 1463–1472, 2011.

[130] A. J. Aspden, M. S. Day, and J. B. Bell, "Turbulence-chemistry interaction in lean premixed hydrogen combustion," *Proc. Combust. Inst.*, **35**, 1321 – 1329, 2015.

[131] A. J. Aspden, J. B. Bell, M. S. Day, and F. N. Egolfopoulos, "Turbulence–flame interactions in lean premixed dodecane flames," *Proc. Combust. Inst.*, **36**, 2005–2016, 2017.

[132] B. Savard and G. Blanquart, "Broken reaction zone and differential diffusion effects in high Karlovitz n-C_7H_{16} premixed turbulent flames," *Combust. Flame*, **162**, 2020 – 2033, 2015.

[133] S. Lapointe and G. Blanquart, "Fuel and chemistry effects in high Karlovitz premixed turbulent flames," *Combust. Flame*, **167**, 294–307, 2016.

[134] B. Bobbitt, S. Lapointe, and G. Blanquart, "Vorticity transformation in high Karlovitz number premixed flames," *Phys. Fluids*, **28**, 015101, 2016.

[135] B. Bobbitt and G. Blanquart, "Vorticity isotropy in high Karlovitz number premixed flames," *Phys. Fluids*, **28**, 105101, 2016.

[136] P. E. Hamlington, A. Y. Poludnenko, and E. S. Oran, "Interactions between turbulence and flames in premixed reacting flows," *Phys. Fluids*, **23**, 125111, 2011.

[137] P. E. Hamlington, R. Darragh, C. A. Briner, C. A. Towery, B. D. Taylor, and A. Y. Poludnenko, "Lagrangian analysis of high-speed turbulent premixed reacting flows: Thermochemical trajectories in hydrogen–air flames," *Combust. Flame*, **186**, 193–207, 2017.

[138] O. Chatakonda, E. R. Hawkes, A. J. Aspden, A. R. Kerstein, H. Kolla, and J. H. Chen, "On the fractal characteristics of low Damköhler number flames," *Combust. Flame*, **160**, 2422–2433, 2013.

[139] Z. M. Nikolaou and N. Swaminathan, "Heat release rate markers for premixed combustion," *Combust. Flame*, **161**, 3073–3084, 2014.

[140] Y. Minamoto, N. Fukushima, M. Tanahashi, T. Miyauchi, T. Dunstan, and N. Swaminathan, "Effect of flow-geometry on turbulence-scalar interaction in premixed flames," *Phys. Fluids*, **23**, 125107, 2011.

[141] M. Tanahashi, Y. Nada, Y. Ito, and T. Miauchi, "Local flame structure in the well-stirred reactor regime," *Proc. Combust. Inst.*, **29**, 2041–2049, 2002.

[142] Y. Shim, S. Tanaka, M. Tanahashi, and T. Miyauchi, "Local structure and fractal characteristics of H_2-air turbulent premixed flame," *Proc. Combust. Inst.*, **33**, 1455–1462, 2011.

[143] Y.-S. Shim, N. Fukushima, M. Shimura, Y. Nada, M. Tanahashi, and T. Miyauchi, "Radical fingering in turbulent premixed flame classified into thin reaction zones," *Proc. Combust. Inst.*, **34**, 1383–1391, 2013.

[144] T. Nilsson, H. Carlsson, R. Yu, and X. Bai, "Structures of turbulent premixed flames in the high Karlovitz number regime: DNS analysis," *Fuel*, **216**, 627–638, 2018.

[145] A. Y. Poludnenko, "Pulsating instability and self-acceleration of fast turbulent flames," *Phys. Fluids*, **27**, 014106, 2015.

[146] P. E. Hamlington, A. Y. Poludnenko, and E. S. Oran, "Intermittency in premixed turbulent reacting flows," *Phys. Fluids*, **24**, 075111, 2012.

[147] C. Towery, A. Poludnenko, J. Urzay, J. O'Brien, M. Ihme, and P. Hamlington, "Spectral kinetic energy transfer in turbulent premixed reacting flows," *Phys. Rev. E*, **93**, 053115, 2016.

[148] J. O'Brien, C. A. Towery, P. E. Hamlington, M. Ihme, A. Y. Poludnenko, and J. Urzay, "The cross-scale physical-space transfer of kinetic energy in turbulent premixed flames," *Proc. Combust. Inst.*, **36**, 1967–1975, 2017.

[149] R. Lamioni, P. E. Lapenna, G. Troiani, and F. Creta, "Flame induced flow features in the presence of Darrieus-Landau instability," *Flow Turbul. Combust.*, **101**, 1137–1155, 2018.

[150] P. E. Lapenna, R. Lamioni, G. Troiani, and F. Creta, "Large scale effects in weakly turbulent premixed flames," *Proc. Combust. Inst.*, **37**, 1945–1952, 2019.

[151] B. Savard and G. Blanquart, "An a priori model for the effective species Lewis numbers in premixed turbulent flames," *Combust. Flame*, **161**, 1547–1557, 2014.

[152] M. Summerfield, S. H. Reiter, V. Kebely, and R. W. Mascolo, "The physical structure of turbulent flames," *J. Jet Prop.*, **24**, 254–255, 1954.

[153] S. B. Pope and M. S. Anand, "Flamelet and distributed combustion in premixed turbulent flames," *Symp. (Int.) Combust.*, **20**, 403–410, 1985.

[154] K. N. C. Bray, "Turbulent transport in flames," *Proc. Royal Soc. London A*, **451**, 231–256, 1995.

[155] M. J. Dunn, A. R. Masri, R. W. Bilger, and R. S. Barlow, "Finite rate chemistry effects in highly sheared turbulent premixed flames," *Flow Turbul. Combust.*, **85**, 621–648, 2010.

References

[156] Y. Minamoto, N. Swaminathan, R. S. Cant, and T. Leung, "Reaction zones and their structure in MILD combustion," *Combust. Sci. Technol.*, **186**, 1075–1096, 2014.

[157] A. J. Aspden, M. S. Day, and J. B. Bell, "Turbulence-flame interactions in lean premixed hydrogen," in *Proceedings of the International Colloquium on the Dynamics of Explosions and Reactive Systems*, 2009.

[158] A. Y. Poludnenko, V. N. Gamezo, and E. S. Oran, "Turbulence-flame interaction and formation of distributed flame," in *Proceedings of the International Colloquium on the Dynamics of Explosions and Reactive Systems*, 2009.

[159] A. Y. Poludnenko and E. S. Oran, "The interaction of high-speed turbulence with flames: Turbulent flame speed," *Combust. Flame*, **158**, 301–326, 2010.

[160] A. J. Aspden, M. S. Day, and J. B. Bell, "Lewis number effects in distributed flames," *Proc. Combust. Inst.*, **33**, 1473–1480, 2011.

[161] Z. Nikolaou and N. Swaminathan, "Direct numerical simulation of complex fuel combustion with detailed chemistry: Physical insight and mean reaction rate modeling," *Combust. Sci. Technol.*, **187**, 1759–1789, 2015.

[162] A. Y. Poludnenko, T. A. Gardiner, and E. S. Oran, "Spontaneous transition of turbulent flames to detonations in unconfined media," *Phys. Rev. Lett.*, **107**, 054501, 2011.

[163] J. F. Driscoll, J. H. Chen, A. Skiba, C. Campbell, E. Hawkes, and H. Wang, "Premixed flames subjected to extreme turbulence: Some questions and recent answers," *Prog. Energy Combust. Sci.*, 76, 100802, 2020.

[164] S. Treichler, M. Bauer, A. Bhagatwala, G. Borghesi, R. Sankaran, H. Kolla, P. S. McCormick, E. Slaughter, W. Lee, A. Aiken, and J. H. Chen, "S3D-legion: An exascale software for direct numerical simulation of turbulent combustion with complex multicomponent chemistry," in T. P. Straatsma, K. B. Antypas, and T. J. Williams, Eds. *Exascale Scientific Applications: Programming Approaches for Scalability, Performance, and Portability*. Boca Raton, FL: Chapman and Hall/CRC, 2017 pp. 257–278.

[165] M. Bauer, S. Treichler, and A. Aiken, "Singe: Leveraging warp specialization for high performance on gpus," **49**, 02 2014, 119–130.

[166] M. J. Berger and P. Colella, "Local adaptive mesh refinement for shock hydrodynamics." *J. Comput. Phys.*, **82(1)**, 64–84, 1989.

[167] A. S. Almgren, J. B. Bell, P. Colella, L. H. Howell, and M. L. Welcome, "A conservative adaptive projection method for the variable density incompressible Navier-Stokes equations," *J. Comput. Phys.*, **142**, 1–46, 1998.

[168] D. Kothe, "Exascale applications: Skin in the game," *Phil. Trans. Royal Soc. London A*, under review, 2019.

[169] J. R. Hertzberg, I. G. Shepherd, and L. Talbot, "Vortex shedding behind rod stabilized flames," *Combust. Flame*, **86**, 1–11, 1991.

[170] W. Meier, X. R. Duan, and P. Weigand, "Investigations of swirl flames in gas turbine model combustor II. turbulence-chemistry interactions," *Combust. Flame*, **144**, 225–236, 2006.

[171] M. S. Sweeney, S. Hochgreb, M. J. Dunn, and R. S. Barlow, "Multiply conditioned analysis of stratification in highly swirling methane/air flames," *Combust. Flame*, **160**, 322–334, 2013.

[172] K. K. Agarwal and R. V. Ravikrishna, "Experimental and numerical studies in a compact trapped vortex combustor: Stability assessment and augmentation," *Combust. Sci. Tech.*, **183**, 1308–1327, 2011.

[173] F. Güthe, J. Hellat, and P. J. Flohr, "The reheat concept: The proven pathway to ultralow emissions and high efficiency and flexibility," *J. Eng Gas Turb. Power*, **131**, 1–7, 2009.

[174] D. A. Pennell, M. R. Bothien, A. Ciani, V. Granet, G. Singla, S. Thorpe, A. Wickstroem, K. Oumejjoud, and M. Yaquinto, "An introduction to the ansaldo GT36 constant pressure sequential combustor," *ASME Turbo Expo: Power for Land Sea and Air*, GT2017-64790, 2017.

[175] K. Aditya, A. Gruber, C. Xu, T. Lu, A. Krisman, M. Bothien, and J. H. Chen, "Direct numerical simulation of flame stabilization assisted by autoignition in a reheat gas turbine combustor," *Proc. Combust. Inst.*, **37**, 2635–2642, 2019.

[176] J. C. Massey, I. Langella, and N. Swaminathan, "A scaling law for the recirculation zone length behind a bluff body in reacting flows," *J. Fluid Mech.*, **875**, 699–724, 2019.

[177] T. M. Wabel, A. W. Skiba, and J. F. Driscoll, "Evolution of turbulence through a broadened preheat zone from conditionally-averaged velocity measurements," *Combust. Flame*, **188**, 13–27, 2018.

[178] M. J. Dunn, A. R. Masri, and R. W. Bilger, "A new piloted premixed jet burner to study strong finite-rate chemistry effects," *Combust. Flame*, **151**, 46–60, 2007.

[179] B. Coriton, M. D. Smooke, and A. Gomez, "Effect of the composition of the hot product stream in the quasi-steady extinction of strained premixed flames," *Combust. Flame*, **157**, 2155–2164, 2010.

[180] B. Coriton, J. H. Frank, and A. Gomez, "Interaction of turbulent premixed flames with combustion products: Role of stoichiometry," *Combust. Flame*, **170**, 37–52, 2016.

[181] S. Chaudhuri, S. Kostka, M. W. Renfro, and B. M. Cetegen, "Blowoff dynamics of bluff body stabilized turbulent premixed flames," *Combust. Flame*, **157**, 790–802, 2010.

[182] Y. M. Marzouk, A. F. Ghoniem, and H. N. Najm, "Dynamic response of strained premixed flames to equivalence ratio gradients," *Proc. Combust. Inst.*, **28**, 1859–1866, 2000.

[183] E. S. Richardson, V. E. Granet, A. Eyssartier, and J. H. Chen, "Effects of equivalence ratio variation on lean, stratified methane-air laminar counterflow flames," *Combust. Theory Model.*, **14**, 775–792, 2010.

[184] B. Coriton, J. H. Frank, A. G. Hsu, M. D. Smooke, and A. Gomez, "Effect of quenching of the oxidation layer in highly turbulent counterflow premixed flames," *Proc. Combust. Inst.*, **33**, 1647–1654, 2011.

[185] R. Zhou and S. Hochgreb, "The behavior of laminar stratified methane/air flames in counterflow," *Combust. Flame*, **160**, 1070–1082, 2013.

[186] E. S. Richardson and J. H. Chen, "Analysis of turbulent flame propagation in equivalence ratio-stratified flow," *Proc. Combust. Inst.*, **36**, 1729–1736, 2017.

[187] A. Bouaniche, N. Jaouen, P. Domingo, and L. Vervisch, "Vitiated high karlovitz n-decane/air turbulent flames: Scaling laws and micro-mixing modeling analysis," *Flow Turbulence Combust.*, **102**, 235–252, 2019.

[188] H. Wang, E. R. Hawkes, B. Savard, and J. H. Chen, "Direct numerical simulation of a high Ka CH4/air stratified premixed jet flames," *Combust. Flame*, **195**, 229–245, 2018.

[189] B. Böhm, C. Heeger, R. L. Gordon, and A. Dreizler, "New perspectives on turbulent combustion: Multi-parameter high-speed planar laser diagnostics," *Flow Turbul. Combust.*, **86**, 313–341, 2011.

[190] C. F. Kaminski and M. B. Long, "Multidimensional diagnostics in space and time," *Appl. Combust. Diag.*, 224–251, 2002.

References

[191] K. Kohse-Höinghaus, R. S. Barlow, M. Aldén, and J. Wolfrum, "Combustion at the focus: Laser diagnostics and control," *Proc. Combust. Inst.*, **30**, 89–123, 2005.

[192] V. Sick, "High speed imaging in fundamental and applied combustion research," *Proc. Combust. Inst.*, **34**, 3509–3530, 2013.

[193] B. Thurow, N. Jiang, and W. Lempert, "Review of ultra-high repetition rate laser diagnostics for fluid dynamic measurements," *Meas. Sci. Technol.*, **24**, 012002, 2012.

[194] G. Damkohler, "The effect of turbulence on the flame velocity in gas mixtures," *Z. Elektrochem. und Angewandte Physikalische Chemie* 46 (1940) 601, English translation, NACA Techn. Memo. No. 1112 1947.

[195] M. Aldén, H. Edner, and S. Svanberg, "Simultaneous, spatially resolved monitoring of C_2 and OH in a C_2H_2/O_2 flame using a diode array detector," *Appl. Phys. B*, **29**, 93–97, 1982.

[196] M. Aldén, H. Edner, G. Holmstedt, S. Svanberg, and T. Högberg, "Single-pulse laser-induced OH fluorescence in an atmospheric flame, spatially resolved with a diode array detector," *Appl. Opt.*, **21**, 1236–1240, 1982.

[197] G. Kychakoff, R. D. Howe, R. K. Hanson, M. C. Drake, R. W. Pitz, M. Lapp, and C. M. Penney, "Visualization of turbulent flame fronts with planar laser-induced fluorescence," *Science*, **224**, 382–384, 1984.

[198] M. Aldén, J. Bood, Z. Li, and M. Richter, "Visualization and understanding of combustion processes using spatially and temporally resolved laser diagnostic techniques," *Proc. Combust. Inst.*, **33**, 69–97, 2011.

[199] Z. Li, B. Li, Z. Sun, X.-S. Bai, and M. Aldén, "Turbulence and combustion interaction: High resolution local flame front structure visualization using simultaneous single-shot PLIF imaging of CH, OH, and CH_2O in a piloted premixed jet flame," *Combust. Flame*, **157**, 1087–1096, 2010.

[200] P. H. Paul and H. N. Najm, "Planar laser-induced fluorescence imaging of flame heat release rate," in *Symp. (Int.) Combust.*, **27**, no. 1, 1998, 43–50.

[201] J. E. Rehm and P. H. Paul, "Reaction rate imaging," *Proc. Combust. Inst.*, **28**, 1775–1782, 2000.

[202] B. Ayoola, R. Balachandran, J. Frank, E. Mastorakos, and C. Kaminski, "Spatially resolved heat release rate measurements in turbulent premixed flames," *Combust. Flame*, **144**, 1–16, 2006.

[203] T. M. Wabel, P. Zhang, X. Zhao, H. Wang, E. Hawkes, and A. M. Steinberg, "Assessment of chemical scalars for heat release rate measurement in highly turbulent premixed combustion including experimental factors," *Combust. Flame*, **194**, 485–506, 2018.

[204] A. C. Eckbreth, *Laser Diagnostics for Combustion Temperature and Species*. Boca Raton, FL: CRC Press, 1996, **3**.

[205] C. Brackmann, J. Nygren, X. Bai, Z. Li, H. Bladh, B. Axelsson, I. Denbratt, L. Koopmans, P.-E. Bengtsson, and M. Aldén, "Laser-induced fluorescence of formaldehyde in combustion using third harmonic Nd: YAG laser excitation," *Spect. Acta Part A: Mol. Bio. Spect.*, **59**, 3347–3356, 2003.

[206] Z. Li, M. Afzelius, J. Zetterberg, and M. Aldén, "Applications of a single-longitudinal-mode alexandrite laser for diagnostics of parameters of combustion interest," *Rev. Sci. Instr.*, **75**, 3208–3215, 2004.

[207] P. H. Paul and J. E. Dec, "Imaging of reaction zones in hydrocarbon–air flames by use of planar laser-induced fluorescence of CH," *Opt. Lett.*, **19**, 998–1000, 1994.

[208] M. G. Allen, R. D. Howe, and R. K. Hanson, "Digital imaging of reaction zones in hydrocarbon–air flames using planar laser-induced fluorescence of CH and C_2," *Opt. Lett.*, **11**, 126–128, 1986.

[209] Z. S. Li, J. Kiefer, J. Zetterberg, M. Linvin, A. Leipertz, X.-S. Bai, and M. Aldén, "Development of improved PLIF CH detection using an Alexandrite laser for single-shot investigation of turbulent and lean flames," *Proc. Combust. Inst.*, **31**, 727–735, 2007.

[210] C. Carter, J. Donbar, and J. Driscoll, "Simultaneous CH planar laser-induced fluorescence and particle imaging velocimetry in turbulent nonpremixed flames," *Appl. Phys. B: Lasers Optics*, **66**, 129–132, 1998.

[211] C. D. Carter, S. Hammack, and T. Lee, "High-speed planar laser-induced fluorescence of the CH radical using the $C^2\Sigma^+ - X^2\Pi(0,0)$ band," *Appl. Phys. B*, **116**, 515–519, 2014.

[212] B. Zhou, J. Kiefer, J. Zetterberg, Z. Li, and M. Aldén, "Strategy for PLIF single-shot HCO imaging in turbulent methane/air flames," *Combust. Flame*, **161**, 1566–1574, 2014.

[213] J. Kiefer, Z. Li, T. Seeger, A. Leipertz, and M. Aldén, "Planar laser-induced fluorescence of HCO for instantaneous flame front imaging in hydrocarbon flames," *Proc. Combust. Inst.*, **32**, 921–928, 2009.

[214] B. Li, D. Zhang, M. Yao, and Z. Li, "Strategy for single-shot CH3 imaging in premixed methane/air flames using photofragmentation laser-induced fluorescence," *Proc. Combust. Inst.*, **36**, 4487–4495, 2017.

[215] B. Li, X. Li, M. Yao, and Z. Li, "Methyl radical imaging in methane–air flames using laser photofragmentation-induced fluorescence," *Appl. Spect.*, **69**, 1152–1156, 2015.

[216] K. Katharina and J. Jeffries Eds., *Applied Combustion Diagnostics, Combustion.* New York: Taylor & Francis, 2002.

[217] C. Brackmann, Z. Li, M. Rupinski, N. Docquier, G. Pengloan, and M. Aldén, "Strategies for formaldehyde detection in flames and engines using a single-mode Nd: YAG/OPO laser system," *Appl. Spect.*, **59**, 763–768, 2005.

[218] T. M. Wabel, A. W. Skiba, and J. F. Driscoll, "Turbulent burning velocity measurements: Extended to extreme levels of turbulence," *Proc. Combust. Inst.*, **36**, 1801–1808, 2017.

[219] C. D. Carter, S. Hammack, and T. Lee, "High-speed flamefront imaging in premixed turbulent flames using planar laser-induced fluorescence of the CH C- X band," *Combust. Flame*, **168**, 66–74, 2016.

[220] J. F. Driscoll, "Turbulent premixed combustion: Flamelet structure and its effect on turbulent burning velocities," *Prog. Energy Combust. Sci.*, **34**, 91–134, 2008.

[221] B. Zhou, C. Brackmann, Z. Li, M. Aldén, and X.-S. Bai, "Simultaneous multi-species and temperature visualization of premixed flames in the distributed reaction zone regime," *Proc. Combust. Inst.*, **35**, 1409–1416, 2015.

[222] B. Zhou, Q. Li, Y. He, P. Petersson, Z. Li, M. Aldén, and X.-S. Bai, "Visualization of multi-regime turbulent combustion in swirl-stabilized lean premixed flames," *Combust. Flame*, **162**, 2954–2958, 2015.

[223] J. A. Sutton and J. F. Driscoll, "Rayleigh scattering cross sections of combustion species at 266, 355, and 532 nm for thermometry applications," *Opt. Lett.*, **29**, 2620–2622, 2004.

[224] K. Watson, K. Lyons, J. Donbar, and C. Carter, "Simultaneous Rayleigh imaging and CH-PLIF measurements in a lifted jet diffusion flame," *Combust. Flame*, **123**, 252–265, 2000.

[225] T. McManus and J. Sutton, "Quantitative planar temperature imaging in turbulent non-premixed flames using filtered Rayleigh scattering," *Appl. Opt.*, **58**, 2936–2947, 2019.

[226] C. Schulz and V. Sick, "Tracer-LIF diagnositcs: Quantitative measurement of fuel concentration, temperature and fuel/air ratio in pratical combustion systems," *Prog. Energy Combust. Sci.*, **31**, 75–121, 2005.

[227] R. Whiddon, B. Zhou, J. Borggren, M. Alden, and Z. S. Li, "Vapor phase tri-methyindium seeding system suitable for high temperature spectroscopy and thermometry," *Rev. Sci. Instr.*, **86**, 093107, 2015.

[228] C. Abram, B. Fond, and Beyrau, "Temperature measurement techniques for gas and liquid flows using thermographic phosphor tracer particles," *Prog. Energy Combust. Sci.*, **64**, 93–156, 2018.

[229] J. D. Miller, N. Jiang, M. N. Slipchenko, J. G. Mance, T. R. Meyer, S. Roy, and J. R. Gord, "Spatiotemporal analysis of turbulent jets enabled by 100-kHz, 100-ms burst-mode particle image velocimetry," *Exp. Fluids*, **57**, 192, 2016.

[230] Z. Wang, P. Stamatoglou, B. Zhou, M. Aldén, X.-S. Bai, and M. Richter, "Investigation of OH and CH_2O distributions at ultra-high repetition rates by planar laser induced fluorescence imaging in highly turbulent jet flames," *Fuel*, **234**, 1528–1540, 2018.

[231] M. Gamba, N. Clemens, and O. Ezekoye, "Volumetric PIV and 2D OH PLIF imaging in the far-field of a low reynolds number nonpremixed jet flame," *Meas. Sci. Technol.*, **24**, 024003, 2012.

[232] G. E. Elsinga, F. Scarano, B. Wieneke, and B. W. van Oudheusden, "Tomographic particle image velocimetry," *Exp. Fluids*, **41**, 933–947, 2006.

[233] F. Scarano, "Tomographic PIV: Principles and practice," *Meas. Sci. Technol.*, **24**, 012001, 2012.

[234] B. Coriton and J. H. Frank, "Experimental study of vorticity-strain rate interaction in turbulent partially premixed jet flames using tomographic particle image velocimetry," *Phys. Fluids*, **28**, 025109, 2016.

[235] J. Weinkauff, D. Michaelis, A. Dreizler, and B. Böhm, "Tomographic PIV measurements in a turbulent lifted jet flame," *Exp. Fluids*, **54**, 1624, 2013.

[236] B. Coriton, A. M. Steinberg, and J. H. Frank, "High-speed tomographic PIV and OH PLIF measurements in turbulent reactive flows," *Exp. Fluids*, **55**, p. 1743, 2014.

[237] D. Ebi and N. T. Clemens, "Simultaneous high-speed 3D flame front detection and tomographic PIV," *Meas. Sci. Technol.*, **27**, 035303, 2016.

[238] R. J. Adrian and J. Westerweel, *Particle Image Velocimetry*. Cambridge: Cambridge University Press, 2011, no. 30.

[239] D. Schanz, S. Gesemann, and A. Schröder, "Shake-the-box: Lagrangian particle tracking at high particle image densities," *Exp. Fluids*, **57**, 70, 2016.

[240] G. Coppola, B. Coriton, and A. Gomez, "Highly turbulent counterflow flames: A laboratory scale benchmark for practical systems," *Combust. Flame*, **156**, 1834–1843, 2009.

[241] J. C. Oefelein, V. Sankaran, and T. G. Drozda, "Large eddy simulation of swirling particle-laden flow in a model axisymmetric combustor," *Proc. Combust. Inst.*, **31**, 2291–2299, 2007.

[242] J. C. Oefelein, "Large eddy simulation of turbulent combustion processes in propulsion and power systems," *Prog. Aero. Sci.*, **42**, 2–37, 2006.

[243] B. Coriton and J. H. Frank, "High-speed tomographic PIV measurements of strain rate intermittency and clustering in turbulent partially-premixed jet flames," *Proc. Combust. Inst.*, **35**, 1243–1250, 2015.

[244] A. M. Steinberg, B. Coriton, and J. H. Frank, "Influence of combustion on principal strain-rate transport in turbulent premixed flames," *Proc. Combust. Inst.*, **35**, 1287–1294, 2015.

[245] B. Coriton and J. H. Frank, "Impact of heat release on strain rate field in turbulent premixed bunsen flames," *Proc. Combust. Inst.*, **36**, 1885–1892, 2017.

[246] G. K. Batchelor, "The effect of homogeneous turbulence on material lines and surfaces," *Proc. Royal Soc. London A*, **213**, 349–366, 1952.

[247] G. Coppola and A. Gomez, "Experimental investigation on a turbulence generation system with high-blockage plates," *Exp. Thermal Fluid Sci.*, **33**, 1037–1048, 2009.

[248] N. Swaminathan and R. W. Grout, "Interaction of turbulence and scalar fields in premixed flames," *Phys. Fluids*, **18**, 045102, 2006.

[249] G. Hartung, J. Hult, C. F. Kaminski, J. W. Rogerson, and N. Swaminathan, "Effect of heat release on turbulence and scalar-turbulence interaction in premixed combustion," *Phys. Fluids*, **20**, 035110, 2008.

[250] T. Sponfeldner, I. Boxx, F. Beyrau, Y. Hardalupas, W. Meier, and A. M. K. P. Taylor, "On the alignment of fluid-dynamic principal strain-rates with the 3D flamelet-normal in a premixed turbulent V-flame," *Proc. Combust. Inst.*, **35**, 1269–1276, 2015.

[251] B. Zhou and J. H. Frank, "Effects of heat release and imposed bulk strain on alignment of strain rate eigenvectors in turbulent premixed flames," *Combust. Flame*, **201**, 290–300, 2019.

[252] C. J. Mueller, J. F. Driscoll, D. L. Reuss, M. C. Drake, and M. E. Rosalik, "Vorticity generation and attenuation as vortices convect through a premixed flame," *Combust. Flame*, **112**, 342–358, 1998.

[253] A. Lipatnikov, S. Nishiki, and T. Hasegawa, "A direct numerical simulation study of vorticity transformation in weakly turbulent premixed flames," *Phys. Fluids*, **26**, 105104, 2014.

[254] TNF, "International workshop on measurement and computation of turbulent non-premixed flames," https://www.sandia.gov/TNF/abstract.html (Accessed April 10, 2019).

[255] K. K. Nomura and G. K. Post, "The structure and dynamics of vorticity and rate of strain in incompressible homogeneous turbulence," *J. Fluid Mech.*, **377**, 65–97, 1998.

[256] J. Jimenez, "Kinematic alignment effects in turbulent flows," *Phys. Fluids A*, **4**, 652–654, 1992.

[257] T. Leung, N. Swaminathan, and P. Davidson, "Geometry and interaction of structures in homogeneous isotropic turbulence," *J. Fluid Mech.*, **710**, 453–481, 2012.

[258] P. E. Hamlington, J. Schumacher, and W. J. Dahm, "Local and nonlocal strain rate fields and vorticity alignment in turbulent flows," *Phys. Rev. E*, **77**, 026303, 2008.

[259] N. A. K. Doan, N. Swaminathan, P. A. Davidson, and M. Tanahashi, "Scale locality of the energy cascade using real space quantities," *Phys. Rev. Fluids*, **3**, 084601, 2018.

[260] B. Böhm, C. Heeger, I. Boxx, W. Meier, and A. Dreizler, "Time-resolved conditional flow field statistics in extinguishing turbulent opposed jet flames using simultaneous high speed PIV/OH-PLIF," *Proc. Combust. Inst.*, **32**, 1647–1654, 2009.

[261] C. Heeger, B. Böhm, S. Ahmed, R. Gordon, I. Boxx, W. Meier, A. Dreizler, and E. Mastorakos, "Statistics of relative and absolute velocities of turbulent non-premixed edge flames following spark ignition," *Proc. Combust. Inst.*, **32**, 2957–2964, 2009.

[262] I. Boxx, M. Stöhr, C. Carter, and W. Meier, "Temporally resolved planar measurements of transient phenomena in a partially pre-mixed swirl flame in a gas turbine model combustor," *Combust. Flame*, **157**, 1510–1525, 2010.

[263] A. Nauert, P. Petersson, M. Linne, and A. Dreizler, "Experimental analysis of flashback in lean premixed swirling flames: Conditions close to flashback," *Exp. Fluids*, **43**, 89–100, 2007.

References

[264] C. F. Kaminski, J. Hult, M. Aldén, S. Lindenmaier, A. Dreizler, U. Maas, and M. Baum, "Spark ignition of turbulent methane/air mixtures revealed by time-resolved planar laser-induced fluorescence and direct numerical simulations," *Proc. Combust. Inst.*, **28**, 399–405, 2000.

[265] A. Dreizler, S. Lindenmaier, U. Maas, J. Hult, M. Aldén, and C. Kaminski, "Characterisation of a spark ignition system by planar laser-induced fluorescence of OH at high repetition rates and comparison with chemical kinetic calculations," *Appl. Phys. B*, **70**, 287–294, 2000.

[266] J. D. Miller, S. R. Engel, T. R. Meyer, T. Seeger, and A. Leipertz, "High-speed CH planar laser-induced fluorescence imaging using a multimode-pumped optical parametric oscillator," *Opt. Lett.*, **36**, 3927–3929, 2011.

[267] J. D. Miller, M. Slipchenko, T. R. Meyer, N. Jiang, W. R. Lempert, and J. R. Gord, "Ultrahigh-frame-rate OH fluorescence imaging in turbulent flames using a burst-mode optical parametric oscillator," *Opt. Lett.*, **34**, 1309–1311, 2009.

[268] K. N. Gabet, R. A. Patton, N. Jiang, W. R. Lempert, and J. A. Sutton, "High-speed CH_2O PLIF imaging in turbulent flames using a pulse-burst laser system," *Appl. Phys. B*, **106**, 569–575, 2012.

[269] N. Jiang and W. R. Lempert, "Ultrahigh-frame-rate nitric oxide planar laser-induced fluorescence imaging," *Opt. Lett.*, **33**, 2236–2238, 2008.

[270] R. A. Patton, K. N. Gabet, N. Jiang, W. R. Lempert, and J. A. Sutton, "Multi-kHz temperature imaging in turbulent non-premixed flames using planar rayleigh scattering," *Appl. Phys. B*, **108**, 377–392, 2012.

[271] S. A. Skeen, J. Manin, and L. M. Pickett, "Simultaneous formaldehyde PLIF and high-speed schlieren imaging for ignition visualization in high-pressure spray flames," *Proc. Combust. Inst.*, **35**, 3167–3174, 2015.

[272] S. A. Kaiser, J. H. Frank, and M. B. Long, "Use of Rayleigh imaging and ray tracing to correct for beam-steering effects in turbulent flames," *Appl. Opt.*, **44**, 6557–6564, 2005.

[273] B. Zhou, A. J. Ruggles, E. Huang, and F. J. H., "Wavelet-based algorithm for correction of beam-steering artefacts in turbulent flow imaging at elevated pressure," *Exp. Fluid*, **60**, 136, 2019.

[274] I. Boxx, C. Heeger, R. Gordon, B. Böhm, M. Aigner, A. Dreizler, and W. Meier, "Simultaneous three-component PIV/OH PLIF measurements of a turbulent lifted, C3H8-Argon jet diffusion flame at 1.5 khz repetition rate," *Proc. Combust. Inst.*, **32**, 905–912, 2009.

[275] A. M. Steinberg, I. Boxx, C. M. Arndt, J. H. Frank, and W. Meier, "Experimental study of flame-hole reignition mechanisms in a turbulent non-premixed jet flame using sustained multi-kHz PIV and crossed-plane OH PLIF," *Proc. Combust. Inst.*, **33**, 1663–1672, 2011.

[276] J. Floyd, P. Geipel, and A. Kempf, "Computed tomography of chemiluminescence (CTC): Instantaneous 3D measurements and phantom studies of a turbulent opposed jet flame," *Combust. Flame*, **158**, 376–391, 2011.

[277] J. Floyd and A. Kempf, "Computed tomography of chemiluminescence (CTC): High resolution and instantaneous 3-D measurements of a matrix burner," *Proc. Combust. Inst.*, **33**, 751–758, 2011.

[278] X. Li and L. Ma, "Capabilities and limitations of 3D flame measurements based on computed tomography of chemiluminescence," *Combust. Flame*, **162**, 642–651, 2015.

[279] S. M. Wiseman, M. J. Brear, R. L. Gordon, and I. Marusic, "Measurements from flame chemiluminescence tomography of forced laminar premixed propane flames," *Combust. Flame*, **183**, 1–14, 2017.

[280] B. R. Halls, D. J. Thul, D. Michaelis, S. Roy, T. R. Meyer, and J. R. Gord, "Single-shot, volumetrically illuminated, three-dimensional, tomographic laser-induced-fluorescence imaging in a gaseous free jet," *Opt. Exp.*, **24**, 10 040–10 049, 2016.

[281] T. Li, J. Pareja, F. Fuest, M. Schütte, Y. Zhou, A. Dreizler, and B. Böhm, "Tomographic imaging of OH laser-induced fluorescence in laminar and turbulent jet flames," *Meas. Sci. Technol.*, **29**, p. 015206, 2017.

[282] L. Ma, Q. Lei, J. Ikeda, W. Xu, Y. Wu, and C. D. Carter, "Single-shot 3D flame diagnostic based on volumetric laser induced fluorescence (VLIF)," *Proc. Combust. Inst.*, **36**, 4575–4583, 2017.

[283] B. R. Halls, et al., "kHz-rate four-dimensional fluorescence tomography using an ultraviolet-tunable narrowband burst-mode optical parametric oscillator," *Optica*, **4**, 897–902, 2017.

[284] M. Winter, J. Lam, and M. Long, "Techniques for high-speed digital imaging of gas concentrations in turbulent flows," *Exp. Fluids*, **5**, 177–183, 1987.

[285] G. Kychakoff, P. H. Paul, I. van Cruyningen, and R. K. Hanson, "Movies and 3-D images of flowfields using planar laser-induced fluorescence," *Appl. Opt.*, **26**, 2498–2500, 1987.

[286] J. Nygren, J. Hult, M. Richter, M. Aldén, M. Christensen, A. Hultqvist, and B. Johansson, "Three-dimensional laser induced fluorescence of fuel distributions in an HCCI engine," *Proc. Combust. Inst.*, **29**, 679–685, 2002.

[287] R. Wellander, M. Richter, and M. Aldén, "Time-resolved (kHz) 3D imaging of OH PLIF in a flame," *Exp. Fluids*, **55**, 1764, 2014.

[288] J. Weinkauff, M. Greifenstein, A. Dreizler, and B. Böhm, "Time resolved three-dimensional flamebase imaging of a lifted jet flame by laser scanning," *Meas. Sci. Technol.*, **26**, 105201, 2015.

[289] K. Y. Cho, A. Satija, T. L. Pourpoint, S. F. Son, and R. P. Lucht, "High-repetition-rate three-dimensional OH imaging using scanned planar laser-induced fluorescence system for multiphase combustion," *Appl. Opt.*, **53**, 316–326, 2014.

[290] J. H. Frank, K. M. Lyons, and M. B. Long, "Technique for three-dimensional measurements of the time development of turbulent flames," *Opt. Lett.*, **16**, 958–960, 1991.

[291] E. Kristensson, Z. Li, E. Berrocal, M. Richter, and M. Aldén, "Instantaneous 3D imaging of flame species using coded laser illumination," *Proc. Combust. Inst.*, **36**, 4585–4591, 2017.

[292] T. Li, J. Pareja, L. Becker, W. Heddrich, A. Dreizler, and B. Böhm, "Quasi-4D laser diagnostics using an acousto-optic deflector scanning system," *Appl. Phys. B*, **123**, 78, 2017.

[293] J. W. Labahn, H. Wu, B. Coriton, J. H. Frank, and M. Ihme, "Data assimilation using high-speed measurements and LES to examine local extinction events in turbulent flames," *Proc. Combust. Inst.*, **37**, 2259–2266, 2019.

[294] H. Wang, E. R. Hawkes, B. Zhou, J. H. Chen, Z. Li, and M. Aldén, "A comparison between direct numerical simulation and experiment of the turbulent burning velocity-related statistics in a turbulent methane-air premixed jet flame at high Karlovitz number," *Proc. Combust. Inst.*, **36**, 2045–2053, 2017.

3 Premixed Combustion Modeling

C. Dopazo, N. Swaminathan, L. Cifuentes, and X.-S. Bai

The flow in practical systems is invariably turbulent and thus the mutual influences of combustion and turbulence are inevitable. The turbulence involves a wide range of length and time scales that interact with thermochemical scales, and these interactions bring fundamental challenges to studying turbulent combustion, as discussed in Sections 2.1 and 2.2. Those associated with the experimental investigations were explored and discussed earlier in Section 2.3 and this chapter focuses on turbulent combustion modeling challenges. Various modeling strategies were developed in past studies to overcome the challenges of turbulent combustion. These approaches for premixed combustion, along with recent advances, are reviewed in this chapter. The modeling of non-premixed and partially premixed combustion are discussed in Chapter 4.

3.1 Introduction

Fuel and air are fully mixed before they enter a combustion zone in premixed flames, which is classically seen as a wave separating hot products from the cold mixture and propagates into the unburnt mixture. Thus, there is a propagation speed for this wave front relative to the flow, which is known as flame speed. This quantity, denoted commonly by S_L for one-dimensional laminar premixed flames, depends on the equivalence ratio, temperature, and pressure of the fuel–air mixture. Its typical values are about 0.4 m/s for stoichiometric hydrocarbon–air mixtures and about 2 m/s for stoichiometric hydrogen–air mixture at room temperature and atmospheric pressure. This (relatively) simple laminar flame shown schematically in Fig. 3.1 depicts important characteristics along with two thermochemical quantities, S_L and flame thickness δ_L, of interest for the turbulent premixed flame modeling point of view.

The phenomenological analysis of Mallard and Le Chattlier [1, 2] is also outlined in Fig. 3.1, which starts by assuming that (1) the heat produced inside the flame of thickness δ_L is simply conducted away to the cold side and (2) the fuel mass flux entering the flame is consumed by it. This analysis helps one to quickly identify how the two thermochemical quantities are related to the thermal diffusivity, $\alpha = \lambda/(\rho_u C_p)$, of the fuel–air mixture. The flame thickness can be estimated by knowing the thermal diffusivity and S_L, both of which can be measured for a given fuel–air mixture. Also, one can define a representative characteristic timescale for the

Premixed Combustion Modeling

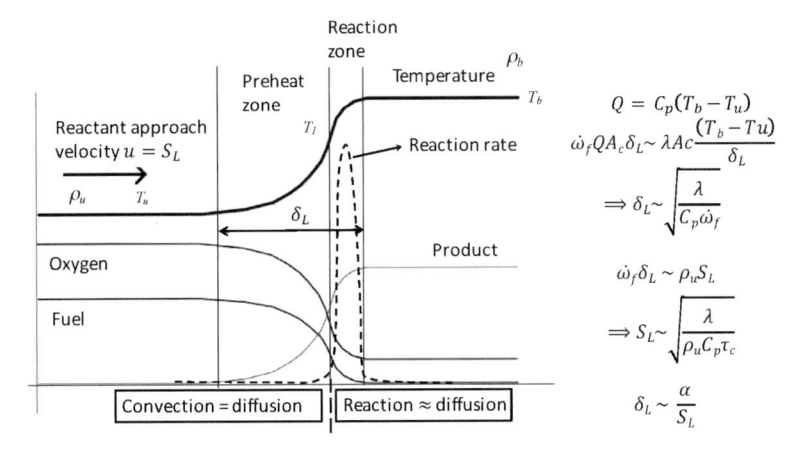

Figure 3.1 Schematic of a one-dimensional laminar premixed flame along with the analysis of Mallard and Le Chattlier [1, 2].

flame as $\tau_c = \delta_L/S_L = \rho_u/\dot{\omega}_f$. The chemical reaction rate, $\dot{\omega}_f$, can be estimated only by knowing S_L and δ_L in this phenomenological analysis.

Sophisticated analyses to estimate the reaction rate were developed in the past by approximating the combustion kinetics to a single irreversible reaction with high activation energy. This analysis splits the flame into two zones: convective-diffusive and reactive-diffusive zones as shown in Fig. 3.1. To outline this analysis briefly, let us assume the mass diffusivity is equal to the thermal diffusivity (unity Lewis number) and under this condition the premixed flame can be investigated using a reaction progress variable, c, that can be defined using a species mass fraction or temperature. For our purpose here, let us take it to be the normalized temperature $c = (T - T_u)/(T_b - T_u)$ and thus its variation across the flame would be similar to the temperature variation shown in Fig. 3.1 but it is bounded between 0 and 1. The progress variable in a steady one-dimensional flame is governed by

$$\rho u \frac{dc}{dx} = \dot{\omega}_c + \frac{d}{dx}\left(\rho \alpha \frac{dc}{dx}\right). \tag{3.1}$$

The variation of c in the preheat zone can be deduced by balancing the convective and diffusive terms, since the reaction rate is negligible there. The methods of activation energy asymptotics can be used to find the variation of c in the reaction zone by balancing the reaction rate with the molecular diffusive flux. These two solutions are then matched at the interface. The details of this analysis can be found in textbooks on combustion, for example [3–5], and this analysis suggests that $S_L \sim T_u^2$ for a small increase in the mixture temperature above the ambient value and $S_L \sim p^{-1/2}$ for the pressure dependence. These results are supported by measurements of the flame speed over a range of thermochemical conditions [5].

Recent protestations to mitigate the impact of combustion on the planet is pushing the boundaries of operational envelops of combustion systems to achieve improved efficiencies and decreased emission of pollutants. Fuel-lean combustion is a promising avenue to achieve these goals [6, 7] and the mixture temperature and pressure may go beyond what has been explored in past combustion studies, and thus one must be cautious in extending the above pressure and temperature dependence of S_L to the conditions of future interest for the following reasons. Although the one-step approximation for combustion kinetics is helpful to gain insights, it has been well accepted that this is inadequate to study formation of pollutants and to find avenues to mitigate this issue. This is also compounded by a change in the role of some of the elementary reactions at elevated temperature and pressure that can lead to different (unexpected) behavior of S_L with p [8, 9]. Thus, further experimental and numerical investigations of laminar premixed flames of various practical fuel and air mixtures at elevated temperature and pressure are worthwhile and helpful in our pursuit to find ways to achieve "greener" (friendlier to the environment) combustion.

In this respect, combustion modeling plays a vital role, as it can help to assess the effectiveness of new combustion techniques before implementing them in practice. As noted in Chapter 1, combustion modeling provides a closure for the mean (in Reynolds-averaged Navier–Stokes [RANS]) or filtered (in large eddy simulation [LES]) reaction rate of the reaction progress variable, $\overline{\dot{\omega}_c}$. Under the assumption of unity Lewis number, this reaction rate is related to the mean fuel consumption rate or the heat release rate.

The choice of combustion modeling depends on the relativity between the characteristic scales of turbulence and combustion. Typical turbulent time scales of interest are its integral (τ_Λ, large) and Kolmogorov (τ_η, viscous or small) scales. The flame timescale $\tau_c = \delta_L/S_L$ is typically used for combustion, with the flame thickness commonly taken to be the thermal thickness $\delta_{th} = 1/|dc/dx|_{\max}$ of the laminar flame shown in Fig. 3.1. When the combustion kinetics is modeled using complex chemistry there could be a number of length and time scales of interest for combustion as discussed in Chapter 5. However, the aforementioned choice is adequate for our immediate purpose here.

Comparing the chemical timescale to the integral and Kolmogorov time scales[1] introduces two nondimensional parameters: Damköhler number Da $= \tau_\Lambda/\tau_c$ and Karlovitz number Ka $= \tau_c/\tau_\eta$, and identifies various regimes of turbulent combustion as discussed in Chapter 2 (see Figs. 2.1, 2.2, and 2.3). Three regimes of broad interest are (1) Da > 1 and Ka $\ll 1$ – corrugated and wrinkled flamelets regime of turbulent combustion; (2) Da > 1; and Ka > 1; and (3) Da < 1 and Ka $\gg 1$ – commonly known as distributed combustion regime. The second regime was called as the thin reaction zone regime by Peters [10] if $1 < $ Ka $ < 100$, where the small-scale turbulence can enter the flame and disturb the preheat zone but not the reaction zone. This implies that δ_L or $\delta_{th} > \eta$, which also means that $\tau_c > \tau_\eta$ and $\delta_r < \eta$, where η is the Kolmogorov length scale and δ_r is the reaction zone (see Fig. 3.1) thickness.

Flamelet combustion is taken to be present if the combustion timescale is much smaller than the characteristic turbulence timescales, which implies that Da > 1 and Ka $\ll 1$. Thus the turbulence scales do not enter into the laminar flame and hence the

---

[1] Integral $\tau_\Lambda = \Lambda/u'$; Kolmogorov $\tau_\eta = (\nu/\varepsilon)^{1/2}$.

structure of the laminar flame is preserved. For the third regime, the turbulence scales are much smaller than the combustion scales and thus the laminar flame structure is expected to be disturbed fully. However, it has been suggested recently that the third combustion regime can be seen as a "distributed flamelets", regime since the local values of Damköhler number are large in the flow regions with substantial heat release and small in regions of weak combustion [11]. Thus, a flamelet-based model can work well for the third regime also.

The objective for the rest of the discussion in this section is to review the recent progress made in the modeling for turbulent premixed combustion and to identify further associated challenges. It is imperative to give a broad perspective on various modeling approaches for turbulent premixed combustion in past studies. The aim is not to give a detailed review on these approaches but rather to describe the general concept involved, and the details can be found in the references provided. The modeling of turbulent premixed flames can be broadly categorized into phenomenological, geometrical, and statistical approaches as suggested in [12].

3.2 Phenomenological Models

3.2.1 Eddy Breakup Model

This is based on the phenomenology of eddy breakup process involved in the energy cascade in turbulent flows. This breakup process brings a cold fuel–air mixture into contact with hot combustion products in premixed combustion occurring at high Damköhler number in large Reynolds number flows [13]. The high Da implies that the combustion is controlled by this turbulent mixing rate, which depends on a characteristic turbulent timescale, τ_t. Thus a model for reaction rate, is written as

$$\overline{\dot{\omega}_c} = \mathcal{A}\,\overline{\rho}\,\frac{\sigma_c^2}{\tau_t}, \tag{3.2}$$

where $\mathcal{A}$ is a model parameter and σ_c^2 is the variance of the reaction progress variable. The turbulent timescale is estimated using the ratio of turbulent kinetic energy, $\widetilde{k}$, to its dissipation rate, $\widetilde{\varepsilon}$, that is, $\tau_t = (\widetilde{k}/\widetilde{\varepsilon})$ for RANS calculations. This timescale for LES is estimated using the subgrid kinetic energy, $\widetilde{k}_{sgs}$ and its dissipation rate $\widetilde{\varepsilon}_{sgs} = A_\varepsilon\,\widetilde{k}_{sgs}^{3/2}/\Delta$, where Δ is the LES filter width and A_ε is a model constant [14–16]. The variance of c needs to be interpreted as the subgrid scale (SGS) variance, which is typically modeled as $\sigma_{c,sgs}^2 = \Delta^2 A_\sigma\,(\nabla\widetilde{c}\cdot\nabla\widetilde{c})$ using the Favre-filtered progress variable gradient. This model for the SGS variance is discussed further in Section 3.5. This reaction rate model is easy to implement in computer codes for practical applications; however, it has been shown that the results have a strong dependence on the numerical solver and grid used for LES [17], which is somewhat less desirable.

The phenomenology of eddy breakup used preceding also suggests that the mean reaction rate is related to the scalar dissipation rate of the progress variable fluctuations. This relationship was demonstrated by Bray [18] using the Bray–Moss–Libby formulationdiscussed in Section 3.4.2, which gave

$$\overline{\dot{\omega}_c}^B = \frac{2}{2C_m - 1} \overline{\rho} \, \widetilde{\epsilon}_c, \qquad (3.3)$$

where $C_m \simeq 0.7$. If one writes $\widetilde{\epsilon}_c \simeq B \, \sigma_c^2/\tau_t$ then the model in Eq. (3.2) is recovered if $A = 2B/(2C_m - 1)$. The foregoing expressions hold for large Da (mixing controlled or fast chemistry) combustion.

The eddy dissipation concept (EDC) originally proposed for non-premixed combustion [19] based on the eddy breakup (EBU) model was modified by Ertesvåg and Magnussen [20] for premixed combustion with finite rate chemistry by assuming that the "fine scales" (where viscous dissipation acts) influence the mixing rate of reactants and products. These fine scales act like a partially or perfectly stirred reactor (Partially Stirred Reactor [PaSR] or Perfectly Stirred Reactor [PSR]) and thus their transient states can be obtained by solving the PSR equation for c, which is

$$\frac{d c^*}{d t} = \frac{-(c^* - \widetilde{c})}{\tau^*} + \frac{\dot{\omega}_c}{\rho}, \qquad (3.4)$$

For the PaSR approach, typically the mixing (first term on the right-hand side of Eq. (3.4)) and the reaction terms are treated sequentially in separate steps using either operator splitting or/and fractional time-stepping techniques. The average resident time of the fluid inside the fine structure is given by $\tau^* = 0.408 \, (\nu/\widetilde{\epsilon})^{1/2}$. These fine structures are assumed to occupy a fraction of the flow field, which is $\gamma^* = 2.138 \, (\nu \widetilde{\epsilon}/\widetilde{k}^2)^{1/4}$ [20]. Now, the mean reaction rate is

$$\overline{\dot{\omega}_c}^E = \overline{\rho} \frac{\gamma^{*2}}{\tau^*} (c^* - \widetilde{c}). \qquad (3.5)$$

Minamoto et al. [21] proposed a possible way to exploit Eqs. (3.3) and (3.5) by defining a parameter $\mathcal{B} = 1/(1 + \text{Da})$ and writing

$$\overline{\dot{\omega}_c} = \mathcal{B} \, \overline{\dot{\omega}_c}^E + (1 - \mathcal{B}) \, \overline{\dot{\omega}_c}^B. \qquad (3.6)$$

In the limit of large Da, the mixing controlled behavior is recovered. When the Da is low the turbulent eddies are expected to alter the flame structure and thus the local combusting eddy can be seen as a reactor, and the aforementioned model recovers this limit. A priori testing conducted by Minamoto et al. [21] using direct numerical simulation (DNS) data shows the potentials of the model in Eq. (3.6). For LES, one can redefine the parameter as $\mathcal{B} = 1/(1 + \text{Da}_\Delta)$, where $\text{Da}_\Delta = \tau_\Delta/\tau_c$ using the SGS timescale $\tau_\Delta = \Delta/u'_\Delta$ instead of the turbulence integral timescale. The LES filter width is Δ and u'_Δ is the velocity scale for the SGS fluctuations. The thickness of the laminar flame is not involved in the aforementioned models because the front is taken to be thin due to large Damköhler number.

3.2.2 Thickened Flame Model

This approach introduced in 1977 by Butler and O'Rourke [22] artificially thickens the laminar flame (flame front) by increasing the diffusivity by a factor F and increasing the chemical timescale, τ_c, by the same factor, ie., $\widehat{\alpha} = F\alpha$ and $\widehat{\tau}_c = F\tau_c$. This is because the flame front, which is of the order of about 0.5 mm (for a fuel–air mixture at room temperature and atmospheric pressure), was hard to resolve in laminar flame computations in 1970s. The simple analysis shown in Fig. 3.1 suggests that the flame thickness $\widehat{\delta_L} \sim \sqrt{\widehat{\alpha \tau_c}} = F\delta_L$ increases by the factor F but the flame speed ($= \sqrt{\widehat{\alpha/\tau_c}}$) remains unchanged. This allows the flame structure to be resolved using practically reasonable grids in 1970s but this is less of a problem today for laminar flames because the computer hardwares and the numerical methods are well advanced. However, it is still the case for turbulent flames at elevated pressure and temperature conditions.

For the one-step reaction, the chemical timescale is inversely related to the pre-exponential factor in the Arrhenius rate expression and thus this factor has to be decreased by the same factor F to keep the flame speed constant. This approach was adopted for LES of turbulent premixed flames by Colin et al. [23]. The thickening of the flame by decreasing the preexponential factor affects the turbulence–chemistry interaction (TCI) and the flame evolution time is also increased artificially by the factor F. These effects are compensated using the flame stretch efficiency function, which is determined empirically using DNS data. Further details can be found in [23]. A number of LES studies have used this approach to investigate turbulent premixed combustion in simple and complex flow configurations and showed a good comparison with measured statistics of flow and flame quantities. These studies are reviewed by Gicquel et al. and the details can be found there [12]. Dynamic evaluation of the efficiency function has also been explored in past studies [24–26]. Also, the thickened flame approach has been combined with other approaches such as tabulated chemistry [27–29] and in situ adaptive tabulation (ISAT) [30] to include multistep combustion kinetics. The ISAT approach introduced by Pope [31] helps to attain computational efficiency when a large chemical kinetic mechanism is used for combustion chemistry in turbulent flame calculations, and this approach is discussed in some detail in the chapter on chemical kinetics (Chapter 5).

3.2.3 Linear Eddy Model

This was proposed by Kerstein to find a modeling approach that does not depend on the combustion regime [32] by using a conceptually and computationally minimal model to represent advection, diffusion, and reaction processes in turbulent combustion. This is done through 1D unsteady reactive-diffusive equation including a term to represent the turbulent stirring, which is a random event. The 1D species mass fraction, Y_i, equation is (see Eq. 1.4 in Chapter 1)

$$\frac{\partial Y_i}{\partial t} = \frac{-1}{\rho}\frac{\partial J_i}{\partial s} + \frac{\dot{\omega}_i}{\rho} + \mathcal{F}_{\text{stir}}, \tag{3.7}$$

where the distance s is along the local normal specified by the gradient of Y_i, the molecular flux is J_i, and $\mathcal{F}_{\text{stir}}$ represents the turbulent stirring events by eddies ranging from the integral to Kolmogorov scales, which is modeled using a "triplet map." This mapping is done stochastically to mimic the random folding and mixing by turbulence.

For LES, these events are at SGS and thus the triplet mapping represents the folding and mixing processes imparted on the scalar fields by eddies of numerical cell size and below. Hence, the foregoing 1D equations are solved in the LEM domain (subgrid) in every LES 3D mesh cell [33, 34]. It is imperative that an equivalent equation for temperature must also be considered. It is clear that there has to be some exchange of information between the LES and LEM domains and also among neighboring LEM domains. The latter part occurs through a splicing procedure described in [35]. All of these procedures bring additional computational burdens and hence LES with LEM can be expensive compared to flamelet-based models. However, some computational savings can be realized using neural network for combustion chemistry [36–38]. Many past studies using this subgrid closure showed some good comparison with measurements for premixed combustion in different regimes, and the details can be found in those studies [39–47]. Also, the LEM in 3D can be used as a postprocessing tool to extract scalar structures using RANS results, as suggested by Grøvdal et al. [48].

3.3   Geometrical Models

The turbulent premixed flame is seen as a thin surface propagating on its own with respect to the local fluid flow, separating unburned reactants from fully burned products when the Damköhler number is large (Da $\gg$ 1). The propagation or displacement of the flame surface occurs because of molecular diffusion and chemical reaction, which are strongly coupled to each other in premixed combustion. Thus, the turbulent premixed combustion may be described if one could track the evolution of the flame surface. Two approaches for this were investigated in many past studies. The first one is known as the flame surface density (FSD) approach and the second one is based on a field or G equation. These two approaches are discussed in the text that follows.

3.3.1   FSD Approach

This approach was first introduced for non-premixed turbulent combustion by Marble and Broadwell [49] based on phenomenological observation and then it was adopted for premixed combustion. The surface density function (SDF), $\Sigma(c = \zeta; \mathbf{x}, t)$, defined by choosing a particular isosurface of $c = \zeta$, is related to the gradient of the progress variable. This relationship was suggested by Batchelor [50] for a generic scalar transport in homogeneous turbulence. The turbulent premixed combustion counterpart was shown rigorously by Pope [51] as

$$\Sigma(\zeta; \mathbf{x}, t) = \langle |\nabla c| \, | c = \zeta \rangle \overline{P}(\zeta; \mathbf{x}, t), \qquad (3.8)$$

where $\overline{P}$ is the probability density function (PDF) of c. Hence, the surface density function is related to the conditional expectation of the scalar gradient, which can be obtained using the joint PDF of c and $|\nabla c|$ as demonstrated by Pope [51], who has remarked that this mathematical relationship was first stated by Bilger. This relationship brings out the important information that the scalar gradient plays a central role in premixed combustion and this observation led to another approach called the scalar dissipation rate based approach, which is elaborated in [52]. The close relationship between Σ and $|\nabla c|$ has been explored in many past studies and is summarized in [53].

The surface density function obeys the transport equation

$$\frac{\partial \Sigma}{\partial t} + \frac{\partial}{\partial x_k}\left[\langle u_k + S_{d,c} n_k\rangle_s \Sigma\right] = \left\langle (\delta_{km} - n_k n_m)\frac{\partial u_k}{\partial x_m}\right\rangle_s \Sigma + \left\langle S_{d,c}\frac{\partial n_k}{\partial x_k}\right\rangle_s \Sigma$$
$$= \langle a_T + S_{d,c}\,\kappa\rangle_s \Sigma = \langle \Phi \rangle_s \Sigma, \tag{3.9}$$

where a_T is the flow[2] tangential strain rate and $\kappa = (\nabla \cdot \mathbf{n})$ is the flame curvature, since n_k is the component of the surface normal $\mathbf{n}$ pointing toward small values of c, in the direction x_k. The five terms of Eq. (3.9) imply the following physical meaning: (1) temporal evolution, (2) convective transport, (3) propagation, (4) flow tangential strain rate, and (5) the mutual influence of displacement speed and curvature respectively. The fifth term, $\kappa\,S_{d,c}$, can also be interpreted as additional tangential strain rate due to the propagation of the curved area element, which is explained further in Section 3.6. The displacement speed, $S_{d,c}$, of the isosurface is defined as

$$S_{d,c} = \frac{\nabla \cdot (\rho \alpha \nabla c) + \dot{\omega}_c}{\rho |\nabla c|}, \tag{3.10}$$

which is typically estimated in terms of burning velocity of freely propagating laminar flame through $\rho\,S_{d,c} = \rho_u S_L$. The tangential strain rate and curvature are induced by the flow and their combined effect is known as stretch, denoted by Φ in Eq. (3.9). The stretch rate denotes the fractional rate of change of surface area per unit volume, that is, $\Phi = d\ln(\delta A)/dt$, where δA is the differential area of interest. This equation was derived by Pope by considering the kinematic balance for the evolution of surface density [54] and also by Candel and Poinsot by applying the transport theorem to an elemental surface area moving with an arbitrary velocity [55]. It is also possible to derive the transport equation using the joint PDF of surface properties as shown in [54]. Alternatively, one can also start with the joint PDF equation for the scalar and its gradient magnitude to get a transport equation for Σ using Eq. (3.8) as shown by Pope [51]. The subscript s in Eq. (3.9) denotes that the quantity is surface averaged, which is defined as

$$\langle Q \rangle_s = \frac{\langle Q|\nabla c|\,|\zeta\rangle}{\langle |\nabla c|\,|\zeta\rangle}, \tag{3.11}$$

for a generic quantity Q.

In the FSD approach, the mean (for RANS paradigm) or filtered (for LES) reaction rate is given by

$$\overline{\dot{\omega}_c} = \rho_u\,S_L\,I_o\,\overline{\Sigma}, \tag{3.12}$$

---

[2] $S_{d,c}\,\kappa$ can be interpreted as the additional strain rate coming from the propagation mechanism or induced by the curvature; see Eq. (3.74).

where $\overline{\Sigma}$ is the FSD (surface density averaged or filtered in an appropriate way) and I_o is a factor, typically of order unity, accounting for the change in the local burning velocity due to strain and curvature effects [56]. This approach translates the reaction rate modeling into a modeling for FSD and there are two ways to do this. One is to use an algebraic expression and the other is to employ a transport equation. Many algebraic models were proposed in past studies for both RANS and LES, which are reviewed in [57–59]. These models are derived by approximating the reaction rate signal as a telegraph signal and the details can be found in [57] for RANS calculation. For example, a simple algebraic closure can be written by equating Eqs. (3.3) and (3.12) and this closure is [60]

$$\overline{\Sigma} = \frac{2}{(2C_m - 1) I_o} \frac{\overline{\rho \tilde{\epsilon}_c}}{\rho_u S_L}. \tag{3.13}$$

The closure models obtained for the scalar dissipation rate, $\tilde{\epsilon}_c = \overline{\rho \alpha (\nabla c'' \cdot \nabla c'')}/\overline{\rho}$, with c'' as the Favre fluctuation of c, in recent studies [52] can be used here. For LES, this dissipation rate would be the filtered dissipation rate written as $\widetilde{N}_c = \widetilde{\alpha}(\nabla \tilde{c} \cdot \nabla \tilde{c}) + \widetilde{\epsilon_{c,\text{sgs}}}$, which is the sum of resolved and SGS contributions. The modeling of the latter part requires extra care and the algebraic closure proposed by Dunstan et al. [61] can be used, for example. A priori analysis of this kind of modeling for $\overline{\dot{\omega}}_c$ using DNS data is explored in [62] and a posteriori analysis has also been reported in a past study [63].

It is also entirely feasible to adapt the EBU type model by approximating the SGS dissipation rate using a simple linear relaxation based model[3] and taking the SGS variance as $\overline{c}(1 - \overline{c})$. This yields

$$\overline{\Sigma} = C_1 \frac{\overline{c}(1 - \overline{c})}{\Delta}, \tag{3.14}$$

where C_1 is a model constant. A variant of this methodology involving the SGS flame wrinkling factor $\Xi_\Delta = \overline{|\nabla c|}/|\nabla \overline{c}|$ was proposed in [65], suggesting that $C_1 = 4\Xi_\Delta \sqrt{6/\pi}$. A transport equation for the SGS flame wrinkling factor can be derived and solved with additional modeling for the generation and removal rates of Ξ_Δ [66–68], which introduces further uncertainties. The simplest way is to treat this factor as unity [65] or use a fractal based algebraic expression

$$\Xi_\Delta = \left(1 + \frac{\Delta}{\eta_i}\right)^{D_f - 2},$$

as has been suggested in [69–72]. The fractal dimension D_f varies typically between 2 and 3 [71, 73, 74] and its scale dependence has also been investigated using dynamic procedure [70]. The inner cutoff length scale η_i can be the Kolmogorov scale or δ_{th} [71] or the inverse of surface-averaged mean flame curvature [69, 70], $|0.5 \langle \nabla \cdot \mathbf{n} \rangle_s|^{-1}$. It seems that a good prediction of $\overline{\dot{\omega}}$ using these models strongly depends on the choices for the exponent and the inner cutoff length scale. These two quantities can also depend on the Karlovitz number, as the fractal nature of the flame

---

[3] Using a linear relaxation based expression for the dissipation rate of the progress variable will lead to serious physical inconsistencies for premixed combustion, as discussed in [64].

surface depends on the combustion regime [75]. The various algebraic SGS closures available for the FSD are evaluated and compared using DNS data in [75] and that study showed that the Karlovitz number dependencies for D_f and η_i are essential to estimate $\overline{\Sigma}$ correctly across the combustion regime.

The algebraic model given in Eq. (3.14) was used in [76] to calculate turbulent premixed flames propagating in a channel with a square obstruction. The computed results showed an underprediction of about 20–30% in peak pressure and also a time lag in the pressure history compared to experimental measurements, which implies that the heat release rate may not have been estimated correctly. Perhaps this could be because of the overcompensation of SGS wrinkling in the aforementioned model. The multiscale analysis showed that turbulent eddies of size smaller than $2\delta_{th}$ contributed less than 10% of the total tangential strain rate [77], which is also confirmed further in [78]. Thus, the LES numerical grid resolving turbulence scales down to a few multiples of δ_{th} would be enough to capture most of the flame straining caused by turbulence through the LES equations and additional modeling may not be required for SGS flame wrinkling (or stretching). Doan et al. [77] noted that eddies larger than $17\delta_{th}$ contributed less than 10% to the total flow tangential strain rate, a_T (see Eq. 3.9). These results suggest that the LES grid does not have to resolve eddies of size δ_{th} or below. This result has a significant implication for LES of turbulent flames in complex flows and geometries of pratical interest (such as gas turbines, for example). The contribution of curvature induced strain rate, $\kappa\, S_{d,c}$, can be positive or negative locally and this is discussed in detail in Section 3.6.

One can also employ a transport equation for $\overline{\Sigma}$ rather than using an algebraic expression. This transport equation can be obtained by either averaging (for RANS) or filtering (for LES) Eq. (3.9). This equation is

$$\frac{\partial \overline{\Sigma}}{\partial t} + \nabla \cdot \left[\langle \mathbf{u} + S_d\, \mathbf{n} \rangle_s \overline{\Sigma} \right] = \langle \Phi \rangle_s \overline{\Sigma} \qquad (3.15)$$

in a generic form. For RANS calculation, this equation takes the following form:

$$\frac{\partial \overline{\Sigma}}{\partial t} + \nabla \cdot (\widetilde{\mathbf{u}}\, \overline{\Sigma}) = -\nabla \cdot \left(\langle \mathbf{u}'' \rangle_s \overline{\Sigma} \right) + \langle \overline{\Phi} \rangle_s \overline{\Sigma} + \langle \Phi'' \rangle_s \overline{\Sigma} - \nabla \cdot \left(\langle S_d\, \mathbf{n} \rangle_s \overline{\Sigma} \right). \qquad (3.16)$$

There will be a similar equation for LES also but the $\overline{\Sigma}$ should be taken as the generalized FSD, $\Sigma_{gen} = \overline{|\nabla c|}$. The surface-averaged quantities with a bar on them, for example $\langle \overline{\Phi} \rangle_s$, should be taken that it is coming from the resolved scales in the LES while the double-primed quantity implies the SGS contribution. For RANS, the meaning implied by these terms is quite straightforward. All the terms on the right-hand side of Eq. (3.16) need to be modeled and these models are elaborated by Cant [58] for both RANS and LES. As noted by Cant, care is required to satisfy the physical realizability of the FSD, that is, $\overline{\Sigma} \geq 0$, while using both algebraic expression or a transport equation. As can be seen in reference [58], the models used for flame stretch always produce (or act as a source, that is, $\langle \Phi'' \rangle_s > 0$) FSD while it is possible to have negative stretch as observed in experiments [79] (through the measurements of scalar dissipation rate which is related to FSD) and DNS data [80]. It is generally thought that the combined effects of curvature and propagation mechanisms can lead to negative stretch if this combined effect is larger than the flow tangential strain rate contribution. Mathematically, this is written as

$$\overline{a_T} < \left| \overline{\langle S_d \kappa \rangle_s} - \overline{\nabla \cdot \langle S_d \mathbf{n} \rangle_s} \right| \tag{3.17}$$

from Eq. (3.9). This condition is written because the tangential strain rate models are always positive semidefinite (i.e., they give $a_T \geq 0$) while the model for the SGS curvature related term has a negative sign [58].

Intriguing effects leading to negative stretch rate are identified by Ruan et al. [80] and these effects originate from the turbulence–scalar interaction process, which is different from the preceding observation on the combined effects of curvature and propagation mechanisms. One can write Eq. (3.9) as

$$\frac{1}{\overline{\Sigma}} \frac{d \overline{\Sigma}}{dt} = \langle \Phi \rangle_s - \frac{\partial \overline{\langle \widetilde{u}_k + \overline{S_d n_k} \rangle_s}}{\partial x_k}, \tag{3.18}$$

where $d \cdot /dt = \partial \cdot /\partial t + (\widetilde{u}_k + \overline{S_d n_k}) \partial \cdot /\partial x_k$. Then using the definition of Φ from Eq. (3.9) in Eq. (3.18) one gets[4]

$$\frac{1}{\overline{\Sigma}} \frac{d \overline{\Sigma}}{dt} = -\overline{\left\langle n_k n_m \frac{\partial u_k}{\partial x_m} \right\rangle_s} - \overline{\left\langle n_k \frac{\partial S_d}{\partial x_k} \right\rangle_s}$$

$$= -\overline{\langle \alpha \cos^2 \theta_\alpha + \beta \cos^2 \theta_\beta + \gamma \cos^2 \theta_\gamma \rangle_s} - \overline{\left\langle n_k \frac{\partial S_d}{\partial x_k} \right\rangle_s}, \tag{3.19}$$

where α, β, and γ are the principal strain rates (eigenvalues of $\partial u_k'' / \partial x_m$) and θ_α, θ_β, and θ_γ are the angles between the surface normal vector $\mathbf{n}$ and the corresponding eigen vectors. The second line of Eq. (3.19) is obtained by using eigen decomposition method for the inner product between the unit normals and the strain tensor. The following observations can be made from Eq. (3.19).

- For a material surface moving with fluid $S_d = 0$ and it is well known that the surface normal aligns with the most compressive strain γ. Thus, $d\overline{\Sigma}/dt = \langle |\gamma| \rangle_s \overline{\Sigma}$, which gives exponential growth for $\overline{\Sigma}$, as has been shown in many past studies, since 1950s. See Section 3.6 for physical insights on this point.
- For a passive surface, which does not affect the flow, propagating with constant S_d, one observes an exponential growth of $\overline{\Sigma}$ with t, as noted by Peters [81] and Kerstein [82]. The situation of constant S_d can lead to cusp (singularity) formation due to the possibility of self-intersection of the propagating surface [54, 83, 84]. The molecular diffusion present in a real situation will influence S_d (see Eq. 3.10) and hence $\partial S_d / \partial x_k$ cannot be zero and thus the possibility for the cusp formation is reduced and $\overline{\Sigma}$ will not grow exponentially.
- There is growing evidence (see Section 2.3) showing that the flame normal aligns with the most extensive strain rate, α, in premixed combustion at large Damköhler

---

[4] See Eq. (3.114).

number. The implications of this are discussed elaborately in Section 3.6. For this situation, Eq. (3.19) yields $d\overline{\Sigma}/dt = -\overline{\langle a \rangle_s \Sigma}$ for unity Lewis number, which suggests that there is a loss of FSD purely from the turbulence–scalar interaction. It is clear that this loss is not from the flame curvature mechanism and this effect is not included in the FSD models proposed in past studies. A model catering to these observations at the SGS level in a satisfactory manner is yet to be developed. The aforementioned effects are compounded by the response of S_d to the curvature in nonunity Lewis number flames [85], which also cannot be ignored for SGS modeling.

It is quite clear from the preceding discussion that the FSD approach translates the modeling of the filtered reaction rate into finding a closure for the FSD, which has been developed in past studies. Many modeling expressions used in this approach need further development to capture many of the turbulent premixed combustion subtleties and thus further research is required.

3.3.2   *G-Equation Model*

As noted earlier, this approach also considers the flame front as a surface separating the unburned gas from burned mixture. The propagation of the front is described using a level-set equation that is valid only on the surface [3]. The level-set equation itself does not contain any information about the thickness of the flame; thus, it has been used as a theoretical model of thin laminar flamelets embedded in turbulence to study the development of wrinkles on the flame surface and turbulent burning velocity in homogeneous turbulent flow [82, 86, 87]. Recently, the level-set equation has been used to study flame instability in laminar and weakly turbulent premixed flames [88, 89], where the effect of hydrodynamic instability and turbulence on the evolution of flame surface and turbulent burning velocity is analyzed. Finite flame thickness has to be taken into account in order to apply the G equation model to industry problems using RANS or LES. This can be achieved by converting the level-set function G to a *distance function*.

3.3.2.1   **Kinematic Equation for Flame Front Propagation**

The propagative form of the instantaneous c equation is written as

$$\frac{\partial c}{\partial t} + \mathbf{u} \cdot \nabla c = S_d |\nabla c|, \tag{3.20}$$

where S_d is the local displacement speed of the iso-scalar surface, $c(\mathbf{x}, t) = \zeta$ (see Eq. 3.10). The diffusive flux in the equation for S_d can be written as

$$\nabla \cdot (\rho \alpha \nabla c) = -\nabla \cdot (\rho \alpha |\nabla c| \mathbf{n}) = -\rho \alpha |\nabla c| \nabla \cdot \mathbf{n} + \mathbf{n} \cdot \nabla (\rho \alpha \mathbf{n} \cdot \nabla c), \tag{3.21}$$

by splitting it into components along the normal and tangent to the surface. As noted earlier for Eq. (3.9), the surface normal unit vector is $\mathbf{n} = -\nabla c/|\nabla c|$, pointing toward the unburned mixture. Now, the displacement speed becomes

$$S_d = \frac{\mathbf{n} \cdot \nabla \left(\rho \alpha \mathbf{n} \cdot \nabla c\right) + \dot{\omega}_c}{\rho |\nabla c|} - \alpha \left(\nabla \cdot \mathbf{n}\right) = S_{d,N} - \alpha \kappa, \quad (3.22)$$

where $\kappa = (\nabla \cdot \mathbf{n})$ is twice the mean curvature of the isosurface $c = \zeta$.

Equation (3.20) can be compared to the level-set G equation

$$\frac{\partial G}{\partial t} + \mathbf{u} \cdot \nabla G = S_d |\nabla G|, \quad (3.23)$$

where $G(\mathbf{x}, t)$ is a level-set function. The level-set $G(\mathbf{x}, t) = G_0$ propagating along $\mathbf{n}$ at a speed of S_d represents the flame surface. The burnt region is $G(\mathbf{x}, t) > G_0$ while $G(\mathbf{x}, t) < G_0$ represents the unburned region. This equation was first introduced by Williams [3] and it can be derived by considering the motion of a fluid element on the flame surface at $\mathbf{x}$ and time t, which is propagating locally at a speed of S_d along $\mathbf{n}$ and convected by the local flow velocity $\mathbf{u}$ [10, 81, 90]. The surface normal is defined, again, as $\mathbf{n} = -\nabla G / |\nabla G|$ and thus

$$\frac{d\mathbf{x}}{dt} = \mathbf{u} + S_d \mathbf{n}. \quad (3.24)$$

The G equation (3.23) can be obtained by differentiating $G(\mathbf{x}, t) = G_0$ with respect to t and making use of Eq. (3.24). Since the G equation is derived to describe the motion of the flame surface it is valid only at the $G(\mathbf{x}, t) = G_0$ surface.

The displacement speed S_d, which is a flame parameter, should be specified to close the G equation. A theoretical expression for S_d has been developed for weakly stretched laminar flamelet propagation [3, 91, 92]

$$S_d = S_L - S_L \kappa \mathcal{L} + \mathcal{L} \mathbf{n} \cdot \nabla \mathbf{u} \cdot \mathbf{n}, \quad (3.25)$$

where S_L is the burning velocity of unstretched planar laminar flame, $\mathcal{L}$ is the Markstein length, and the last term represents the effect of strain on the flame propagation (this term is similar to the alignment term in Eq. 3.19). It is clear that Eq. (3.25) bears a certain similarity to the second part of Eq. (3.22); both involve a term that is proportional to κ and a term that is related to the laminar burning velocity of an unstretched planar flame. In the limit of unstretched flame the two S_d expressions are identical provided that the level-set $G(\mathbf{x}, t) = G_0$ coincides with the isoscalar surface $c(\mathbf{x}, t) = \zeta$.

The equation (3.20) for the isoscalar $c(\mathbf{x}, t) = \zeta$ surface is derived from the transport equation of the reactive scalar, which is generally valid in different regimes of turbulent premixed combustion and also in different, preheat and reaction, zones of the flame provided that S_d on the given isoscalar surface is known. Peters [10, 81] developed a level-set equation valid for both the flamelet regime and the thin reaction zone regime. The equation (3.20) is valid for all regimes of turbulent premixed combustion provided that the local displacement speed of the isoscalar surface is given or modeled properly, which is a challenge. In the broken reaction zone regime at high Karlovitz numbers the reaction zone is highly distorted. The flame front is not well defined; however, the isoscalar surface can be defined. This allows for the use of level-set method to predict flames with extinction holes [93, 94]. Nevertheless, Williams [95] has noted that the G equation approach is strictly valid only in the wrinkled flamelets regime of turbulent premixed combustion (very low turbulence level; see Figs. 2.1 and 2.2 of Section 2.1).

3.3.2.2 Modeling of Flame Front Propagation in RANS and LES

The propagation of an ensemble averaged (in RANS) or filtered (in LES) flame is modeled in the calculations using the level-set approach. For example, the level-set equation describing the filtered flame can be derived using three approaches. The first one is by spatial filtering of the level-set G equation (3.23) (see [96] for details), which follows the methodology used by Peters [10, 81] for ensemble averaged flame (RANS paradigm). The second approach is through kinematic consideration of a fluid element on the filtered flame, which is convected by the filtered flow field and also self-propagating along the local normal direction. This approach is explained in [97, 98]. The third approach is by mathematical rearrangement of the filtered reactive scalar equations in a manner analogous to that used to obtain Eqs. (3.20) to (3.22). The latter approach gives the filtered level-set equation as

$$\frac{\partial \tilde{c}}{\partial t} + \tilde{\mathbf{u}} \cdot \nabla \tilde{c} = S_{d,\text{LES}} |\nabla \tilde{c}|, \tag{3.26}$$

where $\tilde{c}$ is the filtered reaction progress variable, $\tilde{\mathbf{u}}$ is the filtered local fluid velocity, and $S_{d,\text{LES}}$ is the local displacement speed of the filtered isoscalar surface, $\tilde{c}(\mathbf{x},t) = \zeta$. The displacement speed is

$$S_{d,\text{LES}} = \frac{\nabla \cdot J_{c,\text{SGS}} + \nabla \cdot (\overline{\rho \alpha \nabla \tilde{c}}) + \overline{\dot{\omega}_c}}{\overline{\rho} |\nabla \tilde{c}|}, \tag{3.27}$$

where $J_{c,\text{SGS}} = (\overline{\rho \mathbf{u} \tilde{c}} - \overline{\rho \mathbf{u} c})$ is the SGS flux of c, which can be modeled using a SGS scalar flux model, for example, the Smagorinky model [99],

$$J_{c,\text{SGS}} = \overline{\rho} \alpha_T \nabla \tilde{c}, \tag{3.28}$$

where α_T is the SGS diffusivity. The diffusive and SGS flux terms in Eq. (3.27) can be expanded into normal diffusion and curvature-dependent terms similar to Eq. (3.21). This gives

$$S_{d,\text{LES}} = S_{d,\text{LES}}^N - (\alpha + \alpha_T) \nabla \cdot \tilde{\mathbf{n}}, \tag{3.29}$$

where $\tilde{\mathbf{n}} = -\nabla \tilde{c}/|\nabla \tilde{c}|$ is unit normal vector of the resolved surface $\tilde{c}(\mathbf{x},t) = \zeta$. The curvature of this surface is $\nabla \cdot \tilde{\mathbf{n}}$ and the normal component is given by

$$S_{d,\text{LES}}^N = \frac{\tilde{\mathbf{n}} \cdot \nabla \left([\overline{\rho \alpha} + \overline{\rho} \alpha_T] \tilde{\mathbf{n}} \cdot \nabla \tilde{c}\right) + \overline{\dot{\omega}_c}}{\overline{\rho} |\nabla \tilde{c}|}, \tag{3.30}$$

which depends on the molecular diffusion, SGS turbulent transport, and the filtered reaction rate. Thus, it requires modeling and various models have been proposed in past studies [10, 81, 90, 96, 97, 100]. Application of these models to LES of premixed combustion has been reported with some success [94, 96–98, 101]; however, validation of these models has not been sufficient due to limited experimental and DNS data.

3.3.2.3 Level-Set G-Equation Model Coupled with Flamelet Chemistry

The G equation describes the propagation of an isoscalar surface in the flame, while it does not contain any information on the flame structure. It is possible to recover the flame structure through this approach in the following manner. The flame sheet model is the simplest approach, giving

$$\psi(G) = \begin{cases} \psi_u & \text{for } G < G_0 \\ \psi_b & \text{for } G > G_0, \end{cases} \quad (3.31)$$

where ψ_u and ψ_b are the values of reactive scalar ψ in the unburned and burned mixtures respectively. The quantity ψ represents the species mass fractions, density, temperature, etc., and is a function of the level-set G.

In the flamelet regime, $\psi(G)$ can be obtained from the counterflow twin-laminar flame calculation with detailed chemistry for various strain rates of the flame [98, 102]. In this configuration, the flame has a finite thickness, unlike in the flame-sheet model, so the level-set function G should be reconstructed to represent a flame coordinate. This can be done through an reinitialization process, which converts the G function to a distance function,

$$\begin{cases} G = G_0 & \text{for } G = G_0 \\ |\nabla G| = 1 & \text{for } G \neq G_0, \end{cases} \quad (3.32)$$

Without loss of generality, one can set $G_0 = 0$. After the reinitialization the value of $(G - G_0)$ represents the distance to the $G = G_0$ level-set. The $G = G_0$ level can be set at the beginning of the preheat or the inner layer of the laminar flamelet (in the laminar flame results) and the variations of ψ in the laminar flame can then be mapped to $(G - G_0)$ variation that can then be used as a lookup table for the turbulent flame simulations. Since the flamelet is thin in the wrinkled flamelet regime, the location of $G = G_0$ has a negligible impact on the computed distribution of the reactive scalars.

In the LES or RANS framework, computation of the filtered or ensemble averaged scalar quantities requires the PDF of G [93]. If a presumed PDF approach is employed the variance of G, σ_G, is needed. The governing equation for this variance has been developed for RANS [10, 81, 90] and LES [97]. This approach has been shown to be useful for turbulent flame prediction in the flamelet regime. The G equation model using the flamelet chemistry and presumed PDF fails to predicted the broadened distribution of reactive scalars such as CH_2O [103] in the thin reaction zone regime combustion where the preheat zones of the flamelet are broadened by small-scale turbulent eddies. This observation is consistent with the remark by Williams [95] that the G equation approach is strictly valid for the wrinkled flamelet regime of turbulent premixed combustion.

3.4 Statistical Models

The models described in the previous subsection treat the flame as a propagating interface separating burnt products from unburnt mixtures and predominantly use the kinematic aspects of the flame propagation. The statistical models involve the PDF of the reaction progress variable in one form or another to obtain a closure for the mean or filtered reaction rate. Since the reaction rate is a function of species mass fractions and temperature, its mean or filtered value can be written as

$$\overline{\dot{\omega}_j}(\mathbf{x},t) = \int \dot{\omega}_j(\zeta_i) \, \overline{P}(\zeta_i;\mathbf{x},t) \, d\zeta_i, \qquad (3.33)$$

for species j. The symbols ζ_i represent the sample space variable for the vector $\mathcal{Y}_i$ containing the species mass fractions and temperature. The one-point joint PDFs of all the species mass fractions and temperature are denoted using $\overline{P}(\zeta_i;\mathbf{x},t)$, which also varies in space and time. This PDF is the probability density function in the classical sense for the RANS methodology but it is the FDF (filtered density function) representing the stochastic nature of the subgrid fluctuations in LES. This distinction is important to recognize as noted by Howarth and Pope [104]; the FDF is a random quantity and its expected value in the limit of filter width going to zero gives the PDF. The FDF was introduced by Pope [51]. Now, the question is, how to get this PDF or FDF?

This can be obtained by solving a modeled transport equation, known as the transported PDF approach, or the shape of this PDF can be presumed using the first and/or second moments. The first attempt at solving the scalar PDF transport equation was reported in [105]. The mean or variance of the reaction progress variable, or both, are used for the RANS methodology to get the presumed shape. The use of these two approaches for RANS methodology is discussed in many past studies [3, 7, 10, 58, 59, 104, 106]. For LES, the filtered progress variable and SGS variance of the progress variable are used to get the presumed shape or a modeled transport equation for the FDF is solved. Recent advances on these approaches for turbulent premixed combustion are reviewed briefly in the text that follows.

3.4.1 Transported PDF

The transport equations for the joint-scalar PDF and FDF are similar [107]:

$$\overline{\rho}\frac{D\widetilde{P}}{Dt} = -\sum_{j=1}^{N} \frac{\partial \left(\overline{[\nabla \cdot (\rho \mathcal{D}_j \nabla Y_j)]|\zeta_i} \widetilde{P}\right)}{\partial \zeta_j} - \sum_{j=1}^{N} \frac{\partial \left(\overline{\dot{\omega}_j|\zeta_i}\right)\widetilde{P}}{\partial \zeta_j}$$
$$- \nabla \cdot \left[\left(\overline{\rho\mathbf{u}|\zeta_i} - \overline{\rho}\widetilde{\mathbf{u}}\right)\widetilde{P}\right], \qquad (3.34)$$

where $\mathcal{D}_j$ is the molecular diffusivity for scalar j and the quantities inside brackets with an overline imply that they are conditionally filtered (or averaged). This is given for the reaction rate as an example:

$$\left(\overline{\dot{\omega}_j|\zeta_i}\right)\widetilde{P} = \int \dot{\omega}_j\left(\mathbf{x}',t;\mathcal{Y}\right)\mathbf{F}\left(\mathbf{x}',t\right)\mathcal{G}\left(\mathbf{x}'-\mathbf{x}\right)\,d\mathbf{x}', \qquad (3.35)$$

where $\mathbf{F}(\mathbf{x}',t) = \prod_{i=1}^{N}\delta\left[\zeta_i - \mathcal{Y}_i(\mathbf{x}',t)\right]$ is the fine-grained density function and $\mathcal{G}$ is the filter kernel. This transport equation needs modeling for the diffusive (second) and turbulent transport (fourth) terms. The reaction (third) term is closed, which is known to be the biggest advantage of this method. Typically, Eq. (3.34) is solved using the Monte Carlo (MC) method along with the models for the diffusive and turbulent transport terms. The application of this method to turbulent premixed flames is discussed in detail by Lindstedt [108]. As noted by him, it is also quite common to consider a joint PDF of velocity and scalar mass fractions, and sometimes including turbulent frequency [109]. Of course, considering additional quantities for the joint PDF would bring further modeling challenges, so one must be cautious.

For the LES, the turbulent transport flux is typically closed using the gradient hypothesis [110]:

$$\left(\overline{\rho\,\mathbf{u}|\zeta_i} - \overline{\rho}\,\widetilde{\mathbf{u}}\right)\widetilde{P} = -\overline{\rho}\,\nu_t\nabla\widetilde{P}. \qquad (3.36)$$

The conditional scalar diffusion enters the first term on the right-hand side of Eq. (3.34). It synthesizes molecular diffusion enhanced by small-scale vortices and chemical reaction, and it is commonly called the micro-mixing term. The first and simplest micro-mixing closure proposed was the Linear Mean Square Estimation (LMSE) model [105], which implied a linear relaxation of all scalar values toward the mean, namely,

$$\overline{\nabla\cdot\left(\rho\mathcal{D}_j\nabla Y_j\right)|\zeta_i} = -\frac{\overline{\rho}}{2\,\tau_{\text{mix}}}\left(\zeta_j - \widetilde{Y}_j\right), \qquad (3.37)$$

where τ_{mix} is the timescale for the small-scale mixing. This timescale is typically modeled as $\tau_{\text{mix}} = \tau_t/C_\phi$ with $C_\phi \simeq 2$ and turbulent timescale $\tau_t = \widetilde{k}/\widetilde{\epsilon}$. The mixing timescale, τ_{mix}, is related to the scalar dissipation rate and thus the scalar dissipation rate model could be used to deduce the timescale (explained on a physical basis in Section 3.6). The LMSE model can also be interpreted as resulting from the approximation of the Laplacian operator for a small volume or, alternatively, in terms of finite differencing the normal diffusion. This model is deterministic whereas the instantaneous micro-mixing is random. To account for this randomness, Valiño and Dopazo [111] introduced the Binomial Langevin model. A deficiency of the LMSE model is the preservation of the initial PDF shape. To overcome this shortcoming, several models have been developed and tested. The Coalescence and Dispersion (CD) model makes scalar values or MC particles, an efficient way to numerically integrate PDF transport equations, to interact (coalesce) and split-up (disperse), creating two MC particles with scalar values equal to an average (either unweighted or weighted) of the original values. The CD model specialized for turbulent mixing was proposed independently by Dopazo [112] and Janicka et al. [113]. The Euclidean Minimum Spanning Tree (EMST) model selects MC particle pairs for mixing in such a way that particles closer in composition space are more likely to mix than those further away [114]. The Amplitude Mapping Closure (AMC) proposed by Chen et al. [115] maps a Gaussian reference field to a surrogate scalar field, whose known statistics are the same as those associated with the stochastic scalar field of interest. The main drawbacks of all micro-mixing models are discussed in Section 3.6.

Under some conditions, micro-mixing must be influenced by the chemical reaction and its effect must be included in the specification of τ_{mix}, specifically for premixed combustion, as it involves strong coupling between molecular diffusion and chemical reaction. Such a model has been proposed recently by combining the turbulent timescale, τ_t, and a flamelet reaction timescale by using the segregation factor $g = \sigma_c^2/[\widetilde{c}(1-\widetilde{c})]$ [116]

$$\frac{1}{\tau_{mix}} = (1-g)\frac{C_\phi}{\tau_t} + g\frac{\int_0^1 \overline{N_c|\zeta}\, \widetilde{P}\, d\zeta}{\sigma_c^2}, \qquad (3.38)$$

where $\overline{N_c|\zeta}$ is the conditional dissipation rate of c and $\widetilde{P}$ is the marginal PDF of c. This model, tested a priori using DNS data in [116], is obtained by noting that $\tau_{mix} = \sigma_c^2/\widetilde{\varepsilon}_c$ and using a scalar dissipation rate model similar to the bridging model suggested by Champion and his co-workers for the dissipation rate. This bridging model involves g, integral turbulence timescale, τ_t, and a flamelet chemical timescale. This model is summarized in [52]. Many previous LES studies, for example [104, 110, 117–119], discuss the use of RANS τ_{mix} models for LES and details can be found there.

One can also write the conditional diffusion of c in Eq. (3.34) in terms of its conditional scalar dissipation rate, namely,

$$-\frac{\partial\left(\overline{[\nabla\cdot(\rho\alpha\nabla c)]|\zeta}\right)\widetilde{P}}{\partial\zeta} = \nabla\cdot(\alpha\nabla\widetilde{P}) - \frac{\partial^2}{\partial\zeta\partial\zeta}\left[\left(\overline{\rho\,\alpha\nabla c\cdot\nabla c|\zeta}\right)\widetilde{P}\right]. \qquad (3.39)$$

Thus, it is apparent that the conditional scalar dissipation rate, which is related to the scalar mixing frequency, can appear explicitly in Eq. (3.34). However, note that substituting Eq. (3.39) into Eq. (3.34) leads to a physically and mathematically correct PDF transport equation although the negative diffusivity in ζ-space makes its numerical integration unstable.

Numerous studies have been devoted in the recent past to develop closures for the scalar micro-mixing in turbulent premixed combustion. Typically, the mixing frequency required for transported PDF calculations is modeled using the inverse of τ_t for passive scalars. Although this approach worked quite well in many past studies, including the influence of chemical reaction on the mixing frequency was shown to improve the predictions using the LES-FDF approach [118].

In the PDF approach notional particles are employed to solve its transport equation using MC methods. Alternatively, one can employ a Eulerian MC field representation for the scalar fields. Valiño [120] requires that these stochastic fields must remain spatially smooth whereas Sabel'nikov and Soulard [121] allow discontinuities. In either of these approaches, the turbulent reacting flow is represented by a number of notional Eulerian MC fields, for example, the nth field for scalar j, obeying

$$d\zeta_j^n = -\widetilde{\mathbf{u}} \cdot \nabla \zeta_j^n \, dt + \dot{\omega}_j\left(\zeta_j^n\right) dt - \frac{\left(\zeta_j^n - \widetilde{\zeta}_j\right)}{2\tau_{\text{mix}}} dt$$
$$+ \nabla \cdot \left[\left(\mathcal{D}_j + \frac{\nu_t}{\text{Sc}_t}\right) \nabla \zeta_j^n\right] + \sqrt{2\nu_t/\text{Sc}_t} \, \nabla \zeta_j^n \cdot d\mathbf{W}_j^n, \qquad (3.40)$$

where $\mathbf{W}_j$ is the time varying vectorial Wiener process, which is independent of spatial location [120]. The increment $d\mathbf{W}_j^n$ is a normally distributed Markovian random process with zero mean and variance equal to dt. The Favre-averaged quantities are obtained by ensemble averaging over these Eulerian fields and thus the higher the number of these fields the better is the statistical accuracy. About 10 to 30 fields were used in past studies. Examples of this approach applied to turbulent premixed flames can be found in [122–124]. By comparing Eq. (3.40) to the LEM equation, Eq. (3.7), one may see that there are some similarities between these two approaches. The random stirring event, modeled using stochastic triplet mapping, is equivalent to the Wiener process in the Eulerian stochastic fields method.

The deterministic equivalent of the Eulerian stochastic field approach is the multienvironment PDF method proposed in [125, 126] and the details can be found in those studies. The Eulerian field methods are more straightforward to implement into CFD codes than particle-based methods. The transported FDF approach is an active research area and the major challenge for it is still in finding a closure model for the SGS scalar mixing rate containing the close coupling of molecular diffusion, chemical reaction, and turbulence in premixed combustion. A detailed review of the PDF approach can be found in [127].

The aforementioned problem with the mixing timescale is greatly alleviated in presumed shape approaches, which gives the averaged or filtered reaction rate of the progress variable as

$$\overline{\dot{\omega}_c}(\mathbf{x},t) = \overline{\rho} \int_0^1 \frac{\dot{\omega}_c(\zeta)}{\rho} \widetilde{P}(\zeta;\mathbf{x},t) \, d\zeta. \qquad (3.41)$$

The mass-weighted marginal PDF or FDF is obtained by presuming a shape for given values of the first two moments as noted earlier. The reaction rate function, $\dot{\omega}_c(\zeta)$, and the PDF can be obtained in many different ways, thus resulting in different closure strategies, viz., the classical BML, tabulated chemistry, and conditional moment closure approaches.

3.4.2   Bray–Moss–Libby Model

Several reviews of this model are available [57, 58, 128] and its further development for RANS is discussed by Bray [128]. For the sake of completeness, it is reviewed briefly before presenting some ideas for its LES extension, since its use for LES is quite uncommon. This is mainly because it requires another closure to get the filtered reaction and Bray et al. [129] devised a way to overcome this for RANS methodology, which is reviewed by Bray [128].

Premixed Combustion Modeling

The classical BML model presumes the shape of the Reynolds-averaged PDF of c to be

$$\overline{P}(\zeta;\mathbf{x},t) = \alpha(\mathbf{x},t)\,\delta(\zeta) + \beta(\mathbf{x},t)\,\delta(1-\zeta) + \gamma(\mathbf{x},t)\,[H(0) - H(1)]\,f(\zeta;\mathbf{x},t) \tag{3.42}$$

by partitioning the reacting flow into (1) reactants having $0 \leq \zeta \leq \zeta^*$, (2) reaction zones represented by $\zeta^* \leq \zeta \leq (1-\zeta^*)$, and (3) products having $(1-\zeta^*) \leq \zeta \leq 1$. These three categories have probabilities of α, γ, and β respectively. The symbols δ and H represent the Dirac delta and Heaviside functions respectively. The function $f(\zeta;\mathbf{x})$ is the internal part of the PDF and it obeys

$$\int_0^1 f(\zeta;\mathbf{x},t)\,d\zeta = 1 \quad \text{and} \quad \alpha + \beta + \gamma = 1. \tag{3.43}$$

The PDF represented by the above model is shown in Fig. 3.2a. The mass-weighted PDF is related to the Reynolds PDF through $\widetilde{P} = \rho\,\overline{P}/\overline{\rho}$.

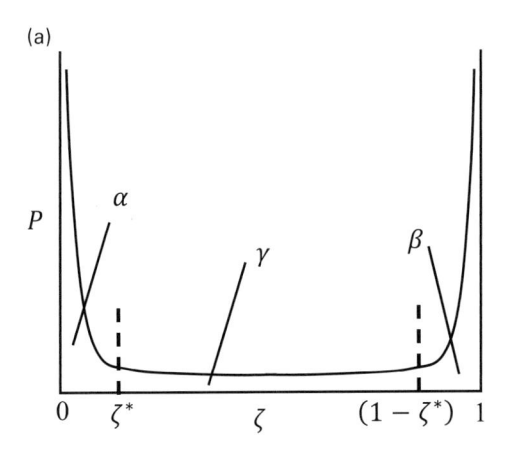

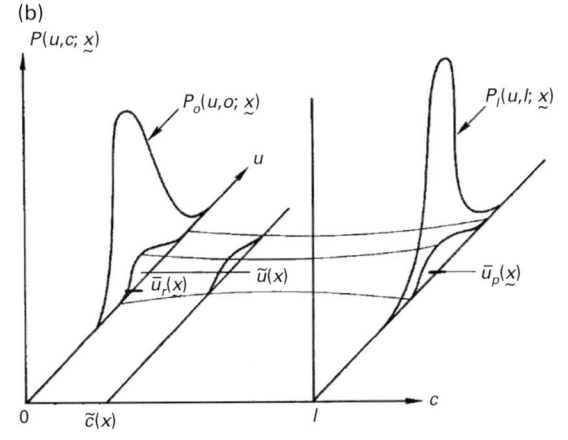

Figure 3.2 Sketch of the BML. (a) PDF of reaction progress variable. (b) Joint PDF of c and velocity.

In the classical bimodal, limit which is acceptable for premixed combustion with Da $\gg 1$, the burning mode part becomes negligible, that is, $\gamma \ll 1$, implying that the flame front is very thin. Under this condition, one gets $\alpha + \beta = 1$ from the second part of Eq. (3.43). If one defines the progress variable using temperature then the ideal gas state equation gives

$$\frac{\rho}{\rho_u} = \frac{1}{1+\tau c}, \quad \text{where} \quad c = \frac{T-T_u}{T_b-T_u} \quad \text{and} \quad \tau = \frac{T_b-T_u}{T_u}, \quad (3.44)$$

when the thermodynamic pressure is constant across the flame. The mean or filtered density can be obtained by averaging or filtering the above equation, which gives $\overline{\rho} = \rho_u/(1+\tau\widetilde{c})$. This can also be obtained by using the PDF (or FDF) through

$$\overline{\rho}(\mathbf{x},t) = \int_0^1 \rho(\zeta)\,\overline{P}(\zeta;\mathbf{x},t)\,d\zeta = \frac{\rho_u}{1+\tau\widetilde{c}}. \quad (3.45)$$

Since $\alpha = 1 - \beta$ for $\gamma \ll 1$ Eq. (3.45) can be used to deduce

$$\beta(\mathbf{x},t) = \frac{(1+\tau)\widetilde{c}}{1+\tau\widetilde{c}} \quad \text{and} \quad \alpha = \frac{1-\widetilde{c}}{1+\tau\widetilde{c}}. \quad (3.46)$$

Now, the second and third moments of the progress variable can be deduced using the presumed shape, if required. The second moment, the Favre variance, is $\sigma_c^2 = \widetilde{c}(1-\widetilde{c})$, which gives a second transport equation for $\widetilde{c}$. Bray [18] showed that the reconciliation of the two $\widetilde{c}$ equations led to an important relationship between the mean reaction and scalar dissipation rates, which is written as Eq. (3.3).

As per Eq. (3.41), the mean reaction rate for the BML model is

$$\overline{\dot{\omega}_c} = \int_{c^*}^{1-c^*} \dot{\omega}_c \frac{\overline{\rho}}{\rho} \widetilde{\gamma}\widetilde{f}\,d\zeta = \int_{c^*}^{1-c^*} \dot{\omega}_c\,\gamma\,f\,d\zeta. \quad (3.47)$$

Hence the mean reaction rate cannot be estimated in the classical BML approach, since $\gamma \approx 0$. Bray et al. [129] showed that

$$\widetilde{\gamma} = \frac{\widetilde{c}(1-\widetilde{c})(1-g)}{I_1 - I_2}, \quad \text{where} \quad I_1 = \int_{c^*}^{1-c^*} \zeta\widetilde{f}\,d\zeta \quad \text{and} \quad I_2 = \int_{c^*}^{1-c^*} \zeta^2\widetilde{f}\,d\zeta \quad (3.48)$$

by using the segregation factor $g = \sigma_c^2/(\widetilde{c}(1-\widetilde{c}))$. Here, the burning mode, $\widetilde{f}$, is undefined and it a requires a model. Within the framework of the BML and in the limit of $g \to 1$, the mass-weighted burning mode can be defined using the gradient of c in a laminar flamelet [129]:

$$\widetilde{f} = \frac{\rho\,C_1}{\overline{\rho}|dc/d\xi|}, \quad \text{where} \quad \xi = \frac{S_L x}{\mathcal{D}}. \quad (3.49)$$

The symbol C_1 is a normalizing constant so that $\int \widetilde{f}\,d\zeta = 1$. Hence, the mean reaction rate in Eq. (3.47) can be evaluated using Eqs. (3.48) and (3.49). Since the limit $g \to 1$ is imposed, this can be viewed as the mean reaction rate when the presumed PDF is deviating from its bimodal limit by a small amount because of finite rate chemistry effects. Nevertheless, this limit can be overcome if one employs a different form for the burning mode part, $\widetilde{f}$. Bray et al. [129] showed that this model worked well for wrinkled and corrugated flamelet regimes of turbulent premixed combustion.

The partitioning illustrated in Fig. 3.2a and discussed earlier can be extended to the joint PDF of c and a velocity component u_k as illustrated in Fig. 3.2b. This PDF is written as

$$\overline{P}(\zeta, u_k; \mathbf{x}, t) = \alpha \, \delta(\zeta) \, f_r(u_k; \mathbf{x}, t) + \beta \, \delta(1 - \zeta) \, f_p(u_k; \mathbf{x}, t)$$
$$+ \gamma \, [H(0) - H(1)] \, f_F(\zeta, u_k; \mathbf{x}, t). \qquad (3.50)$$

From this PDF, one can deduce a simple expression for the turbulent scalar flux

$$\widetilde{u_k'' c''} = \widetilde{c}(1 - \widetilde{c}) \left(\overline{u_{k,p}} - \overline{u_{k,r}} \right) + \mathcal{O}(\gamma) \qquad (3.51)$$

when $\gamma \ll 1$. This expression helped us to realize that the turbulent scalar flux can be a countergradient in turbulent premixed combustion. Many interesting and useful insights can be deduced using this model as demonstrated by Bray and his co-workers in many of their past studies, which are summarized in [57].

One can think of presuming the FDF (or the subgrid PDF) using the aforementioned partitioning approach and it has not been explored fully yet. As noted earlier, estimating the filtered reaction rate would need to be considered carefully since γ is taken to be negligible or it deviates from zero by a small amount. However, the use of the *delta* function to represent the subgrid PDF is quite common for LES of partially premixed combustion. The *beta* function is also used commonly to represent the subgrid PDF of c in both premixed and partially premixed combustion.

The BML approach assumes that the combustion chemistry is typically represented using a single irreversible reaction and thus $\dot{\omega}_c(\zeta)$ in Eqs. (3.41) and (3.47) can be expressed using

$$\dot{\omega}_c(\zeta) = B \rho \, (1 - \zeta) \exp\left[\frac{-\beta (1 - \zeta)}{1 - \widehat{\alpha}(1 - \zeta)} \right], \qquad (3.52)$$

where $\widehat{\alpha} = \tau/(1 + \tau)$ and $\beta = \tau T_a / \left[T_u (1 + \tau)^2 \right]$ with T_a as the activation temperature of the single irreversible reaction with a preexponential factor B. The parameter τ is defined in Eq. (3.44) along with $\zeta(= c)$. If the results of unstrained and strained laminar flame calculations for this quantity, reaction rate, are used then the approach is called as tabulated chemistry approach. Conditional moment closure is also used sometimes to get the filtered or averaged reaction rate.

3.4.3 Conditional Moment Closure Approach

This approach assumes that the fluctuations in various species mass fractions and temperature are related to the fluctuation of a key quantity, which is the progress variable for premixed combustion and mixture fraction for non-premixed combustion. So, transport equations for conditionally averaged or filtered mass fractions and temperature are derived and solved with closure models in this approach. The derivation of these equations starts by simply splitting the mass fraction of species i into its conditional mean and a fluctuation over this mean:

$$\rho Y_i (\mathbf{x},t) = \left(\overline{\rho Y_i | c = \zeta}\right) + \rho Y_i'' = \left(\overline{\rho|\zeta}\right) \widetilde{Q}_i (\zeta;\mathbf{x},t) + \rho Y_i''. \qquad (3.53)$$

Substituting this decomposition into the species mass fraction equation, one gets a transport equation for Q_i as explained by Klimenko and Bilger [130]. The final equation may be written as [130–134]

$$\left(\overline{\rho|\zeta}\right) \frac{\partial \widetilde{Q}_i}{\partial t} + \left(\overline{\rho \mathbf{u}|\zeta}\right) \cdot \nabla \widetilde{Q}_i - \frac{\mathrm{Le}_c}{\mathrm{Le}_i} \left(\overline{\rho N_c|\zeta}\right) \frac{\partial^2 \widetilde{Q}_i}{\partial \zeta^2} = \left(\overline{\dot{\omega}_i|\zeta}\right)$$
$$- \left(\overline{\dot{\omega}_c|\zeta}\right) \frac{\partial Q_i}{\partial \zeta} - \frac{1}{\widetilde{P}} \nabla \cdot \left[\left(\overline{\rho \mathbf{u}'' Y_i''|\zeta}\right) \widetilde{P}\right] + e_{Q_i}, \qquad (3.54)$$

where Le_i is the Lewis number of species i and Le_c is the Lewis number for the reaction progress variable if it is defined using a species mass fraction. If c is defined using temperature then $\mathrm{Le}_c = 1$. The physical meaning of the various terms in Eq. (3.54) is as follows. The first and second terms denote, respectively, the unsteady and convective changes of $\widetilde{Q}_i$. The third term represents the diffusion of the conditional average in the sample space. The fourth term is the chemical reaction rate for species i. The influence of the conditioning variable, which is a reactive scalar, on the evolution of $\widetilde{Q}_i$ is given by the fifth term. The sixth term represents the contribution of conditional fluctuation, Y_i'', to $\widetilde{Q}_i$ evolution. The last term, representing the contributions of the molecular diffusion of $\widetilde{Q}_i$ in physical space and the differential diffusion of mass and heat, is given by

$$e_{Q_i} = \left(\nabla \cdot (\rho \mathcal{D}_i \nabla \widetilde{Q}_i) \Big| \zeta\right) + \left(\rho \mathcal{D}_i \nabla c \cdot \nabla \left(\frac{\partial \widetilde{Q}_i}{\partial \zeta}\right) \Big| \zeta\right)$$
$$+ \left(\frac{\partial \widetilde{Q}_i}{\partial \zeta} \nabla \cdot \left[\left(\frac{\mathrm{Le}_c - \mathrm{Le}_i}{\mathrm{Le}_c}\right) \rho \mathcal{D}_i \nabla c\right] \Big| \zeta\right). \qquad (3.55)$$

The first two terms of e_{Q_i} given in Eq. (3.55) can be neglected for RANS calculations of premixed combustion at large Reynolds number; however, they cannot be neglected for LES since the CMC is for SGS at which the molecular diffusion may not be ignored. The third term represents the differential diffusion effects, which are shown to make weak contribution for LES of premixed flames in the thin reaction zone regime [135]. The role of this term in other regimes of premixed combustion is yet to investigated. When the Lewis number is unity for all the species there is no contribution from the differential diffusion term.

The solution to Eq. (3.54) requires closures for the conditional velocity, conditional dissipation rate, $\overline{N_c|\zeta}$, reaction rates, and the conditional scalar flux. The conditional velocity is commonly assumed to be equal to the unconditional value, $\widetilde{(\mathbf{u}|\zeta)} = \widetilde{\mathbf{u}}$, for LES [135, 136] and a linear model is used for RANS [132, 133]; sometimes it was also neglected [134] depending on the turbulence level. We recommend retaining the convective term on the physical basis and using a simple closure, as has been done in [135, 136] for LES. The conditional scalar flux term is typically estimated using the gradient model for both RANS and LES. The conditional reaction rates are calculated using the Arrhenius rate expression but involving $\tilde{Q}_i$ instead of Y_i, that is,

$$\left(\overline{\dot{\omega}_i|\zeta}\right) = \dot{\omega}_i\left(\tilde{Q}_1, \tilde{Q}_2, \ldots, \tilde{Q}_T\right), \tag{3.56}$$

where $\tilde{Q}_T$ is the conditional temperature. Using DNS data, this closure has been shown to be very good when there are no local extinctions [131, 137]. The filtered reaction rates required for the LES equations are obtained using

$$\overline{\dot{\omega}_c} = \int_0^1 \left(\overline{\dot{\omega}_c|\zeta}\right) \overline{P}(\zeta) \, d\zeta, \tag{3.57}$$

which is an exact relation resulting from Eq. (3.33) after using Bayes' theorem for the joint PDF in that equation.

The closure for the conditional scalar dissipation rate in premixed combustion is tricky because of the strong influence of the reaction rate and its effects on the scalar dissipation rate as shown by Swaminathan and Bray [138]. One plausible modeling for this quantity can be obtained using the identity

$$\tilde{N}_c = \int_0^1 \left(\overline{N_c|\zeta}\right) \tilde{P}(\zeta) \, d\zeta \tag{3.58}$$

and recognizing that the conditional scalar dissipation rate normalized by its value, $\widehat{N}_c$, at the inner reaction zone located at $\widehat{\zeta}$ is a weak function of the stretch rate [139], ie., $\left(\overline{N_c|\zeta}\right)/\widehat{N}_c = \widehat{f}(\zeta)$. Substituting this relation into Eq. (3.58), multiplying the resulting equation by $\widehat{f}$ on both sides and rearranging it gives [139]

$$\left(\overline{N_c|\zeta}\right) = \frac{\tilde{N}_c \widehat{f}}{\int_0^1 \widehat{f} \, \tilde{P} \, d\zeta}, \tag{3.59}$$

where the function $\widehat{f}$ is obtained from the unstrained laminar flame calculated with a complex chemical kinetic mechanism suitable for the fuel–air mixture of interest. Now, the modeling of the conditional scalar dissipation rate is reduced to the modeling of the unconditional mean (for RANS) or filtered (for LES) dissipation rate. This unconditional dissipation rate is written as

$$\overline{\rho N_c} = \overline{\rho}\tilde{N}_c = \overline{(\rho \mathcal{D}(\nabla c \cdot \nabla c))} = \overline{\rho}\mathcal{D}(\nabla\tilde{c} \cdot \nabla\tilde{c}) + \overline{\rho}\tilde{\epsilon}_c$$
$$= \overline{\rho}\mathcal{D}(\nabla\tilde{c} \cdot \nabla\tilde{c}) + \left[\overline{(\rho \mathcal{D}(\nabla c \cdot \nabla c))} - \overline{\rho}\mathcal{D}(\nabla\tilde{c} \cdot \nabla\tilde{c})\right]. \tag{3.60}$$

For RANS $\tilde{N}_c \approx \tilde{\epsilon}_c$, since the contribution from the mean gradient is small but for LES no such approximation can be made and the second term, $\tilde{\epsilon}_c$, denotes the subgrid scalar dissipation rate. The quantity $\tilde{\epsilon}_c$ needs to be modeled for both RANS and LES. As noted for Eq. (3.13), algebraic expressions based on the linear relaxation hypothesis using the SGS turbulence timescale cannot be used to model $\tilde{\epsilon}_c$ in premixed combustion [64]. This remark also holds for the RANS approach. The evolution of this dissipation rate in premixed combustion is influenced by turbulence, chemical reaction, and molecular diffusion. These interactions can be quite different in premixed combustion compared to those in nonreacting flows and thus one cannot simply extend the scalar dissipation rate models from nonreacting flows to premixed combustion. The physical processes influence the dynamics of isoscalar surfaces and their evolution. Various theoretical and numerical investigations of these aspects are discussed in Section 3.6 and recent experimental observations on these dynamics are discussed in Section 2.3.

An algebraic model including these effects was proposed by Dunstan et al. [61], which has been investigated further using DNS [62, 140–142] and LES [11, 63, 135, 136, 143–149]. This algebraic expression is given by

$$\widetilde{\epsilon}_c = \mathcal{F}\left[(2K_c - \tau C_4)\frac{S_L}{\delta_{th}} + C_3\frac{\widetilde{\epsilon}_{sgs}}{\widetilde{k}_{sgs}}\right]\frac{\sigma^2_{c,sgs}}{\beta_c}, \qquad (3.61)$$

where the quantities with a subscript "sgs" imply that they are subgrid quantities and $\mathcal{F} = 1 - \exp(-0.75\Delta/\delta_{th})$, which ensures that $\widetilde{\epsilon}_c \to 0$ when $\Delta \to 0$. The thermochemical parameter is $K_c \approx 0.8\tau$ for a hydrocarbon–air mixture, which also varies with the equivalence ratio. The heat release parameter, τ, is defined in Eq. (3.44). The subgrid kinetic energy and its dissipation rate are denoted by $\widetilde{k}_{sgs}$ and $\widetilde{\epsilon}_{sgs}$ respectively. The other parameters related to the turbulence–scalar interaction are given by

$$C_3 = \frac{1.2\sqrt{Ka_\Delta}}{1 + \sqrt{Ka_\Delta}} \quad \text{and} \quad C_4 = \frac{1.1}{(1 + Ka_\Delta)^{0.4}}, \qquad (3.62)$$

where $Ka_\Delta = \sqrt{\widetilde{\epsilon}_{sgs}\delta_{th}/S_L^3}$. The parameter β_c typically varies in the range $2.5 \le \beta_c \le 7.5$ depending on the value of τ, which is related to the equivalence ratio of the fuel–air mixture. One can also evaluate this parameter dynamically and typical values of β_c were observed to be in the aforementioned range [63, 150]. It is worth noting that the aforementioned parameters are derived using DNS data for the thermochemical processes and their interaction with turbulence in premixed combustion, implying that the aforementioned model for the scalar dissipation rate lacks any tuning constant.

The subgrid variance is obtained by solving its transport equation:

$$\overline{\rho}\frac{D\sigma^2_{c,sgs}}{Dt} \approx \nabla \cdot \left[\left(\overline{\rho\mathcal{D}_c} + \frac{\overline{\rho}\nu_t}{Sc_t}\right)\nabla\sigma^2_{c,sgs}\right] + 2\overline{\rho}\frac{\nu_t}{Sc_t}(\nabla\widetilde{c}\cdot\nabla\widetilde{c})$$
$$- 2\overline{\rho}\,\widetilde{\epsilon}_c + 2\left(\overline{c\,\dot{\omega}_c} - \widetilde{c}\,\overline{\dot{\omega}_c}\right). \qquad (3.63)$$

The last term of the above transport equation is closed using

$$\overline{c\,\dot{\omega}_c} = \int_0^1 \zeta\left(\overline{\dot{\omega}_c|\zeta}\right)\widetilde{P}\,d\zeta. \qquad (3.64)$$

For RANS calculations, $\mathcal{F} = 1$, all the subgrid quantities must be replaced with the corresponding RANS quantities and the symbol Ka will be the Karlovitz number. A value of $\beta_c = 6.7$ was observed to work well for RANS in many past studies [151–157]. The variance required should be obtained using its transport equation, which is similar to the one given earlier for the subgrid variance.

Finally, the mass-weighted subgrid PDF, $\widetilde{P}$, is obtained using the beta function (see Eq. 3.65) with $\widetilde{c}$ and $\sigma^2_{c,sgs}$ calculated at every grid cell through their respective transport equations. The potential numerical issues that may arise while computing the integrals involving this PDF can be avoided using the integration by parts that will lead to the use of the CDF (cumulative distribution function) as discussed in [64]. The potentials of LES-CMC are demonstrated for premixed combustion only in a couple of recent studies [135, 136] and more work is required.

3.4.4 Tabulated Chemistry Approach

One can approximate $\dot{\omega}_c(\zeta)$ appearing in Eq. (3.41) using the results of either unstrained or strained laminar premixed flames. If one uses a strained premixed flame then Eq. (3.41) will involve a double integral to include the effects of all possible values of strain rate (typically going from zero to the extinction strain rate) and the marginal PDF should be replaced with a joint PDF of c and strain rate. This approach was originally proposed by Bradley [158, 159] for RANS. For the sake of simplicity let us ignore the stretch effects for the time being and the marginal PDF of c can be then be obtained using the beta function:

$$\widetilde{P}\left(\zeta;\widetilde{c},\sigma_c^2\right) = \frac{\zeta^{a-1}(1-\zeta)^{b-1}}{\Psi(a,b)} \quad \text{with} \quad \Psi(a,b) = \int_0^1 \zeta^{a-1}(1-\zeta)^{b-1}\,d\zeta,$$

$$\text{where} \quad a = \widetilde{c}\left[\frac{\widetilde{c}(1-\widetilde{c})}{\sigma_c^2} - 1\right] \quad \text{and} \quad b = \frac{(1-\widetilde{c})\,a}{\widetilde{c}}. \tag{3.65}$$

Thus, the averaged reaction rate can be computed a priori and tabulated as function of $\widetilde{c}$ and σ_c^2, which can be looked up during the RANS calculations. Hence, this approach is called a tabulated chemistry approach (TCA). The so-called FGM (flame generated manifold) approach follows these steps using the unstrained freely propagating laminar premixed flame solution for $\dot{\omega}_c(\zeta)$ and defining the progress variable as a linear combination of a few species mass fractions [160]. The TCA assumes that both the large and small turbulence time scales are larger than the chemical timescale, which implies that Da ≫ 1 and Ka ≪ 1, suggesting that the internal structures of the laminar flame are not affected by the turbulent eddies. This condition strictly holds in the wrinkled and corrugated flamelets regimes of turbulent premixed combustion, as noted in Chapter 2. However, DNS calculations of premixed combustion at u'/S_L values as high as 50 showed that the foregoing flamelet assumption holds even at such a large turbulence level [161]. However, the upper limit of the turbulence level for which this flamelet assumption breaks is still an open question.

The predictions using TCA were shown to improve when the stretch effects, parameterized and captured using scalar dissipation rate of the progress variable, were included for RANS calculation of rod-stabilized turbulent "Vee" flames and pilot-stabilized turbulent Bunsen flames [139, 162]. When the straining effects were excluded, the computed flame length was observed to be very much shorter than (nearly one half of) the measured length, suggesting that the fuel is consumed quickly (increased burning rate) in the piloted Bunsen flames [139, 162]. A similar approach was also attempted for filtered reaction rate but using the atomic hydrogen mass fraction to parametrize strained flamelets for building the lookup table [163] for LES of a premixed slot-jet flame simulated using DNS by Sankaran et al. [164]. A priori analysis of this strained flamelet closure for a filtered reaction was also conducted in a past study [165]. In a light of the multiscale analyses reported in [77, 78], it may not be necessary to model subgrid straining effects in an LES resolving 80% to 90% of the turbulent kinetic energy for the following reasons. LES resolves a wide range of turbulent time and length scales and the kinetic energy associated with those scales. Also, the dynamic evolution of and the mutual interactions among the range of turbulent scales including their influences on scalar fields are captured inherently in the LES. If the minimum LES grid cell size is only a few multiples of δ_{th}, as is commonly used for LES, then the residual kinetic energy in the unresolved subgrid scales is small (less than 20%) and these unresolved scales will not affect the flame. Indeed, Doan et al. [77] and Ahmed et al. [78] showed that turbulent eddies of size smaller than $3\delta_{th}$ to $5\delta_{th}$ contribute only less than 10% to the flame straining. Furthermore, LES of a piloted Bunsen flame using strained flamelets led to underprediction of fuel burn rate compared to unstrained flamelet closure [64]. The unstrained flamelet SGS closure has also been shown to work well for multiregime premixed combustion behind a bluff body [11] if one pays close attention to the various SGS closures and their physical and mathematical consistencies, which are discussed next.

3.5 Improvements for the Flamelets: *FlaRe* Approach

The classical way of using Da and Ka is not straightforward for LES, as these two parameters involve integral and Kolmogorov time scales that are unavailable during the LES. But these time scales can be extracted once the LES is finished. So, it is more useful to compare the chemical timescale with local turbulence time scales resolved in the LES to understand if the flamelet could prevail or not. If the chemical timescale at a point in the flow field, $(\mathbf{x}, t)$, is shorter than the resolved scales at that point in time and space then the flame is likely to survive there; otherwise it will extinguish or move to a point where it can survive. Thus, using the global Da and Ka as commonly used for RANS is not quite appropriate for LES and hence simply extending RANS ideas to LES may be inappropriate. Also, extending ideas and concepts developed for passive scalar transport in turbulence cannot be carried over for reactive scalars such as progress variables.

Let us consider Fig. 3.3 showing an arbitrary domain with flame front and flamebrush. This front, which is an instantaneous structure, is resolved only in DNS or in high-speed laser diagnostics and flamebrush is obtained by time-averaging the front over a sufficiently long time. RANS calculations yield flamebrush. The filtered flame computed in LES would be somewhere in between the front and flamebrush depending on the numerical grid used for LES, and when the filtered flame is averaged over sufficiently long time then one gets the flamebrush. The volume shown in Fig. 3.3 is not the numerical cell volume; rather it is arbitrary.

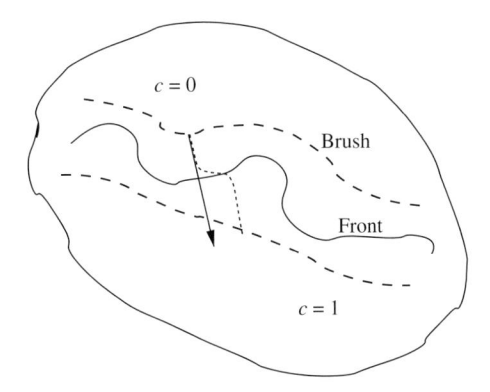

Figure 3.3 A schematic sketch showing flame and flame brush in an arbitrary region of turbulent premixed combustion.

The mixture inside the volume is uniform in space and time with $c = 0$ everywhere if there is no combustion, that is, the reaction rate is zero. Under this condition, there *cannot* be any gradient or fluctuation of c in the domain, even if there is turbulence. Once the combustion is initiated (or occurs) then c will vary from 0 to 1 across the front as shown in the figure, leading to spatial and temporal gradients and this will lead to nonzero $\nabla \tilde{c}$, where $\tilde{c}$ can be seen as Favre-averaged (this will vary over the flame-brush) or Favre-filtered (this will vary over the filtered flame) or instantaneous (varying across the flame front) progress variable. These nonzero gradients will lead to diffusive fluxes and variance of c. So, the reaction rate is the central quantity driving other processes subsequently in turbulent premixed combustion. *This means that the SGS variance and scalar dissipation rate must be modeled in a manner consistent with the reaction rate closure and one cannot, rather shouldn't, use models developed for passive scalars because the passive scalar gradients and variance are generated by turbulence (not combustion!).* It is quite common to see that this important physical aspect of premixed combustion is grossly ignored in the literature by using passive scalar models. This observation applies equally for partially premixed combustion as well. From this discussion, the following points can be noted.

1. *Calculation of filtered reaction rate:*
   Equation (3.41) is rewritten as

$$\overline{\dot{\omega}_c}(\mathbf{x}, t) = \overline{\rho} \int_0^1 \left(\frac{\dot{\omega}_c}{\rho}\right) \tilde{P}\left(\zeta; \tilde{c}, \sigma_{c,\text{sgs}}^2\right) d\zeta,$$

$$= \overline{\rho} \int_0^1 \dot{W}_c \, \tilde{P}\left(\zeta; \tilde{c}, \sigma_{c,\text{sgs}}^2\right) d\zeta = \overline{\rho} \widetilde{\dot{W}_c}. \qquad (3.66)$$

The mean or filtered density, $\overline{\rho}$, must come from the CFD solver and not from the flamelet library. If one uses $\dot{\omega}_c$ rather than $\dot{W}_c$ in the in Eq. (3.66) integral then the flamelet density is used, implying that the left- and right-hand sides of the filtered progress variable equation

$$\overline{\rho}\frac{D\widetilde{c}}{Dt} - \nabla \cdot (\overline{\rho\mathcal{D}\nabla c}) + \nabla \cdot (\overline{\rho\mathbf{u}c} - \overline{\rho}\widetilde{\mathbf{u}}\widetilde{c}) = \overline{\rho}\widetilde{W}_c \qquad (3.67)$$

will have different (inconsistent) densities. This means that the lookup table should supply $\widetilde{W}_c$, an inverse of reaction timescale, *not* the rate of mass or molar production per unit volume for calculations using RANS or LES paradigms. Another subtle, but important, point to note is the definition of c. It is quite common to see in the literature that this variable is defined using a linear combination of species such as CO, CO_2, H_2, etc., with widely differing mass diffusivities (or Lewis number). This yields additional diffusive terms for c, which are typically neglected by ignoring the fact that the molecular diffusion and chemical reactions are closely coupled in premixed (also partially premixed) systems. This is specifically so for LES because combustion is a SGS phenomenon where molecular fluxes cannot be ignored (which is quite commonly done for RANS). Hence, it is recommended to define c using species having similar diffusivities or Lewis numbers.

2. *Matters related to the SGS variance, $\sigma_{c,\text{sgs}}^2$:*

The RANS variance, σ_c^2, is related to the SGS variance, $\sigma_{c,\text{sgs}}^2$, through

$$\sigma_c^2 = \sigma_{c,\text{res}}^2 + \langle \sigma_{c,\text{sgs}}^2 \rangle = \langle (\widetilde{c} - \langle \widetilde{c} \rangle)^2 \rangle + \langle \sigma_{c,\text{sgs}}^2 \rangle, \qquad (3.68)$$

where $\langle \cdot \rangle$ denotes Favre averaging with respect to time; the resolved part, $\sigma_{c,\text{res}}^2$, is obtained using the computed filtered progress variable and the subgrid variance is either modeled algebraically or transported using Eq. (3.63). The algebraic model is (see the discussion below Eq. 3.2)

$$\sigma_{c,\text{sgs,model}}^2 \simeq A_\sigma \Delta^2 (\nabla \widetilde{c} \cdot \nabla \widetilde{c}), \qquad (3.69)$$

with A_σ typically taken to be 0.5 and it can also be evaluated dynamically. This model was originally developed for a mixture fraction [166] but used commonly for progress variables also. The relevance and validity of this model for progress variable variance are discussed in the text that follows. The transport equation for $\sigma_{c,\text{sgs}}^2$, Eq. (3.63), is rewritten as

$$\underbrace{\frac{\partial \overline{\rho}\,\sigma_{c,\text{sgs}}^2}{\partial t}}_{T_0} + \underbrace{\nabla \cdot \left(\overline{\rho}\widetilde{\mathbf{U}}\,\sigma_{c,\text{sgs}}^2\right)}_{T_1} \approx \underbrace{\nabla \cdot \left[\left(\overline{\rho\mathcal{D}_c} + \frac{\overline{\rho}\,\nu_t}{Sc_t}\right)\nabla \sigma_{c,\text{sgs}}^2\right]}_{T_{21}+T_{22}}$$
$$+ \underbrace{2\overline{\rho}\frac{\nu_t}{Sc_t}(\nabla \widetilde{c} \cdot \nabla \widetilde{c})}_{T_3} \underbrace{- 2\overline{\rho}\,\widetilde{\epsilon}_c}_{T_4} + \underbrace{2\overline{\rho}\left(\widetilde{W_c c} - \widetilde{W}_c \widetilde{c}\right)}_{T_5}. \qquad (3.70)$$

The order of magnitude analysis of this equation performed in [64] gives

$$T_0 \sim \mathcal{O}\left(\frac{\rho_u S_L}{\delta_{th}}; \frac{U_{\text{ref}}}{u_\Delta'}\frac{1}{\text{Da}_\Delta}\right) \quad T_1 \sim \mathcal{O}\left(\frac{\rho_u S_L}{\delta_{th}}; \frac{U_{\text{ref}}}{u_\Delta'}\frac{1}{\text{Da}_\Delta}\right)$$

$$T_{21} \sim \mathcal{O}\left(\frac{\rho_u S_L}{\delta_{th}}; \frac{1}{\text{Da}_\Delta \text{Re}_\Delta}\right) \quad T_{22} \sim \mathcal{O}\left(\frac{\rho_u S_L}{\delta_{th}}; \frac{1}{\text{Da}_\Delta}\right)$$

$$T_3 \sim \mathcal{O}\left(\frac{\rho_u S_L}{\delta_{th}}; \frac{1}{\text{Da}_\Delta}\right) \quad T_4 \sim \mathcal{O}\left(\frac{\rho_u S_L}{\delta_{th}}; 1\right) \quad T_5 \sim \mathcal{O}\left(\frac{\rho_u S_L}{\delta_{th}}; 1\right),$$

where $\text{Re}_\Delta = u'_\Delta \Delta/(S_L \delta_{th})$ and $\text{Da}_\Delta = t_{sgs}/t_c = \Delta S_L/(\delta_{th} u'_\Delta)$. The SGS velocity scale is u'_Δ. In the foregoing scaling analysis, the spatial derivatives of filtered quantities, time derivatives, and density are scaled using Δ, Δ/U_{ref}, and ρ_u, respectively. The filtered velocity is scaled using a reference velocity U_{ref}. The molecular diffusivity is scaled using $S_L \delta_{th}$, and the SGS viscosity is scaled as $u'_\Delta \Delta$. The progress variable gradient is produced predominantly by the chemical reaction and thus the appropriate scaling for $\widetilde{\epsilon}_c$ is S_L/δ_{th}. From these scaling relations one concludes that

1. The dissipation and reaction terms are leading order and the dissipation rate model must include contributions from chemical reactions.
2. The subgrid parameters Da_Δ and Re_Δ are likely to be finite and thus the other terms of the $\sigma^2_{c,\text{sgs}}$ transport equation cannot be neglected.
3. If one scales $\widetilde{\epsilon}_c$ using (u'_Δ/Δ) (which means that the scalar gradient and thus the dissipation rate are produced by turbulence) then

$$T_4 \sim \mathcal{O}\left(\frac{\rho_u S_L}{\delta_{th}}; \frac{1}{\text{Da}_\Delta}\right),$$

which implies that T_3 and T_4 are of the same order. Since the dissipation rate is taken to come from turbulence, which is at odds with premixed combustion physics described earlier using Fig. 3.3, a linear relaxation model (see the discussion below Eq. 3.3) can be used for $\widetilde{\epsilon}_c$ and thus

$$\widetilde{\epsilon}_c \simeq \frac{\nu_t}{A_\sigma Sc_t \Delta^2} \sigma^2_{c,\text{sgs}}. \tag{3.71}$$

Using this model in $T_3 \simeq -T_4$, one can deduce Eq. (3.69). Hence, it is clear that the algebraic model for the variance and the linear relaxation model for the dissipation rate should not be used for premixed (also partially premixed) combustion calculations [64]. It is recommended that $\sigma^2_{c,\text{sgs}}$ is obtained using its transport equation, Eq. (3.63), and $\widetilde{\epsilon}_c$ is modeled by Eq. (3.61). This ensures that $\sigma^2_{c,\text{sgs}}$ and $\widetilde{\epsilon}_c$ are modeled consistently with the $\overline{\dot{\omega}}_c$ model and premixed combustion physics. This is the reason for calling the flamelet approach described earlier as Flamelets Revised for consistencies – FlaRe – approach.

These conclusions also hold for RANS, particularly for the terms T_3 to T_5 [167]. The balance of the various terms in the SGS variance equation has been explored using LES results for a dual-swirl GT model burner from DLR and that analysis concurs the aforementioned conclusions [168]. The FlaRe approach has been tested widely using RANS and LES paradigms for premixed and partially premixed combustion, and these studies are reviewed in [142, 147].

Lifted jet turbulent flames are known to be challenging for combustion modeling, specifically the transient evolution of the initial spark in the stratified fuel-air mixture toward the final lift-off height of the flame. The fully established flame can be seen to be a partially premixed flame at the flame base with non-premixed combustion in downstream locations but the initial ignition kernel placed at a downstream location evolves in stratified mixtures. The flame established by the ignition kernel is expected to propagate upstream along the stoichiometric mixture fraction contour toward its final lift-off height. Hence, the set of modeling equations should include $\widetilde{Z}$ and $\sigma_{z,\text{sgs}}^2$ equations for the mixture fraction field.

This case is considered here to highlight the importance of maintaining the modeling and physical consistencies discussed earlier, specifically for the SGS scalar dissipation rate. These lifted flames are investigated experimentally in [169] and numerically using LES-FlaRe in [146, 168] and the LES-Thickened flame model in [170]. The experimental study considered air-diluted (30% air by volume) methane jet having a diameter of $d_j = 5$ mm issuing into ambient air at two bulk mean jet velocities of $U_b = 12.5$ and 25.5 m/s. The ambient disturbance on the jet and flame evolution was avoided by having a coaxial airflow at 0.1 m/s through an annulus having an outer diameter of 200 mm. The initial spark was placed at different downstream locations to study its transient evolution toward the final lift-off location using high-speed movies, which were then used to extract the required information. Further details can be found in the experimental study reported in [169].

Figure 3.4 shows the computed spatial variation of filtered stoichiometric mixture fraction and reaction rate of $\widetilde{c}$ contours, where the mixture fraction is defined using the Bilger mixture fraction and c is the normalized progress variable defined using ($Y_{CO2} + Y_{CO}$). The results obtained using the linear relaxation model in Eq. (3.71) with $(A_\sigma Sc_t) = 0.5$ and SDR model in Eq. (3.61) with $\beta_c = 7.5$ for the subgrid dissipation rate $\widetilde{\epsilon}_c$ are shown for two instants, 1 and 5 ms. These are the elapsed times after initiating the spark. Both cases had the same size initial kernel with the same amount of spark energy as in the experiment. There is almost no difference except for the reduced reaction rate for the linear relaxation model at $t = 1$ ms. The reaction rate becomes weaker and weaker as time goes by (see the contours shown for 5 ms) for this model and this does not help to establish a flame that could propagate toward the final lift-off height. This is clear from the results shown for 5 ms, which also shows the difference in the Z_{st} contour where the kernel is evolving. This difference comes from the flame-induced effects on the flow and turbulent mixing process, which are captured inherently by the LES-FlaRe with the dissipation rate mode in Eq. (3.61).

The variation of kernel diameter during its initial evolution period is shown in Fig. 3.5. This is the diameter of an equivalent sphere having the same volume occupied by the evolving kernel at a given time. The kernel is identified using $\widetilde{c} \geq 0.001$. The difference in d_{kernel} predicted using both dissipation rate models is negligible up to 2 ms and they agree well with measured data available for about 1.3 ms. The diameter starts to grow as t^2 when Eq. (3.61), is used but it grows linearly for the linear relaxation model. By about 7 ms, the filtered reaction rate becomes negligible for the linear relaxation model compared to that predicted

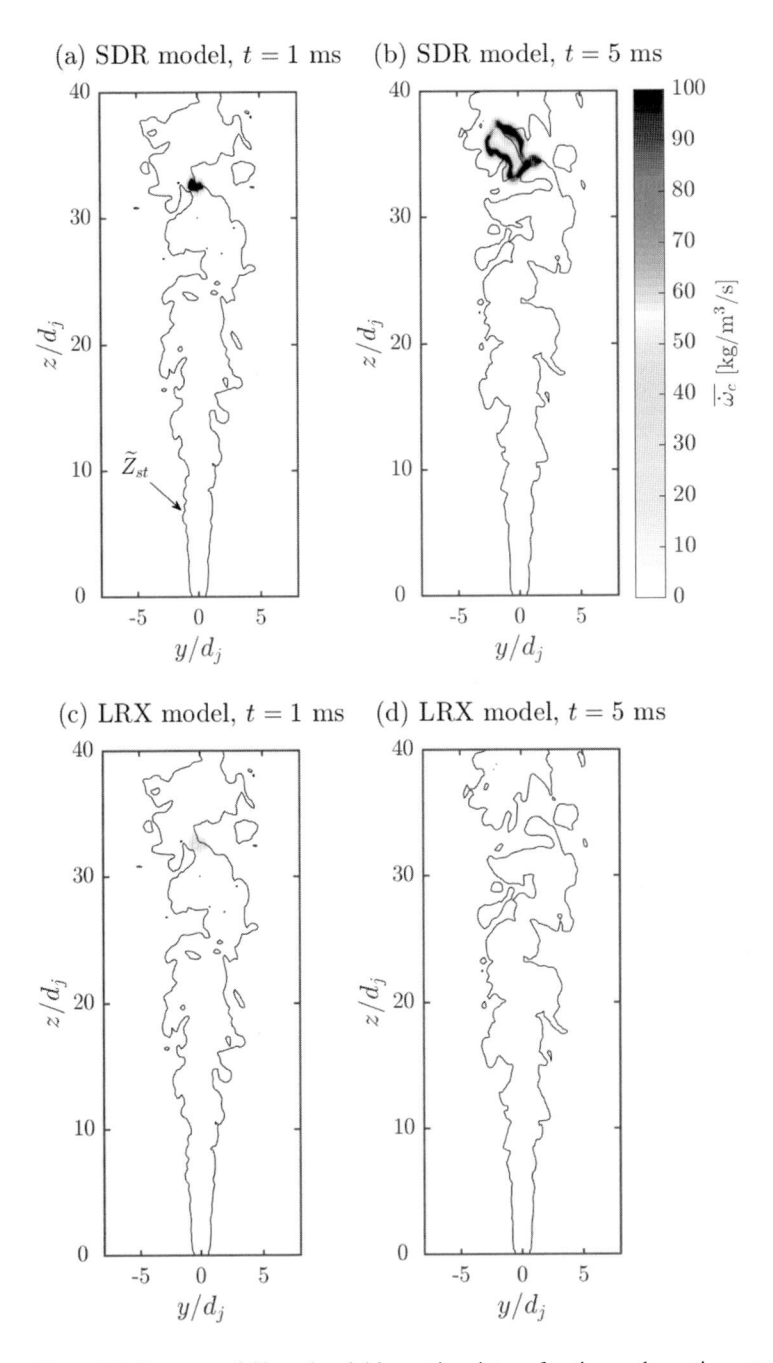

Figure 3.4 Contours of filtered stoichiometric mixture fraction and reaction rate of $\tilde{c}$ at 1 and 5 ms from sparking time. The LES results are shown for two scalar dissipation rate models, SDR in Eqs. (3.61) and linear relaxation (LRX) in Eq. (3.71). This is for the case with $U_b = 25.5$ m/s with initial kernel placed at $30 d_j$.

using Eq. (3.61), as depicted by the reaction rate contours shown in Fig. 3.5. The thickened flame model over predicts the kernel diameter and the growth rate is substantially higher as seen in the figure and this result is taken from [170].

The axial position of the most upstream point of the flame leading edge identified using $\widetilde{T} = 1200$ K is compared to the measurements in Fig. 3.6 for the two flames. The result for the linear relaxation model does not appear in the figure, since the ignition was unsuccessful as depicted in Figs. 3.4 and 3.5. Hence, it is evident that the subgrid closures should maintain physical and mathematical consistencies.

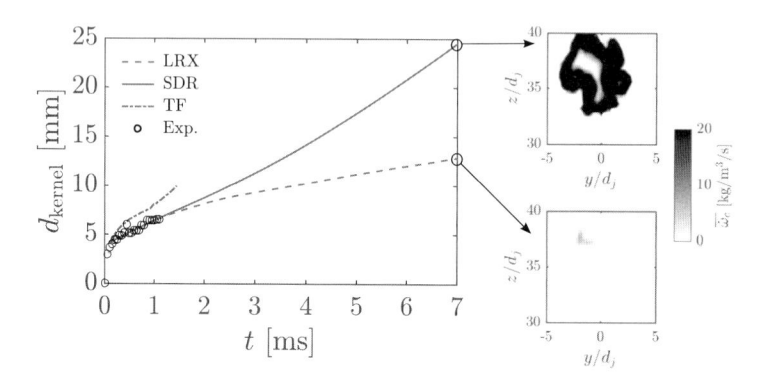

Figure 3.5 Temporal evolution of ignition kernel diameter computed in LES using SDR in Eq. (3.61) and linear relaxation model in Eq. (3.71) for $\widetilde{\varepsilon}_c$. The thickened flame model result from [170] is shown for comparison. The symbols are the experimental data from [169].

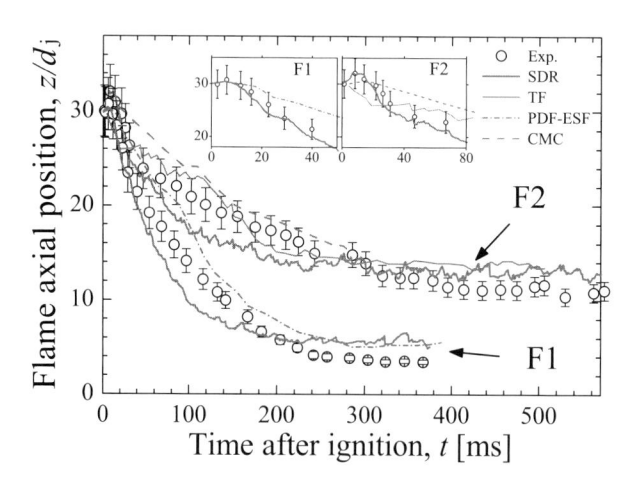

Figure 3.6 Transient evolution of measured (symbols) and computed (lines) flame leading point. The error bars corresponds to the 9% maximum scatter of the experimental data reported in [169]. The flames F1 and F2 have $U_b = 12.5$ and 25.5 m/s respectively with initial sparking at $z = 30d_j$. The thickened flame model result is from [170], the Eulerian stochastic field result is from [122], and the CMC result is from [171].

Nevertheless, one can make the linear relaxation model yield a successful ignition by choosing another value for its parameter (A_σ Sc_t) but that would not be helpful in the view of having a "truly predictive" approach.

The detail of the LES-FlaRe can be found in [146, 168] and these studies demonstrated that the flame leading edge traveled along a spiral path toward their final lift-off height with three distinct stages of evolution; (1) kernel expansion during $0 \leq t \leq 5$ ms (approximately), (2) freely propagating, and (3) stabilizing flame, which were also observed in [170]. The durations of the latter two stages depend on the value of U_b. The thickened flame model result from [170] for the flame F2 and results obtained using the Eulerian stochastic field method in [122] for the flame F1 are shown for comparisons. A temperature value of 1500 K was used for the thickened flame model to detect the flame edge and the difference between 1200 and 1500 K is about one jet diameter. Although the evolution of the flame leading edge shows some dependence on the SGS combustion model the final lift-off heights predicted by these models are very similar. The computational costs, however, for the LES using these three SGS combustion closures differ substantially and the flamelet-based approach (LES-FlaRe) is the most economical one.

3.6 Turbulent Mixing in Premixed Combustion Revisited: Some Fundamental Considerations

It is apparent from the previous sections in this chapter that the physical understanding of the time evolution of spatial scalar heterogeneities is essential to model turbulent mixing and chemical conversion in premixed combustion systems. This is tantamount to comprehending the temporal evolution of scalar gradients of species and temperature. Some simple notions, which might help to grasp and illustrate concepts of mixing-reacting systems, are discussed next.

3.6.1 Characteristic Mixing Times

The transport equation for reaction progress variable, $c(\mathbf{x}, t)$, obeys the conservation equation,

$$\underbrace{\frac{\partial c}{\partial t}}_{c_0/\tau} + \underbrace{u_j \frac{\partial c}{\partial x_j}}_{c_0/\tau_{\text{conv}}} = \underbrace{\frac{1}{\rho} \frac{\partial}{\partial x_j} \left(\rho D \frac{\partial c}{\partial x_j} \right)}_{c_0/\tau_{\text{diff}}} + \underbrace{\frac{\dot{\omega}_c}{\rho}}_{c_0/\tau_c}. \tag{3.72}$$

The orders of magnitude appear underneath every term. The characteristic value of c at a given point and time is c_0, and τ is the local and instantaneous characteristic time for the unsteady term. The convective timescale for an eddy of characteristic size l

and velocity scale u' is $\tau_{conv} = l/u'$. The characteristic diffusive time of a small-scale structure of thickness δ for scalar spatial variation is $\tau_{diff} = \delta^2/\mathcal{D}$, where $\mathcal{D}$ is the molecular diffusivity. The characteristic chemical timescale is τ_c. Different realizations of the turbulent flow will display different characteristic times or, equivalently, the various time scales will have statistical distributions varying in space and time and for different flows and scalar thermochemical systems. In some large-scale structures $\tau \sim \tau_{conv}$, and diffusive and chemical effects may be negligible, with the scalar field being merely convected, whereas for some small-scale structures $c(\mathbf{x}, t)$ may be solely mixed by diffusion at the molecular level and $\tau \sim \tau_{diff}$. Small-scale flow topologiess can also interact with small-scale scalar structures, where $\tau_{conv} \sim \tau_{diff}$. The local chemical term will be important in regions where $\tau_{diff} \sim \tau_c$. The overlapping ranges of the various characteristic times will lead to strong interactions between pairwise mechanisms of convection, molecular diffusion, and chemical conversion.

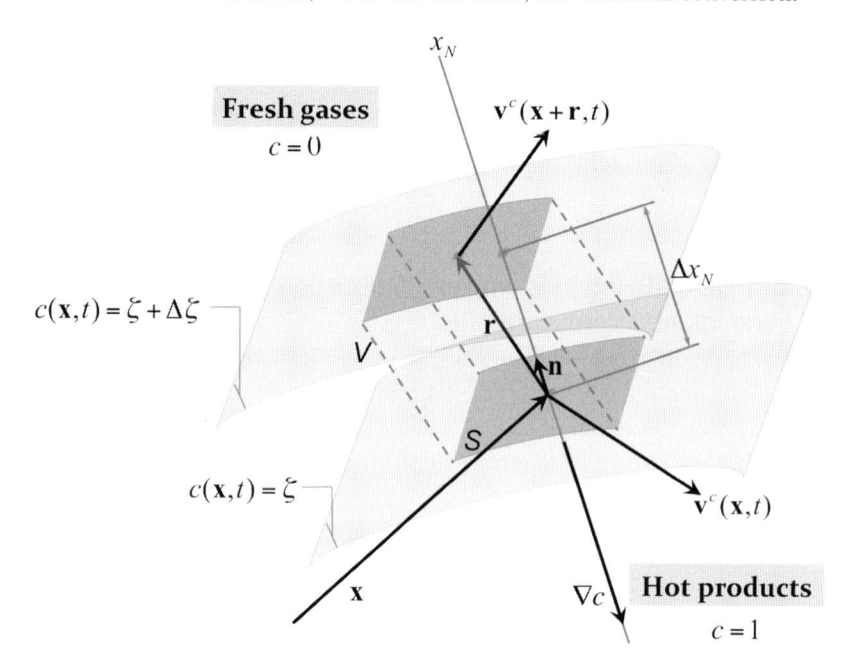

Figure 3.7 Nonmaterial line, surface, and volume elements.

Scalar mixing is tantamount to smooth concentration gradients, which can be addressed by investigating the evolution of isoscalar surfaces. Consider an infinitesimal volume V of a premixed flame bounded by two adjacent isoscalar surfaces $c(\mathbf{x}, t) = \zeta$ and $\zeta + \Delta\zeta$; see Fig. 3.7. Dopazo et al. [172] showed that the relative rate of change of the infinitesimal normal distance, r (for $\mathbf{r} = r\mathbf{n}$) between the two isosurfaces and of an area element S on the isosurface ζ and the volume $V = Sr$ are given by

$$\frac{1}{r}\frac{dr}{dt} = n_i S_{ij} n_j + \mathbf{n} \cdot \nabla S_d = a_N + a_N^a, \tag{3.73}$$

$$\frac{1}{S}\frac{dS}{dt} = \left(\delta_{ij} - n_i n_j\right) S_{ij} + S_d \nabla \cdot \mathbf{n} = a_T + a_T^a \qquad (3.74)$$

$$\frac{1}{V}\frac{dV}{dt} = \nabla \cdot \mathbf{u} + \mathbf{n} \cdot \nabla S_d + S_d \nabla \cdot \mathbf{n}, \qquad (3.75)$$

respectively. The unit vector normal to $c(\mathbf{x}, t) = \zeta$ is $\mathbf{n} = -\nabla c/|\nabla c|$ and the flow strain rate tensor is $S_{ij} = 0.5(\partial u_i/\partial x_j + \partial u_j/\partial x_i)$. The displacement speed of a point $\mathbf{x}$ on the isosurface ζ at time t relative to the fluid is $S_d \mathbf{n} = \mathbf{v}^c - \mathbf{u}$, where $\mathbf{v}^c$ is the absolute velocity of the point and $\mathbf{u}$ is the local flow velocity.

The normal and tangential strain rates induced by the flow, defined in Eqs. (3.73) and (3.74), satisfy the constraint $\nabla \cdot \mathbf{u} = a_N + a_T$. For an area element S located on the x-y plane, $\mathbf{n} = (0, 0, 1)$, $a_N = S_{33} = \partial w/\partial z$, and $a_T = S_{11} + S_{22} = \partial u/\partial x + \partial v/\partial y$. Thus, a_T is named as the area stretch rate induced by the fluid flow and a_N is the flow line stretch rate. The other two terms a_N^a and a_T^a defined in Eqs. (3.73) and (3.74) are additional normal and tangential strain rates induced by thermochemical processes. It is quite clear from their definitions that this additional normal strain rate is related to the variation of S_d along $\mathbf{n}$ and a_T^a is related to the curvature ($\kappa = \nabla \cdot \mathbf{n} = \partial n_k/\partial x_k$). These additional strain rates are also relevant for FSD evolution (see Eqs. 3.18 and 3.19). The displacement speed defined in Eq. (3.10) can be recast into a normal and tangential parts, as written in Eq. (3.22),

$$S_d = S_{d,N} - \mathcal{D}\kappa. \qquad (3.76)$$

A steady one-dimensional laminar premixed flame (see Fig. 3.1) is a good example to illustrate the meaning behind the previous concepts for planar and parallel isoscalar surfaces. An isosurface $c(x, t) = \zeta$, fixed in space, moves with an absolute speed $-u(x)\mathbf{n} + S_d(x)\mathbf{n} = 0$ in the laminar flame and hence $u(x) = S_d(x)$. The absolute velocity of an adjacent isosurface of $c(x + \Delta x) = \zeta + \Delta \zeta$ with $\Delta \zeta > 0$, also vanishes since $-u(x + \Delta x)\mathbf{n} + S_d(x + \Delta x)\mathbf{n} = 0$, which can be recast as $-(u(x) + \Delta x du/dx)\mathbf{n} + (S_d(x) + \Delta x dS_d/dx)\mathbf{n} = 0$. The latter implies $du/dx = dS_d/dx$. As the density decreases across the flame the flow velocity $u(x)$ must increase downstream, that is, $du/dx > 0$ and therefore $dS_d/dx > 0$. This implies that the isosurface displacement speed relative to the fluid must increase downstream to have a steady planar laminar premixed flame. For this simple example, $a_N(x) = du/dx$ is the flow normal strain rate and $a_N^a(x) = dS_d/dx_N = -dS_d/dx$ is the additional normal strain rate. For this planar flame, $\nabla \cdot \mathbf{u} = du/dx$ and $a_T(x) = 0$, $\kappa = 0$ and $a_T^a(x) = 0$. Thus, the distance r between two adjacent isoscalar surfaces remains constant since $d \ln r/dt = 0$ (no stretch rate). Similarly, S and V also remain constant for this simple example. Thus, the propagation velocity S_d results from a balance between the molecular diffusion normal to the flame and chemical conversion and hence $S_d = S_{d,N}$.

The elemental mass m inside the volume V is $m = \rho V$. The relative rate of change of this mass is given by

$$\frac{1}{m}\frac{dm}{dt} = \frac{1}{\rho}\frac{d\rho}{dt} + \frac{1}{V}\frac{dV}{dt}$$
$$= -\nabla \cdot \mathbf{u} + \nabla \cdot \mathbf{u} + \mathbf{n} \cdot \nabla S_d + S_d \kappa$$
$$= \mathbf{n} \cdot \nabla S_d + S_d \kappa = a_N^a + a_T^a \qquad (3.77)$$

after using the overall mass conservation equation and Eq. (3.75). Equation (3.77) yields the mass entrainment rate per unit mass into the nonmaterial volume V due to the additional normal and tangential strain rates. This entrained mass can reduce the thermal energy per unit mass inside the volume V, which can be expressed as

$$\frac{d\,c_p T}{dt} = \frac{1}{m}\frac{dmc_p T}{dt} - \frac{c_p T}{m}\frac{dm}{dt}.$$

The last term in the previous expression represents the dilution of thermal energy by the entrained mass and for large values of dilution, the thermal energy per unit mass $c_p T$ in volume V may be reduced significantly.

The scalar gradient magnitude can be approximated as $|\nabla c| \approx \Delta \zeta / r$, which can be differentiated for constant $\Delta \zeta$ to get

$$-\frac{1}{|\nabla c|}\frac{d|\nabla c|}{dt} = \frac{1}{r}\frac{dr}{dt} = a_N + a_N^a. \qquad (3.78)$$

Swaminathan and Grout [173] and Chakraborty and Swaminathan [174] showed that a scalar gradient is produced when the gradient vector aligns with the most compressive principal strain rate in low Da flames or the gradient is destroyed when the vector aligns with the most extensive principal strain rate in high Da flames. The authors argue that flow dilatation pulls the isoscalar surfaces from each other, dissipating scalar gradients in premixed flames when Da is large; this process is counteracted by chemical reactions, which produce scalar gradients. Equation (3.78) provides a clear explanation to the previous findings. It has been shown that $a_N^a < 0$ in two turbulent premixed flames (with Da equal to 11.48 and 1.16) due to the dominant contribution of the reaction [172, 175]. While $a_N > 0$, in the two flames, $a_N + a_N^a < 0$, with the higher Da case displaying larger values of the total normal strain rate. Recent experimental studies investigating these alignments are discussed in Section 2.3.

Equation (3.78) is a candidate to be considered as a mixing rate, the inverse of a characteristic mixing time, namely [175],

$$\underbrace{-\frac{1}{|\nabla c|}\frac{d|\nabla c|}{dt}}_{\frac{1}{\tau_{\text{mix}}}} = \underbrace{a_N}_{\frac{1}{\tau_\eta}} + \underbrace{\mathbf{n} \cdot \nabla \left[\frac{\mathbf{n} \cdot \nabla (\rho \mathcal{D} \mathbf{n} \cdot \nabla c)}{\rho |\nabla c|}\right]}_{a_N^{\text{nd}} = \frac{1}{\tau_{\text{nd}}} = \frac{\mathcal{D}}{\delta_{\text{nd}}^2}}$$
$$\underbrace{-\mathbf{n} \cdot \nabla (\mathcal{D} \kappa)}_{a_N^{\text{td}} = \frac{1}{\tau_{\text{td}}} = \frac{\mathcal{D}}{\delta_{\text{td}}^2}} + \underbrace{\mathbf{n} \cdot \nabla \left[\frac{\dot{\omega}_c}{\rho |\nabla c|}\right]}_{a_N^{\text{ch}} = \frac{1}{\tau_c}}. \qquad (3.79)$$

The ratio of the first to the last terms on the right-hand side of Eq. (3.79) is of the order of the Karlovitz number, $\mathrm{Ka} = \tau_c/\tau_\eta$, ratio of the characteristic chemical time τ_c to the Kolmogorov timescale, τ_η. Two distinguished limits are of interest here:

1. For $\mathrm{Ka} \ll 1$ the chemical reaction is fast compared to the small-scale turbulence characteristic timescale. Thus, the chemical reaction will dominate over the small-scale flow phenomena. For this situation, the mixing time will be determined by

$$\frac{1}{\tau_{\mathrm{mix}}} = \frac{1}{\tau_{\mathrm{nd}}} + \frac{1}{\tau_{\mathrm{td}}} + \frac{1}{\tau_c}. \quad (3.80)$$

The chemical reaction rate is balanced by the rate of molecular diffusion in *reaction zone* as sketched in Fig. 3.1. This zone will have a thickness equal to $\delta_{\mathrm{nd}} \sim \sqrt{\mathcal{D}\tau_c}$ if the normal diffusion dominates over the tangential diffusion. The thickness of this zone will be $\delta_{\mathrm{td}}^2 \sim \delta_{\mathrm{nd}}/\kappa \sim \mathcal{D}\tau_c$ if tangential diffusion is larger than the normal diffusion. Hence, *the evolution of gradients is controlled by the chemical reaction rate and the mixing time is proportional to τ_c in this limit.*

2. For $\mathrm{Ka} \gg 1$ the chemical reaction is slow compared to the small-scale turbulence characteristic timescale. The chemical term is likely to be negligible and the mixing time is given by

$$\frac{1}{\tau_{\mathrm{mix}}} = \frac{1}{\tau_\eta} + \frac{1}{\tau_{\mathrm{nd}}} + \frac{1}{\tau_{\mathrm{td}}}. \quad (3.81)$$

A *preheat zone* will exist, whose thickness will be of the order $\delta_{\mathrm{nd}} \sim \mathrm{Sc}^{1/4}\eta_{co}$ if the normal diffusion dominates tangential diffusion and the Schmidt number is $\mathrm{Sc} = \nu/\mathcal{D}$ and $\eta_{co} = (\mathcal{D}^3/\widetilde{\varepsilon})^{1/4}$ is the Corrsin–Obukhov length scale [176, 177]. If the tangential diffusion dominate over the normal diffusion then $\delta_{\mathrm{nd}}/\kappa \sim \mathrm{Sc}^{1/2}\eta_{co}^2$. In this limit, the evolution of gradients is controlled by the turbulent and scalar small-scale structures and the mixing time is likely to be proportional to τ_η. Nevertheless, if a flame is present locally then the small-scale turbulence and scalar structures are likely to be influenced by the flame induced effects, specifically at the subgrid level for the LES paradigm.

The dissipation rate of scalar fluctuations due to molecular diffusion, $\epsilon_c = \mathcal{D}|\nabla c|^2$, obeys

$$\frac{1}{\epsilon_c}\frac{d\epsilon_c}{dt} = -2\left(a_N + a_N^a\right), \quad (3.82)$$

when $\mathcal{D}$ is taken to be constant. This equation also confirms that $\left(a_N + a_N^a\right)^{-1}$ is an appropriate characteristic mixing timescale. In summary, the picture emerging from this analysis is that the local Karlovitz number $\mathrm{Ka}(\mathbf{x},t) = \tau_c/\tau_\eta$ determines the mixing/reaction regime. If $\mathrm{Ka} \ll 1$, the characteristic mixing time will be proportional to τ_c in the burning zone of the flame. For $\mathrm{Ka} \gg 1$, the characteristic mixing time will be proportional to τ_η in the preheat zone of the flame, as well as in the fresh reactant and product regions. These remarks should be explored to model micro-mixing in any statistical approach. Taking $\tau_{\text{mix}} \sim \widetilde{k}/\widetilde{\varepsilon}$ seems a gross misrepresentation of the physics of mixing processes in premixed combustion.

3.6.2 Length Scales: Spectral Analysis of a Dynamically Passive Scalar Field Undergoing a Linear Chemical Reaction

Wavenumber spectral analyses of scalar fields have been used in classical theory of turbulent mixing to examine length scale information. Obukhov [176], Corrsin [177], Batchelor [178] and Batchelor et al. [179] characterized scalar spectra for Schmidt numbers less than unity, of order unity, and much larger than unity respectively. Corrsin [180, 181], O'Brien [182, 183] and Pao [184] investigated the effect of linear reaction rate on the spectra of scalar fields undergoing turbulent mixing.

Ulitsky and Collins [185] and Knaus and Pantano [186], among others, have conducted spectral investigations of velocity and scalar fields in premixed and non-premixed combusting flows. The DNS of turbulent flames of Martin and Candler [187] and Jaberi et al. [188] pointed out the essential role played by the pressure dilatation work $p(\nabla \cdot \mathbf{u})$. Kolla et al. [189] used DNS with a realistic finite-rate chemistry to investigate statistically planar premixed flames in a turbulent shear layer, proposed a new density-weighted method, and concluded that pressure dilatation generates turbulent kinetic energy (TKE) over all scales with a spike around scales of the order of the laminar flame thickness. Paes and Xuan [190] examined DNS databases of an n-heptane and air mixture in homogeneous isotropic turbulence representing combustion of heavy hydrocarbons under engine conditions and obtained TKE evolutions in physical and wavenumber spaces to identify effects of heat release on turbulence. The pressure dilatation work $p(\nabla \cdot \mathbf{u})$ dominated TKE budgets during strong heat release over a wide range of wavenumbers. The authors also distinguished viscous dissipation and viscous dilatation terms, whose addition is the correct real viscous dissipation rate of TKE; these terms became stronger with increasing heat release, but were small compared to pressure dilatation.

The TKE transfer spectrum moved energy from small to large scales during the strong heat release period, which may be indicative of an inverse cascade of TKE. Kolla et al. [191] used DNS databases for two different flames with different Re, Da, and Ka, to obtain velocity and scalar dissipation spectra, which deviated from those predicted for turbulent flow of a constant density fluid, and the scalar dissipation spectral data collapsed when normalized by the density-weighted averaged dissipation rates.

The issue of characteristic length scales in chemically reacting turbulent flows will be illustrated here using the seminal ideas introduced by O'Brien [182] and Corrsin [180] on the spectral behavior of a statistically isotropic and stationary scalar field $c(\mathbf{x}, t)$ convected by a solenoidal and stationary turbulence $\mathbf{u}(\mathbf{x}, t)$ and undergoing a linear chemical reaction. The pressure dilatation work is absent in this case and thus results should be interpreted with extreme caution. However, the discussion of the various length scales and spectral limits might be of interest. The scalar $c(\mathbf{x},t)$, not necessarily the reaction progress variable, is governed by

$$\frac{\partial c}{\partial t} + u_j \frac{\partial c}{\partial x_j} = \mathcal{D}\frac{\partial^2 c}{\partial x_j \partial x_j} \pm \frac{c}{\tau_c}, \tag{3.83}$$

where the characteristic chemical reaction timescale is τ_c. The + or − sign corresponds to scalar generation or destruction of c by a chemical reaction.

The spatial autocorrelation of a statistically homogeneous scalar field is defined by

$$R_c(\mathbf{r},t) = \overline{[c(\mathbf{x},t) - \overline{c}(t)][c(\mathbf{x}+\mathbf{r},t) - \overline{c}(t)]}, \tag{3.84}$$

where $\overline{c}(t)$ is the scalar mean, which is a function of time only in homogeneous flows. The spatial Fourier transform of $R_c(\mathbf{r},t)$ is the spatial spectrum, given by

$$\Phi_c(\mathbf{k},t) = \frac{1}{(2\pi)^3} \iiint_{-\infty}^{+\infty} e^{-i\mathbf{k}\cdot\mathbf{r}} R_c(\mathbf{r},t)\, d\mathbf{r}, \tag{3.85}$$

where the wavenumber vector is $\mathbf{k}$. The inverse of $\Phi_c(\mathbf{k},t)$ is

$$R_c(\mathbf{r},t) = \iiint_{-\infty}^{+\infty} e^{i\mathbf{k}\cdot\mathbf{r}} \Phi_c(\mathbf{k},t)\, d\mathbf{k}. \tag{3.86}$$

The 3-D scalar spectrum $E_c(k,t)$ is the spectral density in wavenumber vectors with the same magnitude $k = |\mathbf{k}|$. The scalar spectrum $E_c(k,t)$ is obtained by integrating $\Phi_c(\mathbf{k},t)$ over a sphere of radius k, that is,

$$E_c(k,t) = \frac{1}{2} \oiint \Phi_c(\mathbf{k},t)\, d\sigma(k), \tag{3.87}$$

where $d\sigma(k)$ is an infinitesimal area element on that sphere.

One can easily obtain

$$\frac{\overline{[c(\mathbf{x},t) - \overline{c}(t)]^2}}{2} = \frac{R_c(0,t)}{2} = \frac{1}{2} \iiint_{-\infty}^{+\infty} \Phi_c(\mathbf{k},t)\, d\mathbf{k} = \int_0^\infty E_c(k,t)\, dk \tag{3.88}$$

and $E_c(k,t)$ obeys the equation [180]

$$\frac{\partial E_c(k,t)}{\partial t} = T(k,t) - 2\mathcal{D}k^2 E_c(k,t) \pm \frac{2\,E_c(k,t)}{\tau_c}. \tag{3.89}$$

The term $T(k,t)$ is the scalar convective spectral transfer from small wavenumbers (large-scale structures) to large wavenumbers (small scales). This spectral transfer can be expressed as $T(k,t) = -\partial[s(k,t)E_c]/\partial k$ [184], where sE_c is the scalar spectral flux across a wavenumber k. For stationary and isotropic velocity and scalar fields, the scalar injected at small wavenumbers and is transferred by $T(k,t)$ from small to large wavenumbers, where molecular diffusion and chemical reaction will act. A stationary scalar spectra $E_c(k)$ obeys

$$\frac{ds\,E_c}{dk} = -2\mathcal{D}k^2 E_c \pm \frac{2E_c}{\tau_c}. \tag{3.90}$$

For an *inert scalar* ($\tau_c \to \infty$) with $Sc = \nu/\mathcal{D}$ of order unity, there exists an inertial-convective range with negligible molecular diffusion and therefore one gets

$$\frac{ds E_c}{dk} = 0. \tag{3.91}$$

Relevant parameters in the inertial-convective range are the TKE dissipation rate ε, and the wavenumber k. From purely dimensional considerations, $s(k) = C_1 \varepsilon^{1/3} k^{5/3}$. Hence [192],

$$E_c(k) = K_1 \epsilon_c \varepsilon^{-1/3} k^{-5/3}. \tag{3.92}$$

The inertial-diffusive range obeys the equation

$$\frac{ds E_c}{dk} = -2\mathcal{D}k^2 E_c. \tag{3.93}$$

Once again, with $s(k) = C_2 \varepsilon^{1/3} k^{5/3}$, the solution is

$$E_c(k,t) = K_2 \epsilon_c \varepsilon^{-1/3} k^{-5/3} \exp\left[-\frac{3}{2C_2}\left(\frac{k}{k_{co}}\right)^{4/3}\right]. \tag{3.94}$$

The term $k_{co} = (\varepsilon/\mathcal{D}^3)^{1/4}$ is the Corrsin–Obukhov wavenumber. k_{co} is comparable to the Kolmogorov wavenumber $k_K = (\varepsilon/\nu^3)^{1/4}$ for $Sc \simeq 1$.

For a reacting scalar, the chemical timescale τ_c combined with ε or $\mathcal{D}$ introduces additional characteristic wavenumbers into the system. This creates two possible scenarios for illustrative purpose here.

Case I: $k_{ch} \leq k_{co}$

This limit implies that the reaction zone thickness is greater than or equal to Corrsin–Obukhov diffusive micro-scale $\eta_{co} = (\mathcal{D}^3/\varepsilon)^{1/4}$. A characteristic chemical wavenumber k_{ch} can be determined as $k_{ch} = (\tau_c^3 \varepsilon)^{-1/2}$. The length scale associated to k_{ch} is $\delta_{ch} = (\tau_c^3 \varepsilon)^{1/2} \sim \Lambda \mathrm{Da}^{-3/2}$, where the Damköhler number is $\mathrm{Da} = \tau_\Lambda/\tau_c = (\Lambda/u')/\tau_c$. The condition $k_{ch} \leq k_{co}$ also implies that $\mathrm{Ka}\,\mathrm{Sc}^{1/2} \geq 1$, where Ka is the Karlovitz number. The wavenumber k_{ch} will be located in the inertial range and hence, $s(k) = C\varepsilon^{1/3} k^{5/3}$. The general solution for $E_c(k)$ in dimensionless form is

$$\frac{E_c}{K \epsilon_c \varepsilon^{-1/3} k_{ch}^{-5/3}} = \left(\frac{k}{k_{ch}}\right)^{-5/3}$$

$$\times \exp\left[-\frac{3}{2C}\left(\frac{k}{k_{ch}}\right)^{4/3}\left(\frac{k_{ch}}{k_{co}}\right)^{4/3} \pm \frac{3}{C}\left(\frac{k}{k_{ch}}\right)^{-2/3}\right]. \tag{3.95}$$

The negative and positive signs correspond to scalar generation and consumption respectively. The ratio (k_{co}/k_{ch}) can be rewritten as $k_{co}/k_{ch} = \mathrm{Sc}^{3/4}\mathrm{Ka}^{3/2}$. Figure 3.8 depicts scalar spectra for generated and consumed scalars with $C = 1$, $\mathrm{Sc} = 1$ and four values of Ka. The chemical reaction perturbs the inertial range significantly for the consumed and moderately for the generated scalar. While chemical destruction of $c(\mathbf{x},t)$ implies a monotonic decrease of the spectral content with k, the generated scalar displays a range where $E_c(k)$ can grow, followed by a small range where $E_c(k) \propto k^{-5/3}$. Increasing Ka extends the range of large wavenumber as $\text{Ka} \propto \eta^{-1}$ for fixed τ_c, ν and ε [59].

The following ranges are implicit in the preceding solution: (1) inertial-convective for $k \ll k_{ch}$, (2) inertial-convective-reactive for $k \sim k_{ch}$, and (3) inertial-diffusive for $k \geq k_{ch}$.

Case II: $k_{ch} \gg k_{co}$

The smallest scalar length scales will be determined by τ_c and $\mathcal{D}$, and is $\delta_{ch} \sim (\mathcal{D}\tau_c)^{1/2}$. The characteristic chemical wavenumber is $k_{ch} \sim (\mathcal{D}\tau_c)^{-1/2}$. Now,

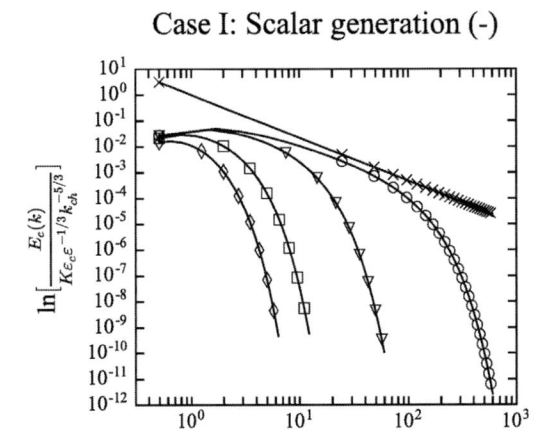

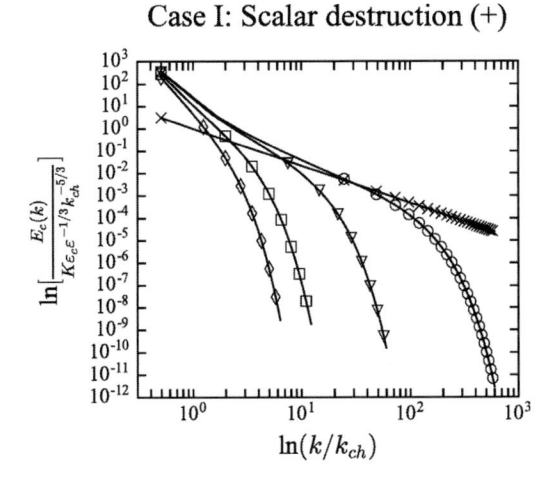

Figure 3.8 Spectra of statistically isotropic and stationary scalar fields, convected by a solenoidal and stationary turbulence, with a linear chemical reaction of generation or destruction. ∘∘ Ka = 21.54; ▽▽ Ka = 4.64; □□ Ka = 1.59; ◇◇ Ka = 1.0; ✕✕ −(5/3) $\ln(k/k_{ch})$.

the condition $k_{ch} \gg k_{co}$ implies that $\mathrm{Ka}\,\mathrm{Sc}^{1/2} \ll 1$. Here also, there will be an inertial-convective range for $k \ll k_{co}$ and an inertial-convective-diffusive range for $k \geq k_{co}$. Using $s(k) = C\varepsilon^{1/3}k^{5/3}$, these two ranges are included in the solution

$$E_c(k) = K\epsilon_c \varepsilon^{-1/3} k^{-5/3} \exp\left[-\frac{3}{2C}\left(\frac{k}{k_{co}}\right)^{4/3}\right]. \tag{3.96}$$

A viscous-diffusive range may appear for $k_{co} \sim k \ll k_{ch}$. The Kolmogorov timescale $\tau_\eta = (\nu/\varepsilon)^{1/2}$ and the wavenumber magnitude define $s(k) = Ck/\tau_\eta$, which leads to the solution

$$E_c(k) = K\epsilon_c \varepsilon^{-3/4} \nu^{3/4} \mathcal{D}^{1/2} \left(\frac{k}{k_B}\right)^{-1} \exp\left[-\frac{3}{2C}\left(\frac{k}{k_B}\right)^{4/3}\right], \tag{3.97}$$

where $k_B = (\varepsilon/\nu\mathcal{D}^2)^{1/4}$ is the Batchelor wavenumber corresponding to the Batchelor length scale $\delta_B \sim (\nu\mathcal{D}^2/\varepsilon)^{1/4}$. The k^{-1} range is similar to that unveiled by Batchelor [178] for $\mathrm{Sc} \gg 1$.

A diffusive-reactive range will occur for $k_{co} \ll k \sim k_{ch}$, with $s(k) = Ck/\tau_c$, yielding

$$E_c(k) = K\epsilon_c \mathcal{D}^{1/2} \tau_c^{3/2} \left(\frac{k}{k_{ch}}\right)^{\frac{\pm 2 - C}{C}} \exp\left[\frac{-1}{C}\left(\frac{k}{k_{ch}}\right)^2\right]. \tag{3.98}$$

When the scalar c is generated (+ sign in the exponent) by the linear chemical reaction one gets $E_c(k) \propto (k/k_{ch})\exp\left[-(k/k_{ch})^2\right]$ if $C = 1$. This suggests that the scalar spectral content will grow for a certain range of k around $k \sim k_{ch}$. On the other hand, if c is consumed by the reaction then $E_c(k) \propto (k/k_{ch})^{-3} \exp\left[-(k/k_{ch})^2\right]$ for $C = 1$.

Dimensionless spectra in the different wavenumber ranges are

$$\frac{E_c(k)}{K\epsilon_c \varepsilon^{-1/3} k_{co}^{-5/3}} = \left(\frac{k}{k_{co}}\right)^{-5/3} \exp\left[-\frac{3}{2C}\left(\frac{k}{k_{co}}\right)^{4/3}\right] \quad \text{for } k \leq k_{co}, \tag{3.99}$$

$$\frac{E_c(k)}{K\epsilon_c \varepsilon^{-1/3} k_{co}^{-5/3}} = \mathrm{Sc}^{1/2} \left(\frac{k}{k_{co}}\right)^{-1}$$

$$\exp\left[-\frac{3}{2C}\mathrm{Sc}^{1/4}\left(\frac{k}{k_{co}}\right)^{4/3}\right] \quad \text{for } k_{co} \sim k \ll k_{ch}, \tag{3.100}$$

$$\frac{E_c(k)}{K\epsilon_c \varepsilon^{-1/3} k_{co}^{-5/3}} = \mathrm{Ka}^{3/2}\mathrm{Sc}^{3/4} \left(\mathrm{Ka}^{1/2}\mathrm{Sc}^{1/4}\right)^{\frac{\pm 2 - C}{C}} \left(\frac{k}{k_{co}}\right)^{\frac{\pm 2 - C}{C}}$$

$$\exp\left[-\frac{1}{C}\mathrm{Ka}\,\mathrm{Sc}^{1/2}\left(\frac{k}{k_{co}}\right)^2\right] \quad \text{for } k_{co} \ll k \sim k_{ch}. \tag{3.101}$$

Premixed Combustion Modeling

The chemical reaction creates large gradients through the enhancement of pre-existing gradients and therefore extends the wavenumber range well beyond k_{co}. This could be considered in a wider sense as a chemical cascade transferring the spectral content of $\overline{c'^2} = \overline{[c(\mathbf{x},t) - \overline{c}(t)]^2}$ from $k \leq k_{co}$ to wavenumbers $k > k_{co}$.

Figure 3.9 shows the spectra in the three k ranges for $C = 1.0$, $Sc = 1.0$, and $Ka = 0.097$. The $k^{-5/3}$ spectrum is apparent. However, for the choice of C, Sc, and Ka values the k^{-1} range is inappreciable and is overshadowed by the exponential decay in Eq. (3.100). A linearly generated scalar can add spectral content at large wavenumbers (around k_{ch}) before molecular diffusion dissipates scalar fluctuations.

The previous spectral treatment has a counterpart in physical space. The conservation equation for the scalar fluctuation $c'(\mathbf{x},t) = c(\mathbf{x},t) - \overline{c}(t)$ convected by isotropic and stationary solenoidal turbulence with zero mean velocity $\overline{\mathbf{u}}(t) = 0$ and scalar field undergoing a linear chemical reaction is

$$\frac{\partial c'}{\partial t} + \mathbf{u} \cdot \nabla c' = \mathcal{D} \nabla^2 c' \pm \frac{c'}{\tau_c}. \tag{3.102}$$

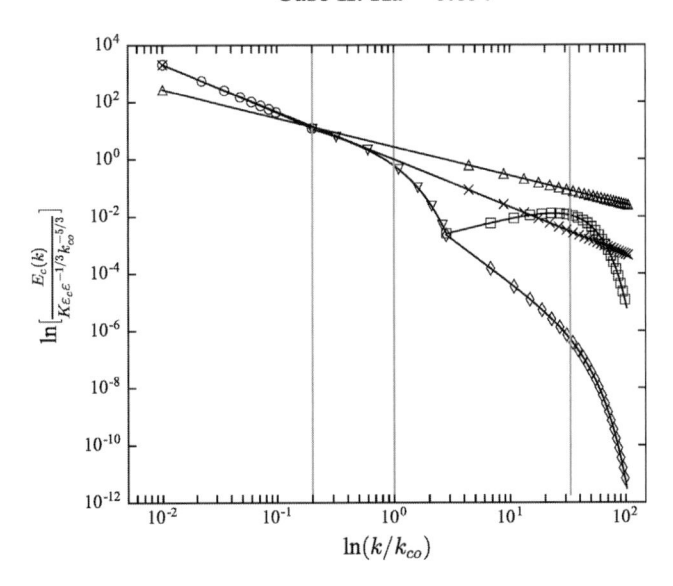

Figure 3.9 Scalar spectra in the three wavenumber ranges (see Eqs. 3.99 to 3.101) for case II. ⊖⊖ $(k \ll k_{co})$; ▽▽ $(k_{co} \approx k \ll k_{ch})$; ▫▫ $(k_{co} \ll k \approx k_{ch})$: generation; ◇◇ $(k_{co} \ll k \approx k_{ch})$: consumption; ✕✕ $-(5/3)\ln(k/k_{co})$; △△ $-\ln(k/k_{co})$.

The mean value, $\overline{c}(t)$, evolves according to

$$\frac{d\overline{c}}{dt} = \pm \frac{\overline{c}}{\tau_c}. \tag{3.103}$$

The Reynolds averaged equations for $(\overline{c'^2}/2)$, $\overline{|\nabla c'|}$ and $\overline{\epsilon}_c$ are, respectively,

$$\frac{d}{dt}\left(\frac{\overline{c'^2}}{2}\right) = -\overline{\epsilon}_c \pm \frac{\overline{c'^2}}{\tau_c}, \tag{3.104}$$

$$\frac{d}{dt}\left(\overline{|\nabla c'|}\right) = -\overline{a_N |\nabla c'|} - \mathcal{D}\overline{|\nabla c'| n_{i,j} n_{i,j}} \pm \frac{\overline{|\nabla c'|}}{\tau_c}, \quad \text{and} \tag{3.105}$$

$$\frac{d\overline{\epsilon}_c}{dt} = -2\overline{a_N \epsilon_c} - 2\mathcal{D}^2 \overline{\frac{\partial |\nabla c'|}{\partial x_j}\frac{\partial |\nabla c'|}{\partial x_j}} - 2\mathcal{D}\overline{\epsilon_c n_{i,j} n_{i,j}} \pm \frac{2\overline{\epsilon}_c}{\tau_c}. \tag{3.106}$$

If the scalar c is linearly generated or consumed then the chemical reaction increases or destroys $(\overline{c'^2}/2)$, $\overline{|\nabla c'|}$ and $\overline{\epsilon}_c$. The average rate of destruction of $(\overline{c'^2}/2)$ is

$$\frac{2}{\overline{c'^2}}\frac{d}{dt}\left(\frac{\overline{c'^2}}{2}\right) = -\frac{2\overline{\epsilon}_c}{\overline{c'^2}} \pm \frac{2}{\tau_c} = -\left(\frac{2\overline{\epsilon}_c}{\overline{c'^2}} \mp \frac{2}{\tau_c}\right). \tag{3.107}$$

Now, the characteristic destruction time for $(\overline{c'^2}/2)$ is

$$\tau_M = \left(\frac{2\overline{\epsilon}_c}{\overline{c'^2}} \mp \frac{2}{\tau_c}\right)^{-1}. \tag{3.108}$$

It is clear that the mixing timescale for an inert scalar is $\tau_M = (\overline{c'^2}/2\overline{\epsilon}_c)$. Also, one sees that the chemical reaction alters the mixing timescale. When there is production of scalar variance (or the scalar itself) by the chemical reaction then $\tau_M \geq (\overline{c'^2}/2\overline{\epsilon}_c)$ and when there is destruction then $\tau_M \leq (\overline{c'^2}/2\overline{\epsilon}_c)$. Hence, one must be cautious in using $(\overline{c'^2}/2\overline{\epsilon}_c)$ as mixing timescale in reacting flows.

Naively, if one assumes that the chemical reaction does not influence the terms involving the strain and molecular diffusivity in Eqs. (3.105) and (3.106) then it can be deduced that $\overline{|\nabla c'|}(t) = \overline{|\nabla c'|}(0) \exp(\pm t/\tau_c)$ and $\overline{\epsilon}_c(t) = \overline{\epsilon}_c(0) \exp(\pm 2t/\tau_c)$. However, it is quite clear that the neglected terms will also feel the effect of chemical reaction as they involve $|\nabla c'|$, which is influenced directly by the reaction. Hence, the combined influence is what matters. If $|2\overline{a_N \epsilon_c}|$ and $\left[2\mathcal{D}^2\overline{(\partial |\nabla c'|/\partial x_j)^2} + 2\mathcal{D}\overline{\epsilon_c n_{i,j} n_{i,j}}\right]$ in Eq. (3.106) are of the same order of magnitude and $|2\overline{\epsilon}_c/\tau_c| \ll |2\overline{a_N \epsilon_c}|$, then $Ka \gg 1$ implying that the characteristic mixing timescale is of the order of Kolmogorov timescale, τ_η. On the other hand, if $|2\overline{\epsilon}_c/\tau_c| \gg |2\overline{a_N \epsilon_c}|$ and $|2\overline{a_N \epsilon_c}| \sim \left[2\mathcal{D}^2\overline{(\partial |\nabla c'|/\partial x_j)^2} + 2\mathcal{D}\overline{\epsilon_c n_{i,j} n_{i,j}}\right]$ then the characteristic mixing timescale is $\delta^2/\mathcal{D} \sim \tau_c$ with $Ka \ll 1$.

3.6.3 Further Relationships Relevant to Modeling

The kinematic form of Eq. (3.20) is

$$\frac{\partial c}{\partial t} + \mathbf{v}^c \cdot \nabla c = 0. \tag{3.109}$$

An equation for $|\nabla c|$ with constant $\mathcal{D}$ is readily obtained:

$$\frac{1}{|\nabla c|}\left(\frac{\partial |\nabla c|}{\partial t} + \mathbf{v}^c \cdot \nabla |\nabla c|\right) = -\left(a_N + a_N^a\right). \tag{3.110}$$

Mixing and reaction can be described in terms of the flame surface density (FSD) [49, 54, 193]. The local and instantaneous isosurface density function is the surface area per unit volume (with dimensions 1/m), namely,

$$\Sigma(\zeta; \mathbf{x}, t) = \Sigma\left[c(\mathbf{x}, t) = \zeta; \mathbf{x}, t\right] = \frac{S\left[c(\mathbf{x}, t) = \zeta; \mathbf{x}, t\right]}{V} = \frac{1}{r(\zeta; \mathbf{x}, t)}. \tag{3.111}$$

Therefore, the widely used FSD is nothing but the inverse of the normal distance, r, between two adjacent surface elements. The variable $\Sigma(\zeta; \mathbf{x}, t)$ has been also defined as

$$\Sigma(\zeta; \mathbf{x}, t) = |\nabla c|\delta[\zeta - c(\mathbf{x}, t)] = \left(\frac{\varepsilon_c}{\mathcal{D}}\right)^{1/2} \delta\left[\zeta - c(\mathbf{x}, t)\right], \tag{3.112}$$

where $\delta\left[\zeta - c(\mathbf{x}, t)\right]$ is the Dirac delta function. The fine-grained PDF of $c(\mathbf{x}, t)$ is defined by $\mathcal{P} = \delta\left[\zeta - c(\mathbf{x}, t)\right]$, which obeys

$$\frac{\partial \mathcal{P}}{\partial t} + \mathbf{v}^c \cdot \nabla \mathcal{P} = 0. \tag{3.113}$$

Combining Eqs. (3.110) and (3.113), a transport equation for $\Sigma(\zeta; \mathbf{x}, t)$ can be obtained

$$\frac{\partial \Sigma(\zeta; \mathbf{x}, t)}{\partial t} + \mathbf{v}^c \cdot \nabla \Sigma = -\left(a_N + a_N^a\right)\Sigma. \tag{3.114}$$

On account of $\Sigma(\zeta; \mathbf{x}, t) = 1/r(\zeta; \mathbf{x}, t)$, Eq. (3.114) is identical to Eq. (3.73) with a sign change. Also, Eq. (3.114) is the same as Eq. (3.19). The above equation for FSD is also identical to the instantaneous and local version of Eq. (3.9), i.e.,

$$\frac{\partial \Sigma}{\partial t} + \nabla \cdot (\mathbf{u} + S_d \mathbf{n})\,\Sigma = (a_T + a_T^a)\Sigma. \tag{3.115}$$

The key role of the total (flow + propagation) normal strain rate, $(a_N + a_N^a)$, in the enhancement or destruction of scalar gradients, hence Σ, is emphasized by Eq. (3.114) as demonstrated by Eq. (3.19), whereas Eq. (3.115) underlines the area element total tangential strain rate, $(a_T + a_T^a)$, customarily used in combustion studies.

3.6.4 Remarks on the G-Equation

Equation (3.23) for the level-set G is valid only at the $G(\mathbf{x}, t) = G_0$ surface (the geometrical flame surface). A model for S_d is [53]

$$S_d = \frac{\rho_u}{\rho}\left(S_L + S_{\kappa, L}\right). \tag{3.116}$$

The term S_L is the unstretched laminar flame burning velocity and $S_{\kappa, L}$ is the curvature induced velocity. On the other hand the governing equation, valid at any isosurface of the reaction progress variable, is

$$\frac{\partial c}{\partial t} + \mathbf{u} \cdot \nabla c = S_d |\nabla c| = \left[\frac{\mathbf{n} \cdot \nabla \left(\rho \mathcal{D} \mathbf{n} \cdot \nabla c \right) + \dot{\omega}_c}{\rho |\nabla c|} - \mathcal{D} \nabla \cdot \mathbf{n} \right] |\nabla c|. \quad (3.117)$$

Equating $\rho_u S_L/\rho = S_{d,N}$ and $\rho_u S_{\kappa,L}/\rho = -\mathcal{D}(\nabla \cdot \mathbf{n})$, given in Eq. (3.22), at the flame surface would imply that $G = c(\mathbf{x},t) = \zeta$. However, identifying the isoscalar surfaces $c(\mathbf{x},t) = \zeta$ to be considered as the flame surface, $G(\mathbf{x},t) = G_0$, is a difficult task. Moreover, while the G approach is of a phenomenological nature, the isosurface method is more general. Difficulties in using the G equation are illustrated in Fig. 3.10, which shows the PDF of the displacement speed. The mean value of S_d is about 10 times the laminar flame speed, while the standard deviation is 13.5. This result is obtained by analyzing DNS data of a piloted turbulent Bunsen flame investigated experimentally by Chen et al [194]. This DNS was run by Cifuentes and Kempf with flame condition: $Re_t = 318$, $Ka = 10$, $Da = 0.6$, $Sc = 0.7$, and $Le = 1.0$.

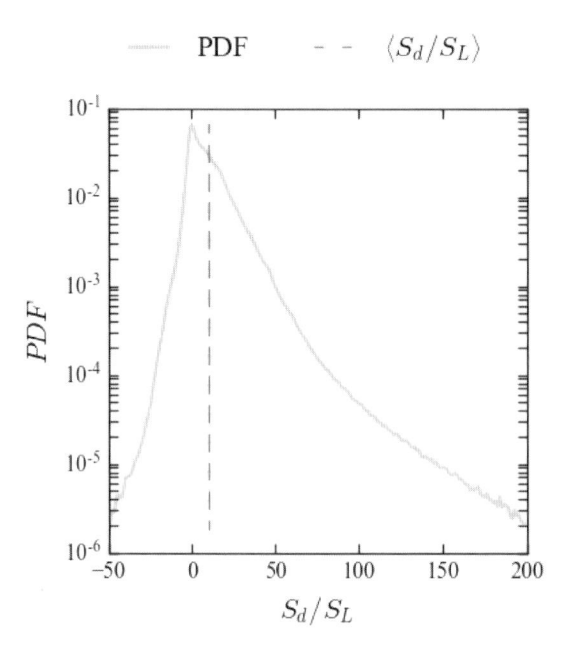

Figure 3.10 Probability density function of the displacement speed, normalized with the laminar flame speed.

3.7 Summary

A description with a critical appraisal of existing models for turbulent premixed combustion is presented. Fundamental physical concepts are revisited and emphasized. Modeling the highly nonlinear and spatially local chemical reaction rates is a great challenge for combustion simulations. In principle, these sources introduce time scales of the various elementary chemical reactions in spatially uniform scalar fields; chemical time scales much smaller and/or much larger than the relevant turbulent time scales will weakly interact with each other. However, in spatially nonuniform cases, a chemical reaction can either enhance or destroy preexisting scalar gradients and, hence, generate new length scales, which interact with those of comparable sizes in the turbulence field. These interactions can become strong when the characteristic time and length scales, and thus also characteristic velocities, of the turbulence are comparable to those of the flame.

When the chemical time and length scales are much smaller than the relevant (both integral and small) scales of turbulence, flames may burn under local laminar flow conditions (flamelet combustion). The turbulence scales will then convect, convolute, and contort the flame. Small-scale turbulence will either interact or not with the flame depending on the relative size of the Kolmogorov microscales compared to those of the flame.

RANS and LES models should account for differences in grids of the computational domain where the chemical conversion is either dominant or negligible. Local models should consider local values of Ka, Da, and Re_t to decide the relative importance of reaction or flow. The traditional definition of the characteristic chemical time, namely, $\tau_c = \rho_u / \dot{\omega}_f$, involves the density in the fresh mixture (where $\dot{\omega}_f = 0$) and $\dot{\omega}_f$ at another location (e.g., the maximum value of $\dot{\omega}_f$). The actual density where the maximum reaction rate occurs can be up to five to seven times smaller than ρ_u and, therefore, the local chemical time can be five to seven times smaller than the commonly estimated τ_c. A similar remark can be made in connection with the definitions of the Damköhler and Karlovitz numbers:

$$\mathrm{Da} = \frac{\tau_\Lambda}{\tau_c} = \left(\frac{S_L}{u'}\right)^2 \frac{u'\Lambda}{\nu_u} = \left(\frac{S_L}{u'}\right)^2 \mathrm{Re}_t = \left(\frac{\Lambda}{\delta_L}\right)^2 \frac{\alpha_u}{\nu_u} \mathrm{Re}_t^{-1}$$

$$\mathrm{Ka} = \frac{\tau_c}{\tau_\eta} = \left(\frac{\delta_L}{\eta}\right)^2 \left(\frac{S_L \delta_L}{\nu_u}\right)^{-1}$$

$$\mathrm{Da}\,\mathrm{Ka} = \frac{\tau_\Lambda}{\tau_\eta} = \mathrm{Re}_t^{1/2}.$$

Hence, one must be cautious in using these parameters to develop physical understanding for model development. Also, the turbulence is modified as the fluid approaches the flame and the local integral and Kolmogorov time scales, where the reactions are occurring, will likely be very different from the commonly used τ_Λ and τ_η ahead of the diffusion-reaction zone. The kinematic viscosity depends on the temperature, and thus the turbulent Reynolds number also changes through the flame. A turbulent flame may become laminar locally if one considers the local rms values of velocity fluctuations, integral length scale, and kinematic viscosity. Thus, a pertinent question is: What is the usefulness of the Reynolds number upstream of the reaction zone to classify or categorize a flame as turbulent if in reality the local conditions at the reaction zone matter?

While conventional definitions of Da, Ka, and Re$_t$ may be useful for turbulent flame taxonomic purposes and for the elaboration of combustion diagrams, local values of those dimensionless parameters, with all variables entering the definition estimated along the computation at a given grid point, would be useful to decide, for example, the relevant mixing timescale for an SGS mixing or combustion model.

The three suggested types of models, phenomenological, geometrical, and statistical, stem from two different viewpoints adopted by the investigators of turbulent combustion systems in the 1970s. On the one hand, statistical approaches were based on traditional formulations in classical turbulence in terms of statistical moments; to overcome the closure problem of the highly nonlinear chemical conversion term the transported scalar PDF method [105] was proposed. On the other hand, some practitioners of large activation energy asymptotics initially examined turbulent combustion as a perturbation of laminar flames in low-intensity turbulence fields; laminar flame speed, S_L, and flame thickness, δ_L, were considered to be relevant parameters, although they were slightly modified by the weak turbulence and systematically entered into model formulations [195].

One of the targets of phenomenological models is to obtain an average of the highly nonlinear reaction rate in terms of known turbulence and thermochemical variables. The EBU model equates rates of reaction, $(\overline{\dot{\omega}_c}/\overline{\rho})$, and of mixing, σ_c/τ_t. The turbulence timescale is $\tau_t = \Lambda/u' \sim (\widetilde{k}/\widetilde{\varepsilon})$, which is the characteristic timescale for the TKE transfer from energy containing eddies to those of neighboring eddies. The transfer time of TKE changes across the TKE spectrum, i.e, τ_t is different from $\tau_I = (l_I/u'_l)$ in the inertial subrange and both also differ from $\tau_\eta = (\eta/u_\eta) = \sqrt{\nu/\widetilde{\varepsilon}}$. What is constant across the TKE spectrum is the rate of TKE transfer, i.e., $\widetilde{\varepsilon} \simeq (u')^3/\Lambda = u_\eta^3/\eta$. Hence, using τ_t to characterize the mixing, which is a small-scale process, timescale seems to be a gross oversimplification.

The EDC model introduces PaSR or PSR notions with a mixing rate proportional to Kolmogorov strain rate. This partially reflects the correct physics on account that in reaction-dominated zones molecular diffusion should balance reaction, whereas in mixing-dominated regions molecular diffusion should balance small-scale convection. In the limit of Ka $\ll$ 1, only the last term on the right side of Eq. (3.4) survives, while for Ka $\gg$ 1 the relaxation term is dominant.

The idea of Minamoto et al. [21] is physically sound, should the local Da or Ka be used. Recast in terms of Ka, it becomes

$$\overline{\dot{\omega}_c} = \frac{\text{Ka}}{1+\text{Ka}} \overline{\dot{\omega}_c}^E + \frac{1}{1+\text{Ka}} \overline{\dot{\omega}_c}^B. \tag{3.118}$$

The two limits of Eq. (3.79) given in Eqs. (3.80) and (3.81) are the correct approximation to the modeling of mixing timescale. In the zones of computational domain where mixing dominates, the local Karlovitz number is Ka $\gg$ 1, and the mixing rate is determined by Eq. (3.81). On the other hand, in reaction-dominated zones Ka $\ll$ 1 and the mixing rate will be determined by Eq. (3.80). Switching from $\overline{\dot{\omega}_c}^B$ or Eq. (3.80) when the local Ka $\ll 1$ to $\overline{\dot{\omega}_c}^E$ or Eq. (3.81) for the local Ka $\gg 1$ during the computations would be the conceptually correct approximation and worth exploring.

The LEM has been successfully applied to many turbulent combustion systems, in spite of the difficulty of physically interpreting the phenomenological triplet map to mimic turbulent mixing.

The physical meaning of the FSD, Σ, as the inverse of the distance between two adjacent isoscalar surfaces is apparent. The transport equation for Σ can be written either in terms of the total tangential strain rate, $\left(a_T + a_T^a\right)$, or in terms of the total normal strain rate, $\left(a_N + a_N^a\right)$. Models for the latter should be explored.

The G equation model describes the propagation of an isoscalar surface in the flame; however, it does not describe the flame structure, since $G(x,t) = G_0 = c(x,t) = \zeta_0$ identifies a single isoscalar surface. If one considers G as the distance function then the flame structure can be extracted by combining it with the flamelet approach. Given that S_d could vary across isosurfaces in the flame, the expression $S_d = S_L - S_L \kappa \mathcal{L} + a_N \mathcal{L}$ can be hardly compared with

$$S_d = \frac{\dot{\omega}_c + \mathbf{n} \cdot \nabla \left(\rho \mathcal{D} \mathbf{n} \cdot \nabla c \right)}{\rho |\nabla c|} - \mathcal{D} \nabla \cdot \mathbf{n}.$$

Setting $S_L \kappa \mathcal{L} = \mathcal{D} (\nabla \cdot \mathbf{n})$ and $S_L + a_N \mathcal{L} = \left(\mathbf{n} \cdot \nabla \left(\rho \mathcal{D} \mathbf{n} \cdot \nabla c \right) + \dot{\omega}_c \right) / \left(\rho |\nabla c| \right)$ is artificial. However, if the turbulence level is very low then the foregoing approximation serves well and thus the G equation approach is strictly valid for the wrinkled flamelet regime of turbulent premixed combustion [95].

Dopazo and OBrien [105] first proposed the transported scalar PDF methodology to deal with highly nonlinear chemical terms in reactive systems. However, modeling the so-called micro-mixing term (i.e., the conditional molecular diffusion) turns out to be a stumbling block for this methodology. However, many models have been developed for nonreacting turbulent flows (e.g., LMSE, binomial Langevin, Coalescence and Dispersion, Mapping Closure, Euclidean Minimum Spanning Tree, among others) and applied to reacting flows also. Some of their main drawbacks have been identified and discussed. Examining details of the turbulent mixing processes in variable density flows with nonnegligible heat release, and, particularly, the role of chemical reactions in modifying scalar gradients, deserves much further research. Inclusion of differential diffusion effects in the mixing models is another important pending challenge.

In the presumed PDF approach, the marginal PDF shape is prescribed for the given first two moments. Typically, the beta function is commonly used and this function gives a bimodal behaviour for large variance (either SGS or RANS) and degenerates to a mono-modal Gaussian shape when the variance is small. The BML approach assumed the PDF to be bimodal when the chemical timescale is very small compared to the relevant turbulent time scales. This approach has led to many useful insights such as countergradient flux in premixed combustion. Recent developments on this approach are presented and discussed. The conditional moment closure method allows including the evolution of finite rate chemistry effects by solving transport equations for conditionally filtered mass fractions of all the chemical species involved in the combustion process. The small-scale mixing denoted using the scalar dissipation rate of the conditioning scalar, the reaction progress variable, needs to be modeled and this limited the progress of CMC for premixed combustion. However, recent advances on the modeling of the scalar dissipation rate of c allowed further development and testing of CMC for premixed combustion. These developments are described in Section 3.4.3.

The filtered reaction rate can be precomputed and stored in a lookup table, which can be used during LES. Although this tabulated chemistry approach has been used in many past studies careful analysis of this approach showed that physical and mathematical inconsistencies can arise if one is not careful in modeling the SGS mixing and reaction processes and their interactions. Improvements to the tabulated chemistry approach have been identified, proposed and tested using DNS and LES in many recent studies. This revised approach is called as FlaRe (flamelets revised for consistencies) and is discussed Section 3.5.

The scalar mixing-reaction problem in turbulent flows is revisited in light of recent findings on the rate of change of nonmaterial line, area, and volume elements. Apart from the flow velocity, $\mathbf{u}$, the normal displacement vector, $S_d \mathbf{n}$, adds strain rates normal and tangential to an isosurface with clear physical meanings. In particular, Eq. (3.79) defines the scalar mixing rate as a balance between the Kolmogorov strain rate, the rates of molecular diffusion in the normal and tangential directions to an isosurface, and the chemical reaction rate. If $Ka \ll 1$ then a balance between molecular diffusion and chemical reaction determines the mixing rate. On the other hand, the balance between flow microscale convection and molecular diffusion dictates the mixing rate for $Ka \gg 1$. The problem of length scales in turbulent combustion is illustrated with a simple example of scalar spectra, à la Corrsin, for isotropic and stationary turbulence with constant density. Recently, several authors have identified the pressure-dilatation term as an important source of TKE in combustion systems. Although this effect is absent in this simple example, our illustration will be pedagogically instructive and, probably, relevant in regions of the computational domain with negligible local heat release. Analytical spectra are presented and discussed for scalars undergoing a linear chemical conversion (either generation and consumption) in the limits of $Ka \ll 1$ and $Ka \gg 1$. The spectra behavior for various wavenumber ranges are described.

Acknowledgments
Figures 3.4 to 3.6 are kindly provided by Dr. Zhi. X. Chen of Cambridge University Engineering Department.

References

[1] E. Mallard, "Speed of flame in air and fire-damp mixtures," *Ann. Mines*, **7**, 355–381, 1875.

[2] E. Mallard and H. L. Chatelier, "Experimental and theoretical researches on the combustion of explosive gaseous mixtures," *Ann. Mines*, **3**, 274–378, 1883.

[3] F. A. Williams, *Combustion Theory*. Redwood City, CA: Addison-Wesley, 1985.
[4] S. R. Turns, *An Introduction to Combustion: Concepts and Application*, 2nd ed. Singapore: McGraw-Hill, 2000.
[5] C. K. Law, *Combustion Physics*. Cambridge: Cambridge University Press, 2006.
[6] D. Dunn-Rankin(Ed), *Lean Combustion Technology and Control*. San Diego, CA: Academic Press, Elsevier Inc., 2008.
[7] N. Swaminathan and K. N. C. Bray (Eds), *Turbulent Premixed Flames*. New York: Cambridge University Press, 2011.
[8] G. Ghiasi, "Simulation of turbulent flames at conditions related to IC engines," Ph.D. dissertation, University of Cambridge, Department of Engineering, 2018.
[9] G. Ghiasi, I. Ahmed, and N. Swaminathan, "Gasoline flame behavior at elevated temperature and pressure," *Fuel*, **238**, 248–256, 2019.
[10] N. Peters, *Turbulent Combustion*. Cambridge: Cambridge University Press, 2000.
[11] I. Langella, N. Swaminathan, and R. W. Pitz, "Application of unstrained flamelet SGS closure for multi-regime premixed combustion," *Combust. Flame*, **173**, 161–178, 2016.
[12] L. Y. M. Gicquel, G. Staffelbach, and T. Poinsot, "Large eddy simulations of gaseous flames in gas turbine combustion chambers," *Prog. Energy Combust. Sci.*, **38**, 782–817, 2012.
[13] D. B. Spalding, "Mixing and chemical reaction in steady confined turbulent flames," *Symp. (International) Combus.*, **13**, 649–657, 1971.
[14] C. Fureby and S. I. Möller, "Large eddy simulation of reacting flows applied to bluff body stabilized flames," *AIAA J.*, **33**, 2339–2347, 1995.
[15] S. I. Möller, E. Lundgren, and C. Fureby, "Large eddy simulation of unsteady combustion," *Proc. Combust. Inst.*, **26**, 241–248, 1996.
[16] E. Giacomazzi, V. Battaglia, and C. Bruno, "The coupling of turbulence and chemistry in a premixed bluff-body flame as studied by LES," *Combust. Flame*, **138**, 320–335, 2004.
[17] K. J. Nogenmyr, C. Fureby, X. S. Bai, P. Peterson, R. Collin, and M. Linne, "Large eddy simulation and laser diagnostic studies on a low swirl stratified premixed flame," *Combust. Flame*, **156**, 25–36, 2009.
[18] K. N. C. Bray, "The interaction between turbulence and combustion," *Proc. Combust. Inst*, **17**, 223–233, 1979.
[19] B. F. Magnussen and B. H. Hjertager, "On mathematical modeling of turbulent combustion with special emphasis on soot formation and combustion," *Symp. (International) Combust.*, **16**, 719–727, 1977.
[20] I. S. Ertesvåg and B. F. Magnussen, "The eddy dissipation turbulence energy cascade model," *Combust. Sci. Technol.*, **159**, 213–235, 2000.
[21] Y. Minamoto, K. Aoki, M. Tanahashi, and N. Swaminathan, "DNS of swirling hydrogen-air premixed flames," *J. Hydrogen Energy*, **40**, 13 604–13 620, 2015.
[22] T. D. Butler and P. J. O'Rourke, "A numerical method for two-dimensional unsteady reacting flows," *Sixteenth Symposium (International) on Combustion, The Combustion Institute, Pittsburgh*, **16**, 1503–1515, 1977.
[23] O. Colin, F. Ducros, D. Veynante, and T. Poinsot, "A thickened flame model for large eddy simulations of turbulent premixed combustion," *Phys. Fluids*, **12**, 1843–1863, 2000.
[24] F. Charlette, C. Meneveau, and D. Veynante, "A power-law flame wrinkling model for LES of premixed turbulent combustion part II: dynamic formulation," *Combust. Flame*, **131**, 181–197, 2002.

[25] G. Wang, M. Boileau, and D. Veynante, "Implementation of a dynamic thickened flame model for large eddy simulations of turbulent premixed combustion," *Combust. Flame*, **158**, 2199–2213, 2011.

[26] P. S. Volpiani, T. Schmitt, and D. Veynante, "Large eddy simulation of a turbulent swirling premixed flame coupling the TFLES model with a dynamic wrinkling formulation," *Combust. Flame*, **180**, 124–135, 2017.

[27] G. Kuenne, A. Ketelheun, and J. Janicka, "LES modeling of premixed combustion using a thickened flame approach coupled with FGM tabulated chemistry," *Combust. Flame*, **158**, 1750–1767, 2011.

[28] H. Wang, Z. Yu, T. Ye, M. Cheng, and M. Zhao, "Large eddy simulation of turbulent stratified combustion using dynamic thickened flame coupled with tabulated detailed chemistry," *App. Math. Model.*, **62**, 476–498, 2018.

[29] F. Proch and A. M. Kempf, "Numerical analysis of the cambridge stratified flame series using artificial thickened flame LES with tabulated premixed flame chemistry," *Combust. Flame*, **161**, 2627–2646, 2014.

[30] S. Emami, K. Mazaheri, A. Shamooni, and Y. Mahmoudi, "LES of flame acceleration and DDT in hydrogen-air mixture using artificially thickened flame approach and detailed chemical kinetics," *Int. J. Hydrogen Energy*, **40**, 7395–7408, 2015.

[31] S. B. Pope, "Computationally efficient implementation of combustion chemistry using in situ adaptive tabulation," *Combust. Theory Model*, **1**, 41–63, 1997.

[32] A. R. Kerstein, "Linear-eddy model of turbulent transport and mixing," *Combust. Sci. Technol.*, **60**, 391–421, 1988.

[33] P. A. McMurtry, S. Menon, and A. R. Kerstein, "A new subgrid model for turbulent combustion: Application to hydrogen-air combustion," *Proc. Combust. Inst.*, **24**, 271–278, 1993.

[34] S. Menon, p. A. McMurtry, and A. R. Kerstein, "A linear eddy mixing model for large eddy simulation of turbulent combustion," in B. Galperin and S. Orszag, Eds., *LES of Complex Engineering and Geophysical Flows*. Cambridge: Cambridge University Press, 1993, pp. 278–314.

[35] S. Menon and A. R. Kerstein, "The linear-eddy model," in *Turbulent Combustion Modeling: Advances, New Trends and Perspectives*. Fluid Mechanics and Its Applications, T. Echekki and E. Mastorakos, Eds. Dordrecht: Springer Science+Business Media, 2011, **95**, pp. 221–247.

[36] A. S. Baris and S. Menon, "Turbulent premixed flame modeling using artificial neural networks based chemical kinetics," *Proc. Combust. Inst.*, **32**, 1605–1611, 2009.

[37] A. S. Baris and S. Menon, "Linear eddy mixing based tabulation and artificial neural networks for large eddy simulations of turbulent flames," *Combust. Flame*, **157**, 62–74, 2010.

[38] A. S. Baris, E. R. Hawkes, and S. Menon, "Large eddy simulation of extinction and reignition with artificial neural networks based chemical kinetics," *Combust. Flame*, **157**, 566–578, 2010.

[39] T. Smith and S. Menon, "Model simulations of freely propagating turbulent premixed flames," *Proc. Combust. Inst.*, **26**, 299–306, 1996.

[40] V. Chakravarthy and S. Menon, "Large eddy simulation of turbulent premixed flames in the flamelet regime," *Combust. Sci. Technol.*, **162**, 175–222, 2000.

[41] V. Sankaran and S. Menon, "Structure of premixed flame in the thin-reaction-zones regime," *Proc. Combust. Inst.*, **28**, 203–210, 2000.

[42] V. Sankaran and S. Menon, "Subgrid combustion modeling of 3-D premixed flames in the thin-reaction-zone regime," *Proc. Combust. Inst.*, **30**, 575–582, 2005.

[43] S. Undapalli and S. Menon, "LES of premixed and non-premixed combustion in a stagnation point reverse flow combustor," *Proc. Combust. Inst.*, **32**, 1537–1544, 2009.

[44] S. Srinivasan and S. Menon, "From flamelet to distributed/broken reaction zone regimes: investigations using the linear eddy mixing model," *AIAA Paper*, **2012-4111**, 1–15, 2012.

[45] S. Srinivasan and S. Menon, "Linear eddy mixing model studies of high Karlovitz number turbulent premixed flames," *Flow Turb. Combust.*, **93**, 189–219, 2014.

[46] S. Srinivasan, R. Ranjan, and S. Menon, "Flame dynamics during combustion instability in a high-pressure, shear-coaxial injector combustor," *Flow Turb. Combust.*, **94**, 237–262, 2015.

[47] R. Ranjan, B. Muralidharan, Y. Nagaoka, and S. Menon, "Subgrid-scale modeling of reaction-diffusion and scalar transport in turbulent premixed flames," *Combust. Sci. Technol.*, **188**, 1496–1537, 2016.

[48] F. Grøvdal, S. Sannan, J.-Y. Chen, A. R. Kerstein, and T. Løvås, "Three-dimensional linear eddy modeling of a turbulent lifted hydrogen jet flame in a vitiated co-flow," *Flow Turbulence Combust.*, **101**, 993–1007, 2018.

[49] F. Marble and J. E. Broadwell, "The coherent flame model of non-premixed turbulent combustion." *Project Squid TRW-9-PU (Project Squid Headquarters*, Purdue University, West Lafayette, IN)" 1977.

[50] G. K. Batchelor, "The effect of homogeneous turbulence on material lines and surfaces," *Proc. Royal Soc. Lond. A*, **231**, 349–366, 1952.

[51] S. B. Pope, "Computations of turbulent combustion: Progress and challenges," *Proc. Combust. Inst.*, **23**, 591–612, 1990.

[52] N. Chakraborty, M. Champion, A. Mura, and N. Swaminathan, "Scalar-dissipation-rate approach," in N. Swaminathan and K. N. C. Bray, Eds., *Turbulent Premixed Flames*. Cambridge: Cambridge University Press, 2011, pp. 74–102.

[53] L. Vervisch, V. Moureau, P. Domingo, and D. Veynante, "Flame surface density and the G equation," in N. Swaminathan and K. N. C. Bray, Eds., *Turbulent Premixed Flames*. Cambridge: Cambridge University Press, 2011, pp. 60–74.

[54] S. B. Pope, "The evolution of surfaces in turbulence," *Int. J. Engng. Sci.*, **26**, 445–469, 1988.

[55] S. M. Candel and T. J. Poinsot, "Flame stretch and the balance equation for the flame area," *Combust. Sci. Tech.*, **70**, 1–15, 1990.

[56] K. N. C. Bray, "Studies of the turbulent burning velocity," *Proc. R. Soc. Lond. A*, **431**, 315–335, 1990.

[57] K. N. C. Bray and P. A. Libby, "Recent developments in the BML model for turbulent premixed combustion," in P. A. Libby and F. A. Williams, Eds., *Turbulent Reacting Flows*. New York: Academic Press, 1994, pp. 116–151.

[58] S. R. Cant, "RANS and LES modelling of premixed turbulent combustion," in T. Echekki and E. Mastorakos, Eds., *Turbulent Combustion Modelling: Advances, New Trends and Perspectives*. Dordrecht: Springer Science+Business Media B.V., 2011, pp. 63–90.

[59] T. J. Poinsot and D. Veynante, *Theoretical and Numerical Combustion*. Philadelphia: Edwards Inc., 2001.

[60] K. N. C. Bray and N. Swaminathan, "Scalar dissipation and flame surface density in premixed turbulent combustion," *C. R. Mecanique*, **334**, 466–473, 2006.

[61] T. D. Dunstan, Y. Minamoto, N. Chakraborty, and N. Swaminathan, "Scalar dissipation rate modelling for large eddy simulation of turbulent premixed flames," *Proc. Combust. Inst.*, **34**, 1193–1201, 2013.

[62] Y. Gao, N. Chakraborty, and N. Swaminathan, "Algebraic closure of scalar dissipation rate for large eddy simulations of turbulent premixed combustion," *Combust. Sci. Technol.*, **186**, 1309–1337, 2014.

[63] I. Langella, N. Swaminathan, Y. Gao, and N. Chakraborty, "Assessment of dynamic closure for premixed combustion large eddy simulation," *Combust. Theory Modelling*, **19**, 628–656, 2015.

[64] I. Langella and N. Swaminathan, "Unstrained and strained flamelets for LES of premixed combustion," *Combust. Theory Model.*, **20**, 410–440, 2016.

[65] M. Boger, D. Veynante, H. Boughanem, and A. Trouvé, "Direct numerical simulation analysis of flame surface density concept for large eddy simulation of turbulent premixed combustion," *Proc. Combust. Inst.*, **27**, 917–925, 1998.

[66] H. G. Weller, G. Tabor, A. D. Gosman, and C. Fureby, "Application of a flame-wrinkling LES combustion model to a turbulent mixing layer," *Proc. Combust. Inst.*, **27**, 899–907, 1998.

[67] C. Fureby, "A computational study of combustion instabilities due to vortex shedding," *Proc. Combust. Inst.*, **28**, 783–791, 2000.

[68] C. Fureby, "Large eddy simulation of combustion instabilities in a jet engine afterburner model," *Combust. Sci. and Tech.*, **161**, 213–243, 2000.

[69] F. Charlette, C. Meneveau, and D. Veynante, "A power-law flame wrinkling model for les of premixed turbulent combustion part I: non-dynamic formulation and initial tests," *Combust. Flame*, **131**, 159–180, 2002.

[70] F. Charlette, C. Meneveau, and D. Veynante, "A power-law flame wrinkling model for les of premixed turbulent combustion part II: dynamic formulation," *Combust. Flame*, **131**, 181–197, 2002.

[71] R. Knikker, D. Veynante, and C. Meneveau, "A priori testing of a similarity model for large eddy simulations of turbulent premixed combustion," *Proc. Combust. Inst.*, **29**, 2105–2111, 2002.

[72] R. Knikker, D. Veynante, and C. Meneveau, "A dynamic flame surface density model for large eddy simulations of turbulent premixed combustion," *Phys. Fluids*, **16**, L91–L94, 2004.

[73] O. Colin, F. Ducros, D. Veynante, and T. Poinsot, "A thickened flame model for large eddy simulations of turbulent premixed combustion," *Phys. Fluids*, **12**, 1843–1863, 2000.

[74] K. Hiraoka, Y. Minamoto, M. Shimura, Y. Naka, N. Fukushima, and M. Tanahashi, "A fractal dynamic SGS combustion model for large eddy simulation of turbulent premixed flames," *Combust. Sci. Technol.*, **188**, 1472–1495, 2016.

[75] N. Chakraborty and M. Klein, "A priori direct numerical simulation assessment of algebraic flame surface density models for turbulent premixed flames in the context of large eddy simulation," *Phys. Fluids*, **20**, 085 108 1–14, 2008.

[76] M. P. Kirkpatrick, S. W. Armfield, A. R. Masri, and S. S. Ibrahim, "Large eddy simulation of a propagating turbulent premixed flame," *Flow, Turbulence Combust.*, **70**, 1–19, 2003.

[77] N. A. K. Doan, N. Swaminathan, and N. Chakraborty, "Multiscale analysis of turbulence-flame interaction in premixed flames," *Proc. Combust. Inst.*, **36**, 1929–1935, 2017.

[78] U. Ahmed, N. A. K. Doan, J. Lai, M. Klein, N. Chakraborty, and N. Swaminathan, "A priori direct numerical simulation modeling of scalar dissipation rate transport in head-on quenching of turbulent premixed flames," *Phys. Fluids*, **30**, 105 102,1–13, 2018.

[79] F. O'young and R. W. Bilger, "Scalar gradient and related qunatities in turbulent premixed flames," *Combust. Flame*, **109**, 682–700, 1997.

[80] S. Ruan, N. Swaminathan, and Y. Mizobuchi, "Investigation of flame stretch in turbulent lifted jet flame," *Combust. Sci. Technol.*, **186**, 243–272, 2014.

[81] N. Peters, "The turbulent burning velocity for large-scale and small-scale turbulence," *J. Fluid Mech.*, **384**, 107–132, 1999.

[82] A. R. Kerstein, "Simple derivation of Yakhot's turbulent premixed flamespeed formula," *Combust. Sci. and Tech.*, **60**, 163–165, 1988.

[83] W. Kollmann and J. H. Chen, "Dynamics of the flame surface area in turbulent non-premixed combustion," *Proc. Combust. Inst.*, **25**, 1091–1098, 1994.

[84] W. Kollmann and J. H. Chen, "Pocket formation and the flame surface density equation," *Proc. Combust. Inst.*, **27**, 927–934, 1998.

[85] D. Bradley, "Problems of predicting turbulent burning rates," *Combust. Theory Model.*, **6**, 361–382, 2002.

[86] M. Kardar, G. Parisi, and Y. C. Zhang, "Dynamic scaling of growing interfaces," *Phys. Rev. Letter*, **56**, 889–892, 1986.

[87] H. G. Im, T. S. Lund, and J. H. Ferziger, "Large eddy simulation of turbulent front propagation with dynamic subgrid models," *Phys. Fluids*, **9**, p. 3826, 1997.

[88] F. Creta and M. Matalon, "Propagation of wrinkled turbulent flames in the context of hydrodynamic theory," *J. Fluid Mech.*, **680**, 225–264, 2011.

[89] N. Fogla, F. Creta, and M. Matalon, "The turbulent flame speed for low-to-moderate turbulence intensities: Hydrodynamic theory vs. experiments," *Combust. Flame*, **175**, 155–169, 2017.

[90] N. Peters, "A spectral closure for premixed turbulent combustion in the flamelet regime," *J. Fluid Mech.*, **242**, 611–629, 1992.

[91] P. Pelce and P. Clavin, "Influence of hydrodynamics and diffusion upon the stability limits of laminar premixed flames," *J. Fluid Mech.*, **124**, 219–237, 1982.

[92] M. Matalon and B. J. Matkowsky, "Flames as gasdynamic discontinuities," *J. Fluid Mech.*, **124**, 239–259, 1982.

[93] K. J. Nogenmyr, P. Petersson, X. S. Bai, A. Nauert, J. Olofsson, and C. Brackman, "Large eddy simulation and experiments of stratified lean premixed methane/air turbulent flames," *Proc. Combust. Inst.*, **31**, 1467–1475, 2007.

[94] K. J. Nogenmyr, C. Fureby, X. S. Bai, P. Petersson, R. Collin, and M. Linne, "Large eddy simulation and laser diagnostic studies on a low swirl stratified premixed flame," *Combust. Flame*, **156**, 25–36, 2009.

[95] F. A. Williams, "Some recent studies in turbulent combustion," in *Smart Control of Turbulent Combustion*, A. Yoshida, Ed. Tokyo: Springer Verlag, 2001, 1–12.

[96] H. Pitsch and L. D. D. Lageneste, "Large eddy simulation of turbulent combustion using a level-set approach," *Proc. Combust. Inst.*, **29**, 2001–2008, 2002.

[97] H. Pitsch, "A consistent level set formulation for large-eddy simulation of premixed turbulent combustion," *Combust. Flame*, **143**, 587–598, 2005.

[98] P. Wang and X. S. Bai, "Large eddy simulation of turbulent premixed flames using level-set G-equation," *Proc. Combust. Inst.*, **30**, 583–591, 2005.

[99] J. Smagorinsky, "General circulation experiments with the primitive equations: I. The basic equations," *Mon. Weather Review*, **91**, 99–164, 1963.

[100] V. Moureau, B. Fiorina, and H. Pitsch, "A level set formulation for premixed combustion LES considering the turbulent flame structure," *Combust. Flame*, **156**, 801–812, 2009.

[101] Z. Tan and R. D. Reitz, "An ignition and combustion model based on the level-set method for spark ignition engine multidimensional modeling," *Combust. Flame*, **145**, 1–15, 2006.

[102] P. Nilsson and X. S. Bai, "Level-set flamelet library approach for premixed turbulent combustion," *Exp. Thermal Fluid Sci.*, **21**, 87–98, 2000.

[103] B. Li, E. Baudoin, R. Yu, Z. W. Sun, Z. S. Li, X. S. Bai, M. Aldén, and M. S. Mansour, "Experimental and numerical study of a conical turbulent partially premixed flame," *Proc. Combust. Inst.*, **32 II**, 1811–1818, 2009.

[104] D. C. Haworth and S. B. Pope, "Transported probability density function methods for Reynolds-Averaged and Large-Eddy simulations," in T. Echekki and E. Mastorakos, Eds., *Turbulent Combustion Modelling: Advances, New Trends and Perspectives*. Dordecht: Springer Science+Business Media B.V., 2011, pp. 119–142.

[105] C. Dopazo and E. E. O'Brien, "An approach to the autoignition of a turbulent mixture," *Acta Astronaut*, **1**, 1239–1266, 1974.

[106] P. A. Libby and F. A. Williams, Eds., *Turbulent Reacting Flows*. Academic Press, New York, 1994.

[107] F. Gao and E. E. O'Brien, "A large-eddy simulation scheme for turbulent reacting flows," *Phys. Fluids A.*, **5**, 1282–1284, 1993.

[108] R. P. Lindstedt, "Transported probability density function methods for premixed turbulent flames," in N. Swaminathan and K. N. C. Bray, Eds., *Turbulent Premixed Flames*. Cambridge: Cambridge University Press, 2011, pp. 102–135.

[109] D. H. Rowinski and S. B. Pope, "PDF calculations of piloted premixed jet flames," *Combust. Theory Model.*, **15**, 245–266, 2011.

[110] P. J. Colucci, F. A. Jaberi, P. Givi, and S. B. Pope, "Filtered density function for large eddy simulation of turbulent reacting flows," *Phys. Fluids*, **10**, 499–515, 1998.

[111] L. Valiño and C. Dopazo, "A binomial langevin model for turbulent mixing," *Phys. Fluids A*, **3**, 3034–3037, 1991.

[112] C. Dopazo, "Relaxation of initial probability density function in the turbulent convection of scalar fields," *Phys. Fluids*, **22**, 20–30, 1979.

[113] J. Janicka, W. Kolbe, and W. Kollmann, "Closure of the transport-equation for the probability density-function of turbulent scalar fields," *J. Non-Equil. Thermodyn.*, **4**, 47–66, 1979.

[114] S. Subramaniam and S. B. Pope, "A mixing model for turbulent reactive flows based on Euclidean minimum spanning trees," *Combust. Flame*, **115**, 487–514, 1998.

[115] H. Chen, S. Chen, and R. H. Kraichnan, "Probability distribution of a stochastically advected scalar field," *Phys. Rev. Lett.*, **63**, 2657–2660, 1989.

[116] M. Kuron, Z. Ren, E. R. Hawkes, H. Zhou, H. Kolla, and J. H. Chen, "A mixing timescale model for TPDF simulations of turbulent premixed flames," *Combust. Flame*, **177**, 171–183, 2017.

[117] D. H. Rowinski and S. B. Pope, "Computational study of lean premixed turbulent flames using RANSPDF and LESPDF methods," *Combust. Theory Model.*, **17**, 610–656, 2013.

[118] H. Wang, T. Pant, and P. Zhang, "LES/PDF modeling of turbulent premixed flames with locally enhanced mixing by reaction," *Flow Turbulence Combust.*, **100**, 147–175, 2018.

[119] H. Turkeri, X. Zhao, S. B. Pope, and M. Muradoglu, "Large eddy simulation/probability density function simulations of the cambridge turbulent stratified flame series," *Combust. flame*, **199**, 24–45, 2019.

[120] L. Valiño, "Field Monte Carlo formulation for calculating the probability density function of a single scalar in a turbulent flow," *Flow Turbulence Combust.*, **60**, 157–172, 1998.

[121] V. Sabel'nikov and O. Soulard, "Rapidly decorrelating velocity-field model as a tool for solving one-point Fokker-Planck equations for probability density functions of turbulent reactive scalars," *Phys. Rev. E*, **72**, 016 301-1 – 22, 2005.

[122] W. P. Jones and V. N. Prasad, "LES-pdf simulation of a spark ignited turbulent methane jet," *Proc. Combust. Inst.*, **33**, 1355–1363, 2011.

[123] W. P. Jones, A. J. Marquis, and F. Wang, "Large eddy simulation of a premixed propane turbulent bluff body flame using the Eulerian stochastic field method," *Fuel*, **140**, 514–525, 2015.

[124] M. A. Picciani, E. S. Richardson, and S. Navarro-Martinez, "Resolution requirements in stochastic field simulation of turbulent premixed flames," *Flow Turbulence Combust.*, **101**, 1103–1118, 2018.

[125] R. O. Fox, *Computational Models for Turbulent Reacting Flows*. Cambridge: Cambridge University Press, 2003.

[126] Q. Tang, W. Zhao, M. Bockelie, and R. O. Fox, "Multi-environment probability density function method for modelling turbulent combustion using realistic chemical kinetics," *Combust. Theory Model.*, **11**, 889–907, 2007.

[127] D. C. Howarth, "Progress in probability density function methods for turbulent reacting flows," *Prog. Energy Combust. Sci.*, **36**, 168–259, 2010.

[128] K. N. C. Bray, "Laminar flamelets and the Bray, Moss, and Libby model," in N. Swaminathan and K. N. C. Bray, Eds., *Turbulent Premixed Flames*. Cambridge: Cambridge University Press, 2011, pp. 41–60.

[129] K. N. C. Bray, M. Champion, P. A. Libby, and N. Swaminathan, "Finite rate chemistry and presumed pdf models for premixed turbulent combustion," *Combust. Flame*, **146**, 665–673, 2006.

[130] A. Y. Klimenko and R. W. Bilger, "Conditional moment closure for turbulent combustion," *Prog. Energy Combust. Sci.*, **25**, 595–687, 1999.

[131] N. Swaminathan and R. W. Bilger, "Analyses of conditional moment closure for turbulent premixed flames," *Combust. Theory Modelling*, **5**, 241–260, 2001.

[132] S. Amzin, N. Swaminathan, J. W. Rogerson, and J. H. Kent, "Conditional moment closure for turbulent premixed flames," *Combust. Sci. Technol.*, **184**, 1743–1767, 2012.

[133] S. Amzin and N. Swaminathan, "Computations of turbulent lean premixed combustion using conditional moment closure," *Combust. Theory Modelling*, **17**, 1125–1153, 2013.

[134] S. M. Martin, J. C. Kramlich, G. Kosály, and J. J. Riley, "The premixed conditional moment closure method applied to idealized lean premixed gas turbine combustors," *J. Eng. Gas Turb. Power*, **125**, 895–900, 2003.

[135] D. Farrace, K. Chung, M. Bolla, Y. M. Wright, K. Boulouchos, and E. Mastorakos, "A LES-CMC formulation for premixed flames including differential diffusion," *Combust. Theory Modelling*, **22**, 411–431, 2018.

[136] D. Farrace, K. Chung, S. S. Pandurangi, Y. M. Wright, K. Boulouchos, and N. Swaminathan, "Unstructured LES-CMC modelling of turbulent premixed bluff body flames close to blow-off," *Proc. Combust. Inst.*, **36**, 1977–1985, 2017.

[137] N. Swaminathan, R. W. Bilger, and B. Cuenot, "Relationship between turbulent scalar flux and conditional dilatation in premixed flames with complex chemistry," *Combust. Flame*, **126**, 1764–1779, 2001.

[138] N. Swaminathan and K. N. C. Bray, "Effect of dilatation on scalar dissipation in turbulent premixed flames," *Combust. Flame*, **143**, 549–565, 2005.

[139] H. Kolla and N. Swaminathan, "Strained flamelets for turbulent premixed flames, I: Formulation and planar flame results," *Combust. Flame*, **157**, 943–954, 2010.

[140] Y. Gao, N. Chakraborty, T. Dunstan, and N. Swaminathan, "Assessment of Reynolds averaged Navier-Stokes modelling of scalar dissipation rate transport in turbulent oblique premixed flames," *Combust. Sci. Technol.*, **187**, 1584–1609, 2015.

[141] J. Lai and N. Chakraborty, "A priori direct numerical simulation modeling of scalar dissipation rate transport in head-on quenching of turbulent premixed flames," *Combust. Sci. Technol.*, **188**, 1440–1471, 2016.

[142] G. Ghiasi, N. A. K. Doan, N. Swaminathan, B. Yennerdag, Y. Minamoto, and M. Tanahashi, "Assessment of SGS closure for isochoric combustion of hydrogen-air mixture," *J. Hydrogen Energy*, **43**, 8105–8115, 2018.

[143] I. Langella, N. Swaminathan, Y. Gao, and N. Chakraborty, "LES of premixed combustion: Sensitivity to SGS velocity modelling," *Combust. Sci. Technol.*, **189**, 43–78, 2017.

[144] T. Ma, Y. Gao, A. M. Kempf, and N. Chakraborty, "Validation and implementation of algebraic LES modelling of scalar dissipation rate for reaction rate closure in turbulent premixed combustion," *Combust. Flame*, **161**, 3134–3153, 2014.

[145] D. Butz, Y. Gao, A. M. Kempf, and N. Chakraborty, "Large eddy simulations of a turbulent premixed swirl flame using an algebraic scalar dissipation rate closure," *Combust. Flame*, **162**, 3180–3196, 2015.

[146] Z. X. Chen, S. Ruan, and N. Swaminathan, "Large eddy simulation of flame edge evolution in a spark-ignited methane-air jet," *Proc. Combust. Inst.*, **36**, 1645–1652, 2017.

[147] I. Langella, Z. X. Chen, N. Swaminathan, and S. K. Sadasivuni, "Large-eddy simulation of reacting flows in industrial gas turbine combustor," *J. Propulsion Power*, **34**, 1269–1284, 2018.

[148] J. C. Massey, I. Langella, and N. Swaminathan, "Large eddy simulation of a bluff body stabilised premixed flame using flamelets," *Flow Turbulence Combust.*, **101**, 973–992, 2018.

[149] Z. X. Chen, N. Swaminathan, M. Stöhr, and W. Meier, "Interaction between self-excited oscillations and fuel-air mixing in a dual swirl combustor," *Proc. Combust. Inst.*, **37**, 2325–2333, 2019.

[150] Y. Gao, N. Chakraborty, and N. Swaminathan, "Dynamic closure of scalar dissipation rate for large eddy simulations of turbulent premixed combustion: A direct numerical simulations analysis," *Flow Turbulence Combust.*, **95**, 775–802, 2015.

[151] S. Ruan, N. Swaminathan, and O. Darbyshire, "Modelling of turbulent lifted jet flames using flamelets: a priori assessment and a posteriori validation," *Combust. Theory Modelling*, **18**, 295–329, 2014.

[152] Z. Chen, S. Ruan, and N. Swaminathan, "Simulation of turbulent lifted methane jet flames:effects of air-dilution and transient flame propagation," *Combust. Flame*, **162**, 703–716, 2015.

[153] H. Kolla, J. W. Rogerson, N. Chakraborty, and N. Swaminathan, "Scalar dissipation rate modelling and its validation," *Combust. Sci. Technol.*, **181**, 518–535, 2009.

[154] O. Darbyshire and N. Swaminathan, "A presumed joint PDF model for turbulent combustion with varying equivalence ratio," *Combust. Sci. Technol.*, **184**, 2036–2067, 2012.

[155] I. Ahmed and N. Swaminathan, "Simulation of spherically expanding turbulent premixed flames," *Combust. Sci. Technol.*, **185**, 1509–1540, 2013.

[156] I. Ahmed and N. Swaminathan, "Simulation of turbulent explosion of hydrogen-air mixtures," *J. Hydrogen Energy*, **39**, 9562–9572, 2014.

[157] S. Ruan, N. Swaminathan, M. Isono, T. Saitoh, and K. Saitoh, "Simulation of premixed combustion with varying equivalence ratio in gas turbine combustor," *J. Propulsion Power*, **31**, 861–871, 2015.

[158] D. Bradley, "How fast can we burn?" *Proc. Combust. Inst*, **24**, 247–262, 1992.

[159] D. Bradley, P. H. Gaskell, and X. J. Gu, "Application of a Reynolds stress, stretched flamelet, mathematical model to computations of turbulent burning velocities and comparison with experiments," *Combust. Flame*, **96**, 221–248, 1994.

[160] J. A. van Oijen and L. P. H. de Goey, "Modelling of premixed laminar flames using flamelet–generated manifolds," *Combust. Sci. Tech.*, **161**, 113–137, 2000.

[161] A. Y. Poludnenko and E. S. Oran, "The interaction of high-speed turbulence with flames: global properties and internal flame structure," *Combust. Flame*, **157**, 995–1011, 2010.

[162] H. Kolla and N. Swaminathan, "Strained flamelets for turbulent premixed flames II: Laboratory flame results," *Combust. Flame*, **157**, 1274–1289, 2010.

[163] E. Knudsen, H. Kolla, E. R. Hawkes, and H. Pitsch, "LES of a premixed jet flame DNS using a strained flamelet model," *Combust. Flame*, **160**, 2911–2927, 2013.

[164] R. Sankaran, E. R. Hawkes, J. H. Chen, T. Lu, and C. K. Law, "Structure of a spatially developing turbulent lean methane-air Bunsen flame," *Proc. Combust. Inst.*, **31**, 1291–1298, 2007.

[165] P. Trisjono, K. Kleinheinz, E. R. Hawkes, and H. Pitsch, "Modeling turbulence-chemistry interaction in lean premixed hydrogen flames with a strained flamelet model," *Combust. Flame*, **174**, 194–207, 2016.

[166] C. D. Pierce and P. Moin, "A dynamic model for subgrid-scale variance and dissipation rate of a conserved scalar," *Phys. Fluids*, **10**, 3041–3044, 1998.

[167] N. Chakraborty and N. Swaminathan, "Effects of Lewis number on scalar variance transport in premixed flames," *Flow Turbulence Combust.*, **87**, 261–292, 2011.

[168] Z. Chen, "Simulation of partially premixed turbulent flames," Ph.D. dissertation, University of Cambridge, Department of Engineering, Cambridge, UK, https://doi.org/10.17863/CAM.7488, 2016.

[169] S. F. Ahmed and E. Mastorakos, "Spark ignition of lifted turbulent jet flames," *Combust. Flame*, **146**, 215–231, 2006.

[170] G. Lacaze, E. Richardson, and T. Poinsot, "Large eddy simulation of spark ignition in a turbulent methane jet," *Combust. Flame*, **156**, 1993–2009, 2009.

[171] H. Zhang, A. Giusti, and E. Mastorakos, "LES/CMC modelling of ignition and propagation in a non-premixed methane jet," *Proc. Combust. Inst.*, **37**, 2125–2132, 2019.

[172] C. Dopazo, L. Cifuentes, J. Hierro, and J. Martin, "Micro-scale mixing in turbulent constant density reacting flows and premixed combustion," *Flow Turbulence Combust.*, **96**, 547–571, 2016.

[173] N. Swaminathan and R. W. Grout, "Interaction of turbulence and scalar fields in premixed flames," *Phys. Fluids*, **18**, 045 102–1 – 9, 2006.

[174] N. Chakraborty and N. Swaminathan, "Influence of the Damköhler number on turbulence–scalar interaction in premixed flames I: Physical insight," *Phys. Fluids*, **19**, 045 103–1 – 10, 2007.

[175] C. Dopazo and L. Cifuentes, "The physics of scalar gradients in turbulent premixed combustion and its relevance to modeling," *Combust. Sci. Technol.*, **188**, 1376–1397, 2016.

[176] A. Obukhov et al., "Am obukhov, izv. akad. nauk sssr, ser. geogr. geofiz. 13, 58 1949." *Izv. Akad. Nauk SSSR, Ser. Geogr. Geofiz.*, **13**, p. 58, 1949.

[177] S. Corrsin, "On the spectrum of isotropic temperature fluctuations in an isotropic turbulence," *J. App. Phys.*, **22**, 469–473, 1951.

[178] G. K. Batchelor, "Small-scale variation of convected quantities like temperature in turbulent fluid part 1. general discussion and the case of small conductivity," *J. Fluid Mech.*, **5**, 113–133, 1959.

[179] G. K. Batchelor, I. D. Howells, and T. A. Townsend, "Small-scale variation of convected quantities like temperature in turbulent fluid part 2. the case of large conductivity," *J. Fluid Mech.*, **5**, 134–139, 1959.

[180] S. Corrsin, "The reactant concentration spectrum in turbulent mixing with a first-order reaction," *J. Fluid Mech.*, **11**, 407–416, 1961.

[181] S. Corrsin, "Further generalization of Onsager's cascade model for turbulent spectra," *Phys. Fluids*, **7**, 1156–1159, 1964.

[182] E. O'Brien, On the Statistical Behavior of a Dilute Reactant in Isotropic Turbulence. PhD dissertation, The Johns Hopkins University," 1960.

[183] E. O'Brien, "Lagrangian history direct interaction equations for isotropic turbulent mixing under a second-order chemical reaction," *Phys. Fluids*, **11**, 2328–2335, 1968.

[184] Y. Pao, "Statistical behavior of a turbulent multicomponent mixture with first-order reactions," *AIAA J.*, **2**, 1550–1559, 1964.

[185] M. Ulitsky and L. Collins, "Application of the eddy damped quasi-normal markovian spectral transport theory to premixed turbulent flame propagation," *Phys. Fluids*, **9**, 3410–3430, 1997.

[186] R. Knaus and C. Pantano, "On the effect of heat release in turbulence spectra of non-premixed reacting shear layers," *J. Fluid Mech.*, **626**, 67–109, 2009.

[187] M. Martın and G. Candler, "Effect of chemical reactions on decaying isotropic turbulence," *Phys. Fluids*, **10**, 1715–1724, 1998.

[188] F. A. Jaberi, D. Livescu, and C. K. Madnia, "Characteristics of chemically reacting compressible homogeneous turbulence," *Phys. Fluids*, **12**, 1189–1209, 2000.

[189] H. Kolla, E. R. Hawkes, A. R. Kerstein, N. Swaminathan, and J. H. Chen, "On velocity and reactive scalar spectra in turbulent premixed flames," *J. Fluid Mech.*, **754**, 456–487, 2014.

[190] P. Paes and Y. Xuan, "Numerical investigation of turbulent kinetic energy dynamics in chemically-reacting homogeneous turbulence," *Flow, Turbulence Combust.*, 1–20, 2018.

[191] H. Kolla, X.-Y. Zhao, J. H. Chen, and N. Swaminathan, "Velocity and reactive scalar dissipation spectra in turbulent premixed flames," *Combust. Sci. Technol.*, **188**, 1424–1439, 2016.

[192] H. Tennekes and J. Lumley, *A First Course in Turbulence*. MIT MA: Press, 1972.

[193] L. Vervisch, E. Bidaux, K. Bray, and W. Kollmann, "Surface density function in premixed turbulent combustion modeling. similarities between probability density function and flame surface approaches," *Phys. Fluids*, **7**, 2496–2503, 1995.

[194] Y. Chen, N. Peters, G. Schneemann, N. Wruck, U. Renz, and M. Mansour, "The detailed flame structure of highly stretched turbulent premixed methane-air flames," *Combust. Flame*, **107**, 1996.

[195] F. A. Williams, "An approach to turbulent flame theory," *J. Fluid Mech.*, **40**, 401–421, 1970.

4 Non-premixed and Partially Premixed Combustion Modeling

C. Fureby and X.-S. Bai

4.1 Introduction and Background

Turbulent non-premixed and partially premixed combustion are of practical importance in many different applications, including internal combustion engines, liquid fueled gas turbines, turbojet engines, furnaces, fires, and ramjet and scramjet engines. Most non-premixed problems involve mixing and subsequent reaction of fuel and oxidizer. While non-premixed flames, i.e., conventional diffusion flames, have drawn a large amount of attention from the combustion community in the past, partially premixed combustion has recently gained a widespread interest from various industry branches. This is primarily the result of the recent strengthening of the emission legislation around the world. Although premixed flames (see Chapter 3) would be preferred from the point of view of an emission, achieving complete premixing in industrial applications is very challenging. Another practical aspect is the fear of flashback, which needs to be prevented in many applications. The comparatively high level of nitrogen oxides (NOx) that usually results from diffusion combustion is the reason the industry has turned to partially premixed regimes and/or to staged combustion schemes. Issues of mixing times in practical combustion applications mean that although the overall air–fuel ratio may be designed to reduce NOx, the nonhomogeneous mixing produces local partial premixing.

Figure 4.1 presents some examples of turbulent non-premixed combustion, including traditional jet diffusion flames [1, 2], low-swirl stratified flames [3], supersonic high-temperature stratified jet-in-cross-flow combustion [4], and spray combustion in an aero gas turbine combustor [5]. Based on collected information from many different experimental sources, turbulent non-premixed jet flames appear to consist of wrinkled to highly wrinkled reaction layers, called flamelets, but as the turbulence intensity increases, these flamelets become increasingly shredded [6]. Cellular-like structures are also sometimes observed in turbulent non-premixed jet flames, and are related to the wrinkling and distortion of the stoichiometric isosurface at which combustion is usually considered to occur [7]. In the supersonic high-temperature stratified jet-in-cross-flow combustion case the instantaneous flame structure seems to consist of a set of broken/fragmented and thickened flamelets. The flame in the aero gas turbine combustor is not clearly visible, but often consists of a short, wrenched and wrinkled tube falling back on itself around the combustor centerline to allow the central recirculation region to develop. Since many gas turbine combustors are annular, these flames usually interact with neighboring flames, with the flow and pressure distribution potentially creating thermoacoustic oscillations and instabilities. Accordingly, turbulent non-premixed combustion can take on a variety of characteristic features due to the surrounding conditions, and is therefore not very easily described.

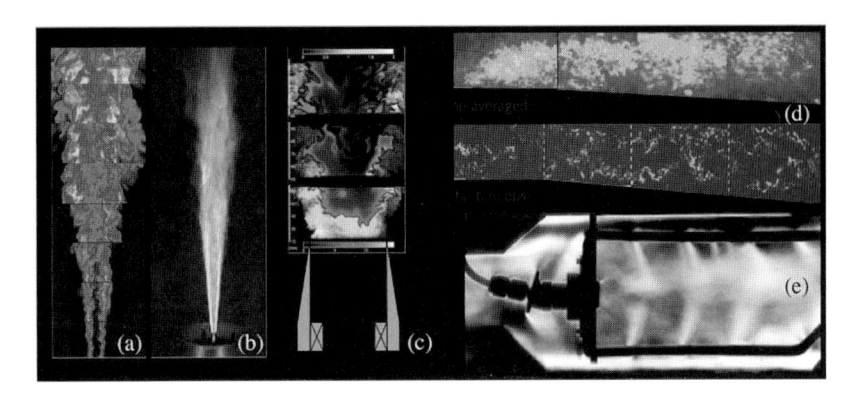

Figure 4.1 Turbulent non-premixed and partially premixed combustion. (a, b) Experimental images from jet diffusion flames [1, 2]. (c) Experimental images from a low-swirl flame [3]. (d) Mean (upper) and instantaneous (lower) experimental CH PLIF images of supersonic high-temperature stratified jet-in-cross-flow combustion [4]. (e) Instantaneous Schlieren images of aeronautical gas turbine combustion [5].

In this chapter we will discuss the physics and modeling of turbulent non-premixed and partially premixed combustion with a goal to provide an overview of the state-of-the-art knowledge in physics and computational modeling. An introductory description of the fundamental aspects of turbulent non-premixed and partially premixed combustion, in terms of different characteristic scales and combustion diagrams, will be followed by a brief but broad overview of different modeling capabilities, suitable for turbulent non-premixed and partially premixed combustion. Thereafter, a comparison between experimental data and a number of large eddy simulation (LES) predictions of turbulent non-premixed and partially premixed combustion will be presented. This will serve the purpose of describing how well turbulent non-premixed and partially premixed combustion can be modeled using modern methodologies. Following this, a more elaborate description of the intrinsic flow and combustion physics will be provided, facilitated by the LES predictions. Finally, some concluding remarks will be offered.

4.2 Regime Diagrams for Turbulent Non-premixed and Partially Premixed Combustion

Characterizing turbulent non-premixed combustion is considerably more difficult than turbulent premixed combustion (see Section 2.1), since non-premixed flames do not propagate, i.e., they do not exhibit a characteristic speed. Additionally, there is no obvious length scale, and the flame thickness is governed by the topology of mixing layers between the fuel and the oxidizer. Therefore, in order to develop a regime diagram for non-premixed combustion, it is necessary to identify relevant variables that influence the flame structure. Since many chemical reactions occur in a real flame, it is difficult to define a unique characteristic chemical time scale, τ_c, based on the chemical kinetics. We may, however, proceed, following the discussion on turbulent premixed combustion in Chapter 2, and simply assume that $\tau_c = \delta_u/s_u = \alpha/s_u^2$, in which $\delta_u = \alpha/s_u$ is the laminar flame thickness, α is the thermal diffusion coefficient, and s_u the laminar flame speed. The principal fluid mechanical process in non-premixed combustion is the mixing between fuel and oxidizer, which is characterized by the scalar dissipation rate at stoichiometric conditions, $\chi_{st} = (\alpha|\nabla z|^2)_{st}$, in which z is the mixture fraction and the subscript st denotes the evaluation at stoichiometric conditions. The mixture fraction, z, measures the degree of mixing, and is typically normalized to 1 in the fuel stream and 0 in the oxidizer stream. The inverse of the scalar dissipation rate, χ, is a time-scale representative of the mixing and defined by $\tau_\chi = \chi_{st}^{-1}$. As $\tau_\chi = \ell_\chi/v_\chi = \chi_{st}^{-1}$ and $\ell_\chi = \alpha/v_\chi$, it is possible to define the mixing related length and velocity scales as $\ell_\chi = \tau_\chi v_\chi = \sqrt{\alpha/\chi_{st}}$ and $v_\chi = \ell_\chi/\tau_\chi = \sqrt{\alpha\chi_{st}}$, respectively. From this we may define a Damköhler (Da) number, $Da_\chi = \tau_\chi/\tau_c$, which can be further expressed in terms of the mixing length and velocity scales, ℓ_χ and v_χ, as $Da_\chi = (\ell_\chi/\delta_u)/(v_\chi/s_u)$. A difficulty with this definition, however, is that it is not associated with the same length and velocity scales as used in the integral Re number, $Re_I = \ell_I v'/\nu$, characterizing the flow. This can be regulated by simply expanding the definition of Da_χ by the ratios ℓ_χ/ℓ_I and v'/v_χ, and rearranging terms so that $Da_\chi = (v'/v_\chi)Da_I(\ell_\chi/\ell_I)$ or, alternatively, $Da_I = (v_\chi/v')Da_\chi(\ell_I/\ell_\chi)$, in which we can recognize the conventional Da number $Da_I = (\ell_I/\delta_u)/(v'/s_u)$. If we further assume that the mixing scales are on the order of the Kolmogorov scales, ℓ_K and v_K, such that $Da_I = (\ell_I/\ell_K)Da_\chi(v_K/v')$, using the relations between the integral and Kolmogorov scales ($\ell_I/\ell_K = Re_I^{3/4}$ and $v'/v_K = Re_I^{1/4}$), we finally arrive at the expression $Da_I = Da_\chi Re_I^{1/2}$. Furthermore, from the definitions of Da and Re_I, using the approximation $\delta_u = \nu/s_u$ it follows that $v'/s_u = Da_\chi^{-2/3}(\ell_I/\delta_u)^{1/3}$. Figure 4.2a shows a non-premixed combustion diagram based on the logarithms of Da_I and Re_I [8], and alternatively, Fig. 4.2b shows a non-premixed combustion diagram based on the velocity and length scale ratios, similar to that employed for premixed combustion in Chapter 3.

The flame is said to be laminar for $Re < 1$, and can otherwise be considered turbulent. For $\ell_I/\delta_u < 1$, all turbulent eddies penetrate the reaction zone to enhance the intensity of the transport processes in the combustion front, causing intense mixing and reaction. This regime is called the *distributed reaction regime*. Based on DNS of flame/vortex interactions of Cuenot and Poinsot [8], the remaining flamelet regime can be further subdivided into quenching flamelets, unsteady flamelets, and flamelets, depending on how the vortices interact with the flame. More precisely, the direct

Non-premixed and Partially Premixed Combustion Modeling

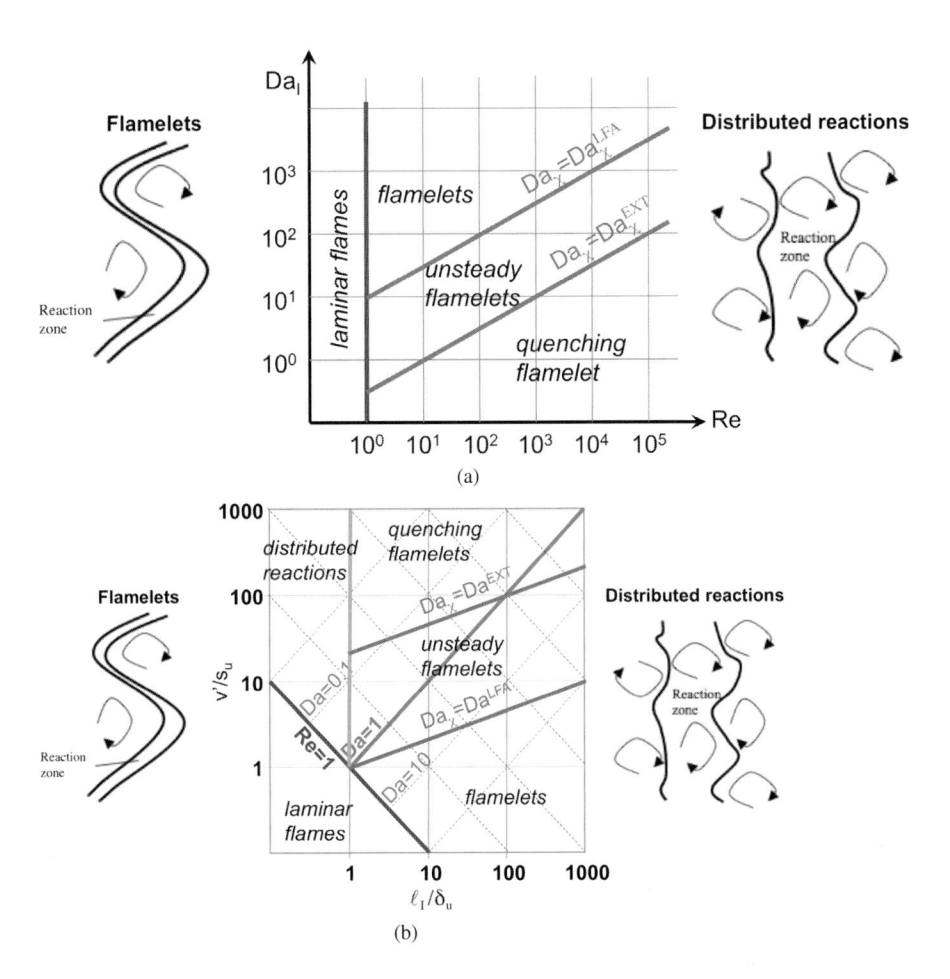

Figure 4.2 Turbulent non-premixed combustion diagrams (center) together with sketches of two different generic turbulent flames.

numerical simulation (DNS) results [8] are able to distinguish between quenching flamelets, unsteady flamelets, and flamelets, and depending on the value of $Da_\chi = (\ell_\chi/\delta_u)/(v_\chi/s_u)$, which may also be expressed as $Da_\chi = (\ell_I/\delta_u)^{1/2}/(v'/s_u)^{3/2}$, different combustion regimes can be distinguished. In the *flamelet regime*, defined by $Da_\chi < Da^{LFA}$, in which Da^{LFA} is the critical Da number for the laminar flamelet assumption (LFA) to be valid, the non-premixed turbulent flame behaves like a laminar flame element, a flamelet, embedded in turbulence. In the *unsteady flamelet regime*, defined by $Da^{LFA} < Da_\chi < Da^{EXT}$, in which Da^{EXT} is the critical Da number for extinction, unsteady effects were observed to wrinkle and otherwise distort the laminar flame element. In the *quenching flamelets regime*, defined by $Da_\chi < Da^{EXT}$, the non-premixed flames tend to gradually blow out with increasing turbulence intensity.

4.3 Theoretical Models of Non-premixed and Partially Premixed Combustion

In this section we discuss the fundamental theoretical modeling elements required to model non-premixed and partially premixed combustion based on the representation of the combustion chemistry and the associated reaction mechanism. The effects of turbulence will be addressed separately in Section 4.4. Premixed and non-premixed combustion will be discussed first, followed by the combined and more complex field of partially premixed combustion. Premixed flames are only briefly discussed here for convenience of discussion of partially premixed flames. A detailed discussion on premixed flames is given in Chapters 2 and 3.

4.3.1 Premixed Combustion

In premixed combustion, the fuel and oxidizer are mixed prior to entering the reaction zone. As the premixed flame is a reaction wave propagating from burned to fresh gases, the classical approach to this modeling situation is based on a reaction coordinate, commonly referred to as the reaction progress variable, c. The progress variable can be defined with the help of any quantity such as temperature, enthalpy, or mass fractions, provided it is bounded by a single value in the unburnt gas (zero) and another in the burned gas (unity). Across the flame, the intermediate values describe the progress of the reaction to turn the fresh gas into burned gas penetrating the flame sheet. Although its definition is not unique, the choice of c should be guided so that (1) c should evolve in a monotonic manner from fresh to burnt gases and the species derivative in c-space should stay moderate for tabulation accuracy, and that (2) the reactive scalars from which c is constructed should all evolve on comparable time scales. Various definitions for the progress variable have been employed for, e.g., hydrocarbon combustion [9–12]. A more generalized definition is provided by Najafi-Yazdi et al. [13], in which $c = \sum_{i=1}^{N} \omega_i Y_i$ where Y_i are the mass fractions and ω_i weighting coefficients obtained from singular value decompositions of the matrix of flamelet solutions. Under the assumption of species independent Schmidt numbers or unity Lewis number the transport equations for the species mass fractions becomes

$$\partial_t(\rho c) + \nabla \cdot (\rho \mathbf{v} c) = \nabla \cdot (D_c \nabla c) + \dot{w}_c, \tag{4.1}$$

in which D_c is the diffusivity of c, and $\dot{w}_c = \sum_{i=1}^{N} \omega_i \dot{w}_i$ with $\dot{w}_i$ being the species reaction rates. The right-hand side of Eq. (4.1) represents the reaction diffusion unbalance, where the reaction source term, $\dot{w}_c$, can be obtained either from its definition, using tabulated values of $\dot{w}_i$, or from a model. The simplest model, assuming an infinitely thin flame, suggests that $\dot{w}_c$ can be expressed as $\dot{w}_c = \rho_u s_u \delta(c - c_0)$, in which δ is the Dirac function, s_u the laminar flame speed, and $c = c_0$ the location of the flame front.

Alternatively, following Williams [14], a level-set method can be used to describe the evolution of the flame front as an interface between the unburned and burned gases. The level-set function G is a scalar field defined such that the flame front position is at $G = G_0$, and that G is negative in the unburned mixture. The instantaneous and local G-equation may be derived by considering the instantaneous flame surface.

An implicit representation of this surface can be given as $G(\mathbf{x},t) - G_0 = 0$, which defines the level-set function G. Here, $\mathbf{x}$ is the vector of space coordinates. By differentiating this expression one finally obtains

$$\partial_t G + (d\mathbf{x}_f/dt) \cdot \nabla G = 0, \quad (4.2)$$

where $\mathbf{x}_f$ is the flame front location. If the curvature radius of the instantaneous flame front is locally larger than the flame thickness, δ_u, the flame front propagation speed is given by

$$d\mathbf{x}_f/dt = \mathbf{v} + s_u \mathbf{n}, \quad (4.3)$$

in which $\mathbf{n} = -\nabla G/|\nabla G|$ is the flame normal vector into the unburned mixture. Note that in Section 4.2, s_u is the laminar flame speed of a planar unstretched premixed flame. Here, it is the modeled local displacement speed, cf. Eq. (3.24) of Chapter 3, neglecting the effect of flame stretch. Combining Eqs. (4.2) and (4.3) yields the instantaneous G-equation,

$$\partial_t G + v \cdot \nabla G = s_u |\nabla G|, \quad (4.4)$$

which is a level-set Hamilton–Jacobi equation. Since this equation has been derived from (4.2) and (4.3), which both only describe the flame surface, also (4.4) is valid at the flame surface only. The remaining G-field is arbitrary and commonly defined to be a distance function.

Under the assumption of equal D_i the species transport equations can therefore be replaced by Eq. (4.1) or (4.2), and at the same time introduce the flamelet library, $Y_i = Y_i^{lib}(c)$, or, $Y_i = Y_i^{lib}(G)$, which describes the relation between the species mass fractions, Y_i, and the reaction progress variable, c, or the level set function, G. Flamelet libraries can be constructed for each equivalence ratio, ϕ, and inlet temperature, T_0, and so forth. so that a more generalized flamelet library, $Y_i = Y_i^{lib}(c, \phi, T_0, \ldots)$, is obtained. Moreover, for premixed combustion simulations, both the c- and G-equations are used in the context of Reynolds-averaged Navier–Stokes (RANS), large-eddy simulation (LES), and direct numerical simulation (DNS) but supplemented with relevant modeling to represent the effects of the averaging (for RANS) or depletion of eddy scales (for LES).

4.3.2 Non-premixed Combustion

In non-premixed combustion the fuel and oxidizer enter the reaction zone via separate streams, so that the fuel and oxidizer must mix before combustion can take place. In most combustion applications, the chemical reaction rates are fast compared to the mixing rates so that the reactions are completed as soon as the reactants are mixed. The classical approach to this modeling situation is based on describing the mixing using a conserved scalar. The fast chemistry assumption further implies that the instantaneous species mass fractions and temperature are functions only of the conserved scalar. The functions are nonlinear, however, and the central problem in the conserved scalar equilibrium (CSE) model, e.g., [15, 16], becomes linking the species mass fractions, Y_i, and temperature, T, to those of the conserved scalar.

There are several scalar variables that are conserved under chemical reactions that all can be used as a basis for describing the mixing and subsequent reactions in a non-premixed reactant flow. As we usually assume the chemical reactions to be fast compared to the mixing, it is sufficient to then consider a single step global reaction of the form

$$\nu_F F + \nu_{O_2} O_2 \Rightarrow \nu_{CO_2} CO_2 + \nu_{H_2O} H_2O, \quad (4.5)$$

where F denotes the (hydrogen or hydrocarbon) fuel and ν_k are the stoichiometric coefficients. For all irreversible, single-step reaction mechanisms, the reaction rate, $\dot{w}$, is related to the species reaction rates, $\dot{w}_i$, through $\dot{w} = \dot{w}_i/(M_i \nu_i)$, which is the same for all species i. For a single-step irreversible reaction mechanism the species transport equations becomes

$$\partial_t(\rho Y_i) + \nabla \cdot (\rho \mathbf{v} Y_i) = \nabla \cdot (D_i \nabla Y_i) + \dot{w}_i, \quad (4.6)$$

with $i = \{F, O_2, CO_2, H_2O\}$, in which $\dot{w}_i = (M_i \nu_i)\dot{w}$. Note also that $Y_{N_2} = 1 - \Sigma_{i=1} Y_i$. By dividing Eq. (4.6) by $M_i \nu_i$, while assuming the species and diffusivities are equal and constant, Eq. (4.6) becomes

$$\partial_t(\rho \alpha_i) + \nabla \cdot (\rho \mathbf{v} \alpha_i) = \nabla \cdot (D_i \nabla \alpha_i) + \dot{w}_i/(M_i \nu_i), \quad (4.7)$$

in which $\alpha_i = Y_i/(M_i \nu_i)$. If we now select a reference species, say $i = O_2$, then we can introduce the coupling functions $\beta_{i,O_2} = \alpha_i - \alpha_{O_2}$ so that Eq. (4.7) becomes

$$\partial_t(\rho \beta_{i,O_2}) + \nabla \cdot (\rho \mathbf{v} \beta_{i,O_2}) = \nabla \cdot (D_i \nabla \beta_{i,O_2}). \quad (4.8)$$

The coupling functions, β_{i,O_2}, must be normalized so that the conditions in the fuel and oxidizer streams are satisfied, resulting in the coupling functions $z_{i,O_2} = (\beta_{i,O_2} - \beta_{i,O_2}^O)/(\beta_{i,O_2}^F - \beta_{i,O_2}^O)$, in which superscripts F and O denote fuel and oxidizer streams, respectively. The coupling function z_{i,O_2} then also by virtue of Eq. (4.8) satisfy the convection–diffusion equations,

$$\partial_t(\rho z_{i,O_2}) + \nabla \cdot (\rho \mathbf{v} z_{i,O_2}) = \nabla \cdot (D_i \nabla z_{i,O_2}). \quad (4.9)$$

Since the equations (4.9) are of the same form and have the same initial and boundary conditions, they collapse to a single equation for the mixture fraction, $z = z_{i,O_2}$,

$$\partial_t(\rho z) + \nabla \cdot (\rho \mathbf{v} z) = \nabla \cdot (D_i \nabla z). \quad (4.10)$$

The intrinsic flame structure, i.e. how Y_i depends on the mixture fraction z, can be derived by inserting the coupling functions $\beta_{i,O_2} = \alpha_i - \alpha_{O_2}$, in which $\alpha_i = Y_i/(M_i \nu_i)$, into the coupling functions $z_{i,O2} = (\beta_{i,O_2} - \beta_{i,O_2}^O)/(\beta_{i,O_2}^F - \beta_{i,O_2}^O)$ for pairs of O_2 and other species. Hence,

$$\begin{aligned} s Y_{C_n H_m} - Y_{O_2} + Y_{O_2}^O &= z(s Y_{C_n H_m}^F + Y_{O_2}^O), \\ p Y_{CO_2} - Y_{O_2} + Y_{O_2}^O &= (z-1) Y_{O_2}^O, \\ q Y_{H_2O} - Y_{O_2} + Y_{O_2}^O &= (z-1) Y_{O_2}^O, \end{aligned} \quad (4.11)$$

with $s = M_{O_2}\nu_{O_2}/M_F\nu_F$, $p = M_{O_2}\nu_{O_2}/M_{CO_2}\nu_{CO_2}$ and $q = M_{O_2}\nu_{O_2}/M_{H_2O}\nu_{H_2O}$ being the stoichiometric ratios for each pair of species. From Eq. (4.11) it also follows that $z_{st} = 1/(\Theta + 1)$ is the stoichiometric mixture fraction value, in which $\Theta = sY_F^F/Y_{O_2}^O$. From relations (4.11), we can extract the dependence of Y_i on z. This dependence is commonly referred to as a flamelet library (Fig. 4.3), and describes the intrinsic structure of a laminar flame that can be used as basis for formulating turbulent combustion models, as will be elaborated on later.

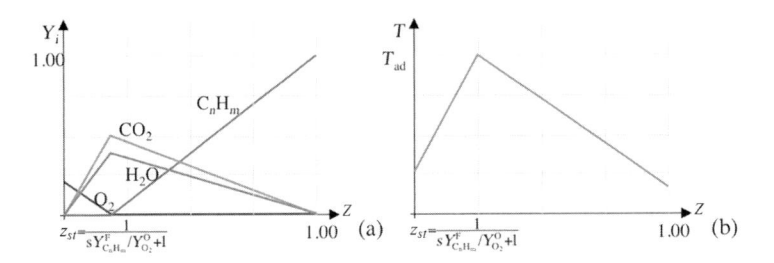

Figure 4.3 Schematic representation of the species dependence on the mixture fraction, z, in the conserved scalar equilibrium model.

Under the assumption of equal D_i, the species transport equations can therefore be replaced by Eq. (4.10), and at the same time we introduce the flamelet library $Y_i = Y_i^{lib}(z)$ that describes the relation between the species mass fractions, Y_i, and the mixture fraction, z. In the same way as described for premixed flames flamelet libraries can be constructed for different inlet temperatures, T_0, and so forth. so that a more generalized flamelet library, $Y_i = Y_i^{lib}(z, T_0, \ldots)$, is obtained. Moreover, for non-premixed combustion simulations, the z-equations are used in the context of RANS, LES, and DNS but supplemented with relevant modeling to represent the effects of the averaging (for RANS) or depletion of eddy scales (for LES).

4.3.3 Partially Premixed and Stratified Combustion

The main challenge in the modeling of partially premixed and stratified combustion is to achieve a dual description of varying premixed fronts (i.e., defined for a nonuniform distribution of local equivalence ratios) and trailing diffusion flames. Premixed and non-premixed turbulent combustion are two idealized regimes that lead to flames with different properties. The premixed front is thin and propagates, whereas the diffusion flame is mixing controlled and does not propagate. In light of Sections 4.3.2 and 4.3.3, we thus need to incorporate both z and c (or alternatively G) in our model of partially premixed and stratified combustion. Focusing here on the reaction progress variable, c, this may thus be defined as $c = Y_c/Y_c^{Eq}$ in which $Y_c = Y_{CO} + Y_{CO_2} + Y_{H_2} + Y_{H_2O}$, where Y_c^{Eq} denotes the equilibrium value of Y_c. A transport equation for the progress variable, c, may then be consistently derived from the fuel mass-fraction transport equation, using the definition of c and the mixture fraction equation (4.10), such that

$$\partial_t(\rho z) + \nabla \cdot (\rho \mathbf{v} z) = \nabla \cdot (D \nabla z),$$
$$\partial_t(\rho c) + \nabla \cdot (\rho \mathbf{v} c) = \nabla \cdot (D \nabla c) + \dot{w}_c + \frac{2}{Y_c^{Eq}(z)} \frac{dY_c^{Eq}(z)}{dz} D (\nabla c \cdot \nabla z) \quad (4.12)$$
$$+ \frac{c}{Y_c^{Eq}(z)} \frac{d^2 Y_c^{Eq}(z)}{dz^2} D (\nabla z \cdot \nabla z).$$

The left-hand side of Eq. (4.12) contains the unsteady and convective terms and the right-hand side contains diffusion, $\nabla \cdot (D \nabla c)$, chemical reactions, $\dot{w}_c$, and two additional terms, the third term and the fourth term, respectively. These two terms are not present in (4.1) and arise due to the normalization of Y_c. The advantage of using c instead of Y_c is that the variable is normalized so closures developed for homogenous premixed combustion can be used [17], but one has to consider the terms accounting for the variable stoichiometry of the mixture. The third term contains the cross-scalar dissipation rate between the progress variable, c, and the mixture fraction, z, $\rho \chi_{cz} = D \nabla c \cdot \nabla z = D|\nabla c||\nabla z|\mathbf{n}_c \cdot \mathbf{n}_z$; which represents the transport of reactants across the isosurface of z, whereas the fourth term contains the scalar dissipation rate, $\rho \chi_{zz} = D \nabla z \cdot \nabla z = D|\nabla z|^2$, which represents the rate of scalar mixing at the smallest scales and has a significant effect on extinction and ignition in non-premixed combustion. Furthermore, $\mathbf{n}_c = -\nabla c/|\nabla c|$ and $\mathbf{n}_z = -\nabla z/|\nabla z|$ are the unit normal vectors to c and z, respectively. With the definitions of χ_{zz} and χ_{cz} introduced, (4.12) becomes

$$\partial_t(\rho c) + \nabla \cdot (\rho \mathbf{v} c) = \nabla \cdot (D \nabla c) + \dot{w}_c + \frac{2}{Y_c^{Eq}(z)} \frac{dY_c^{Eq}(z)}{dz} \rho \chi_{cz} + \frac{c}{Y_c^{Eq}(z)} \frac{d^2 Y_c^{Eq}(z)}{dz^2} \rho \chi_{zz}. \quad (4.13)$$

The second term of the right-hand side of (4.13) signifies premixed (with mixture stratification) mode combustion, the third term signifies non-premixed mode combustion, and the fourth term denotes mixed mode combustion resulting from interactions of ∇c and ∇z. When the flame front is assumed thinner than any other flow scales, the transport of reactants through the flame is supposed to be one-dimensional in the direction $-\mathbf{n}_c$, and transport phenomena in the other directions are negligible. For partially premixed flames propagating along a given isomixture fraction surface, $\mathbf{n}_z$ points in a direction almost perpendicular to $\mathbf{n}_c$, and hence $\mathbf{n}_z \cdot \mathbf{n}_c \approx 0$. Under a flamelet hypothesis, one may then assume that effects resulting from interactions between varying premixed flamelets, defined for various values of z, are kept small compared to those leading to flamelet propagation in the normal direction $\mathbf{n}_c$ and the hypothesis $\chi_{cz} \approx 0$ may be formulated for the propagating part of the turbulent flame – a hypothesis that is not valid for the diffusion flame branches. If a propagation-based model for $\dot{w}_c$ is preferred, i.e., $\dot{w}_c = \rho_u s_u |\nabla c|$, then

$$\partial_t(\rho c) + \nabla \cdot (\rho \mathbf{v} c) = \nabla \cdot (D \nabla c) + \rho_u s_u \mathbf{n}_c |\nabla c| + \frac{2}{Y_c^{Eq}(z)} \frac{dY_c^{Eq}(z)}{dz} D|\nabla z||\mathbf{n}_z||\nabla c|$$
$$+ \frac{c}{Y_c^{Eq}(z)} \frac{d^2 Y_c^{Eq}(z)}{dz^2} D \nabla z \cdot \nabla z. \quad (4.14)$$

From Eq. (4.14), we find that the third term on the right-hand side modifies the flame propagation speed and the propagation direction, and the gradient of z might accelerate or decelerate the flame, depending on which direction it propagates.

For modeling purposes, it is interesting to compare the third term to the flame speed term (the second term), i.e., $D|\nabla z|/\rho_u s_u \approx 2\delta_u/\delta_z$, where δ_u and δ_z are the flame thickness, and the z-layer thickness, respectively. Hence, if z is close to homogeneity, δ_z is large compared to δ_u and the third term is negligible. Otherwise, the effect of the third term cannot be neglected a priori.

4.4 Modeling of Non-premixed and Partially Premixed Combustion

In this section, we will briefly describe a number of different LES methodologies that can be used, and have been used, to study non-premixed and partially premixed combustion. This summary is by no means complete but is intended to provide some guideline to which methods can be applied in different flame situations. Based on the previous theoretical discussion, it is clear that both premixed and non-premixed combustion are limiting cases, and the models discussed in this section will be models that can be used to simulate the full span of combustion regimes, including non-premixed, premixed, partially premixed, and stratified combustion. In the following, we will only discuss the combustion modeling. The modeling of the unresolved transport terms, resulting from the low pass filtering of the transport equation is assumed to be performed using any of the conventional subgrid models summarized by Sagaut [18].

4.4.1 Flamelet Models

Flamelet models can be considered an extension of infinitely fast chemistry models, assuming that the flame is thin compared to the length scales of the flow, and the flame behaves like an interface between fuel and oxidizer (in non-premixed combustion) [19], or between reactants and products (in premixed combustion) [20]. Due to the scale separation, it is convenient to decouple the flow simulation from that of the chemistry, which can be separately modeled using zero- and one-dimensional laminar flames, resulting in the flamelet library, Y_i^{lib}, which can be combined with c, G and/or z, and with information about how the turbulence influences mixing, self-ignition and fully developed turbulent combustion by means of other parameters such as the scalar dissipation rate, χ_{zz}, cross-scalar dissipation rate χ_{xz}, or the flame wrinkling, Ξ.

A flamelet model for all the aforementioned combustion regimes can be formulated based on the mass, momentum, and energy equations together with the transport equations for z and c, i.e., Eqs. (4.12) and (4.13). Several different approaches have been used, e.g., [9, 21, 22], to develop LES models based on (4.12) and (4.13), and here we will review these methods. The governing equations, including Eqs. (4.12) and (4.13), are first low-pass filtered, resulting in

$$\partial_t(\bar{\rho}\tilde{z}) + \nabla \cdot (\bar{\rho}\tilde{\mathbf{v}}\tilde{z}) = \nabla \cdot (D\nabla\tilde{z} - \mathbf{b}_z),$$

$$\partial_t(\bar{\rho}\tilde{c}) + \nabla \cdot (\bar{\rho}\tilde{\mathbf{v}}\tilde{c}) = \nabla \cdot (D\nabla\tilde{c} - \mathbf{b}_c) + \overline{\dot{w}_c} + \overline{\frac{2}{Y_c^{Eq}(z)}\frac{dY_c^{Eq}(z)}{dz}\rho\chi_{cz}},$$

$$+ \overline{\frac{c}{Y_c^{Eq}(z)}\frac{d^2 Y_c^{Eq}(z)}{dz^2}\rho\chi_{zz}}, \tag{4.15}$$

in which the bar and tilde denote filtered and density-weighted filtered variables, whereas $\mathbf{b}_z$ and $\mathbf{b}_c$ denote the unresolved transport terms associated with z and c, respectively.

The premixed combustion part of the c-equation (4.15) can according to [9, 21] be modeled using a conventional joint probability density function (PDF) approach,

$$\overline{\dot{w}_c} = \int_0^1 \int_0^1 \dot{w}_c(\zeta,\eta) P(\zeta,\eta) d\zeta d\eta, \tag{4.16}$$

where ζ and η are sample space variables for c and z respectively, and $P(\zeta,\eta)$ is the subgrid joint PDF, whereas Duwig and Fureby [22] instead propose to use a propagation formulation using a flame wrinkling or flame surface density factor. The flamelet reaction rate $\dot{w}_c(\zeta,\eta)$ can be obtained either from diffusion flamelets parameterized using c [9] or from unstrained planar laminar premixed flames computed over the whole range of z spanning the flammability limits [21]. The joint PDF employed by Langella et al. [21] assumes statistical independence between ζ and η so that $P(\zeta,\eta) \approx P(\zeta)P(\eta)$, which is quite common in LES. The marginal PDFs of ζ and η are modeled using beta functions $P(\zeta) = \beta(\zeta; \tilde{z}, \widetilde{z''^2})$ and $P(\eta) = \beta(\eta; \tilde{c}, \widetilde{c''^2})$, in which $\widetilde{z''^2}$ and $\widetilde{c''^2}$ are the subgrid variances of z and c, respectively. The variances $\widetilde{z''^2}$ and $\widetilde{c''^2}$ can be modeled either using algebraic expressions, e.g., [23], or using modeled transport equations as in [21].

The assumption of statistical independence of c and z has been examined within the RANS framework using DNS and experimental data [24]. It was shown that the assumption is reasonable for certain flames, e.g., rod-stabilized flames, whereas the assumption is not valid for Bunsen flames, in which the covariance effect of c and z is important. For the latter case, Darbyshire and Swaminathan proposed to use two marginal PDFs obtained using the beta functions for the given values of means and variances, combined together as per a prescribed covariance using copulas [24, 25]. Within the context of LES, the importance of the covariance effect of c and z at the subgrid level is, however, an open question and it depends on whether the grid is sufficiently fine to resolve the local flame structure, which can be satisfied for atmospheric conditions (lab-scale flames) but may not be for elevated pressures and temperatures [26]. Nevertheless, the agreement between the measured and computed averaged temperatures using the statistical independent marginal PDF of c and z at an aero-combustor exit is acceptable [27].

The mixed mode combustion part of the c-equation (4.15) have been shown to be an order of magnitude smaller than the other two in previous studies by Domingo et al. [28], but are included in the model developed by Duwig and Fureby [22], as the propagation formulation adopted by these authors facilitates including this term. When not using a propagation formulation, this term can be neglected for consistency with the joint PDF closure and for the sake of simplicity, since including this term would introduce additional modeling.

The non-premixed part of the c-equation (4.15) can be modeled using the marginal PDF of η following the ideas and concepts of [15, 29] such that

$$\overline{\frac{c}{Y_c^{Eq}(z)} \frac{d^2 Y_c^{Eq}(z)}{dz^2} \rho \chi_{zz}} = \bar{\rho} \tilde{c} \tilde{\chi}_{zz} \int_0^1 \frac{1}{Y_c^{Eq}(\eta)} \frac{d^2 Y_c^{Eq}(\eta)}{d\eta^2} P(\eta) d\eta, \qquad (4.17)$$

in which $\tilde{\chi}_{zz} = \tilde{D}|\nabla \tilde{z}|^2 + c_z(\nu_k/\Delta^2)\widetilde{z''^2}$ where Δ is the filter width, is the density weighted modeled scalar dissipation rate of the mixture fraction, in which the second term is the modeled subgrid contribution.

The flamelet approach described here allows for efficient tabulation of the filtered reaction rates as a function of the four parameters $\tilde{z}$, $\tilde{c}$, $\widetilde{z''^2}$, and $\widetilde{c''^2}$, that are provided by the LES. The reaction rates can be retrieved from look-up tables during the simulation. The look-up tables are typically generated for freely propagating one-dimensional planar laminar flames and can be obtained for any reaction global, skeletal, or detailed mechanism. Some elaborate details on the flamelets based modeling for partially premixed combustion can be found in Chen et al. [27].

4.4.2 Thickened Flame Model

The thickened flame model was developed for modeling of turbulent premixed flames, cf. Chapter 3, Section 3.2.2. This model can be used as a submodel for the premixed flame parts of a partially premixed flame, e.g., the leading lean and rich premixed flames of a lifted non-premixed flames. O'Rourke and Bracco [30] were the first to note that the laminar flame speed, $s_u \propto \sqrt{D\dot{w}/\rho^2}$, and flame thickness, $\delta_u \propto D/(\bar{\rho} s_u)$, can be used to rescale δ_u while preserving s_u. This can be used to artificially thicken a turbulent, marginally resolved, flame so that it can be resolved on an LES grid while maintaining s_u. Colin et al. [31] developed an LES model from these observations by reducing $\dot{w}$ by a factor F and increasing D with the same factor F, such that

$$\partial_t(\bar{\rho}\tilde{Y}_i) + \nabla \cdot (\bar{\rho}\tilde{\mathbf{v}}\tilde{Y}_i) = \nabla \cdot (FD_i \nabla \tilde{Y}_i) + \tfrac{1}{F} M_i \Sigma_{j=1}^M P_{ij} \dot{w}_j. \qquad (4.18)$$

For a thickened flame, the flame front wrinkling, Ξ_ℓ, is underestimated by a factor $E = \Xi_\ell/\Xi_\Delta$, in which Ξ_Δ is the resolved flame wrinkling. Following [31], this underestimation can be corrected by increasing s_u by E, so that the turbulent flame speed, s_t, becomes $s_t = Es_u$. This may in turn be accomplished by multiplying both D and $\dot{w}$ by the factor E such that

$$\partial_t(\bar{\rho}\tilde{Y}_i) + \nabla \cdot (\bar{\rho}\tilde{\mathbf{v}}\tilde{Y}_i) = \nabla \cdot (FED_i \nabla \tilde{Y}_i) + \tfrac{E}{F} M_i \Sigma_{j=1}^M P_{ij} \dot{w}_j. \qquad (4.19)$$

The flame thickening factor was originally considered to be of the form $F = \Delta/\delta_u$, which may be as large as 10, and can also vary over the flame. This formulation is, however, not ideal away from the reaction zone, where mixing dominates, and a modified expression for the flame thickening factor, F, was proposed by Legier et al. [32], such that

$$F = 1 + (F_{max} - 1)\Omega = 1 + (F_{max} - 1)\tanh(\beta \dot{w}/\dot{w}_{max}), \qquad (4.20)$$

in which $F_{max} = \max(\Delta/\delta_u, 1)$, $\dot{w}_{max}$ is the maximum of $\dot{w}$, and β is a parameter controlling the thickness of the transition layer between thickened and nonthickened zones. The limitation of F_{max} to unity ensures that the flame is not thinned when the grid is sufficiently fine.

The choice of E is somewhat arbitrary but requires a model that assumes that the internal structure of the flame is not altered by the turbulence. Many options are available but most often the model of Charlette et al. [33] is employed, in which

$$E = [1 + \min(\Delta/\delta_u, \Gamma_\Delta v'/s_u)]^\gamma, \qquad (4.21)$$

in which $\Gamma_\Delta = \Gamma_\Delta(\Delta/\delta_u, v'/s_u, Re_\Delta)$ is the ITNFS function [34], describing the net straining effect of all turbulent scales smaller than Δ, and $\gamma = 0.5$. More recently, a dynamic version of the LES-TF model have been developed by Wang et al. [35], who proposed to evaluate F dynamically, using information at the scales Δ and 2Δ, respectively. Although the LES thickened flame model was originally developed for global combustion chemistry, it can be used for skeletal and comprehensive combustion chemistry.

4.4.3 Localized Time Scales Combustion Models

Another versatile LES combustion modeling approach is based on a systematic generalization of the eddy dissipation concept (EDC) model of Magnussen et al. [36, 37], the fractal model (FM) of Giacomazzi et al. [38], and the partially stirred reactor (PaSR) model of Sabelnikov and Fureby [39, 40]. In all these three models, combustion is assumed to take place in fine-structure regions characterized by intense chemical activity, mixing, and vorticity embedded in a surrounding region of lower levels of vorticity and chemical activity as postulated by Corrsin [41], and observed using DNS by Tanahashi et al. [42]. The low-pass filtered reaction rates, $\bar{\dot{w}}_i$, can generally be expressed in terms of a PDF [43], so that $\bar{\dot{w}}_i = \int \wp(\Psi, \tilde{\Psi}, \ldots) \dot{w}_i(\Psi) d\Psi$, in which Ψ denotes the vector of dependent variables $\{\rho, T, Y_i\}$. As a consequence of the parting into reacting fine structures and surroundings, the PDF can be formulated as $\wp(\Psi) = \gamma^* \delta(\Psi^*) + (1 - \gamma^*) \delta(\Psi^0)$, in which δ is the Dirac delta function, γ^* the reacting volume fraction, whereas the superscripts (*) and (^{0}) denotes the fine structures and surroundings, respectively. This finally results in that the low-pass filtered species transport equations can be modeled as

$$\partial_t(\bar{\rho}\tilde{Y}_i) + \nabla \cdot (\bar{\rho}\tilde{\mathbf{v}}\tilde{Y}_i) = \nabla \cdot (D_i \nabla \tilde{Y}_i - \mathbf{b}_i) + \gamma^* \dot{w}_i(\bar{\rho}, T^*, Y_i^*)$$
$$+ (1 - \gamma^*)\dot{w}_i(\bar{\rho}, T^0, Y_i^0). \qquad (4.22)$$

To close Eq. (4.22) the reacting volume fraction, γ^*, and the dependent variables, $\{T, Y_i\}$, at the states of the reacting fine structures (*) and surrounding (^{0}), need to be provided. To accommodate this, each LES cell is presumed to consist of reacting fine structures and surroundings interacting through local mass and energy balance equations of the form $\bar{\rho}(Y_i^* - Y_i^0)/\tau^* = \dot{w}_i(Y_i^*, T^*)$ and $\bar{\rho}\Sigma_{i=1}^N(Y_i^*h_i^* - Y_i^0 h_i^0)/\tau^* = \Sigma_{i=1}^N h_{i,f}^\theta \dot{w}_i(Y_i^*, T^*)$, in which τ^* is the subgrid residence time. By noticing that $\tilde{Y}_i = \gamma^* Y_i^* + (1-\gamma^*)Y_i^0$ and $\tilde{T} = \gamma^* T^* + (1-\gamma^*)T^0$, these balance equations become

$$\begin{cases} \bar{\rho}(Y_i^* - \tilde{Y}_i) = (1-\gamma^*)\tau^* \dot{w}_i(Y_i^*, T^*), \\ \bar{\rho}\Sigma_{i=1}^N(Y_i^* h_i^*(T^*) - \tilde{Y}_i \tilde{h}_i(\tilde{T})) = (1-\gamma^*)\tau^* \Sigma_{i=1}^N h_{i,f}^\theta \dot{w}_i(Y_i^*, T^*). \end{cases} \quad (4.23)$$

The reacting volume fraction, γ^*, and the residence time, τ^*, must finally be modeled to obtain a closed set of equations. This may be achieved in different ways, resulting in the subdivision and recovery of the EDC, FM, and PaSR models, respectively.

In the EDC model [36, 37], γ^* and τ^* are both estimated from a theoretical model of the energy cascade consistent with the Kolmogorov hypothesis [44]. Using this cascade model [37], the modeled subgrid dissipation, ε, can be equated with the dissipation at the first level of the cascade process, $\varepsilon^{(1)}$, to provide expressions of the fine structure velocity, $v^* \approx v_K$, and length scales, $\ell^* \approx 2.5\ell_K$, respectively. Further assuming that the reacting volume fraction coincides with the fine structure intermittency factor [41], $\gamma^* = (v^*/v^{(1)})^3 = (v^*/k^{1/2})^3$, one can estimate the reacting volume fraction as $\gamma^* = 1.02(\nu/\Delta v')^{3/4}$. From the mass exchange between the fine structures, the subgrid residence time can be estimated as $\tau^* = 1.24(\Delta \nu/v'^3)^{1/2}$.

In the FM model, γ^* is estimated using an assumed fractal-like behavior of the fine structures. This results in that $\gamma^* = \gamma_N(\Delta/\ell_K)^{D_3-2}$, in which $\gamma_N = N_K/N_T$ is the ratio of the number N_K of Kolmogorov (ℓ_K) scales to the total number of scales, N_T, generated locally and D_3 is the local fractal dimension. To estimate γ_N, a model must be adopted to control the fractal generation process, and in [38] an analytical fit of the form $\gamma_N \approx 1 - [(0.36(\Delta/\ell_K - 1))/(1 + 0.0469(\Delta/\ell_K - 1)^{2.7})]$ is used. The fractal dimension is estimated using box-counting so that $D_3 = 3 - [\log(\pi)/\log(\Delta/\ell_K)]$. Combining these expressions, we find that γ^* is a function of the ratio of Δ to ℓ_K which has an asymptotic value of $\gamma^* = 1/\pi \approx 0.318$ for high cell Re numbers, $Re_\Delta = |\mathbf{v}|\Delta/\nu$.

In the PaSR model [40], the reacting volume fraction is estimated using spatiotemporal intermittency, and is modeled as $\gamma^* \approx \tau_c/(\tau^* + \tau_c)$. The chemical time-scale is modeled as $\tau_c \approx \delta_u/s_u$, where δ_u and s_u are the laminar flame thickness and flame speed, respectively. The modeling of τ^* is based on that the fine structure area-to-volume ratio is provided by the dissipative length scale, $\ell_D = (\nu\Delta/v')^{1/2}$, governed by the viscosity, ν, and the subgrid velocity stretch, v'/Δ, where $v' = \sqrt{2k/3}$ is the subgrid velocity fluctuations, and that the velocity associated with these structures is the Kolmogorov velocity, $v_K = (\nu\varepsilon)^{1/4}$, so that $\tau^* = \ell_D/v_K$. By combining the expressions for ℓ_D and v_K, using the Kolmogorov scales, $\ell_K = (\nu^3/\varepsilon)^{1/4}$ and $\tau_K = (\nu/\varepsilon)^{1/2}$, in which $\varepsilon = v'^3/\Delta$ is the subgrid dissipation, results in that the residence time is $\tau^* = \sqrt{\tau_K \tau_\Delta}$, in which $\tau_\Delta = \Delta/v'$ is the subgrid time scale. This was noted [45] to correctly represent the statistics of ε, and hence also of the small-scale mixing, which is essential to the onset of chemical reactions.

4.4.4 Transported PDF LES Combustion Models Based on Eulerian Stochastic Fields Models

In the Eulerian stochastic field (ESF) PDF method, the N_P notional particles used in the Lagrangian particle Eulerian mesh PDF method [46] are replaced by N_ϕ notional Eulerian fields that evolve according to stochastic partial differential equations (PDEs). The system of stochastic PDEs is devised such that its one-point one-time Eulerian joint PDF evolves according to the PDF transport equation,

$$\frac{\partial(\bar{\rho}P)}{\partial t} + \nabla \cdot [\bar{\rho}\tilde{v}P] = \nabla \cdot [(D_t)\nabla P]$$
$$+ \frac{\partial}{\partial \psi_\alpha}\left[c_m \frac{D_t}{\Delta^2}(\psi_\alpha - \langle\phi_\alpha\rangle)P\right] + \frac{\partial}{\partial \psi_\alpha}[\bar{\rho}s_\alpha(\psi)P], \quad (4.24)$$

in which P is the probability density function, D_t is the sum of molecular and turbulent diffusivity, c_m is a modeling coefficient, s_α are the source terms of the underlying (species) transport equations, Ψ is the sample space of the array of scalars ϕ, and $\langle\rangle$ denotes the averaging of values. The number of PDEs to be solved is equal to the product of the number of composition variables and the number of fields, $N_P \times N_\phi$, and the stochastic PDEs of the ESF method can be solved using (essentially) standard Eulerian computational fluid dynamics (CFD) methods.

Two independent developments of ESF-PDF method have appeared in the literature. One originated with Valino [47] and another was introduced by Sabelnikov and Soulard [48]. Different physical and mathematical arguments are invoked in the two developments. The Valino approach is based on the Ito stochastic calculus, assuming that the fields are smooth at the mesh resolution. Sabelnikov and Soulard have developed the stochastic field method using the alternative Stratonovich stochastic calculus (for further details see [49]).

Development and application of the Valino ESF approach has been the subject of several publications, e.g., [50, 51]. Here, we will summarize the ESF method as implemented following the approach introduced by Valino [48, 50, 51]. With superscript (n) referring to any one of the fields, the ESF equation corresponding to (4.24) can be written as

$$d\xi_\alpha^n = -(\bar{\rho}\tilde{v} \cdot \nabla \xi_\alpha^n)dt + \nabla \cdot (D_t \cdot \nabla \xi_\alpha^n)dt + \bar{\rho}\sqrt{2D_t/\bar{\rho}}\nabla \xi_\alpha^n \cdot d\mathbf{W}^n$$
$$- \tfrac{1}{2}c_d(\bar{\rho}/\tau_k)(\xi_\alpha^n - \tilde{\phi}_\alpha)dt + \bar{\rho}\dot{w}_\alpha^n dt, \quad (4.25)$$

in which $\mathbf{W}^n$ is a vector-valued Wiener process that varies in time, but is independent of spatial location. The first three terms on the right-hand side of (4.25) correspond, respectively, to advection, chemical reactions, and mixing. The terms involving the apparent turbulent diffusivity, D_t, correspond to a gradient transport model for turbulent velocity fluctuations.

In the absence of the term involving $d\mathbf{W}^n$, Eq. (4.25) could be recast as a PDE by dividing by dt and taking the limit $dt \rightarrow 0$. A key point in the Valino formulation is that each field is smooth at the grid scale,. The formulation of the stochastic term is consistent with this, since the same random increment is applied uniformly in each coordinate direction for each field. It is important that the numerical implementation of the stochastic term remains consistent with the intended interpretation of the stochastic PDE. An implementation that is consistent with the Valino development (using the Ito stochastic calculus) is to approximate $d\mathbf{W}^n$ at the beginning of each time-step, Δt, and to advance explicitly in time. A usual approach is to approximate $d\mathbf{W}^n$ by $\eta t^{1/2}$, where η is a dichotomic vector. By construction, estimates of density-weighted mean quantities are obtained by ensemble averaging over the fields, $\tilde{f} \approx \sum_{n=1}^{N_F} f(\xi^{(n)})/N_F$.

There are other fundamental limitations and practical implementation issues with ESF. The requirement that each field must remain spatially smooth, according to Valino, limits the class of mixing models that can be implemented. This excludes mixing models based on stochastic algorithms at the local cell level. Implementation of advanced mixing models in ESF thus remains an open area for research. Care is required at inflow/outflow boundaries when fluctuations are large, relative to the mean, and it is not clear what boundary condition should be enforced on a plane or axis of symmetry. While mean values must respect spatial symmetries, it is not evident that each individual field should be symmetric. The usual practice, and the one that is usually adopted, is to enforce symmetry for each stochastic field. The main motivation for Eulerian stochastic fields methods compared to Lagrangian particle based methods is the relative ease of implementation of Eulerian stochastic fields methods in Eulerian CFD codes.

4.4.5 Conditional Moment Combustion Model

The conditional moment closure (CMC) model was developed independently by Klimenko [52], and Bilger [53]. The CMC model calculates conditional moments at a fixed location within the flow, using modeled transport equations for the conditional moments of the scalars with no assumptions on the small-scale structure of reaction zones or on the relative time scale of chemistry and the turbulence. The conditional mean, Q_i, of the species mass fraction Y_i is defined by $Q_i(\mathbf{x},t) = \tilde{Y}_i(\mathbf{x},t)|_{z=\eta}$, where η is the sample space of the mixture fraction, $z = z(\mathbf{x},t)$. This also implies that $\tilde{Y}_i(\mathbf{x},t) = Q_i(\eta;\mathbf{x},t) + Y_i''(\mathbf{x},t)$, where Y_i'' are the fluctuations with $\tilde{Y}''|_{z=\eta} = 0$. The conditional mean $Q_i(\eta)$ is a nonrandom function of the variables $\mathbf{x}$ and t, and by inserting $\tilde{Y}_i = Q_i(\eta) + Y_i''$ into the species transport equations and differentiating, using Fick's law,

$$\rho \partial_t Q_i + \rho \mathbf{v} \cdot \nabla Q_i - D_i \nabla^2 z \frac{\partial^2 Q}{\partial \eta^2} - \nabla \cdot (D_i \nabla Q_i) - D_i (\nabla z) \cdot \left(\nabla \cdot \frac{\partial Q}{\partial \eta} \right)$$
$$+ \rho \partial_t Y_i'' + \rho \mathbf{v} \cdot \nabla Y'' - \nabla \cdot (D_i \nabla Y_i'') = \dot{w}_i. \tag{4.26}$$

The scalar dissipation term will here be denoted by $N = D\nabla^2 z$, and taking the conditional expectation of (4.26) with respect to $z(\mathbf{x},t) = \eta$ results in

$$\overline{\rho|_\eta} \partial_t Q_i + \overline{\rho|_\eta \mathbf{v}|_\eta} \cdot \nabla Q_i = \overline{N|_\eta} \frac{\partial^2 Q}{\partial \eta^2} + \overline{\dot{w}_i|_\eta} + e_Q + e_Y, \tag{4.27}$$

in which the source terms are defined by

$$e_Q = \overline{(\nabla \cdot (D\nabla Q_i) + D(\nabla z) \cdot (\nabla \cdot \frac{\partial Q}{\partial \eta}))|_\eta},$$
$$e_Y = -\overline{(\rho \partial_t Y_i'' + \rho \mathbf{v} \cdot \nabla Y_i'' - \nabla \cdot (D\nabla Y_i''))|_\eta}. \quad (4.28)$$

respectively. Several unclosed terms appear in the transport equation (4.27) including the source terms e_Q and e_Y, the conditional velocity $\overline{\mathbf{v}|_\eta}$, and the conditional scalar dissipation rate, $\overline{N|_\eta}$.

The e_Q-term can usually be neglected under high Reynolds number assumptions [54], but needs to be modeled if differential diffusion effects are to be included [55]. Similarly, the e_Y-term involves all the fluctuations around the conditional mean, must not be omitted in the presence of significant differential diffusion effects, but can be set to zero in the absence of differential diffusion and if a suitable set of conditioning variables is selected. In practice, a gradient diffusion approximation for e_Y is often employed [56]. The importance of e_Y, and the accuracy of its closure, strongly depends on flow and flame conditions, particularly for lifted flames where flame propagation is the dominating stabilization mechanism [57].

Several models for the conditional velocity, $\overline{\mathbf{v}|_\eta}$, have been proposed [58, 59], over the years. Based on the work of Swaminathan and Bilger [58], and Li and Bilger [60], using a combination of DNS and experimental data, it appears that the linear model, $\overline{\mathbf{v}|_\eta} = \tilde{\mathbf{v}} + [\widetilde{\mathbf{v}''z''}/\widetilde{z''^2}](\eta - \tilde{z})$, in which $\widetilde{\mathbf{v}''z''}$ is approximated by a gradient model such that $\widetilde{\mathbf{v}''z''} = -D_k \nabla \tilde{z}$ [54], and the PDF gradient model, $\overline{\mathbf{v}|_\eta} = \tilde{\mathbf{v}} + D_k \nabla (\ln P(\eta))$ in which $D_k \approx \mu_k/Sc_z$ [59], both capture these data sufficiently enough to be useful models for the conditional velocity, $\overline{\mathbf{v}|_\eta}$.

The conditional scalar dissipation rate term, $\overline{N|_\eta}$, plays a key role in CMC, especially near the reaction zone where $\partial^2 Q / \partial \eta^2$ is significant [52, 60]. Mell et al. [61], performed an evaluation of models for $\overline{N|_\eta}$ using DNS of homogeneous turbulence. This analysis suggests that the amplitude mapping closure (AMC) model provides good predictions. He and Rubinstein [62], however, concluded that the AMC model was incorrect in terms of its asymptotic behavior for unsymmetric binary mixing, and instead proposed the mapping closure approximation (MCA) model based on the AMC model. The MCA model was successful for some cases but includes a time-dependent function that has not been developed successfully. Devaud et al. [63] suggested a new method for modeling $\overline{N|_\eta}$ derived from the PDF transport equation and presented good agreement with inhomogeneous DNS data and also disclosed some discrepancies of the Girimaji model [64]. The AMC model [61] can be comprehensively summarized as

$$\overline{N|_\eta} = G(\eta) \bar{N} / \int_0^1 G(\eta) P(\eta) d\eta,$$
$$G(\eta) = \exp(-2[erf^{-1}(2\eta - a)]^2), \quad (4.29)$$
$$\bar{N} = \tilde{\chi}/2,$$

in which erf is the error function, and $\tilde{\chi} = c_D(\varepsilon/k)\widetilde{z''^2}$ is the unconditional Favre mean scalar dissipation rate that can be obtained from the conventional equality of the timescales for the velocity and the mixture fraction. Following Beguier et al. [65], $C_D = 2$.

In CMC, the PDF of the mixture fraction, z, is usually modeled by either a clipped Gaussian PDF or a beta PDF, also requiring knowledge of the mixture fraction variance, $\widetilde{z''^2}$, that can be obtained either from a modeled transport equation or estimated using multiple levels of filtering. The most common choice in practical simulations is the beta PDF.

The key to successful CMC modeling is the accuracy of the closure of the conditionally averaged chemical source term, $\overline{\dot{w}_i|_\eta}$. The assumption of relatively small fluctuations around the conditional mean allows for simple first order closure of the reaction rate term, i.e., $\overline{\dot{w}|_\eta} = \dot{w}_i(Q, \eta)$. This closure has proven adequate for many applications, but more complex closures may become necessary if fluctuations become significant. Note that the CMC equation (4.27) remains valid even for large fluctuations; however, the accuracy of the closures will suffer and different modeling strategies will need to be employed. Rather unexpectedly, improved closures of the e_Y-term may not be necessary, however, accurate modeling of $\overline{\dot{w}_i|_\eta}$, is necessary [66]. The latter can be achieved by multiple conditioning or by second order closures [67].

4.4.6 Linear Eddy Combustion Model

A more comprehensive combustion model is the linear eddy mixing (LEM) model of Kerstein [68], and Menon et al. [69]. This approach has been developed in the past decade to offer a closure directly at the subgrid scales for all combustion modes. LEM is a stochastic approach aimed at simulating, rather than modeling, the effects of turbulence on the chemistry, and it is not limited by the scale separation hypothesis [70]. The parameters controlling the LEM turbulent mixing model require only the validity of the Re number independence in the limit of large Re numbers, which is a safe assumption for most flow of engineering interest. Due to this extended validity range, the LEM model can be expected to perform in any combustion mode, and to be able to accurately handle flames near to, or even outside, the flammability limits.

In LEM, the scalar equations are not filtered, and instead the large-scale advection, turbulent mixing by eddies smaller than the grid size, molecular diffusion and chemical reaction are resolved at their appropriate length and time scales inside each LES cell. While the LES filtered conservation equations for mass, momentum, and energy are numerically integrated on the LES grid, the evolution of the species fields is tracked using a two-scale, two-time approach. For any scalar an exact and unfiltered Eulerian transport equation can be written as

$$\rho \partial_t Y_i = -\rho \mathbf{v} \cdot \nabla Y_i - \nabla \cdot (D \nabla Y_i) + \dot{w}_i, \qquad (4.30)$$

where the first right-hand-side term represents the total convection, the second is the molecular diffusion and the last term is the unfiltered chemical reaction source term.

The velocity vector can be decomposed as $v = \tilde{\mathbf{v}} + \mathbf{v}'_f + \mathbf{v}''$, where $\tilde{\mathbf{v}}$ represents the LES resolved velocity, $\mathbf{v}'_f$ is the subgrid velocity normal to the control volume surface, obtained from k, and $\mathbf{v}''$ is the small-scale velocity fluctuation, unresolved at the LES level. Using the decomposition of $\mathbf{v}$ the terms in (4.30) equations characterizing the large and small scale processes can thus be written

$$\rho(Y_i^* - Y_i^n)/\Delta t_{\text{LES}} = -\rho \tilde{\mathbf{v}} \cdot \nabla Y_i^n - \rho \mathbf{v}'_f \cdot \nabla Y_i^n, \tag{4.31}$$

$$Y_i^{n+1} = Y_i^* - \int_t^{t+\Delta t_{\text{LES}}} \frac{1}{\rho}[\rho \mathbf{v}'' \cdot \nabla Y_i^n - \nabla \cdot (D_i \nabla Y_i^n) - \dot{w}_i]dt.$$

In the above, Δt_{LES} is the LES time step, Y_i^n and Y_i^{n+1} are consecutive time values of Y_i and Y_i^* is an intermediate solution, after the large-scale convection is completed. In Eq. (4.31), the first term under the integral represents the subgrid stirring, the second is the subgrid molecular diffusion and the last accounts for the combustion chemistry. The molecular diffusion and the chemical reaction contribution to the small-scale transport are resolved on a one-dimensional grid inside each LES cell at a resolution much finer than the LES resolution, and approaching the Kolmogorov scale. The 1D computational domain is aligned in the direction of the flame normal inside each LES cell, ensuring an accurate representation of flame normal scalar gradients [68]. On this domain (denoted the LEM domain) molecular diffusion (term I), chemical reactions (term II), diffusion of heat via species molecular diffusion (term III), heat diffusion (term IV) and combustion heat release (term V) are resolved, according to

$$\rho^{\text{LEM}} \partial_t Y_k^{\text{LEM}} + F_k^{stirr} - \underbrace{\nabla_s \cdot (D_k \nabla_s Y_k^{\text{LEM}})}_{I} = \underbrace{\dot{w}_k}_{II},$$

$$\rho^{\text{LEM}} c_p \partial_t T^{\text{LEM}} + F_T^{stirr} - \underbrace{\sum_{k=1}^{N}(\rho c_p D_k \nabla_s Y_k^{\text{LEM}} \nabla_s T^{\text{LEM}})}_{III} \tag{4.32}$$

$$- \underbrace{\nabla_s \cdot (D_T \nabla_s T^{\text{LEM}})}_{IV} = - \underbrace{\sum_{k=1}^{N}(h_k \dot{w}_k)}_{V},$$

Here, the superscript LEM indicates values at the subgrid level, LEM, s is the spatial coordinate along the LEM domain, and F_k^{stirr} and F_T^{stirr} represent the effect of the subgrid turbulence on the species k mass fraction field, respectively on the temperature field.

The small-scale turbulent stirring (F_k^{stirr} and F_T^{stirr}) is implemented explicitly on the same grid using stochastic rearrangement events that mimic the action of an eddy upon the scalar field using a method known as triplet mapping and designed to recover the 3D inertial range scaling laws [68, 71]. The location of this stirring event is chosen from a uniform distribution. The frequency at which stirring events occur is given by

$$\lambda = \frac{54}{9} \frac{v Re_\Delta \left[(\Delta/\lambda_K)^{5/3} - 1\right]}{c_k \Delta^3 \left[1 - (\lambda_K/\Delta)^{4/3}\right]}, \tag{4.33}$$

where c_k stands for the scalar turbulent diffusivity, set to 0.067 [72]. The eddy size, ℓ, ranges from the Kolmogorov scale, l_K, to the grid size, Δ with a distribution given in [68],

$$f(\ell) = \frac{(5/3)\ell^{-8/3}}{l_K^{-5/3} - \Delta^{-5/3}}, \qquad (4.34)$$

where the Kolmogorov scale is determined as $l_K = N_{l_K} \Delta \text{Re}_\Delta^{-4/3}$ and N_{l_K} is an empirical constant that reduces the effective range of scales between the integral length scale and Kolmogorov scale but without altering the turbulent diffusivity [71].

Once the LES computations are completed at a given time step, LEM domain cells (and/or cell fractions) are exchanged between the LES cells in a manner that accounts for the mass fluxes across the LES cell faces. The volume of each LEM cell is modified after each LEM step in order to account for the volumetric expansion due to the temperature change induced by the chemical heat release. After this is completed, the LEM domain is regridded to maintain a constant cell volume over the entire LEM domain. Finally, the subgrid scalar fields in each LES cell are ensemble averaged to obtain the LES-resolved scalar field, Y_i, which is used in the LES energy equation and equation of state. Further details on the LEM–LES numerical implementation and the underlying assumptions can be found in the above cited references.

4.5 Discussion

The above models are only a few examples of the models for non-premixed flames and partially premixed flames reported in the literature. Among the models that are not discussed above, one can mention the Lagrangian PDF methods [43, 73] and the multiple mapping conditioning (MMC) approach [74, 75] that combines the advantages of the PDF and the conditional moment closure methods. Both Lagrangian PDF and MMC methods have been applied to simulate turbulent flames, including turbulent jet flames [76, 77], lifted flames, and flames with local extinction and reignition [78, 79]. Due to the limited length of this text, it is a formidable task to list all models developed. It should be mentioned that different approaches have their own merits in modeling different turbulent combustion processes. One may classify the various models for turbulent combustion into two groups, *general* models and *specific* models. General models are those coupling directly with finite rate chemistry, such as the Lagrangian PDF, PDF–ESF, EDC, PaSR, and LEM. Some of these models have been discussed in modeling both premixed flames (Chapter 3) and non-premixed flames (Section 4.4 of the present chapter). General models are aiming at modeling of multi-modes combustion processes, e.g., including non-premixed flames, premixed flames, and ignition processes, by integrating directly the finite rate chemistry. However, these models are often computational expensive, and the computational costs increase with the number of species and reactions considered in the finite-rate chemistry model. These models also employ model parameters that have to be verified for specific problems, e.g., the mixing constant and the number of stochastic fields in the PDF-ESF model. Specific models, such as the stationary flamelet-library models, have the advantage of pre-tabulation of chemical species or reaction rates in the flamelet coordinates, e.g., mixture fraction, reaction progress variable, or scalar dissipation rate. Thus, the computational cost is independent of the chemical reaction mechanisms used. The validity of specific models may be limited to the regimes of flames based on which the models were developed.

4.6 Case Studies

In the following section, we present LES results for three different cases to demonstrate the predictive capabilities of LES in general and the differences between different LES combustion models. The cases are presented in increasing level of complexity, and for additional information about the different cases we refer to the references cited below.

4.6.1 The KAUST Diffusion Flame Burner

Turbulent diffusion flames are often prone to lift-off due to near burner local extinction. At high fuel jet velocities, the flame may easily be blown off. Elbaz and Roberts [80] developed a diffusion flame burner to investigate the local extinction and flame stability (hereafter referred to as the KAUST burner). Figure 4.4a shows a sketch of the burner and the measurement systems. The burner is made up of two sections, a straight concentric tube and a quartz quarl section in which methane and air are mixed and combustion could take place. The swirling airflow was generated via four tangential air inlets to the outer air tube. The tangential air streams mixed with the axial air upstream of the burner, where the swirling airflow surrounded the central fuel tube.

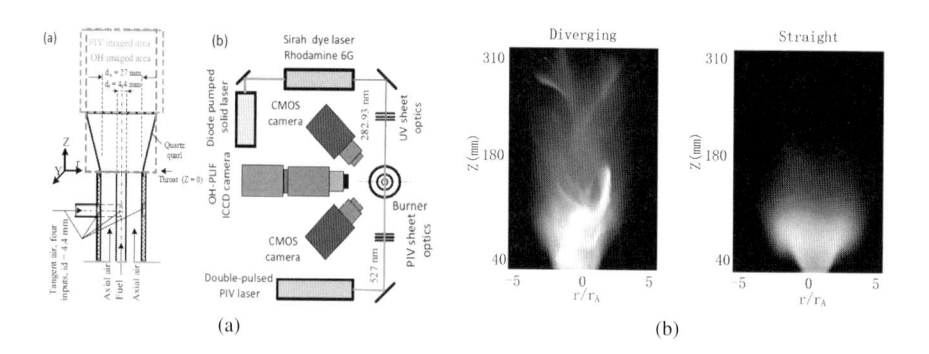

Figure 4.4 Schematic illustration of the KAUST diffusion flame burner and the experimental window (a), the laser system (b), and photos of the flames with a diverging quarl and straight quarl.

It was demonstrated that the critical fuel jet velocities for both lift-off and blow-out for the burner with a diverging quarl are significantly higher than the corresponding ones for the burner with a straight quarl [81]. Figure 4.4 shows photographs of the flames with the different quarls. The fuel (methane) jet velocity is 8.76 m/s and the bulk flow velocity of the air at the plane of the fuel jet exit ($z = 0$) is 4.85 m/s. In the diverging quarl case, the flame showed a luminescence region up to 310 mm above the fuel tube exit. The flame with the diverging quarl showed a yellow color of the flame luminescence indicating soot formation in the flame, whereas the flame in the burner with the straight quarl showed a blue (nonsooty) tulip shape with the luminescence region much smaller than that of the diverging quarl flame. Simultaneous PIV/OH-PLIF measurements were conducted for the region above the quarl exit. The OH measurements were extended to cover the full quarl. As shown in Fig. 4.5, the OH PLIF signal is found in thin layers in both flames indicating flamelet type of reaction zone structure in both flames. The OH layer in the diverging quarl burner is shown to penetrate deep into the quarl, whereas it is just slightly into the quarl of the straight quarl burner. The time-averaged axial velocity from PIV data showed a significant difference in the two flames with the peak axial velocity in the diverging quarl burner shifting toward larger radial position than that in the straight quarl burner.

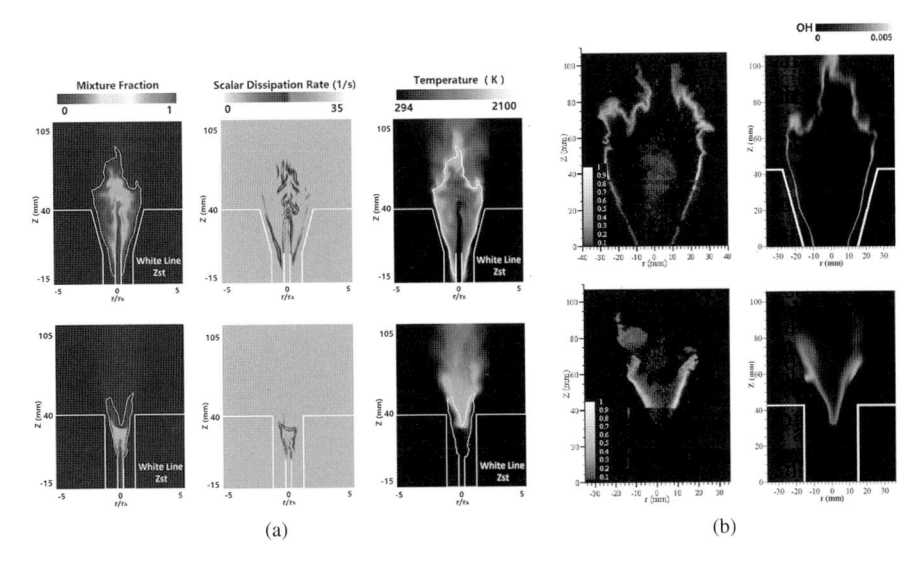

Figure 4.5 Instantaneous distribution of mixture fraction (first column from left), scalar dissipation rate (second column from left), temperature (third column from left), OH mass fraction (first column from right) from LES, and OH PLIF intensity from experiments (second column from right). The stoichiometric mixture fraction is marked with white lines. The upper row shows results for the diver-ging quarl while the bottom row for the straight quarl.

LES with the eulerian stochastic field (ESF) model was used to study flame structures [82], together with the SG35 reaction mechanism [83]. As shown in Fig. 4.5, the OH results from the LES-ESF model is in good agreement with the OH PLIF results. The LES prediction of the mean velocity field in the straight quarl burner is in good agreement with the PIV results (Fig. 4.6), whereas the LES can also replicate the trend of peak-shifting of the mean axial velocity toward larger radial position in the diverging quarl burner, however, with some discrepancy in the quantitative comparison of the mean velocity profile with the PIV experiments. This discrepancy is likely attributed to the uncertainties in the boundary conditions of the LES [82]. Comparison of the mean axial velocity in the two burners showed that the diverging quarl burner exhibits a negative axial velocity in the quarl and in the airflow tube (Fig. 4.3), which allows the fuel to be convected toward the airflow tube; compare with the mixture fraction field in Fig. 4.5. The axial velocity in the airflow tube and the quarl of the straight quarl burner is positive, which results in a narrower fuel flow downstream of the fuel nozzle, with a high gradient of mixture fraction and hence, a high scalar dissipation rate. This leads to local flame extinction near the burner exit and the flame is lifted off the fuel nozzle in the straight quarl burner. The lift-off of the flame allows a partial premixing of air into the fuel stream, which is the reason for the blue (nonsooty) color flame luminosity in the straight quarl burner. In the diverging quarl burner, the flame is stabilized in the airflow tube due to the recirculating flow induced by the diverging quarl geometry. The central fuel flow is not partially premixed with the airflow, which explains the sooty yellow color luminosity of the diverging quarl flame. The axial velocity along the centerline of the burner is higher near the exit of the quarl in the diverging quarl burner than that in the straight quarl burner, which is the reason for the longer flame in the diverging quarl burner.

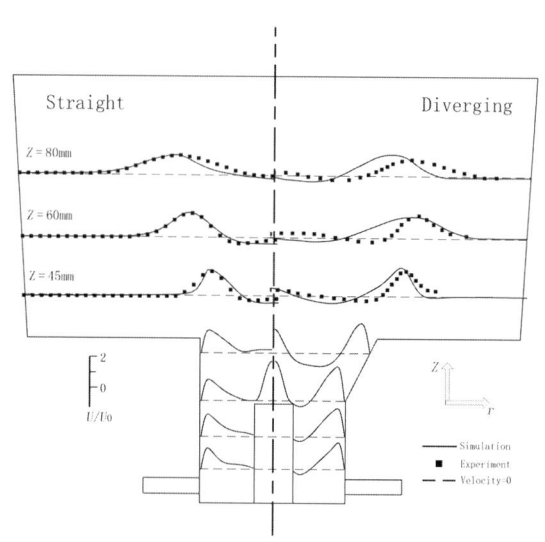

Figure 4.6 Time-aeveraged axial velocity profiles from PIV and LES for the straight quarl case (left) and the diverging quarl case (right).

In the most recent experiments of the KAUST burner flame series, local flame extinction and reignition have been observed in the diverging quarl burner at high fuel velocities [81]. The LES-ESF model has shown to be able to replicate the local extinction and reignition process [81]. It is expected that other models, such as stationary flamelet models [84] and CMC models [52–54] that are well tested for the prediction of Sandia and Sydney flame series [12, 56, 85–87], could also predict the attached diffusion flame in the diverging quarl. The application of these models to the lifted flames in the straight quarl burner, and more challengingly the local extinction and reignition process in diverging quarl burner at high fuel jet velocities will be very interesting and it remains to be further investigated.

4.6.2 Stratified Flames: The Lund Low-Swirl Burner

The experimental low swirl burner considered here is based on that of Bedat and Cheng [88], but with a modified swirl arrangement [89, 90], for studying lean premixed low swirl stabilized turbulent flames. Perspective and top views of the low-swirl burner are shown in Figs. 4.7a and 4.7b, whereas Figs. 4.7c and 4.7d show schematics of the midplane in combination with experimental and computed images of the flame. Fuel and air are fed into a settling chamber, below the burner, which contains a perforated grid. The fuel–air-mixture is then transported to the coflow arrangement and the burner. Two perforated plates are located upstream of the $D = 50$ mm swirler, consisting of an outer part, with eight swirl-vanes, and an inner part, which consists of a perforated plate that allows for a certain portion of the flow to pass through the swirler section without gaining swirl. Downstream of the swirler, a nozzle discharges the swirling mixture into a coflow of air of 0.35 m/s. Two operational points, at a fuel–air ratio $\phi \approx 0.62$, have been experimentally studied [3, 89–92], at 27 kW and 40 kW, respectively. The swirl-number, S, is around 0.55, resulting in characteristic velocities of $v_1 \approx 15$–20 m/s, $v_2 \approx 5$–10 m/s, and $v_3 \approx 0.35$ m/s, for the two operational points corresponding to Re numbers of 20,000 and 30,000, respectively.

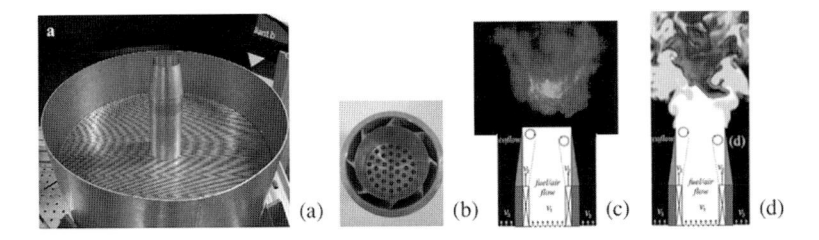

Figure 4.7 (a) The low-swirl burner mounted in the coflow arrangement. (b) The swirl arrangement and a schematic drawing of the burner induced flow field combined with (d) and experimental image of the flame and (d) a computational image of the flame from [91].

The low-swirl burner has been extensively investigated, computationally and experimentally, to characterize the flame and to assist validating computational models. Experimental investigations on this rig have been conducted using single-shot particle Image velocimetry (PIV) and laser Doppler anemometry (LDA) velocity measurements [3, 89, 90], simultaneous acetone and OH-planar laser-induced fluorescence (PLIF) measurements [91], simultaneous PIV and OH-PLIF measurements [89, 90], and Rayleigh thermometry measurements [3, 89]. Recently, rotational coherent anti-Stokes Raman spectroscopy (RCARS) measurements have been carried out along the center axis of the burner [92], and at cross sections through the flame at different heights [93]. Rayleigh scattering, LDA, and different PLIF measurements have been performed for turbulent flame speed measurements [88, 93, 94], and recently PIV and OH-PLIF measurements were made to investigate fuel effects on the flame structures and dynamics [95]. The low-swirl burner has been studied using LES with various combustion models and different simplified reaction mechanisms [3, 96–99], as well as DNS [95].

Figure 4.8 shows LES predictions of the low-swirl burner using the PaSR, TFM and ESF combustion models at an equivalence ratio of $\phi = 0.65$ and LES-ESF model predictions at gradually increasing equivalence ratios, $\phi = 0.70$ and 0.80. Included are also experimental OH images of the lifted flame cup for the different equivalence ratios. The computational images show volumetric renderings of the fuel mass fraction in green, Y_{CH4}, and the temperature, T. The flame is lifted above the burner and stabilized in its central part, where the fuel mass fraction is sufficiently high. In the proximity of the burner exit plane, the fuel–air flow stream undergoes its first Kelvin–Helmholtz instability, evident by the shedding structures in the shear layer. Further downstream the flow becomes turbulent, resulting in a rapid mixing between the central flow stream and the ambient air. The leading flame front is weakly wrinkled, due to the low turbulence intensity in the center region of the flow field, where the Karlovitz (Ka) number is about unity, whereas in the flame trailing edge region Ka $\approx$ 30, positioning the flame in the thin reaction zone regime with associated small scale turbulence interaction within the flame. From the experimental images, flame wrinkling at scales smaller than the LES grid can be seen; however, coexistence of fuel and OH within a thick reaction zone is not observed, therefore implying that the flame is flamelet like and that the flamelet assumption used in the numerical modeling is valid. The flow exiting the burner, including the inner turbulent nonswirling part and the annular shear layers, couples directly to the flame dynamics and the flame wrinkling. The flame lift-off height is well predicted by all models tested. Furthermore, the

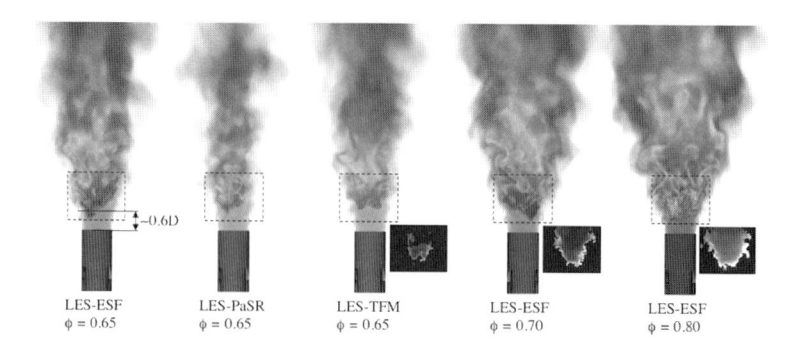

Figure 4.8 Low-swirl burner: LES results using, from left to right, the PaSR, TFM, and ESF combustion models at an equivalence ratio of $\phi = 0.65$ and LES-ESF model predictions at $\phi = 0.70$ and 0.80. Included are also experimental OH images of the flame cup for the different equivalence ratios.

variation of lift-off height with ϕ is also well predicted, clearly suggesting that this LES methodology works well.

Figure 4.9 presents quantitative comparisons of the time-averaged axial velocity, $\langle v_x \rangle$, and temperature, $\langle T \rangle$, between the LES-PaSR, TFM, and ESF models, and the LES G-equation based flamelet model [3], with experimental data from [89] and [93], respectively, at $\phi = 0.65$. From the $\langle v_x \rangle$ profiles, we observe generally reasonable agreement between LES predictions and experimental data but with the LES G-equation flamelet model showing wider $\langle v_x \rangle$ profiles from the burner nozzle and downwards. Regarding the temperature distribution, the $\langle T \rangle$ profiles show that the LES-TFM model underpredicts the temperature close to the burner but shows better agreement with the experimental profiles further away from the burner. The LES-PaSR behaves similarly, but shows better agreement with the experimental data closer to the burner. The LES-ESF shows best overall agreement with the experimental data.

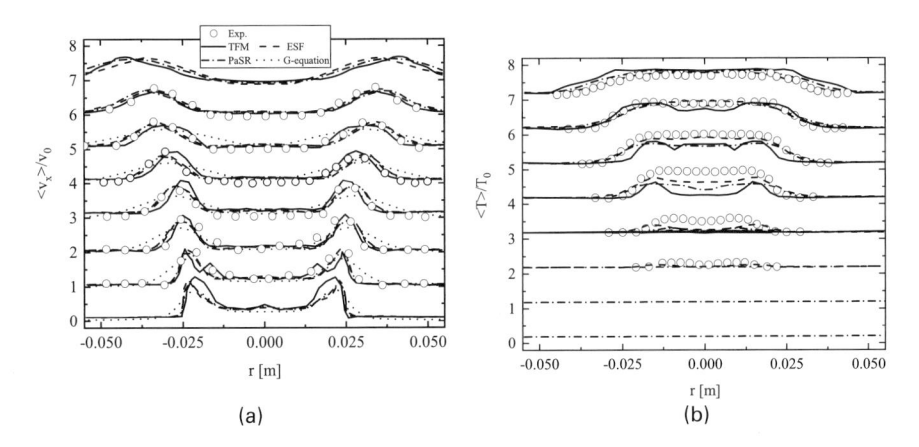

Figure 4.9 Low-swirl burner: Time-averaged (a) axial velocity, $\langle v_x \rangle$, and (b) temperature, $\langle T \rangle$, profiles at $x/D = 0.0, 0.2, 0.4, 0.6, 0.8, 1.0, 1.2$, and 1.4 at $\phi = 0.65$. PIV and CARS data are from Refs.[89] and [93], respectively, and LES predictions from finite-rate chemistry models and G-equation based flamelet model [3].

4.6.3 Non-Premixed and Partially Premixed Flames: The Turchemi Model Combustor

The Turchemi model combustor of Fig. 4.10 is a single-burner laboratory version of a Siemens SGT-100 power generation gas turbine engine, composed of a full-scale single burner attached to an optical combustion chamber, and installed in a high-pressure test rig at DLR Stuttgart, burning methane CH_4. One nonreacting case and four reacting cases were experimentally investigated in [100–102]. The most commonly studied case was operated at 3.0 bar, with an air temperature of 685 K, an air mass flow, with panel cooling, of 0.1749 kg/s, a fuel mass flow of 0.0062 kg/s, and a thermal power of 335 kW. The burner consists of a radial air intake with 12 rectangular channels each with multiple small holes to inject the methane resulting in an initially strongly heterogeneous air and fuel stream. The air and fuel streams mix as they flow through a prechamber of 46 mm length and 86 mm diameter, until they discharge into the combustor, resulting in an M-shaped flame (Fig. 4.10b). Extensive experimental studies were performed including PIV and one-dimensional simultaneous Raman scattering (1DR), from which the temperature can be reconstructed.

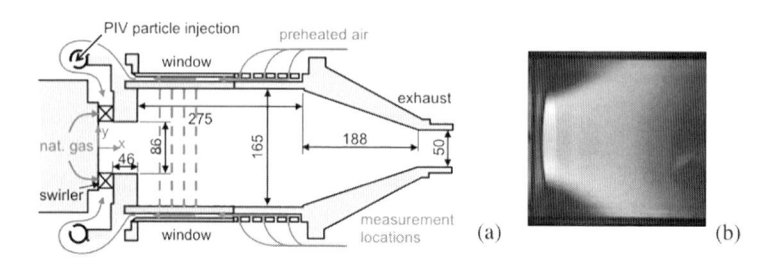

Figure 4.10 The Turchemi combustor. (a) Schematic of the SGT-100 burner and optical combustion chamber in high-pressure test rig at DLR Stuttgart and (b) a typical flame image.

Several LES studies of this case have been performed, including those of Sadasivuni et al. [103], Bulat et al. [104], Bulat et al. [105], Filosa et al. [106], Jaravel et al. [107], Langella et al. [108], and Fedina et al. [109], using different LES combustion models and reaction mechanisms. Fedina et al. [109] compared the performance of two reduced mechanisms, SG35 [83] and SR39 [110], and two global mechanisms, WD2 [111] and JL4 [112]. Figure 4.11 presents results using the PaSR model [39, 40], and the skeletal SG35 mechanism from [109], in terms of instantaneous volumetric renderings of the CH_4 mass-fraction, Y_{CH4}, and heat-release, Q, representative of the luminous flame, and contours of the axial velocity, v_x, and temperature, T, on two half-planes, together with an isosurface of the second invariant of the velocity gradient tensor, λ_2, colored by v_x. The fuel discharges into the 12 radial swirlers, where it starts to mix with the radially supplied air before entering the prechamber at a fixed angle. This results in a strongly swirling flow, with most of the fuel flanking the pre-chamber wall. The fuel continues to mix with the air until the now well-mixed, wrinkled and furrowed, M-shaped fuel–air mixture discharges into the combustor, having a protracted inner V-structure extending well into the prechamber. The flame develops around the M-shaped fuel–air cloud, and is composed of an inner, wrinkled or furrowed, V-shaped flame located in the inner shear layer of the discharging fuel–air mixture, and an outer swirling flame, taking the shape of a truncated wrenched cone located in the outer shear layer of the discharging fuel–air mixture. The two flame elements connect at the swirling annular flame tip, $\sim 0.6D$ downstream of the burner exit depending on the LES combustion model. The heat-release results from multiple thin layers of interpenetrating reaction structures and layers of intermediate and radical species being also sensitive to the strain rate.

Figure 4.12 provides a quantitative comparison of the time-averaged and rms fluctuations of the axial velocity, v_x, and temperature, T, across the combustion chamber at the cross sections $x/D = 1.21$, 1.44, 1.66, and 2.00, respectively, comparing predictions from some of the aforementioned LES combustion models. More

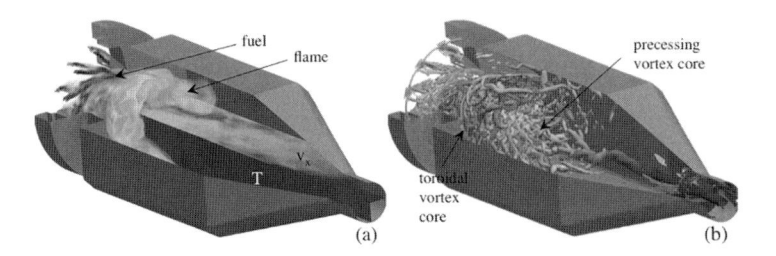

Figure 4.11 The Turchemi combustor. (a) Instantaneous volumetric rendering of CH_4 (fuel) and heat-release rate, $\dot{Q}$, (flame) together with contours of the axial velocity (vx) and temperature (T), and (b) an isosurface of the second invariant of the velocity gradient tensor, λ_2, colored by the axial velocity for LES-PaSR with the SG35 reaction mechanism.

specifically, we compare the LES-TFM-ARC predictions of Jaravel et al. [107], the LES-PaSR/TFM/SF-SG25 predictions of Fedina et al. [109], the LES-SF-JL4 predictions of Bulat et al. [104], and the LES flamelet predictions of Langella et al. [108], with the experimental data [100–102]. Concerning $\langle v_x \rangle$ we generally find good agreement across the whole suite of LES models but with LES-TFM-ARC [107], LES-TFM-SG25 [109], and LES-SF-JL4 [104], predicting a too strong exit vortex core; the LES-SF-JL4, predicting a too narrow jet; and the LES-PaSR-SG25, predicting the strongest central recirculation zone. The axial rms-velocity fluctuations, v_x^{rms}, are similarly predicted by all LES models. Concerning $\langle T \rangle$ the differences due to the choice of LES model is substantially larger: Here, only the LES-TFM-ARC, LES-SF-SG25, and the LES-PaSR-SG25, show a sufficiently accurate agreement with the experimental data at all four cross sections. The LES flamelet predictions of Langella et al., the LES-SF-JL4 predictions of Bulat et al., and the LES-TFM-SG25, predictions of Fedina et al., all overpredict $\langle T \rangle$ in the outer shear layer, and predicts a significantly shorter flame than observed experimentally, and with the LES-TFM-ARC, LES-SF-SG25, and LES-PaSR-SG25 prediction. The rms-temperature fluctuations, T^{rms}, are reasonably well predicted by the LES-TFM-ARC, LES-SF-SG25, and the LES-PaSR-SG25 models but are underpredicted by the LES flamelet

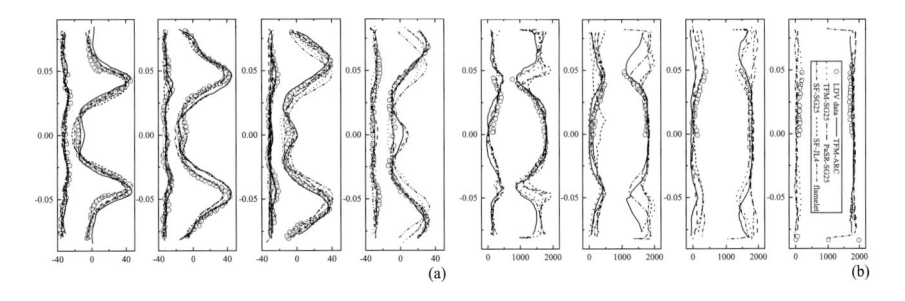

Figure 4.12 The Turchemi combustor, Case A. Comparison of measured and predicted time-averaged (right set of lines in each panel) and rms fluctuations (left set of lines in each panel) of the axial velocity (a) and temperature (b) for different LES simulations at $x/D = 1.21$, 1.44, 1.66, and 2.00. (Symbols) experimental data [100–102], lines are from different chemical kinetic models: LES-TFM-ARC, Jaravel et al. [107], LES-PaSR-SG25, Fedina et al. [109], LES-TFM-SG25, Fedina et al. [109], LES-SF-SG25, Fedina et al. [109], LES-SF-JL4, Bulat et al. [104], and LES-flamelet, Langella et al. [108].

predictions of Langella et al. [108], the LES-SF-JL4 predictions of Bulat et al. [104], and the LES-TFM-SG25 predictions of Fedina et al. [109].

4.6.4 Non-Premixed And Partially Premixed Flames: Prediction of Thermoacoustically Unstable Flames

Prediction of thermoacoustically unstable flames and flames close to the lean blowoff (LBO) limit is important for the design of combustion systems, for example, gas turbine engines. There are complex mechanisms giving rise to thermoacoustically unstable combustion and it is a challenging task for LES to replicate the physical process. High-fidelity experiments are needed to provide data for validation of numerical modeling of such combustion process. In this regard, a well-known partially premixed combustion system is the gas turbine model combustor developed by the German Aerospace Center, DLR. The system contains two swirler generators and a combustion chamber having a square cross section with an internal area of 85 × 85 mm^2 and a length of 114 mm. Figure 4.13a shows the schematic of the burner and the combustion chamber. The dry ambient air supplied at the bottom of the plenum flows through two swirlers having the same rotational direction before exiting into the combustion chamber through concentric circular and annular nozzles of diameter of 15 and 25 mm, respectively. Extensive measurements using laser diagnostics for three operating conditions were made, including thermoacoustically stable and unstable conditions, and a flame close to blow-off [113, 114]. The thermoacoustically stable flame has a higher air flow speed and a lower global equivalence ratio than the thermoacoustically unstable flame. The stable flame has been investigated using LES [115–119]. The thermoacoustically unstable flame has been investigated using LES

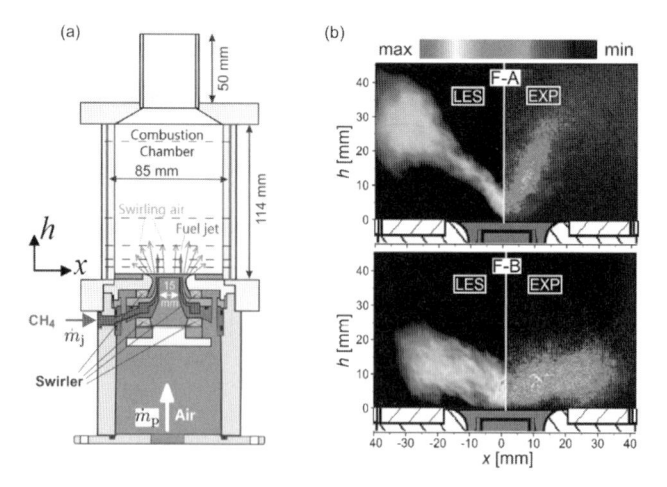

Figure 4.13 The DLR gas turbine model combustor. (a) Schematic of the combustion system. (b) Mid-plane time-averaged CH fields obtained from LES [119] and measurements [114] for the thermoacoustically stable flame (F-A) and the thermoacoustically unstable flame (F-B).

and a flamelet model [118, 119], and the close-to-blowoff flame, which has the lowest global equivalence ratio (of 0.55) among the three flame cases, was recently simulated using LES [120].

In the experiments, it was observed that although the mean flow fields in the thermoacoustically stable and unstable flames are very similar, the shapes of the flames are very different (Fig. 4.13b). A transition of a V-shaped flame in the thermoacoustically stable flame condition (Flame F-A) to a flat flame in the unstable condition (F-B) is found, suggesting a significant change of the important convective delay in the thermoacoustic feedback loop. Chen et al. [118, 119] carried out LES of these flames using a presumed PDF flamelet model based on the mixture fraction and a reaction progress variable, Eqs. (4.15) and (4.16). Since the combustion process involves both premixed and non-premixed combustion modes, the filtered reaction rate is decomposed into a premixed flame part and a non-premixed flame part. The premixed flame part is computed using the flamelet reaction rate obtained from 1D calculations of a freely propagating laminar flames for different mixture fractions covering the flammable range. For the non-premixed part, an algebraic model [29] [also see Eq. (4.17)] was used. It was shown that the presumed PDF flamelet LES model can replicate the different flame shapes (Fig. 4.13b), and the flame dynamics observed in the experiments. The LES results provided a physical explanation for the transition of flame shapes. It was found that the different impedances experienced by the pressure oscillations propagating through the two swirling injector passages with different internal geometries resulted in a significantly varying air mass flow split between the two air passages during a thermoacoustic oscillation cycle. This caused a periodic variation of the radial momentum of the fuel jets injected between the two swirling air flows. The resulting flapping of the fuel jets created an enhanced radial fuel–air mixing that led to the flattened flame shape in the thermoacoustically unstable flame case [119]. It was found that the precessing vortex core (PVC) generated in the combustion chamber due to the swirling inflow, which provided an anchoring mechanism for both the stable and the unstable flames, was strongly influenced by the pressure oscillations in the unstable flame case, being axially stretched and compressed at high and low pressures, respectively [118].

The flame close to blow-off was also successfully simulated using the presumed PDF flamelet model discussed earlier [120]. The LES identified two distinct stages for the flame evolution. In the first stage, the flame was anchored by its stable and robust root located near the centerline in the near field of the burner. In the second stage, entrainment of the inflammable mixture into the flame region within the internal recirculation zone, which was generated due to vortex breakdown of the swirling flow, led to the loss of the flame root and initiated a lift-off event. These two stages were observed to switch from one to the other, in both the LES and the experiments. It is expected that the two-stage flame evolution process will be more sensitive to the heat loss and the chemical kinetics, e.g., the competition between chain branching and chain terminating reactions, when the global equivalence ratio is decreased. Flamelet models accounting for these detailed chemistry effects should be developed to properly predict the complete blow-off process.

4.7 Concluding Remarks

Turbulent non-premixed and partially premixed combustion exhibits multiple modes, e.g., diffusion flames, edge flames, local extinction holes, and reignition. When complex fuels are involved, involving, e.g., two-stage ignition chemistry, low-temperature chemistry (LTC) may play an important role in the combustion and emission process. Recent application of DNS to non-premixed combustion of a DME jet flame [121] has shown the presence of a penta-brachial flame structure including an LTC branch, a triple-flame branch, and a high-temperature branch upstream of the triple flames. At elevated temperature and high-pressure conditions, the problem becomes more challenging since the coupling of ignition, diffusion flame, and premixed flame propagation becomes more involved and experiments are difficult to perform, whereas numerical simulations are limited due to the lack of chemical kinetic mechanisms for high-pressure conditions. The physics of turbulent non-premixed and partially premixed combustion requires further investigation using advanced experimental methods and high-fidelity numerical simulations such as DNS and LES. Sophisticated experiments on non-premixed and partially premixed flames have been reported in the TNF workshop series [122], and high-quality experiments on spray non-premixed combustion under internal combustion engine conditions have been reported in the Engine Combustion Network (ECN) series [123]. It is expected that collaboration within the TNF and ECN frameworks will continue to play a pivotal important role in improving the understanding and modeling of non-premixed and partially premixed combustion.

Comprehensive understanding of the fundamental flame structures is essential for the development of CFD models for predicting the multimode combustion process. In principle, combustion can be modeled using regime or mode-independent *general* models, based on finite-rate chemistry. This approach is being used in numerical simulation of turbulent combustion in internal combustion engines, gas turbines, and supersonic combustion for fundamental investigation. However, for engineering design simulations the computational cost of using such models is often too high to be useful, in particular, when complex fuels and relatively detailed chemical kinetic mechanisms are employed. Thus, improvement of the model efficiency for general models for multimode combustion is needed. One recent effort in this direction is the work of Jangi et al. [124], in which the time-consuming Lagrangian PDF method is accelerated using a chemistry coordinate mapping (CCM) approach that integrates the chemical reaction rates of the Monte Carlo particles within a low-dimensional thermodynamically identical phase space. Analysis of DNS data is expected to provide insight into the identification of low-dimensional phase space. For single-mode combustion processes, e.g., classical diffusion flames, high-efficiency specially developed models are desirable for design simulations. Examples of such models are stationary and unsteady flamelet models, and flamelet/progress variable approach [9], CMC models, and so forth, as documented in the TNF workshop series [122], and the flamelet generated manifold (FGM) models that were first developed for premixed flames [125]. Progress of specially designed models for single-mode combustion has been made in the past (Section 4.6) and it is desirable to extend these models to multimode combustion.

References

[1] V. Bergmann, W. Meier, D. Wolff, and W. Stricker, "Application of spontaneous Raman and Rayleigh scattering and 2D LIF for the characterization of a turbulent CH4/H2/N2 jet diffusion flame," *Appl. Phys. B*, **66**, 489–502, (1998).

[2] W. Meier, R. S. Barlow, Y.-L. Chen, and J.-Y. Chen, "Raman/Rayleigh/LIF measurements in a turbulent CH4/H2/N2 jet diffusion flame: experimental techniques and turbulence–chemistry interaction," *Combust. Flame*, **123**, 326–343, (2000).

[3] K.-J. Nogenmyr, C. Fureby, X.-S. Bai, P. Petersson, R. Collin, and M. Linne, "Large eddy simulation and laser diagnostic studies on a low swirl stratified premixed flame," *Combust. Flame*, **156**, 25–36, (2009).

[4] D. Micka and J. Driscoll, "Stratified jet flames in a heated (1364k) cross-flow with autoignition," in *49th AIAA Aerospace Sciences Meeting including the New Horizons Forum and Aerospace Exposition*, 2011, p. 321.

[5] S. Woo and I. Jeung, "A study on the flow characteristics of gas turbine engine combustor on/off the operation envelope," in *Asia-Pacific Conference on Combution*, **5**, 2005, 161–164.

[6] A. Ratner, J. Driscoll, J. Donbar, C. Carter, and J. Mullin, "Reaction zone structure of non-premixed turbulent flames in the intensely wrinkled regime," *Proc. Combust. Inst.*, **28**, 245–252, (2000).

[7] S. Burke and T. Schumann, "Diffusion flames," *Indust. Engin. Chem.*, **20**, 998–1004, (1928).

[8] B. Cuenot and T. Poinsot, "Effects of curvature and unsteadiness in diffusion flames. implications for turbulent diffusion combustion," in *Symp. (International) Combust.*, **25**, no. 1. Elsevier, 1994, 1383–1390.

[9] C. D. Pierce and P. Moin, "Progress-variable approach for large-eddy simulation of non-premixed turbulent combustion," *J. Fluid Mech.*, **504**, 73–97, (2004).

[10] J. van Oijen and L. De Goey, "Modelling of premixed counterflow flames using the flamelet-generated manifold method," *Combust. Theory Model.*, **6**, 463–478, (2002).

[11] B. Fiorina, R. Baron, O. Gicquel, D. Thevenin, S. Carpentier, N. Darabiha, et al., "Modelling non-adiabatic partially premixed flames using flame-prolongation of ILDM," *Combust. Theory Model.*, **7**, 449–470, (2003).

[12] M. Ihme and H. Pitsch, "Prediction of extinction and reignition in nonpremixed turbulent flames using a flamelet/progress variable model: 2. Application in LES of Sandia flames D and E," *Combust. Flame*, **155**, 90–107, (2008).

[13] A. Najafi-Yazdi, B. Cuenot, and L. Mongeau, "Systematic definition of progress variables and intrinsically low-dimensional, flamelet generated manifolds for chemistry tabulation," *Combust. Flame*, **159**, 1197–1204, (2012).

[14] J. D. Buckmaster, *The mathematics of combustion*. SIAM, 1985.

[15] R. W. Bilger, "The structure of diffusion flames," *Combust. Sci. Technol.*, **13**, 155–170, (1976).

[16] R. W. Bilger, "Turbulent flows with non-premixed reactants," in P. A. Libby and F. A. Williams, Eds., *Turbulent Reacting Flows, Topics in Applied Physics*, **44**, 1980, p. 65.

[17] P. Domingo, L. Vervisch, and K. Bray, "Partially premixed flamelets in LES of nonpremixed turbulent combustion," *Combust. Theory Model.*, **6**, 529–551, (2002).

[18] P. Sagaut, *Large eddy simulation for incompressible flows*. Berlin/Heidelberg: Springer Verlag, 2001.

[19] N. Branley and W. Jones, "Large eddy simulation of a turbulent non-premixed flame," *Combust. Flame*, **127**, 1914–1934, (2001).

[20] E. Hawkes and R. Cant, "Implications of a flame surface density approach to large eddy simulation of premixed turbulent combustion," *Combust. Flame*, **126**, 1617–1629, (2001).

[21] I. Langella, Z. X. Chen, N. Swaminathan, and S. K. Sadasivuni, "Large-eddy simulation of reacting flows in industrial gas turbine combustor," *J. Propul. Power*, **34**, 1269–1284, (2018).

[22] C. Duwig and C. Fureby, "Large eddy simulation of unsteady lean stratified premixed combustion," *Combust. Flame*, **151**, 85–103, (2007).

[23] S. S. Girimaji and Y. Zhou, "Analysis and modeling of subgrid scalar mixing using numerical data," *Phys. Fluids*, **8**, 1224–1236, (1996).

[24] O. R. Darbyshire and N. Swaminathan, "A presumed joint pdf model for turbulent combustion with varying equivalence ratio," *Combust. Sci. Technol.*, **184**, 1019–1042, (2012).

[25] Z. X. Chen, S. Ruan, and N. Swaminathan, "Simulation of turbulent lifted methane jet flames: Effects of air-dilution and transient flame propagation," *Combust. Flame*, **162**, 703–716, (2015).

[26] Z. X. Chen, N. A. K. Doan, S. Ruan, I. Langella, and N. Swaminathan, "A priori investigation of subgrid correlation of mixture fraction and progress variable in partially premixed flames," *Combust. Theory Model.*, **22**, 862–882, (2018).

[27] Z. X. Chen, I. Langella, and N. Swaminathan, "The role of CFD in modern jet engine combustor design," in *Advances in Jet Engines, DOI: http://dx.doi.org/10.5772/intechopen.88267*. IntechOpen, 2019, 1–28.

[28] P. Domingo, L. Vervisch, and J. Réveillon, "DNS analysis of partially premixed combustion in spray and gaseous turbulent flame-bases stabilized in hot air," *Combust. Flame*, **140**, 172–195, (2005).

[29] Z. Chen, S. Ruan, and N. Swaminathan, "Large eddy simulation of flame edge evolution in a spark-ignited methane–air jet," *Proc. Combust. Inst.*, **36**, 1645–1652, (2017).

[30] P. O'Rourke and F. Bracco, "Two scaling transformations for the numerical computation of multidimensional unsteady laminar flames," *J. Comput. Phys.*, **33**, 185–203, (1979).

[31] O. Colin, F. Ducros, D. Veynante, and T. Poinsot, "A thickened flame model for large eddy simulations of turbulent premixed combustion," *Phys. Fluids*, **12**, 1843–1863, (2000).

[32] J.-P. Legier, T. Poinsot, and D. Veynante, "Dynamically thickened flame LES model for premixed and non-premixed turbulent combustion," in *Proceedings of the Summer Program*, **12**. Center for Turbulence Research Stanford, CA, 2000.

[33] F. Charlette, C. Meneveau, and D. Veynante, "A power-law flame wrinkling model for LES of premixed turbulent combustion part I: Non-dynamic formulation and initial tests," *Combust. Flame*, **131**, 159–180, (2002).

[34] C. Meneveau and T. Poinsot, "Stretching and quenching of flamelets in premixed turbulent combustion," *Combust. Flame*, **86**, 311–332, (1991).

[35] G. Wang, M. Boileau, and D. Veynante, "Implementation of a dynamic thickened flame model for large eddy simulations of turbulent premixed combustion," *Combust. Flame*, **158**, 2199–2213, (2011).

[36] B. Magnussen, "On the structure of turbulence and a generalized eddy dissipation concept for chemical reaction in turbulent flow," in *19th Aerospace Sciences Meeting*, 1981, p. 42.

[37] I. S. Ertesvåg and B. F. Magnussen, "The eddy dissipation turbulence energy cascade model," *Combust. Sci. Technol.*, **159**, 213–235, (2000).

[38] E. Giacomazzi, C. Bruno, and B. Favini, "Fractal modelling of turbulent combustion," *Combust. Theory Model.*, **4**, 391–412, (2000).

[39] M. Berglund, E. Fedina, C. Fureby, J. Tegnér, and V. Sabel'Nikov, "Finite rate chemistry large-eddy simulation of self-ignition in supersonic combustion ramjet," *AIAA J.*, **48**, 540–550, (2010).

[40] V. Sabelnikov and C. Fureby, "LES combustion modeling for high re flames using a multi-phase analogy," *Combust. Flame*, **160**, 83–96, (2013).

[41] S. Corrsin, "Turbulent dissipation fluctuations," *Phys. Fluids*, **5**, 1301–1302, (1962).

[42] M. Tanahashi, M. Fujimura, and T. Miyauchi, "Coherent fine-scale eddies in turbulent premixed flames," *Proc. Combust. Inst.*, **28**, 529–535, (2000).

[43] S. B. Pope, "Pdf methods for turbulent reactive flows," *Prog. Energy Combust. Sci.*, **11**, 119–192, (1985).

[44] S. B. Pope, *Turbulent Flows* Cambridge: Cambridge University Press, 2000.

[45] P. Yeung, S. Pope, and B. Sawford, "Reynolds number dependence of Lagrangian statistics in large numerical simulations of isotropic turbulence," *J. Turbul.*, p. N58, (2006).

[46] F. Gao and E. E. OBrien, "A large-eddy simulation scheme for turbulent reacting flows," *Phys. Fluids A*, **5**, 1282–1284, (1993).

[47] L. Valiño, "A field Monte Carlo formulation for calculating the probability density function of a single scalar in a turbulent flow," *Flow Turbul. Combust.*, **60**, 157–172, (1998).

[48] V. Sabelnikov and O. Soulard, "White in time scalar advection model as a tool for solving joint composition pdf equations," *Flow Turbul. Combust.*, **77**, 333–357, (2006).

[49] S. Karlin and H. E. Taylor, *A second course in stochastic processes*. Philadelphia: Elsevier, 1981.

[50] R. Mustata, L. Valiño, C. Jiménez, W. Jones, and S. Bondi, "A probability density function Eulerian Monte Carlo field method for large eddy simulations: application to a turbulent piloted methane/air diffusion flame (Sandia D)," *Combust. Flame*, **145**, 88–104, (2006).

[51] W. P. Jones and V. N. Prasad, "Large eddy simulation of the Sandia Flame Series (D–F) using the Eulerian stochastic field method," *Combust. Flame*, **157**, 1621–1636, (2010).

[52] A. Y. Klimenko, "Multicomponent diffusion of various admixtures in turbulent flow," *Fluid Dynamics*, **25**, 327–334, (1990).

[53] R. W. Bilger, "Conditional moment closure for turbulent reacting flow," *Phys. Fluids A*, **5**, 436–444, (1993).

[54] A. Y. Klimenko and R. W. Bilger, "Conditional moment closure for turbulent combustion," *Prog. Energy Combust. Sci.*, **25**, 595–687, (1999).

[55] A. Kronenburg and R. W. Bilger, "Modelling differential diffusion in nonpremixed reacting turbulent flow: model development," *Combust. Sci. Technol.*, **166**, 195–227, (2001).

[56] E. Richardson, N. Chakraborty, and E. Mastorakos, "Analysis of direct numerical simulations of ignition fronts in turbulent non-premixed flames in the context of conditional moment closure," *Proc. Combust. Inst.*, **31**, 1683–1690, (2007).

[57] C. B. Devaud and K. N. C. Bray, "Assessment of the applicability of conditional moment closure to a lifted turbulent flame: first order model," *Combust. Flame*, **132**, 102–114, (2003).

[58] N. Swaminathan and R. W. Bilger, "Analyses of conditional moment closure for turbulent premixed flames," *Combust. Theory Model.*, **5**, 241–260, (2001).

[59] P. Colucci, F. Jaberi, P. Givi, and S. Pope, "Filtered density function for large eddy simulation of turbulent reacting flows," *Phys. Fluids*, **10**, 499–515, (1998).

[60] J. D. Li and R. W. Bilger, "Measurement and prediction of the conditional variance in a turbulent reactive-scalar mixing layer," *Phys. Fluids A*, **5**, 3255–3264, (1993).

[61] W. E. Mell, V. Nilsen, G. Kosály, and J. Riley, "Investigation of closure models for nonpremixed turbulent reacting flows," *Phys. Fluids*, **6**, 1331–1356, (1994).

[62] G.-W. He and R. Rubinstein, "The mapping closure approximation to the conditional dissipation rate for turbulent scalar mixing," *J. Turbul.*, **4**, 2–2, (2003).

[63] C. B. Devaud, R. W. Bilger, and T. Liu, "A new method of modeling the conditional scalar dissipation rate," *Phys. Fluids*, **16**, 2004–2011, (2004).

[64] S. S. Girimaji, "On the modeling of scalar diffusion in isotropic turbulence," *Phys. Fluids A*, **4**, 2529–2537, (1992).

[65] C. Beguier, I. Dekeyser, and B. Launder, "Ratio of scalar and velocity dissipation time scales in shear flow turbulence," *Phys. Fluids*, **21**, 307–310, (1978).

[66] A. Kronenburg, "Double conditioning of reactive scalar transport equations in turbulent nonpremixed flames," *Phys. Fluids*, **16**, 2640–2648, (2004).

[67] A. Kronenburg and A. Papoutsakis, "Conditional moment closure modeling of extinction and re-ignition in turbulent non-premixed flames," *Proc. Combust. Inst.*, **30**, 759–766, (2005).

[68] A. R. Kerstein, "Linear-eddy modeling of turbulent transport. II: Application to shear layer mixing," *Combust. Flame*, **75**, 397–413, (1989).

[69] S. Menon, P. McMurtry, and A. Kerstein, "A linear eddy mixing model for large eddy simulation of turbulent combustion," in B. Galperin and S. A. Orszag, Eds., *Large Eddy Simulations of Complex Engineering and Geophysical Flows*, Cambridge: Cambridge University Press, 1993, pp. 287–314.

[70] N. Peters, *Turbulent Combustion*. Cambridge: Cambridge University Press, 2000.

[71] T. M. Smith and S. Menon, "One-dimensional simulations of freely propagating turbulent premixed flames," *Combust. Sci. Technol.*, **128**, 99–130, (1997).

[72] V. Chakravarthy and S. Menon, "Linear-eddy simulations of reynolds and schmidt number dependencies in scalar mixing," *Phys. Fluids*, **13**, 488–499, (2001).

[73] S. Pope, "Lagrangian pdf methods for turbulent flows," *Ann. Rev. Fluid Mech.*, **26**, 23–63, (1994).

[74] A. Y. Klimenko and S. Pope, "The modeling of turbulent reactive flows based on multiple mapping conditioning," *Phys. Fluids*, **15**, 1907–1925, (2003).

[75] M. Cleary and A. Y. Klimenko, "A generalised multiple mapping conditioning approach for turbulent combustion," *Flow, Turbul. Combust.*, **82**, p. 477, (2009).

[76] R. R. Cao and S. B. Pope, "The influence of chemical mechanisms on PDF calculations of nonpremixed piloted jet flames," *Combust. Flame*, **143**, 450–470, (2005).

[77] P. Vaishnavi and A. Kronenburg, "Multiple mapping conditioning of velocity in turbulent jet flames," *Combust. Flame*, **157**, 1863–1865, (2010).

[78] H. Wang and S. B. Pope, "Lagrangian investigation of local extinction, re-ignition and auto-ignition in turbulent flames," *Combust. Theory Model.*, **12**, 857–882, (2008).

[79] S. K. Ghai, S. De, and A. Kronenburg, "Numerical simulations of turbulent lifted jet diffusion flames in a vitiated coflow using the stochastic multiple mapping conditioning approach," *Proc. Combust. Inst.*, **37**, 2199–2206, (2019).

[80] A. M. Elbaz and W. L. Roberts, "Investigation of the effects of quarl and initial conditions on swirling non-premixed methane flames: Flow field, temperature, and species distributions," *Fuel*, **169**, 120–134, (2016).

[81] A. M. Elbaz, S. Yu, X. Liu, X. Bai, I. Khesho, and W. L. Roberts, "An experimental/numerical investigation of the role of the quarl in enhancing the blowout limits of swirl-stabilized turbulent non-premixed flames," *Fuel*, **236**, 1226–1242, (2019).

[82] X. Liu, A. M. Elbaz, C. Gong, X. Bai, H. Zheng, and W. L. Roberts, "Effect of burner geometry on swirl stabilized methane/air flames: A joint LES/OH-PLIF/PIV study," *Fuel*, **207**, 533–546, (2017).

[83] M. D. Smoke and V. Giovangigli, "Formulation of the premixed and nonpremixed test problems," in *Reduced kinetic mechanisms and asymptotic approximations for methane-air flames*. Berlin/Heidelberg: Springer-Verlag, 1991, 1–28.

[84] N. Peters, "Laminar diffusion flamelet models in non-premixed turbulent combustion," *Prog. Energy Combust. Sci.*, **10**, 319–339, (1984).

[85] H. Pitsch and H. Steiner, "Large-eddy simulation of a turbulent piloted methane/air diffusion flame (Sandia flame D)," *Phys. Fluids*, **12**, 2541–2554, (2000).

[86] A. Kempf, R. Lindstedt, and J. Janicka, "Large-eddy simulation of a bluff-body stabilized nonpremixed flame," *Combust. Flame*, **144**, 170–189, (2006).

[87] A. Kronenburg and M. Kostka, "Modeling extinction and reignition in turbulent flames," *Combust. Flame*, **143**, 342–356, (2005).

[88] B. Bedat and R. Cheng, "Experimental study of premixed flames in intense isotropic turbulence," *Combust. Flame*, **100**, 485–494, (1995).

[89] P. Petersson, J. Olofsson, C. Brackman, H. Seyfried, J. Zetterberg, M. Richter, M. Aldén, M. A. Linne, R. K. Cheng, A. Nauert *et al.*, "Simultaneous PIV/OH-PLIF, Rayleigh thermometry/OH-PLIF and stereo PIV measurements in a low-swirl flame," *App. Opt.*, **46**, 3928–3936, (2007).

[90] K.-J. Nogenmyr, P. Petersson, X.-S. Bai, A. Nauert, J. Olofsson, C. Brackman, H. Seyfried, J. Zetterberg, Z. Li, M. Richter *et al.*, "Large eddy simulation and experiments of stratified lean premixed methane/air turbulent flames," *Proc. Combust. Inst.*, **31**, 1467–1475, (2007).

[91] K.-J. Nogenmyr, P. Petersson, X.-S. Bai, C. Fureby, R. Collin, A. Lantz, M. Linne, and M. Alden, "Structure and stabilization mechanism of a stratified premixed low swirl flame," *Proc. Combust. Inst.*, **33**, 1567–1574, (2011).

[92] A. Bohlin, E. Nordström, H. Carlsson, X.-S. Bai, and P.-E. Bengtsson, "Pure rotational CARS measurements of temperature and relative O2-concentration in a low swirl turbulent premixed flame," *Proc. Combust. Inst.*, **34**, 3629–3636, (2013).

[93] H. Carlsson, E. Nordström, A. Bohlin, P. Petersson, Y. Wu, R. Collin, M. Aldén, P.-E. Bengtsson, and X.-S. Bai, "Large eddy simulations and rotational CARS/PIV/PLIF

measurements of a lean premixed low swirl stabilized flame," *Combust. Flame*, **161**, 2539–2551, (2014).

[94] R. Cheng, I. Shepherd, B. Bedat, and L. Talbot, "Premixed turbulent flame structures in moderate and intense isotropic turbulence," *Combust. Sci. Technol.*, **174**, 29–59, (2002).

[95] M. Day, S. Tachibana, J. Bell, M. Lijewski, V. Beckner, and R. K. Cheng, "A combined computational and experimental characterization of lean premixed turbulent low swirl laboratory flames: I. methane flames," *Combust. Flame*, **159**, 275–290, (2012).

[96] K. Nogenmyr, X. Bai, C. Fureby, P. Petersson, R. Collin, M. Linne, and M. Aldén, "A comparative study of LES turbulent combustion models applied to a low swirl lean premixed burner," in *46th AIAA Aerospace Sciences Meeting and Exhibit*, 2008, p. 513.

[97] E. Knudsen and H. Pitsch, "A general flamelet transformation useful for distinguishing between premixed and non-premixed modes of combustion," *Combust. Flame*, **156**, 678–696, (2009).

[98] D. Kang, F. Culick, and A. Ratner, "Combustion dynamics of a low-swirl combustor," *Combust. Flame*, **151**, 412–425, (2007).

[99] M. Mansour and Y.-C. Chen, "Stability characteristics and flame structure of low swirl burner," *Exp. Thermal Fluid Sci.*, **32**, 1390–1395, (2008).

[100] U. Stopper, M. Aigner, W. Meier, R. Sadanandan, M. Stöhr, and I. S. Kim, "Flow field and combustion characterization of premixed gas turbine flames by planar laser techniques," *J. Eng. Gas Turb. Power*, **131**, p. 021504, (2009).

[101] U. Stopper, M. Aigner, H. Ax, W. Meier, R. Sadanandan, M. Stöhr, and A. Bonaldo, "PIV, 2D-LIF and 1D-Raman measurements of flow field, composition and temperature in premixed gas turbine flames," *Exp. Thermal Fluid Sci.*, **34**, 396–403, (2010).

[102] U. Stopper, W. Meier, R. Sadanandan, M. Stöhr, M. Aigner, and G. Bulat, "Experimental study of industrial gas turbine flames including quantification of pressure influence on flow field, fuel/air premixing and flame shape," *Combust. Flame*, **160**, 2103–2118, (2013).

[103] S. K. Sadasivuni, G. Bulat, V. Sanderson, and N. Swaminathan, "Application of scalar dissipation rate model to Siemens DLE combustors," in *ASME Turbo Expo 2012: Turbine Technical Conference and Exposition*. American Society of Mechanical Engineers, 2012, 361–370.

[104] G. Bulat, W. Jones, and A. Marquis, "Large eddy simulation of an industrial gas-turbine combustion chamber using the sub-grid PDF method," *Proc. Combust. Inst.*, **34**, 3155–3164, (2013).

[105] G. Bulat, E. Fedina, C. Fureby, W. Meier, and U. Stopper, "Reacting flow in an industrial gas turbine combustor: LES and experimental analysis," *Proc. Combust. Inst.*, **35**, 3175–3183, (2015).

[106] A. Filosa, B. Noll, M. Di Domenico, and M. Aigner, "Numerical investigations of a low emission gas turbine combustor using detailed chemistry," in *50th AIAA/ASME/SAE/ASEE Joint Propulsion Conference*, 2014, p. 3916.

[107] T. Jaravel, E. Riber, B. Cuenot, and G. Bulat, "Large eddy simulation of an industrial gas turbine combustor using reduced chemistry with accurate pollutant prediction," *Proc. Combust. Inst.*, **36**, 3817–3825, (2017).

[108] I. Langella, Z. X. Chen, N. Swaminathan, and S. K. Sadasivuni, "Large-eddy simulation of reacting flows in industrial gas turbine combustor," *J. Propul. Power*, **34**, 1269–1284, (2018).

[109] E. Fedina, C. Fureby, G. Bulat, and W. Meier, "Assessment of finite rate chemistry large eddy simulation combustion models," *Flow Turbul. Combust.*, **99**, 385–409, (2017).

[110] E. Sher and S. Refael, "A simplified reaction scheme for the combustion of hydrogen enriched methane/air flame," *Combust. Sci. Technol.*, **59**, 371–389, (1988).

[111] C. K. Westbrook and F. L. Dryer, "Simplified reaction mechanisms for the oxidation of hydrocarbon fuels in flames," *Combust. Sci. Technol.*, **27**, 31–43, (1981).

[112] W. Jones and R. Lindstedt, "Global reaction schemes for hydrocarbon combustion," *Combust. Flame*, **73**, 233–249, (1988).

[113] W. Meier, X. Duan, and P. Weigand, "Investigation of swirl flames in a gas turbine model combustor: II. turbulence-chemistry interactions," *Combust. Flame*, **144**, 225–236, (2006).

[114] P. Weigand, W. Meier, X. Duan, W. Stricker, and M. Aigner, "Investigation of swirl flames in a gas turbine model combustor: I. flow field, structures, temperature, and species distributions," *Combust. Flame*, **144**, 205–224, (2006).

[115] Y. C. See and M. Ihme, "Large eddy simulation of a partially premixed gas turbine model combustor," *Proc. Combust. Inst.*, **35**, 1225–1234, (2015).

[116] A. Benim, S. Iqbal, W. Merier, F. Joos, and A. Wiedermann, "Numerical investigation of turbulent swirling flames with validation in a gas turbine model combustor," *App. Thermal Eng.*, **110**, 202–212, (2017).

[117] A. Donini, R. Bastiaans, J. A. van Oijen, and L. P. H. de Goey, "A 5-D implementation of FGM for the large eddy simulation of a stratified swirled flame with heat loss in a gas turbine combustor," *Flow Turbul. Combust.*, **98**, 887–922, (2017).

[118] Z. X. Chen, I. Langella, N. Swaminathan, M. Stohr, W. Meier, and H. Kolla, "large eddy simulation of a dual swirl gas turbine combustor: flame/flow structures and stabilization under thermoacoustically stable and unstable conditions," *Combust. Flame*, **203**, 279–300, (2019).

[119] Z. X. Chen, N. Swaminathan, M. Stohr, and W. Meier, "Interaction between self-excited oscillations and fuel-air mixing in a dual swirl combustor," *Proc. Combust. Inst.*, **37**, 2325–2333, (2019).

[120] J. C. Massey, Z. X. Chen, and N. Swaminathan, "Lean flame root dynamics in a gas turbine model combustor," *Combust. Sci. Technol.*, **191**, 1019–1042, (2019).

[121] Y. Minamoto and J. H. Chen, "DNS of a turbulent lifted DME jet flame," *Combust. Flame*, **169**, 38–50, (2016).

[122] "TNF Workshop," https://tnfworkshop.org, last accessed July 1, 2019.

[123] "Engine Combustion Network," https://ecn.sandia.gov, last accessed July 1, 2019.

[124] M. Jangi, X. Zhao, D. C. Haworth, and X.-S. Bai, "Stabilization and liftoff length of a non-premixed methane/air jet flame discharging into a high-temperature environment: An accelerated transported pdf method," *Combust. Flame*, **162**, 408–419, (2015).

[125] J. A. van Oijen, A. Donini, R. J. M. Bastiaans, J. H. M. ten Thije Boonkkamp, and L. P. H. De Goey, "State-of-the-art in premixed combustion modeling using flamelet generated manifolds," *Prog. Energy Combust. Sci.*, **57**, 30–74, (2016).

5 Chemical Kinetics

E. J. K. Nilsson, C. Fureby, and A. Aspden

5.1 Introduction

In the abstract of a paper in the 31st Proceedings of the Combustion Institute (2007), C. K. Law made a statement regarding the development of combustion science: "being aided by the rapid advance in computational capability, combustion has entered the era of quantitative predictability" [1]. One focus of that paper are the possibilities and challenges associated with incorporation of realistic chemical kinetics in combustion modeling. The decade that has passed since the publication of that paper has indeed shown rapid progress in development of comprehensive kinetic mechanisms as well as reduced kinetic mechanisms. Another overview of the field from the same period is the 2004 publication by Hilbert et al. [2], who reviewed the literature on implementation of explicit chemical schemes in computational fluid dynamics (CFD). They point out that a main driver for the development of these schemes is the need to accurately predict pollutant formation. Another motivation is that for flames near extinction the chemistry becomes important. Several recent works [3, 4] highlight the rapid development during the last decade, pointing out that improved fundamental understanding of both the chemical and physical processes involved in combustion, their interplay, as well as increased computational capacity, has moved the field significantly forward. The challenges remain and, in addition, the increasing number of proposed renewable fuels demands an extension of kinetic mechanisms and experimental datasets. Also, as turbulent combustion simulations with explicit chemistry become an important tool for engineering applications the demand on simplified kinetic schemes increase.

The use of numerical simulations to predict and investigate turbulent combustion in furnaces, internal combustion, gas turbine, or rocket engines, and so forth relay heavily on the description of the chemical kinetics. Until not long ago the most dominant part of the modeling was believed to be the modeling of the mean or filtered reaction rates in Reynolds-averaged Navier–Stokes (RANS) and large eddy simulation (LES), respectively, but recently, this has proved to be not always the case for LES. Several studies, e.g., [5, 6], using finite rate chemistry LES clearly point out the underlying reaction mechanism as the most important part of the LES, more important than the subgrid flow and reaction rate models.

The main challenges in modeling of turbulent combustion systems with explicit chemistry is the wide range of spatial and temporal scales, the large amount of species and reactions, and the interactions between reactivity and diffusion [7].

Using a flamelet approach the assumption is made that the chemical time scales are much smaller than the turbulent time scales. In these models the coupling between chemistry and turbulent mixing are parameterized using one or two variables, which result in a computationally feasible system that, however, is not sufficient to capture relevant combustion phenomena for some applications. For descriptions of combustion regimes sensitive to chemistry and chemistry–turbulence interactions PDF-type models have been shown to be more appropriate. To be computationally feasible these models require a statistical description of the smallest scales and incorporation of a reduced set of chemical reactions [7]. An important task is thus to produce chemical kinetics schemes of tractable size and good capacity to predict combustion in relevant ranges of conditions.

Generally, a high combustion efficiency, i.e., complete conversion of fuel to final products to yield maximum energy output, comes hand-in-hand with large emission of the pollutant nitrogen oxides (NOx) efficiently produced at high temperatures. The trade-off between efficient combustion processes and clean exhaust is one of the driving forces in development of new fuels and improvement of combustion devices. For accurate assessment of air quality and health effects resulting from combustion emissions it is not enough to predict the harmful major combustion products like NOx and carbon monoxide (CO). Minor constituents like aldehydes, important byproducts in biofuel combustion, can be highly toxic and need to be accurately predicted. Also, soot formation in combustion process and the resulting formation of particles in the atmosphere are essential for human health assessment [4]. Gas phase pollutants produced in fossil fuel combustion are mainly the toxic CO, the greenhouse gas carbon dioxide (CO_2), NOx, and hydrocarbons (HC) that are unburnt fuel or fuel fragments [8].

The biofuels in practical use today are mainly ethanol and fatty acid methyl esters (FAME) biodiesel. These compounds have oxygen atoms in their molecular structure, which result in additional oxygenated pollutants such as the toxic formaldehyde and acetaldehyde [9]. The biofuels are advantageous from a global environmental perspective since they contribute less to the net increase of greenhouse gases compared to the fossil fuels [10]. On a local and regional perspective biofuel combustion can be considered positive for air quality, since less soot is formed and in general smaller amounts of CO and HC are released. On the other hand, NOx may increase and the oxygenated pollutants are toxic in small amounts and also result in increased formation of ground-level ozone [11, 12]. Emissions from biofuel combustion modify air composition compared to fossil fuel combustion, and whether the overall effect is positive for air quality depends on a range of factors such as temperature and presence of other pollutants. From energy security and global warming perspectives the transition to biofuel combustion is necessary, which mean that the changes in air composition is inevitable [10]. Research in pollutant formation from biofuel combustion is therefore essential to allow development of clean and efficient combustion devices. Modeling of turbulent combustion of biofuels, with models of high enough quality to predict pollutant formation, are an important part of the continued development. The scientific community has during the past decade significantly advanced the understanding of the chemistry governing biofuel combustion, but the understanding is not complete and it has not to a significant extent been used to generate chemical kinetics mechanisms small enough for CFD.

One weakness in the CFD scientific community is the noncritical use of kinetic mechanisms, often as a result of lack of communication between experts on kinetics and CFD modeling. Many kinetic mechanisms, in particular the ones that are small enough to be suitable for CFD, are applicable only over a very limited range of conditions. Use of the kinetic mechanisms outside the range for which it is developed might lead to erroneous results [3]. Importance of chemical description has been reviewed by Hilbert et al. [2], and in the case of LES investigated by Fureby et al. [5, 6], Franzelli et al. [13], and Jaravel et al. [14].

As outlined above, there is a need to implement explicit chemical description in turbulent combustion simulations. The present chapter aims at introducing the reader to methods and principles of chemical kinetics modeling by giving a brief description of key concepts, recent developments in the field and indications of future needs. The chemistry governing various combustion phenomena will be outlined and typical mechanisms for natural gas and common liquid fuels, fossil and renewable, will be presented.

The structure of this chapter is as follows. The basic concepts of chemical kinetics are introduced in Section 5.2, followed by presentation of the chemistry that determines ignition, flame propagation, and extinction. Section 5.2 incorporates a brief introductions to experimental data, with a focus on interpretation of experimental data from a modeler's perspective, with indications about uncertainties. This section is in particular a source of references to important reviews covering experimental data for combustion properties of important fuels. The different types of kinetic mechanisms and typical development approaches are explained in Section 5.3, with particular attention on reduced mechanisms small enough to be incorporated in LES, with capacity to predict pollutant formation. Section 5.4 gives an overview of available combustion chemistry mechanisms for practical fuels: natural gas, heavy liquid fuels, and ethanol. It is not a complete review of all existing mechanisms but rather a guide among the most commonly used and recent mechanisms of different levels of complexity. The implementation of explicit chemistry in LES is given in Section 5.5, where several cases are presented and discussed. The final section summarizes and concludes the current state of this research and identifies some potential directions for future development.

5.2 Combustion Chemistry

Two processes affect chemical composition: reaction and molecular diffusion [7]. Molecular mixing occurs at very small scales and is of significant importance for some combustion phenomena. The effects of mixing on the chemistry and global combustion characteristics, and the implications it has for implementation of explicit chemistry in turbulent combustion modeling, is mentioned in Section 5.5. The basic theory and main features of elementary chemical reactions governing the combustion characteristics are outlined in the present section. Then, the chemistry of the distinct phenomena, viz., ignition [15], flame propagation [16], and extinction [17] are discussed along with comments on experimental systems.

To isolate different combustion phenomena, they are studied in simple laboratory geometries, from which information can be extracted in three levels of detail [1]. The global level consists of the combustion characteristics such as ignition delay time, laminar burning velocity, and extinction strain rate, which are the parameters describing the global flame behavior. On the detailed level, temperature and chemical composition of reactive mixtures or flames are investigated. The third level is investigation of the system response to variation in kinetic or other (temperature, pressure) parameters. In this section experiments commonly used for mechanism validation are presented, outlining their strengths and weaknesses. Key references, where the experimental approaches have been evaluated in detail, are given. One goal of the present section is to give the reader an idea on why and how the different experimental datasets are used in mechanism development and validation. An important part of this is to indicate the uncertainties in the experiments.

5.2.1 Chemical Kinetics

5.2.1.1 Basic Theory and Parameterization

A chemical process like the combustion of a hydrocarbon fuel to produce carbon dioxide and water can be summarized by an overall reaction, here exemplified by that of methanol:

$$2\,CH_3OH + 3\,O_2 \longrightarrow 2\,CO_2 + 4\,H_2O \qquad R1$$

In reality a sequence of reactions, in the case of methanol on the order of hundreds of reactions, take place in the process. Each separate reaction step is called an *elementary reaction*, which is defined as the conversion of reactants to products without observable intermediates. An elementary bimolecular reaction of significant importance in most reaction systems involving hydrogen or hydrocarbon fuels is that of a hydrogen atom with molecular oxygen, given below as reaction R2. This reaction generally promotes reactivity in the system, since it is chain branching, i.e., two radicals are produced from one radical and a stable specie, and thus the radical pool increase.

$$H + O_2 \longrightarrow OH + O \qquad R2$$

A majority of the relevant elementary chemical reactions are of second order (bimolecular), meaning that two species react to form products, as exemplified by reaction R2. The rate of change, r, of a reactant, A, is determined by the reaction rate coefficient, k, and the concentration of reactants:

$$r = \frac{d[A]}{dt} = -k[A][B]. \qquad (5.1)$$

The reaction order is defined by the number of entities participating in molecular encounters required for the products to form. Some second-order reactions involve a collision with a third body, commonly denoted M, that does not change chemically in the process. The collision partner M could be any specie in the reaction mixture, likely N_2 or O_2, in a regular combustion system with air as oxidizer. All collisions with a neutral body do not result in reaction and some species are more efficient collision partners than others. In connection with every reaction involving M it is thus necessary to specify the collisional efficiency of each component that may act as M. Reaction R3

$$H + O_2 + M \longrightarrow HO_2 + M \qquad \text{R3}$$

is the third-body reaction corresponding to the reaction R2, and later in this section we will return to these two reaction as an example of how the relative importance of them change as pressure increase.

The rate of the three-body reaction is, of course, also dependent on the presence of collision partners, M. The reaction rate coefficient is dependent on several separate steps, first the production of an excited intermediate [AB]*, and the two possible pathways for the intermediate to dissociate back to reactants or to be collisionally relaxed and form stable product:

$$r = \frac{d[A]}{dt} = -k[A][B][M]. \qquad (5.2)$$

First-order or unimolecular reactions involve chemical transformation of one reactant and could either be an isomerization process, see reaction R4, or a decomposition into products as in reaction R5. In the case of a unimolecular reaction the collision partner will excite the molecule, which will either relax back by a second collision, isomerize, or decompose to get rid of the excess energy.

$$CH_3O_2 \longrightarrow CH_2OOH \qquad \text{R4}$$
$$CH_2O + M \longrightarrow CO + H_2 + M \qquad \text{R5}$$

Combustion systems involve a wide range of temperature variation and accurate representation of the chemistry must include temperature dependent reaction rate coefficients. The extended Arrhenius equation is the standard representation used in combustion kinetic schemes and it is given by

$$k = A\, T^n \exp\left(-\frac{E_a}{RT}\right), \qquad (5.3)$$

where A is the preexponential factor, E_a is the activation energy, R is the universal gas constant, and T is the temperature. For many reactions the temperature dependence is sufficiently described by the classic Arrhenius expression without the factor T^n in Eq. (5.3). For a bimolecular reaction the reaction rate constant, k, has a unit of cm^3 molecule^{-1} s^{-1}.

Next, we will briefly introduce the expressions used to parameterize the pressure dependence in kinetic mechanisms. Part of the pressure dependence can intuitively be explained by noting that the pressure increases the concentration of M and thus reactions involving a collision partner will also increase as per Eq. (5.2). In the case of a collisionally excited molecule, the pressure dependence is governed by a competition between the dissociation to products and deexcitation back to the original molecule. At pressures below the so-called low-pressure limit, every molecule excited above the dissociation threshold will indeed dissociate and the reaction rate is represented by the Arrhenius form given in Eq. (5.3) for k_0. In the high-pressure limit, the pressure independent rate constant is denoted using k_∞ and the molecule may undergo a number of collisions in this limit even after the energy of the molecule reached above the dissociation threshold. The most complex behavior is in the intermediate pressure regime, called the *fall-off region*, where the reaction rate is resulting from a competition between dissociation and collisional energy transfer. Using a unimolecular reaction as an example the effective rate constant is commonly represented by the following relation:

$$k = k_\infty \left(\frac{P_r}{1 + P_r} \right) F, \tag{5.4}$$

where the reduced pressure P_r is given by

$$P_r = \frac{k_0 [\text{M}]}{k_\infty}. \tag{5.5}$$

The parameter F, sometimes called the blending factor, can be set to 1, which is the simple Lindeman approach, but it is more accurately described with a more complex formulation like the commonly used Troe formalism. Further detail on the parameterization can be found in, for example, the kinetics book by Turanyi and Tomlin [18]. Important insights on pressure dependence is also given in the overview article by Klippenstein [19].

5.2.1.2 Experimental and Computational Kinetics

Reaction kinetics can be investigated in dedicated experimental studies or using quantum chemistry calculations. To give a complete description of a chemical reaction for incorporation in a combustion kinetic mechanism, the temperature and possible pressure dependences, as outlined in the previous section, need to be determined. Furthermore, if competing product channels exist then those channels need to be identified and quantified.

Dedicated chemical kinetics experiments is a science on its own right and will not be discussed here, but we like to point out that in particular at high temperatures and/or pressures the experiments are technically very challenging. Quantum chemical calculations are valuable to complement or extend experimental studies, but the accuracy of the results depend on a number of factors such as type of reactants, choice of method, and so forth, which is also outside the scope of this discussion. For more information about the use of quantum chemistry calculations in combustion kinetics we recommend the overviews given by Miller, Pilling and Troe [20], Klippenstein [19], and the references therein.

Chemical kinetic reaction rate data have been experimentally determined for more than a century using a range of methods, and the amount of data available in the open literature is immense. As chemistry started to be commonly considered in combustion research the first extensive data evaluations summarizing the present state of understanding of reaction kinetic, where compiled, mainly by Tsang and co-workers in the late 1980s [21–24]. In these evaluations, all available data for chemical reactions are presented and recommendations are made on reaction rate expressions and product formation. Since then several updated evaluations have been published, most notable is the one by Baulch et al. [25] in 2005. In recent year quantum chemistry calculations have become important in chemical kinetics research and in the last evaluation calculation data are incorporated when needed to resolve experimental discrepancies or provide data in regimes where no experiments exist.

Here we present some reaction rate expressions for reactions of methanol, to illustrate the use of experiments, calculations, and data evaluations. Figure 5.1a shows plots of reaction rate expressions for reactions of methanol with hydrogen atom, reactions R6a and R6b, and peroxy radical, reactions R7a and R7b, both giving the isomeric products hydroxymethyl (CH_2OH) and methoxy radical (CH_3O). We start by considering hydrogen abstraction by the hydrogen atom:

$$CH_3OH + H \longrightarrow CH_2OH + H_2 \qquad \text{R6a}$$

$$CH_3OH + H \longrightarrow CH_3O + H_2 \qquad \text{R6b}$$

$$CH_3OH + HO_2 \longrightarrow CH_2OH + H_2O_2 \qquad \text{R7b}$$

$$CH_3OH + HO_2 \longrightarrow CH_3O + H_2O_2 \qquad \text{R7b}$$

In the 1984 evaluation, Tsang presented an expression for the overall rate of reaction R6, i.e., the loss of the reactants, methanol and hydrogen. However, the understanding of product formation was not sufficient at that time. Whether abstraction occurs on the hydroxyl group or the carbon depends on bond strengths and complex aspects related to stability of intermediates. Early on it was often assumed that the hydroxyl hydrogen was target for H-abstraction and thus the main product was the methoxy radical in reaction R6b. In subsequent evaluations, the overall reaction was given similar expressions even though there was some disagreement at high temperatures. Both overall reaction rate and branching were experimentally determined for a limited range of temperatures around 1000 K showing that the reaction R6a was the dominant product channel at this temperature. Reaction rates calculated theoretically were lower than those from experiments and gave a slightly different temperature dependence but essentially the same trend with respect to branching to the two product channels.

Figure 5.1b presents an early experimental study at low temperatures, that later proved to be essentially in agreement with a theoretical study. However, another theoretical study performed at the same time but using different methods under predicts the experimental data. The two theoretical studies are essentially in agreement concerning the minor product channel. This points out the strengths and weaknesses

5.2.2 Global Combustion Characteristics

The combustion processes proceed starting with an ignition following flame propagation and finally extinction of the flame. These events are governed by chemical reactivity, each sensitive to a particular subset of reactions. In the present subsection the important chemistry will be explained and the experiments used to investigate the phenomena will be outlined. The experimental methods can be divided into different classes, the first being experiments of homogeneous mixtures in various reactors and shock tubes, where the chemical phenomena are completely isolated. The second class includes experiments where transport processes are also important, such as premixed flame propagation and extinction. A final class that has got some attention recently is non-premixed extinction experiments, considered by some researchers to be essential for development of mechanisms applicable in turbulent combustion simulations.

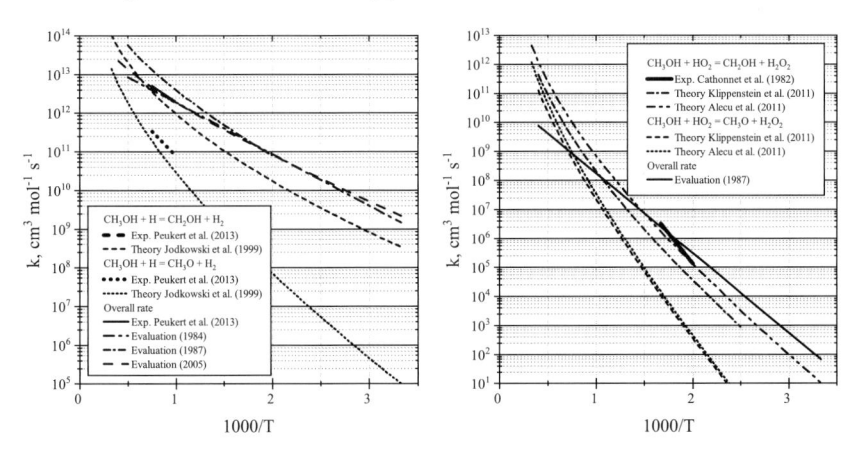

Figure 5.1 Reaction rate constant expression for the reactions of methanol with H and HO_2 including data from experimental studies [26, 27], theoretical calculations [28–30], and data evaluations [21, 25, 31].

5.2.2.1 Ignition

A homogeneous mixture of a fuel, an oxidizer and possibly an inert will ignite when the temperature and pressure conditions are such that the fuel and oxidizer undergo a hydrogen abstraction reaction to form a fuel radical and hydroperoxyl,

$$R-CH_3 + O_2 \longrightarrow R-CH_2 + HO_2 \qquad \text{R8}$$

These kinds of reactions are slow and endothermic, but initiate a chain of reactions as two reactive radicals are produced. These radicals can react in a large number of ways, but the most important consideration is that the following reactions are rapid and to a large extent exothermic. The produced fuel radical can undergo reaction with yet another O_2 by either abstraction reaction R9, or addition mechanism given by reaction R10, the later can have various products but the alkylperoxy route, reaction R10, dominate under many conditions,

$$R-CH_2-CH_2+O_2 \longrightarrow R-CH=CH_2+HO_2 \qquad R9$$
$$R-CH_2+O_2 \longrightarrow R-CH_2O_2 \qquad R10$$

The alkylperoxy radical can be involved in further reactions, but at low temperatures it mainly isomerizes to form hydroperoxyalkyl,

$$R-CH_2OO \longrightarrow R-CH-O-OH \qquad R11$$

All these peroxy species have a rich low-temperature chemistry with many possible decomposition and reaction products, making low-temperature chemistry highly complex. For further detail on the low-temperature chemistry and autoignition we refer to the review by Zador et al. [15].

Again, an important point is rapid production of reactive radicals, resulting in an ignition event. A large fraction of the reactions is hydrogen abstraction by OH, H, O, or another radical. On most fuels hydrogens are available to be abstracted from different positions, and Fig. 5.2 demonstrates how this results in a larger variation in products from an oxygenated fuel compared to an alkane, by showing fuel radicals from propane and propanol. For propane molecule, the abstraction can occur on the central carbon or one of the terminal carbons, while for both n-propanol and isopropanol there are four unique positions for hydrogen abstraction. The relative importance of the different products channels is largely determined by the bond strength in the different C–H bonds and the O–H bond. For small oxygenates the presence of the oxygen containing group has a large effect on the overall reactivity of the molecule, while for longer chained species the overall reactivity and product distribution start to resemble that of the corresponding alkane.

In general, the time for a gas mixture to ignite, known as the ignition delay time, decreases with increasing temperature. For long chained hydrocarbon fuels at low temperature there is, however, a reverse trend – an increase in ignition delay time with increase in temperature. This trend holds for a temperature range of 50–100 K and at slightly higher temperatures it goes back to the "normal" variation. The region with reverse trend is said to be the negative temperature coefficient (NTC) region, and it divides the low-temperature kinetics and the high-temperature kinetics. The chemistry occurring in the NTC-region was explained by Prince et al. [32, 33]. In summary, it is a result of a peculiar temperature dependence of hydroperoxyalkyl peroxy radicals in a temperature region where it can either dissociate back to its reactants or decompose into a new peroxy intermediate. The chemistry underlying the NTC behavior involve many reactions that are of negligible importance at higher temperatures, and are often not included in reduced kinetic mechanisms. However, since low-temperature chemistry is becoming increasingly utilized in combustion devices the need for mechanisms that can accurately predict the NTC behavior is increasing.

The importance of both accurate pressure dependencies in the chemical reactions and description of transport phenomena is apparent when considering the hydrogen

Chemical Kinetics 209

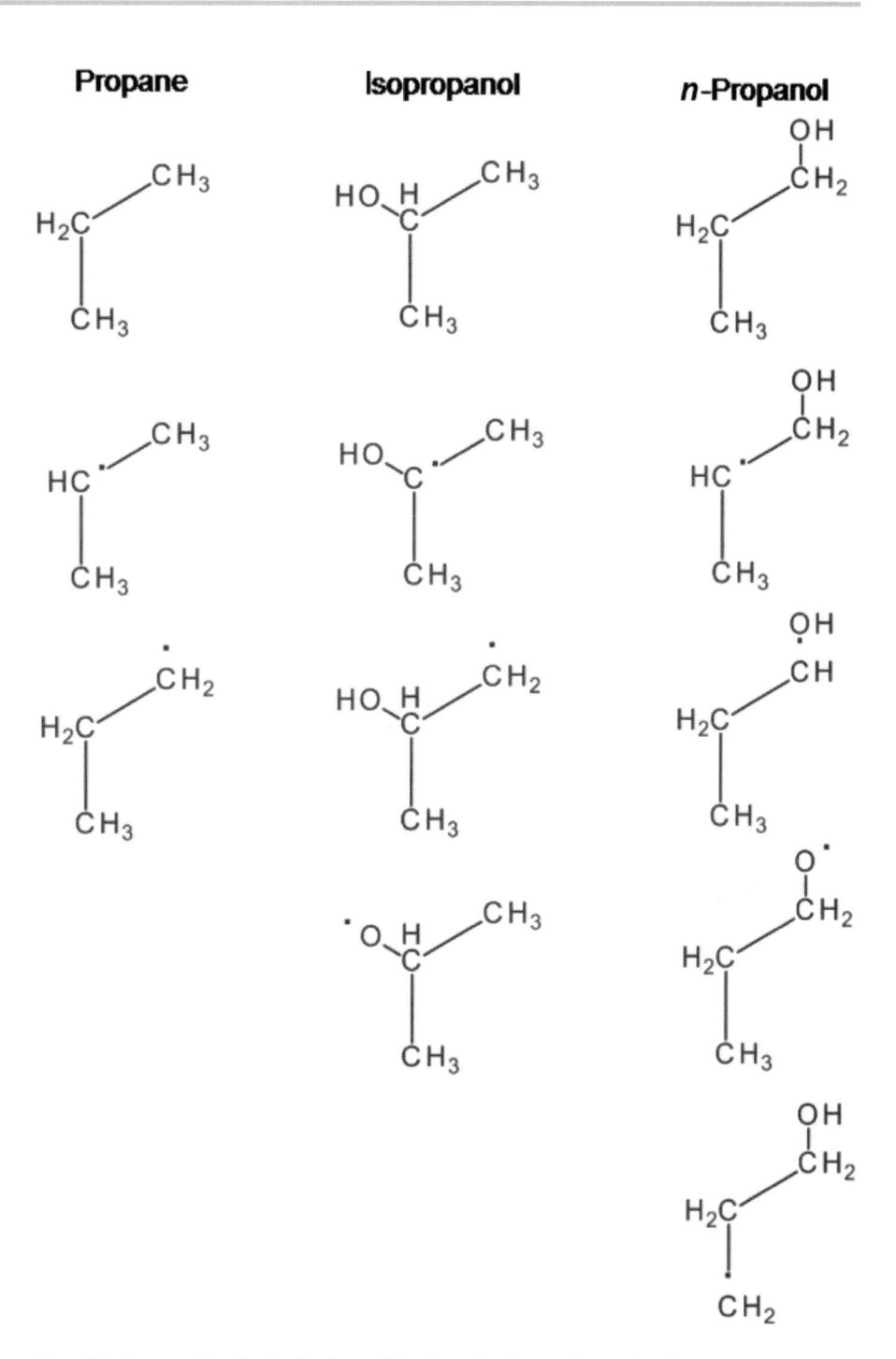

Figure 5.2 Intermediate fuel radicals resulting from hydrogen abstraction from propane, isopropanol, and n-propanol.

ignition, as an example, given by Law [1]. Hydrogen gas counterflowing with heated air at 1 atm and at 10 atm are shown to occur by very different mechanisms; at the low-pressure radical runaway is the key process while at the higher pressure thermal runaway in the presence of diffusive transport.

Ignition delay times are most commonly determined using shock tubes, but rapid compression machines (RCMs) and counterflow flame configurations are important for collecting data over a broad range of conditions. A shock tube is highly suitable for measuring ignition delay time at high temperatures and over a wide range of pressures. In a shock tube, a gas mixture is subject to a shock wave that instantaneously increase the pressure and temperature resulting in an ignition event. The experiment is supposedly to be zero-dimensional (in space) and homogeneous, which means that the ignition event is controlled solely by chemical kinetics. However, this method has limitations that limits its applicability; at some conditions the ignition event may not be truly homogeneous and boundary layer effects can affect the results. This means that the experiments require careful design and the processing of data and evaluation of uncertainties are crucial to assess the value of the data. An in-depth analysis of these aspects, including the choice of definition of ignition delay time, has been presented by Davidson and Hanson [34]. An important message from that analysis is that ignition delay times from shock tubes are less reliable if they are shorter than about 50 μs or longer than several milliseconds, and the user of shock tube data should be mindful of this while comparing modeling and experiments.

Low temperature ignition delays can be measured in RCMs, which is essentially a device simulating a single compression stroke of an internal combustion engine [35, 36]. Non-premixed ignition of fuel counterflowing with a heated oxidizer jet is another method used for investigating ignition phenomena [37, 38] and is suitable for development of detailed kinetic mechanisms.

5.2.2.2 Flame Propagation

Laminar flames are sustained by chemical reactions at high temperatures and thus laminar burning velocity is a flame property governing information on high temperature chemistry [39]. Capturing correct propagation speed and structure of laminar flames are also essential targets for mechanism development because these well-defined flames serve as the basis for fundamental principles of flame theory and thus are a key to understanding and modeling of complex combustion phenomena such as flame front instabilities, extinction and turbulence–chemistry interaction [1]. For a brief introduction to the fundamental principles governing premixed as well as diffusion flames we refer the reader to the review by Law [1].

Flame chemistry is governed mainly by reactions of small molecules and radicals involving O, H, and only one C. Chain branching reactions such as the one given in R12

$$H + O_2 \longrightarrow OH + O \qquad \text{R12}$$

strongly promote reactivity and therefore increase flame propagation speed. Figure 5.3 presents the sensitive reactions from a detailed and a reduced mechanisms for methane–air flames and one can see that the reaction R12 is the most important one. The most sensitive reactions to lower reactivity are the H-consuming reactions. The reaction with methyl radical is for rich flames and reaction with molecular oxygen to produce the less reactive HO_2 radical is for lean flames.

Chain carrying reactions produce one new radical while consuming one radical, but can still have a strongly promoting effect if the produced radical is important for a chain branching reaction. This is the case for the following two reactions producing an H radical that feed reaction R12; both are among the most sensitive reactions listed in Fig. 5.3.

$$HCO(+M) \longrightarrow CO + H(+M) \qquad \text{R13}$$

$$CO + OH \longrightarrow CO_2 + H \qquad \text{R14}$$

Recombination reactions, such as that given in R15, decrease reactivity

$$H + OH(+M) \longrightarrow H_2O(+M) \qquad \text{R15}$$

Flame propagation is to a large extent a competition between the production and destruction of reactive radicals. A large radical pool increases the reactivity and therefore also the laminar burning velocity.

Laminar burning velocity is mostly determined using burner stabilized flames, spherical flames or counterflow flames, Egolfopoulos et al. [40] give an overview of these methods and few details on additional methods. These methods should not be seen as competitive but complementary to each other. Burner stabilized flames using the heat flux method have the advantage that laminar burning velocity is determined on an unstretched laminar flame and therefore no stretch corrections need to be applied, a thorough evaluation of this method was recently performed by Alekseev et al. [41]. However, burner stabilized flames are useful only at atmospheric pressure

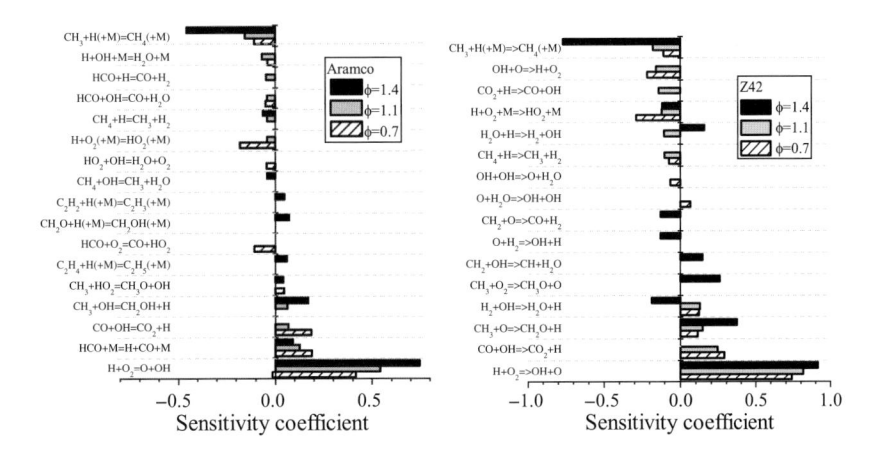

Figure 5.3 Sensitivity coefficients for methane–air flames at 298 K and 1 atm, at three different equivalence ratios. Left panel is the detailed Aramco mechanism and right-hand side is the skeletal Z42 mechanism.

or slightly higher, up to about 5 bar [42]. Outwardly propagating flame methods have the advantage that they can cover a wider range of pressures but it is very important to consider correct treatment of stretch effects and other sources of uncertainties as discussed by Chen [43].

Experimentally determined laminar burning velocities have been reported and reviewed in several comprehensive works in recent years, for example the combined experimental review and modeling update by Ranzi et al. [44] for hydrocarbon and oxygenated fuels and the extensive review of hydrogen, alkanes, and alcohols by Konnov et al. [16].

5.2.2.3 Extinction

Extinction is a high temperature flame phenomena, just like flame propagation, but it has been shown that the controlling kinetics may be notably different [17, 45]. Thus, validating a kinetic mechanism by capturing the laminar burning velocity data does not automatically imply that the chemistry governing extinction phenomena are accurately described. As pointed out by Esposito and Chelliah [46] accurate modeling of non-premixed extinction is essential for a kinetic mechanism used for combustion device simulations. An additional complication is that for fuel molecules larger than C_2, extinction is very sensitive to fuel diffusion, which means that the transport description in the simulations become increasingly important. While extinction of flames burning smaller fuel molecules such as methane and methanol are sensitive to the same reactions as the laminar burning velocity, larger fuels like iso-octane can have dominant sensitivity to reactivity of fuel fragments.

Since extinction of non-premixed flames in a counterflow burner is sensitive to chemical kinetics and transport, these are valuable for investigation of the coupling of these properties. Extinction of flames of large fuel components are particularly sensitive to the mass diffusivity, a result of these large molecules diffusing slowly and thus their transport to the reaction zone is a rate-limiting step.

5.3 Chemical Kinetic Mechanisms

This section gives some detail about the definitions, properties, and construction of mechanisms of different complexity. Different types of mechanisms are described, as outlined in Fig. 5.4, essentially following the definitions of Hilbert et al. [2] and Fiorina et al. [47].

In a perspective article published in 2018, Kohse-Höinghaus [4] makes the distinction between engineering and scientific models for chemical kinetics and points out that the reduced (engineering) mechanisms should be systematically obtained from the detailed (scientific) ones. The reduced schemes need to focus on the relevant fundamental backbone and leave out details of minor significance for the target application.

Reduced mechanisms commonly build on a mechanism of higher complexity and thus a significantly larger number of species and reactions. Recent successful methodologies do, however, not rely completely on detailed mechanisms [48, 49],

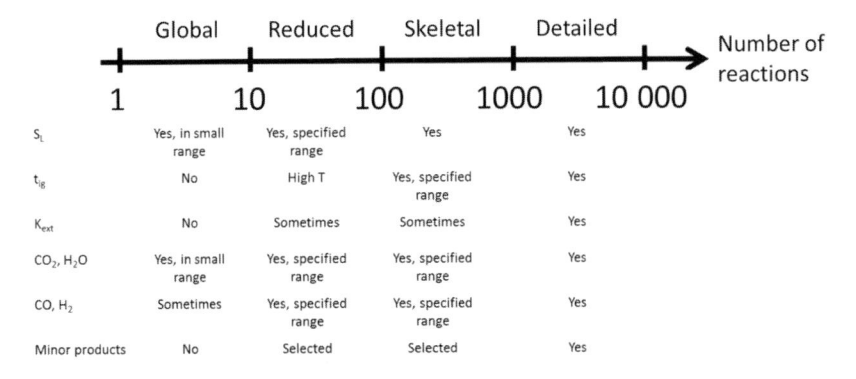

	Global	Reduced	Skeletal	Detailed	
	1	10	100	1000	10 000 → Number of reactions
S_L	Yes, in small range	Yes, specified range	Yes	Yes	
t_{ig}	No	High T	Yes, specified range	Yes	
K_{ext}	No	Sometimes	Sometimes	Yes	
CO_2, H_2O	Yes, in small range	Yes, specified range	Yes, specified range	Yes	
CO, H_2	Sometimes	Yes, specified range	Yes, specified range	Yes	
Minor products	No	Selected	Selected	Yes	

Figure 5.4 Schematic showing the size range for different types of mechanisms, as explained in this section, including indications on typical performance for various combustion characteristics.

as explained later in this section. The chemical fidelity of the reduced mechanisms deduced from detailed mechanisms inevitably depends on the starting mechanism. This means that the first step in a successful mechanism reduction is creation or selection of a comprehensive detailed mechanism. The comprehensiveness of a mechanism can be defined as its ability to reproduce combustion properties related to ignition, propagation, and extinction of a flame, over a range of temperatures and pressures. Additionally, it should be able to handle various fuel/oxidizer ratios and accurately predict concentrations of products over the full range of conditions. Another aspect of comprehensiveness is the hierarchical nature of combustion mechanisms, meaning that a mechanism for a larger fuel will contain the complete submechanisms for combustion of smaller fuels that are intermediates of the combustion of the larger fuel [1]. A simple example of this is that all comprehensive hydrocarbon mechanisms should contain a complete mechanism for CO combustion.

5.3.1 Complete or Detailed Mechanisms

In *complete chemical reaction mechanisms* all possible reactions of fuel and all intermediates are included and represented by accurate reaction rate constants describing the temperature and pressure dependencies. A truly complete mechanism should consist of only elementary reactions. In reality no such mechanisms exist since our understanding is not complete, but this specific term "complete mechanism" can be used for mechanisms constructed with the intention to make them complete. These schemes are also commonly called as *detailed mechanisms* but this term also include schemes where simplifications have been made on purpose, for example, by excluding some minor reaction paths. A complete mechanism is, in theory, expected to be comprehensive, i.e., to accurately reproduce all possible combustion cases at all relevant conditions. Detailed mechanisms are commonly developed to be as comprehensive as possible and a measure of true comprehensiveness is that a mechanism should be accurate for the full fuel hierarchy. This means that a mechanism for combustion of,

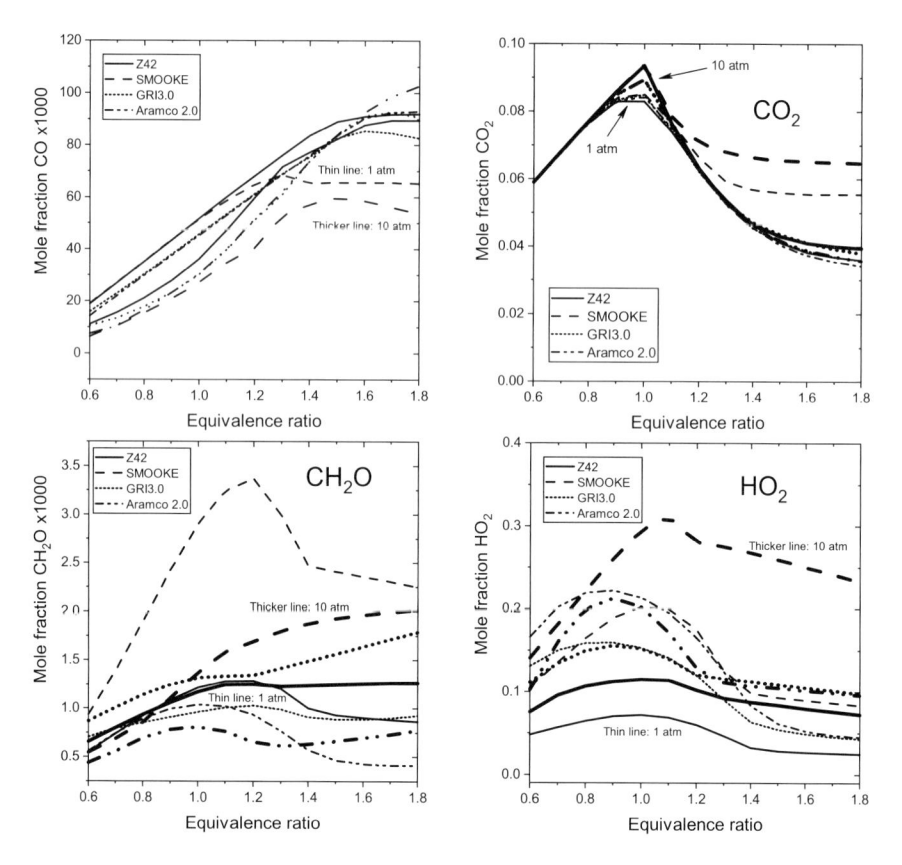

Figure 5.5 Formation of CO, CO_2, CH_2O, and HO_2 in laminar flames of methane, at 1 and 10 atm, calculated using kinetic mechanisms of different complexities.

for example, the C_3 fuel propane does accurately predict combustion characteristics of all smaller, C_0–C_3, fuels. Examples of recently developed comprehensive mechanisms are Aramco 2.0 and the mechanism of Konnov et al. [50–53]. As implied, these mechanisms include all the current knowledge of combustion kinetics of the included fuels, but there are, however, some differences. An important difference is that Konnov et al. chose not to modify any rate parameters, while others adjust reaction rate constants to give better agreement with the experimental data [54], commonly called as "tuning." It is often suggested that tuning within the limits of the uncertainty of experimental data is justified [55].

5.3.2 Skeletal Mechanisms

As expressed by Xin et al. [56] a detailed mechanism consist of species that are critical, nonessential, or marginal for a given simulation. The importance of species can be evaluated by species flux or sensitivity analysis, as explained by Turanyi and Tomlin [18], the first one being the cheapest method while the second one accurately quantifies species importance but for a larger computational cost. *Skeletal mechanism* is a term that include a range of schemes, from detailed to quite small, that have in common that they consist of an intact sequence of elementary reactions from fuel to final products carbon dioxide and water for hydrocarbon fuels. The name is most often used to denote highly simplified schemes where only the most relevant reaction sequences describing particular combustion phenomena are included, and in those cases the comprehensiveness is restricted.

A range of methods have been developed to extract skeletal mechanisms from detailed ones, for example principal component analysis (PCA) and direct relation graph (DRG) methods. The DRG method was first introduced by Lu and Law in 2005 [57] and has rapidly developed into a widely used method with several extensions to improve its capacity, such as DRGEP [14] and DRGASA [56] that include implementation of error propagation and sensitivity analysis respectively. DRG map coupled species to achieve a consistent smaller set of reactions from an extensive mechanism. The methods developed subsequently use different approaches to further identify relatively unimportant species. Comparison of the DRG-based methods was done by Qiu et al. [58], who concluded that DRG combined with DRGEP was the optimal method. Nagy and Turányi have compared their own SEM-CM method with the DRG-based methods and concluded that SEM-CM gives the smallest mechanism for a given simulation error [59].

5.3.3 Reduced Mechanisms

Reduced mechanisms include simplifications in the reaction path, for example, by lumping several elementary reactions together. Many reduction strategies usually based on a combination of skeletal reduction and time scale analysis are developed to deduce reduced mechanisms. An example of this is the ARC approach presented by Jaravel et al. [14] to produce a mechanism for LES of a gas turbine combustor. They create a skeletal mechanism using DRGEP and then select QSSA species in the reduced set, using the level of importance (LOI) criterion [60]. Methods based on separation of time scales have been proposed by Lam et al. [61, 62] and Maas and Pope [63].

Many reduction procedures relay on one combustion phenomena, commonly flame propagation or ignition. Esposito and Chelliah [46] used PCAS to extract reduced mechanisms based on the three properties, viz., ignition, flame propagation, and extinction separately and then combined them into a single mechanism. They concluded that non-premixed extinction is the phenomenon that results in the largest skeletal mechanism, i.e., the level of reduction is constrained by this property.

A recent approach for development of reduced kinetic mechanisms was outlined by Gao et al. [64] and then further refined and presented as HyChem by Goldin et al. [49, 65]. The method treats the fuel breakdown in a semiglobal fashion and couples it to a more detailed C_0–C_4 reaction subset. The fuel breakdown relies on the assumption that pyrolysis of larger hydrocarbons results in a similar set of smaller hydrocarbon segments and that this pyrolysis is fast compared to oxidation of the fuel fragments. As a part of development of HyChem several sets of development targets were tried for the optimization, to allow evaluation of their importance for the performance of the final mechanism [65]. It was clear that using only pyrolysis data as targets resulted in a mechanism that could not accurately predict ignition delay times, while a combination of ignition delay and flame speed targets was successful. Including laminar burning velocity calculations result in a significantly longer time for the optimization process but it is evident that it is necessary to achieve a versatile mechanism capable of capturing all of the three important laminar combustion characteristics.

The C_0–C_4 base for the HyChem mechanism was developed from the USC-II mechanism [66] using DRG/DRGASA and consists of 47 species and 263 reactions [65]. Fuel breakdown consists of a decomposition step and oxidation by six radicals. Application of this method of lumped fuel cracking was used to produce mechanisms for several hydrocarbons, out of which n-butane was selected for implementation in DNS [64]. The results strongly indicate that the approach is valid and that focus on the smaller species in a reduced mechanism yields good predictive capability with a comparably low computational cost, but it is pointed out that for large fuels it is applicable mainly for high temperature applications. A similar methodology has independently been developed by Zettervall et al. for applications on

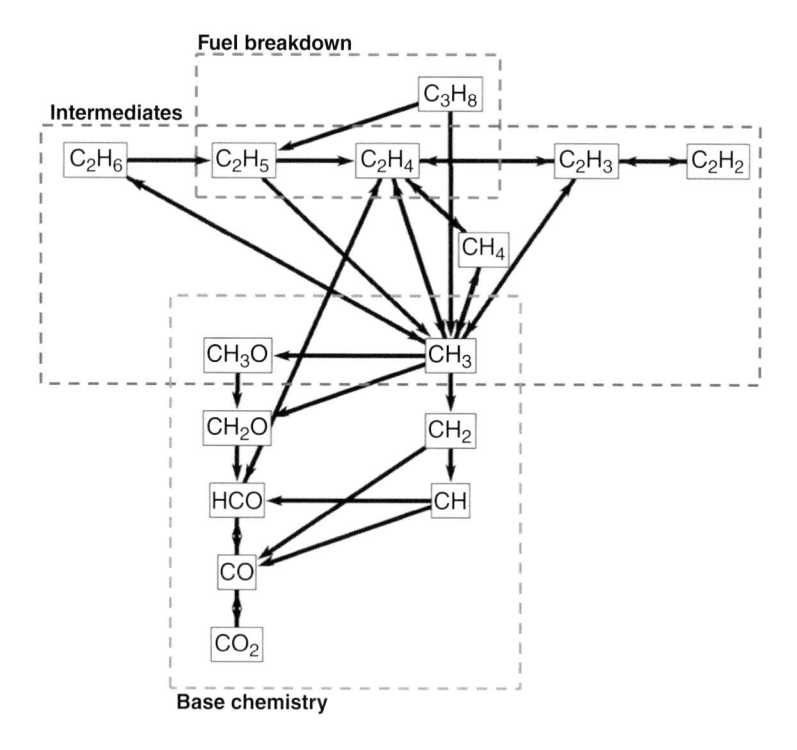

Figure 5.6 Reaction path diagrams showing the hydrocarbon compounds in the combustion of propane/air, modeled for a laminar flame at stoichiometric conditions using the Z66 mechanism by Zettervall et al. [69].

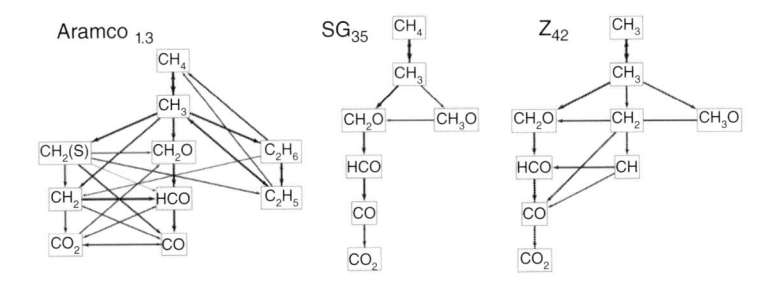

Figure 5.7 Reaction path diagrams showing the 10 most important carbon containing species in stoichiometric methane–air flames for the detailed mechanism Aramco 1.3 and the reduced schemes SG35 and Z42.

methane [67], ethylene [68], propane [69], and kerosene [58] combustion. Zettervall et al. use a smaller base mechanism and the resulting mechanisms are significantly smaller than the HyChem mechanisms.

The approach of Zettervall et al. [48, 69] is shown schematically in Fig. 5.6 for propane as a reaction path diagram showing three distinct parts. These parts include a very simple fuel decomposition subset, the restricted set of intermediate hydrocarbon reactions, and a more extensive base subset incorporating the C/H/O reactions. The alkane mechanisms by Zettervall et al. [58–61] have a common base mechanism for the C/H/O chemistry, with 40 irreversible reactions.

Figure 5.7 exemplifies the differences among mechanisms used to simulate methane–air flames using reaction paths for up to 10 most important reactions on the path from fuel to carbon dioxide. The panel to the left show 10 most important carbon-containing species for the detailed Aramco 1.3 mechanism [54] and recombination paths resulting in C_2 species are also included. The reduced mechanisms SG35 [70] and Z42 [67] include only C_1 species. SG35 is a commonly used reduced mechanism that shows good performance for flames at lean conditions, while showing large overpredictions of reactivity at stoichiometric and rich conditions. Zettervall et al. used SG35 as a starting point and added a set of seven reactions including CH and CH_2, which significantly improved the capacity of the mechanism to predict flame characteristics at richer conditions.

5.3.4 Global Mechanisms

Global mechanisms are comparatively crude simplifications that include one to four reactions involving the fuel and oxidizer, final products, and sometimes a few intermediate species like CO. The reaction rate expressions are tuned to give accurate estimations of laminar flame propagation, fuel breakdown, and production of major species, within a limited range of conditions. These simplified mechanisms have limited capability in simulations since they do not capture the chemical kinetic effects governing re-ignition and extinction processes. For complex and computationally demanding systems the global mechanisms are still the only affordable possibility and there is a continuous development of improved global schemes for various applications. Compared to tabulation methods the global mechanisms can be advantageous, since they are computationally cheaper and give accurate combustor exit temperatures and formation of the main pollutants such as CO and NOx.

The global one- and two-step mechanisms by Westbrook and Dryer (WD1 and WD2), developed in the early 1980s [38, 39], are possibly the most used global mechanisms. Other well-known global mechanisms are the four-step mechanisms by Jones and Lindstedt (JL4) [40] and the mechanism by the group of Peters [41]. The four-step schemes include in the range five to nine species, one of them being molecular hydrogen, which Seshadri et al. [41] showed to significantly improve the global schemes at moderately rich conditions.

5.3.5 Mathematical Stiffness

The computational time taken for a given kinetic mechanism is dependent not only on number of reactions and species, but also the numerical stiffness of the mechanism. The main limiting factor is the range of time scales involved in the mechanism. The shortest timescales are those of highly reactive radicals requiring small integration steps [1] while the longest time scale is related to the oxidation of CO requiring long integration time. Another limiting effect can be reversible reactions, with fast reverse rates leading to equilibrium [39].

5.4 Mechanisms for Natural Gas, Heavy Liquid Fuels, and Ethanol

Her, we present typical mechanisms for natural gas and common liquid fuels, detailed and reduced mechanisms that are either well known to the CFD community, and new mechanisms that have shown promising results.

5.4.1 Natural Gas and/or Methane

Methane, CH_4, is the likely fuel for which the largest number of reduced mechanisms have been constructed, and likely the most common fuel used as a target for mechanisms development. Table 5.1 lists a selection of detailed, reduced, and global mechanisms for methane–air combustion. The Aramco mechanism [54] is the largest and most extensively validated mechanism, used widely in detailed chemistry modeling and by many considered as state of the art for detailed chemical kinetics for several fuels. The other detailed mechanism included here is the well-known GRI-mech 3.0 [71], which is largely outdated but still in use to some extent. The three mid-sized mechanisms SG35, SR39, and Z42 (see Table 5.1) have been used for LES with finite rate chemistry. Z42 is an extended and improved version of SG35, with better performance at rich conditions and in several other aspects. The global mechanisms, WD1, WD2, and JL4, have been used in numerous CFD studies during the past 30 years, and are still of importance in cases where more extensive chemistry cannot be implemented.

Table 5.1 Selected mechanisms for CH$_4$–air combustion

Name	Reference	Mech.	Fuel	No. of species	No. of reactions
Aramco	[54]	Detailed	C$_0$-C$_4$		325 rev
GRI 3.0	[71]	Detailed	CH$_4$, NG	53	35 irrev
SG35	[70]	Reduced	CH$_4$	16	39 irrev
SR39	[72, 73]	Reduced	CH$_4$ + H$_2$	17	42 irrev
Z42	[67, 74]	Reduced	CH$_4$	18	6 irrev
JL4	[75]	Global	CH$_4$, C$_3$H$_8$	6	1/3 irrev
WD1/WD2	[76, 77]	Global	CH$_4$, C$_3$H$_8$	4/5	

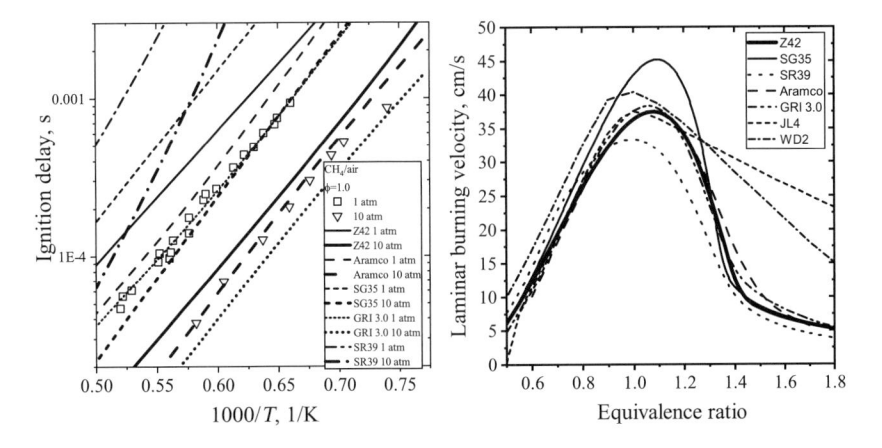

Figure 5.8 Ignition delay time (left) and laminar burning velocity (right) for methane–air flames. Ignition delay at two pressures plotted together with experimental data from Hu et al. [78]. Laminar burning velocity at 1 atm and 298 K initial gas temperature.

An interesting aspect for reduced methane mechanisms is whether to include the C$_2$ chemistry or not. In reality, CH$_3$ will combine to give C$_2$H$_6$, specifically at rich conditions. This is the first step toward soot formation and thus of significant interest for some applications. One approach to this is to include a global step for C$_2$H$_6$ chemistry, as was done by Sher and Refael [72, 73]. The mechanism of Smooke and Giovangigli, probably the most widely used mechanism for methane combustion, does not include C$_2$ chemistry [70].

Figure 5.8 presents ignition delay times and laminar burning velocities for the mechanisms in Table 5.1. The detailed mechanisms are in good agreement with experimental data, and as expected the Aramco mechanisms is the one yielding the best overall agreement. The reduced mechanism predicts longer ignition delay times, for SR35 by more than an order of magnitude. Z42 predicts higher reactivity than its mother mechanism SG35, which indicates that the modifications are valid, but it still ignites a little slowly at lower pressure. For the laminar burning velocity the detailed kinetic mechanisms are in satisfactory agreement with experimental results, with quite good agreement at lean conditions and larger scatter in experimental results

Table 5.2 Selected mechanisms for combustion of heavy fuels

Name	Reference	Mech.	Fuel	No. of species	No. of reactions
Westbrook	[86]	Detailed	FAME		
LLNL	[87]	Detailed	Diesel	2885	11, 754
LLNL	[88]	Reduced	Diesel	163	887
Ranzi	[44]	Large skeletal	Kerosene		
HyChem	[49, 84, 85]	Reduced	Jet fuels	Several versions	
Z65	[48]	Reduced	Kerosene/$C_{12}H_{23}$	18	65
Kundu	[89]	Reduced	Kerosen/$C_{12}H_{23}$	16	23

and between the mechanisms at rich conditions. The reduced mechanisms are also in general agreement at lean conditions, while showing significant differences around stoichiometric conditions and peak laminar burning velocities.

5.4.2 Heavy Liquid Fuels

Heavy hydrocarbon fuels consist of straight and branched alkanes, alkenes, and cyclic compounds. In a compression ignition (CI) engine, the autoignition is of crucial importance and need to be understood and modeled properly. The widely used fractions of fossil oil used as fuels, petrol, diesel, and aviation fuels, consist of a large number of different alkanes and aromatics. It is not possible to model these explicitly, but instead surrogate mixtures of typical fuel components are used [79]. The situation is the same for the biodiesel compounds that commonly consists of fatty acid methyl esters (FAMEs), with different composition depending on the source [79, 80].

Detailed mechanisms for fossil fuel and biodiesel fuel surrogates have been developed by the group at Lawrence Livermore, including works of Westbrook, Pitz, and co-workers [81, 82] and efforts to produce reduced schemes [83]. The CRECK group in Milano has also developed several mechanisms of different levels of complexity.

As mentioned in Section 5.3.3, two groups, Zetterval et al. [48] and the HyChem group [49], have recently developed methodologies of a semiglobal type where a complex fuel mixture is represented by a composite molecule with properties that are averages of the fuel mixture [49, 84, 85]. These methods are necessary for creation of small mechanisms for heavy fuels, since there are no complete mechanisms that can be used for reduction. These mechanism are listed in Table 5.2.

5.4.3 Alcohols

Alcohols with up to five carbon atoms are proposed as transportation fuels since they have physicochemical properties suitable for modern engines and can be produced from biomass [90]. The current state of knowledge on alcohol combustion chemistry was reviewed in 2014 by Sarathy et al. [90], including an overview of experimental studies conducted up to that time and discussions on reaction classes of particular

Table 5.3 Selected mechanisms for ethanol–air combustion

Name	Reference	Mech.	No. of species	No. of reactions
Aramco 2.0	[54]	Detailed		
Marinov	[91]	Detailed	56	351
DRG-EtOH	[92]	Reduced	24	26

relevance to alcohol kinetics. The dominating use of alcohols in combustion applications is transportation engines, in particular spark ignition (SI) engines, while there is also some emerging interest of alcohol combustion for gas turbine applications. The alcohol in practical use today is ethanol for road transport engines, and the following discussion gives a brief overview of ethanol combustion kinetics including detailed and reduced kinetic mechanisms available currently.

The majority of kinetic mechanisms constructed for ethanol combustion target SI engine conditions, suggesting that important validation targets are premixed laminar flame speeds and homogeneous ignition delay times [90]. For validation of mechanisms for CI engines also non-premixed flame configurations are included as validation targets. Ideally, validation for all cases should be performed over a wide range of equivalence ratios and temperatures, and at elevated pressures; this might, however, not always be possible due to a lack of experimental data.

A large number of detailed mechanisms for ethanol combustion have been published since the first one was presented by Natarajan and Bhaskaran in 1981; see Table 5.3. One of the most well-known and extensively used mechanism was published by Marinov [91] in 1999. The Marinov mechanism was for some time treated as a benchmark for ethanol combustion, but it is important to note its limitations, i.e., lack of validation at low temperatures and engine-relevant pressures. A more recent and extensively validated detailed mechanism including ethanol is Aramco, which has been validated for lower temperatures in reactor studies and higher temperatures for ignition delays.

A significant challenge in construction of detailed and reduced mechanisms for alcohol fuels is the implementation of branching ratios for decomposition and abstraction reactions. As exemplified for propane/propanol in Fig. 5.2, the presence of an OH group increases the number of unique abstraction sites and thus the number of different fuel radicals. For a reduced mechanism to be sufficiently small for use in CFD simulations it is necessary to remove some of the reaction paths or lump them together. Ideally this is done based on fundamental knowledge on the importance of the different channels at the target conditions. A complication is the fact that the dominating site for abstraction often changes with temperature. Ethanol does, however, lack significant low-temperature reactivity; as a result, hydrogen abstraction from the fuel does not favor the path leading to the peroxyradical chemistry but mainly goes to formation of acetaldehyde and HO_2.

The ethanol combustion chemical mechanism includes numerous oxygenated intermediates, radicals, and stable species, with various functionalities. Many of these are highly reactive and are not present in end products emitted from the engine. The most important oxygenates that need to be considered in the emissions on the basis of environment and health impacts are the aldehydes acetaldehyde and formaldehyde. A good kinetic mechanism, either detailed or reduced, therefore needs to predict the concentrations of these oxygenates accurately.

5.5 Incorporating Chemistry in Combustion Modeling

The importance of finite rate chemistry in large-eddy simulation (LES) is illustrated in this section by briefly showing and discussing selected results from different studies emphasizing the importance of the chemical kinetics for turbulent flames. The cases selected are high-Re number practical cases computed using finite rate chemistry LES models typically employing dynamic subgrid models such as the LDKM model [93], advanced combustion closures such as the PaSR or SF models [94, 95], and global to skeletal reaction mechanisms using simplified transport models based on Fick's and Hirschfelder and Curtiss' laws [96]. Even significant differences between detailed reaction mechanisms have been observed [2].

5.5.1 Turbulence–Chemistry Interactions in LES

Mixing and chemical reactions occur in scales smaller than convection. Large-eddy simulations require closure models for filtered reaction rates and recently Aspden et al. [97] evaluated factors affecting correlation terms in the models. Finite-rate chemistry is incorporated in LES by solving filtered transport equation for the molecular components, where the reaction rates can be computed explicitly using filtered Arrhenius reaction rates. Aspden et al. [97] concluded that flame response to turbulence is a smaller effects on finite-rate chemistry than the filtering operation. The difference did, however, decrease with increasing Karlovitz number. The important radical species OH, H, and O were found to be less impacted by the filtering than other radicals. These nonlinear effects in response are considered as the main challenges.

5.5.2 The Turchemi Combustor: Methane

The Turchemi model combustor shown in Fig. 5.9 is a single-burner laboratory version of Siemens SGT-100 engine composed of a full-scale single burner attached to an optical combustion chamber and installed in a high-pressure test rig at DLR Stuttgart, burning methane CH_4. One nonreacting case and four reacting cases were experimentally investigated in [98–100]. The most commonly investigated case was operated at 3 bar with an air temperature of 685 K; an air mass flow rate, including panel cooling air, of 0.1749 kg/s; a fuel mass flow rate of 0.0062 kg/s; and a thermal power of 335 kW. The radial burner has multiple fuel injection holes and provides proper initial mixture of fuel and air, which flows through a prechamber of 46 mm length and 86 mm exit diameter. The mixture enters the combustion chamber, where the M-shaped flame establishes in regions of shear layers between internal and external flow zones. Extensive experimental studies were performed including PIV; one-dimensional simultaneous Raman scattering (1DR), from which the temperature can be reconstructed; OH* chemiluminescence (OH*); and OH PLIF (OH).

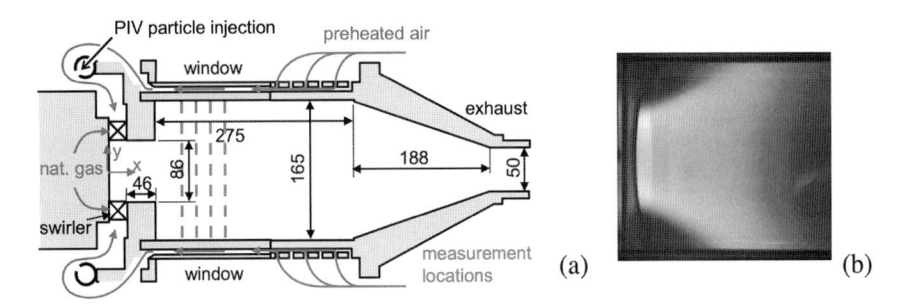

Figure 5.9 The Turchemi combustor. (a) Schematic of the SGT-100 burner and optical combustion chamber in high-pressure test rig at DLR Stuttgart and (b) a typical flame image.

Several LES studies of this case have been performed including those of Sadasivuni et al. [101], Bulat et al. [102], Bulat et al. [103], Filosa et al. [104], Jaravel et al. [14], Langella et al. [105], and Fedina et al. [5], using different LES combustion models and reaction mechanisms. Fedina et al. [5] compared the performance of two reduced mechanism, SG35 [70] and SR39 [72, 73], and two global mechanisms, WD2 [76, 77] and JL4 [75]. Figure 5.10a presents results from [5] for the SG35 mechanism, in terms of instantaneous volumetric renderings of the CH$_4$ mass fraction, Y_{CH4}, and heat release, Q, representative of the luminous flame, and contours of the axial velocity, v_x, and temperature, T, on two half-planes, together with an isosurface of the second invariant of the velocity gradient tensor, λ_2, colored by v_x, from the LES prediction. The fuel discharges into the 12 radial swirlers where it starts to mix with the radially supplied air before entering the prechamber at a fixed angle. This results in a strongly swirling flow, with most of the fuel flanking the prechamber wall, where it continues to mix with the air until the now well-mixed, wrinkled and furrowed, M-shaped fuel–air mixture discharges into the combustor, having a protracted inner V-structure extending well into the prechamber. The flame develops around the M-shaped fuel–air cloud, and is composed of an inner, wrinkled or furrowed, V-shaped flame located in the inner shear-layer of the discharging fuel–air mixture, and an outer swirling flame, taking the shape of a truncated wrenched cone located in the outer shear layer of the discharging fuel–air mixture. The two flame elements connect at the swirling annular flame tip, located at about $0.6D$ downstream of the burner exit depending on the LES combustion model. The heat release results from multiple thin layers of interpenetrating reaction structures and layers of intermediate and radical species being also sensitive to the strain rate.

Figure 5.11 provides a quantitative comparison of the time-averaged and rms fluctuations of the axial velocity, v_x, and temperature, T, across the combustion chamber at the cross sections $x/D = 1.21$, 1.44, 1.66, and 2.00, respectively, from Bulat

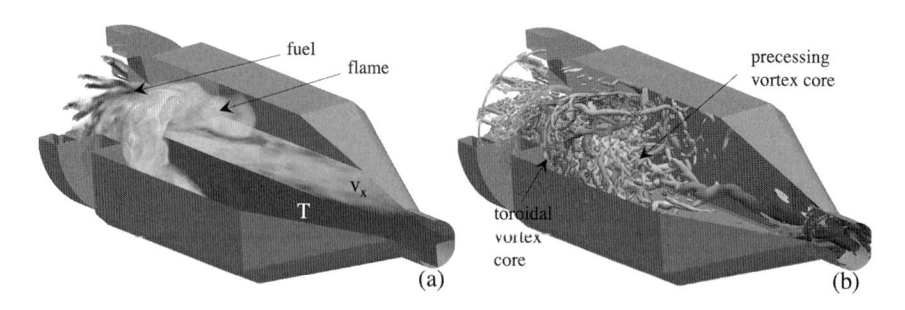

Figure 5.10 The Turchemi combustor. (a) Instantaneous volumetric rendering of CH_4 (fuel) and heat-release, Q, (flame) together with contours of the axial velocity (v_x) and temperature (T), and (b) an isosurface of the second invariant of the velocity gradient tensor, λ_2, colored by the axial velocity for LES-PaSR with the SG35 reaction mechanism.

et al. [102], comparing reaction mechanisms and from Fedina et al. [5], comparing LES combustion models. Regarding the comparison of reaction mechanism SG35, SR39, JL4, and WD2 in Fig. 5.11a one may conclude that $\langle v_x \rangle$ is rather insensitive to the reaction mechanism, which is perhaps not that surprising since the volumetric expansion (related to heat release), which is similar for all reaction mechanisms investigated, mainly influences the flow and hence $\langle v_x \rangle$. The averaged temperature, $\langle T \rangle$, on the other hand appears more sensitive with the flame-shape making an unambiguous impact on $\langle T \rangle$. In general we find that the largest deviations between the experimental data and the LES predictions occur for the two global WD2 and JL4 reaction mechanisms whereas both skeletal SG35 and SR39 reaction mechanisms show improved overall agreement. This is evident in particular for the SG35 reaction mechanism. From kinetic studies, one knows that SR39 overpredict the ignition delay time, τ_{ign}, and underpredict the extinction strain rate, σ_{ext}, whereas SG35, and the more recent Z42 mechanism, both agree well with the data. On the contrary, the global reaction mechanisms, WD2 and JL4, underpredict τ_{ign} and overpredict σ_{ext}. Regarding the comparison of LES combustion models in Fig. 5.11b, using the SG35 reaction mechanism, one finds similar or smaller influences of the LES combustion model, with the LES-SF and LES-PaSR models showing the best agreement with the experimental data provided that the SG35 (or Z42) reaction mechanisms are employed. The largest differences occur at $x/D = 1.44$, which is approximately at the far end of the flame, where small differences in flame speeds makes a large impact on the flame size, shape, and structure. All in all, one finds that the influence of the reaction mechanism appears larger than the influence of the LES combustion model.

5.5.3 The Volvo Bluff Body Combustor: Propane

The VOLVO bluff body combustor studied in [106–108] consists of a rectilinear channel, with a rectangular cross section, divided into an inlet section, a combustor section and an exhaust as shown in Fig. 5.12a. In the inlet section the incoming air is distributed over the cross section by a critical orifice plate that also separates

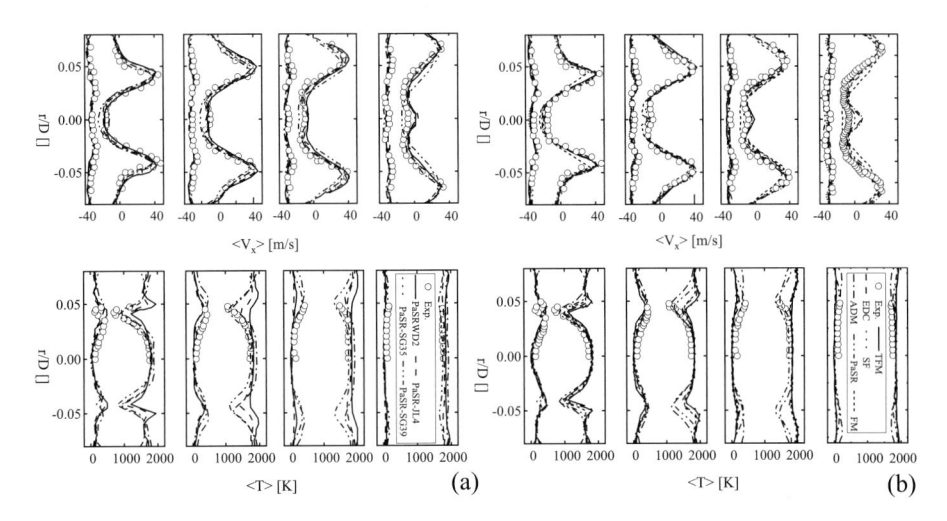

Figure 5.11 The Turchemi combustor, Case A. (a) Comparison of measured and predicted time-averaged (right set of lines in each panel) and rms fluctuations (left set of lines in each panel) of the axial velocity (upper panel) and temperature (lower panel) for different reaction mechanisms at $x/D = 1.21$, 1.44, 1.66 and 2.00. (+) experimental data, [98–100], lines are LES results from different chemical kinetic models, LES-PaSR-WD2, LES-PaSR-JL4, LES-PaSR-SR39, and LES-PaSR-SG35. (b) Comparison of measured and predicted time-averaged (right set of lines in each panel) and rms fluctuations (left set of lines in each panel) of the axial velocity (upper panel) and temperature (lower panel) for different LES combustion models at $x/D = 1.21$, 1.44, 1.66 and 2.00. (+) experimental data, [98–100], lines are results from different LES models, LES-PaSR, LES-EDC, LES-FM, LES-TFM, LES-SF and LES-ADM.

the combustor acoustically from the air-supply system. Propane (C_3H_8) is injected and premixed with the air just upstream of the critical orifice plate by a multiorifice injector, and a honeycomb screen is employed to control the turbulence level. The combustor is of modular design with the walls split into several interchangeable sections with the top and bottom walls being water cooled and the side walls being air cooled to accommodate the quartz windows for optical access as shown in Fig. 5.12. The premixed C_3H_8–air flame is stabilized after a triangular-shaped flameholder of height $h = 0.040$ m, and the combustor discharges into a large diameter circular duct in which far downstream water mist is used to cool the exhausts. Experimental data using gas analysis [108], laser Doppler velocimetry (LDV) [107], coherent anti-Stokes Raman scattering (CARS) [106], and high-speed and schlieren imaging were collected for two operating conditions, Cases I and II, described in [106–108], Both operating conditions were for an air mass flow rate of $m_0 = 0.6$ kg/s at a pressure of $p_0 = 101$ kPa and an equivalence ratio of $\phi = 0.61$, but at temperatures of $T_0 = 288$ K and 600 K respectively. This resulted in inflow velocities of $v_0 = 17.6$ m/s and 36.6 m/s, with Reynolds numbers of 50, 118 and 29, 732, respectively.

Several LES studies of this case have been performed in the past [6, 69, 95, 109–115]. This case has also been selected as the initial test case for the Model Validation for Propulsion (MVP) series of workshops [116], involving more than 20 participating research groups aiming at evaluating the predictive capabilities of LES combustion models.

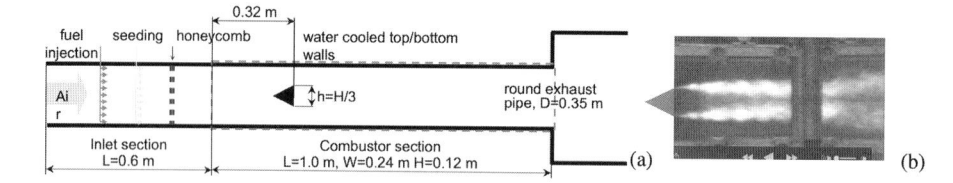

Figure 5.12 Schematic (a) of the VOLVO bluff-body combustor and (b) high-speed video image of Case I, showing symmetric flame flapping dominated by Kelvin–Helmholz instabilities.

Figure 5.13 presents perspective views of Cases I and II, respectively, using the skeletal Z66 [69] reaction mechanism in terms of volume renderings of the axial velocity, v_x, together with an isosurface of the second invariant of the velocity gradient tensor, λ_2; heat release, Q; and OH, CH$_2$O, and HCO mass fractions. For both cases v_x reveal an extended central recirculation zone downstream of the flameholder compared to the nonreacting flow followed by an unsteady and gradually increasing v_x caused by the volumetric expansion from the exothermicity. The LES results further show that spanwise, $\widetilde{\omega}_3$, vortex structures are simultaneously shed off the upper and lower corners of the flameholder whereas longitudinal, $\widetilde{\omega}_{12}$, vortex structures form in the regions between neighboring $\widetilde{\omega}_3$ vortex structures, resulting in two initially symmetric vortex braids that gradually widen and lose their symmetry due to vortex–vortex, barocliniqe torque and viscosous effects. The flames stabilize in the vortex braids behind the flameholder due to a combination of incessant ignition of the premixed fuel–air mixture in the shear layers trailing off the flameholder and the recirculation of hot combustion products in the flameholder wake. The heat-release and OH, CH$_2$O, and HCO species distribution are thus also initially symmetric but gradually develop a staggered pattern. The lower ratio of burnt to unburnt mixture temperatures, T_b/T_u, in Case II ($T_b/T_u = 3.5$) compared to that in Case I (6.3) decrease the exothermicity and volumetric expansion, whereas the higher burnt temperature, T_b, of Case II compared to Case I increases the reaction rates of the post-flame reactions, while the reactions in the flame zone progress with approximately the same rates as in Case I. This results in the lower values of T_b/T_u and promote asymmetric vortex shedding with transition from symmetric to asymmetric vortex shedding moving upstream with decreasing T_b/T_u. These effects combine to modify the combustion chemistry in the pre-heat and post-flame zones.

Figure 5.14 presents profiles of the (a) mean axial velocity, $\langle \widetilde{v}_x \rangle$, and the (b) rms axial velocity fluctuations, $\widetilde{v}_x^{rms}$, at $x/h = 0.95$, 3.75, and 9.40, between LES using the global WD2 [76, 77] and JL4 [75], and skeletal Z66 [69] and H73 [117] C$_3$H$_8$–air reaction mechanisms for Case I (upper panels) and Case II (lower panels) and experimental LDV data [107]. All LES show a gradual transition from a U-shaped $\langle \widetilde{v}_x \rangle$ profile just downstream of the flameholder at $x/h = 0.95$ via a V-shaped $\langle \widetilde{v}_x \rangle$ profile in the recovery region at $x/h = 3.75$ to an almost flat $\langle \widetilde{v}_x \rangle$ profile halfway along the combustor at $x/h = 9.40$. This transition is due to the volumetric expansion caused by the exothermicity and is related to the reaction mechanism. In the near-wake, at

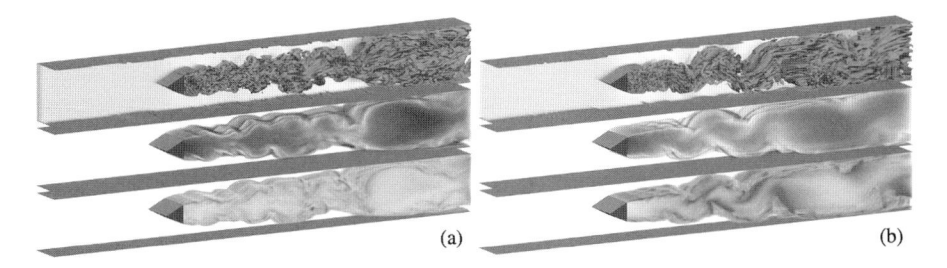

Figure 5.13 The Volvo validation rig combustor, simulated with the Z66 mechanism. Perspective views from the side in terms of (top) volume renderings of the axial velocity, v_x, combined with an isosurface of the second invariant of the velocity gradient tensor, λ_2, (middle) heat-release, Q, and (bottom) mass fractions of OH (in gray), C_2H_5 (green), CH_2O (blue), HCO (red), and CO (orange) for (a) Case I ($v_{in} = 17$ m/s, $T_{in} = 288$ K) and (b) Case II (36 m/s, 600 K).

$x/h = 0.95$, the WD2 mechanism results in a more V-shaped velocity profile with a marginally too strong velocity recovery, whereas the JL4 mechanism underpredicts the velocity recovery somewhat compared to the Z66 and H73 reaction mechanisms, which both result in distinct U-shaped velocity profiles that are in good agreement with the LDV data. Furthermore, all four mechanisms overpredict the velocity fluctuations in the near-wake of the flameholder. In the recovery region, at $x/h = 3.75$, the JL4 mechanism overpredicts the velocity reversal for Case I and underpredicts it for Case II. WD2 is found to underpredict the velocity reversal in both cases, whereas both Z66 and H73 are found to be in reasonably good agreement with the LDV data for both cases. Concerning $\widetilde{v}_x^{rms}$, it is somewhat surprisingly noticed that the best agreement between LES predictions and LDV data is obtained for the JL4 mechanism. The Z66 and H73 mechanism predictions group reasonably well together, showing acceptable agreement with the LDV data whereas the WD2 mechanism result in a somewhat too wide profile with too high fluctuation levels. In the fully developed turbulent combustion region, $x/h = 9.40$, all four mechanisms results in similar, quite flat, $\langle \widetilde{v}_x \rangle$ profiles in good overall agreement with the experimental data. For $\widetilde{v}_x^{rms}$ profiles the best agreement between LES predictions and LDV data is again obtained for the JL4 mechanism, with the Z66 and H73 mechanism predictions following closely, whereas the WD2 mechanism again results in a rather too wide profile with substantially too high fluctuation levels.

Figure 5.15 shows profiles of the (a) mean temperature, $\langle \widetilde{T} \rangle$, and the (b) rms temperature fluctuations, $\widetilde{T}^{rms}$, at $x/h = 0.95$, 3.75 and 8.75, between LES using the global WD2 and JL4, and skeletal Z66 and H73 C_3H_8–air reaction mechanisms for Case I (upper panels) and Case II (lower panels) and experimental gas analysis data [108], and CARS data [106]. Satisfactory agreement is found for all LES model predictions, resulting in a broadening of the high-temperature flame region with increasing distance downstream of the flameholder consistent with the experimental data, in particular the CARS data. Comparing the gas analysis and CARS $\langle \widetilde{T} \rangle$ profiles demonstrates that the temperature profiles from the gas analysis are wider than those from CARS,

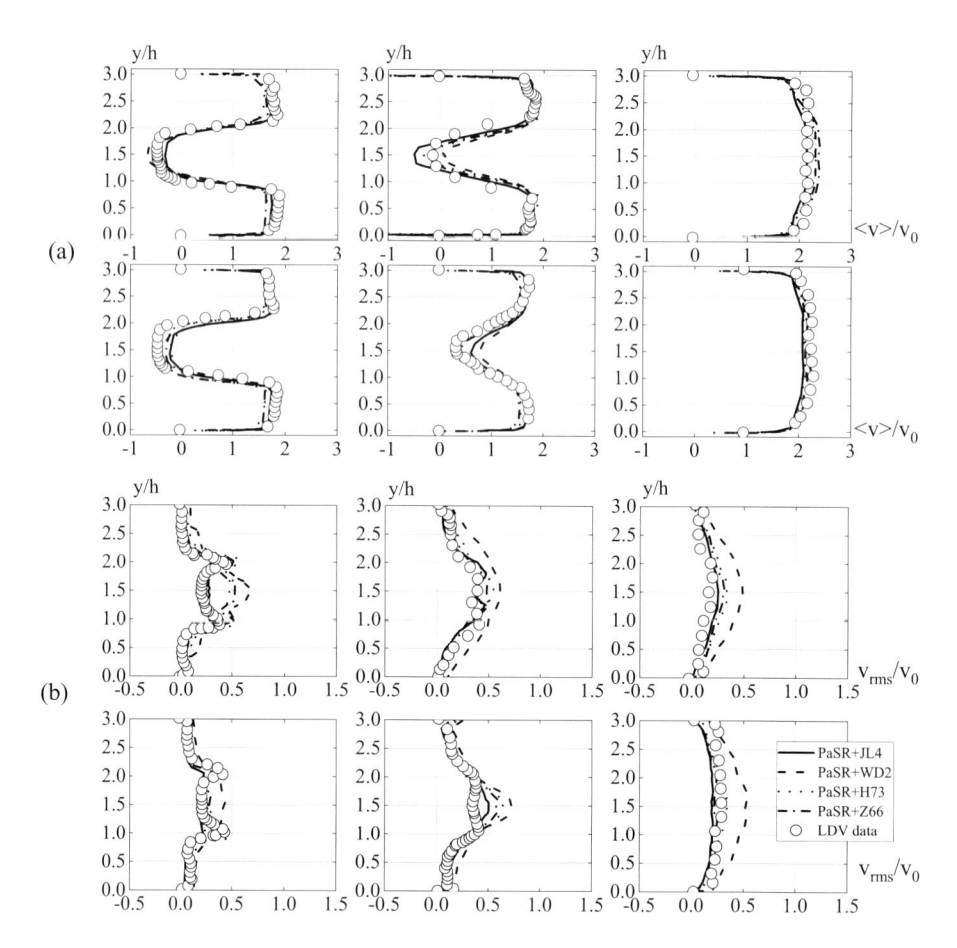

Figure 5.14 The Volvo validation rig combustor. Mean axial velocity (a) and rms velocity fluctuations (b) at $x/h = 0.95$, 3.75 and 9.40, respectively for Case I (upper panels) and Case II (lower panels). (○) LDV data ($x/h = 0.95$, 3.75, and 9.40), [107], and LES predictions on the medium grid using LES-PaSR+WD2, LES-PaSR+JL4, LES-PaSR+Z66, and LES-PaSR+H73.

in particular for Case II. This issue has been debated previously, and the community consensus is that the CARS data is more reliable since it is nonintrusive and direct. The LES model predictions generally agree more satisfactorily with the CARS data than with the gas analysis data, and with the $\langle \widetilde{T} \rangle$ profiles from the Z66 and H73 mechanisms showing very good agreement with the CARS data for all three cross sections and both cases. It should, however, be noted that all LES predictions tend to predict a marginally too narrow flame compared to the gas analysis and CARS data in the near-wake at $x/h = 0.95$ and a too wide flame compared to the CARS data at the most downstream location in the fully developed turbulent combustion region at $x/h = 8.75$ for Case II. Furthermore, the WD2 mechanism tends generally to underpredict the $\langle \widetilde{T} \rangle$ in almost all cross sections. The $\widetilde{T}^{rms}$ profiles from the CARS data are reasonably well captured by the JL4, Z66, and H73 mechanisms with some overestimates of $\widetilde{T}^{rms}$ in the shear layers at $x/h = 0.95$. The WD2 mechanism, however, results in a slightly different

$\widetilde{T}^{\text{rms}}$ profile shape, in particular in the near-wake, at $x/h = 0.95$, and overpredicts the fluctuation levels in the recovery region, at $x/h = 3.75$.

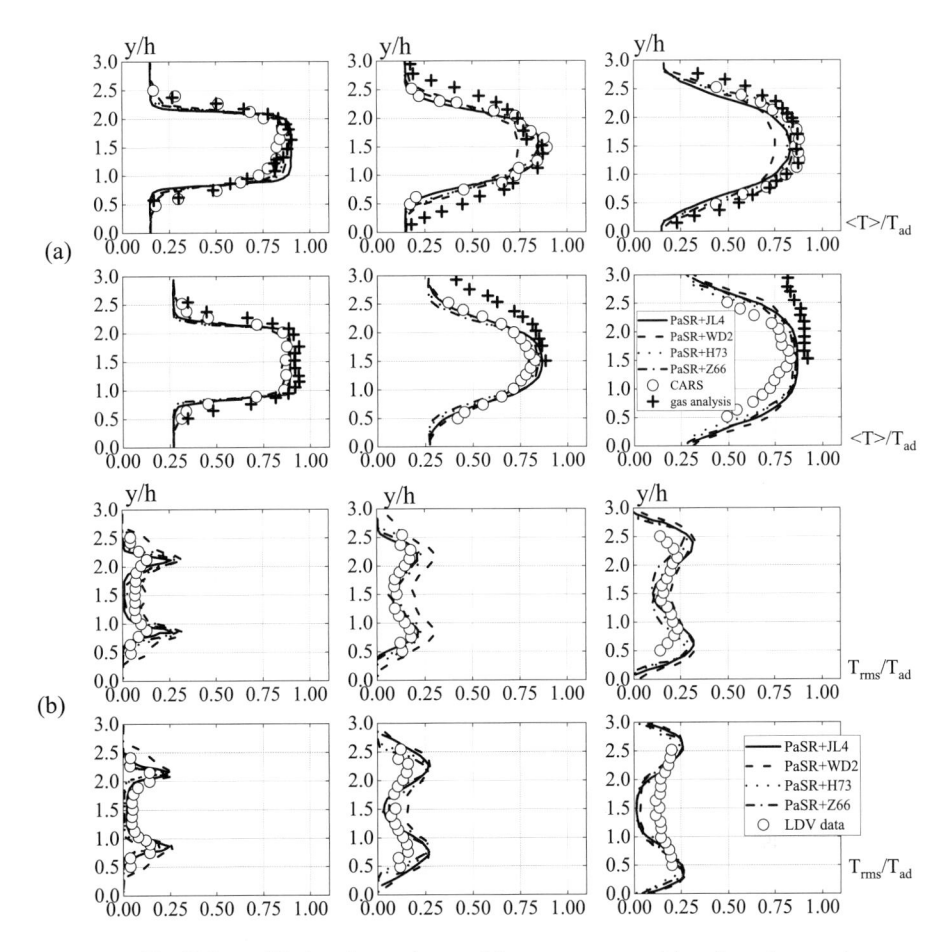

Figure 5.15 The Volvo validation rig combustor. Mean temperature (a) and rms temperature fluctuations (b) at $x/h = 0.95$, 3.75 and 9.40, respectively for Case I (upper panels) and Case II (lower panels). (+) gas analysis data, [107], ($x/h = 3.75$ and 8.75), and (o) CARS data, [106], ($x/h = 0.95$, 3.75, and 8.75), and LES predictions on the medium grid using LES-PaSR+WD2, LES-PaSR+JL4, LES-PaSR+Z66, and LES-PaSR+H73.

5.5.4 A Supersonic Cavity Stabilized Combustor: Ethylene (and Hydrogen)

Driscoll and Micka's supersonic cavity stabilized combustor [117–120] consists of a 2D $Ma = 2.2$ nozzle feeding a constant-area isolator, at the end of which a cavity is mounted followed directly by a $4°$ diverging combustor section and a large-area circular exhaust. Room temperature H_2 (or H_2 and C_2H_4) was injected sonically through a single 2.49 mm diameter injection port located at 44.5 mm upstream of the leading edge of the cavity on the combustor centerline. An electric heater together with an

H$_2$-fueled vitiator were employed to attain air stagnation temperatures ranging between $T_0 = 1040$ and 1500 K at stagnation pressures of $p_0 = 460$ and 590 kPa. Make-up oxygen was added to maintain O$_2$ mole fraction of 0.21 in the vitiator products. Here, we consider four cases characterized by a stagnation pressure of $p_0 = 590$ kPa and stagnation temperatures of $T_0 = 1130$ K, 1240 K, 1370 K, and 1500 K, respectively. For all the cases the H$_2$ temperature is 288 K at an equivalence ratio of $\phi = 0.26$. The experimental data sets comprise images of the flame luminosity, wall-pressure data along the isolator, cavity and combustor and OH laser-induced fluorescence (OH-LIF). From the experiments two combustion modes were discovered: cavity stabilized and jet wake stabilized combustion, with the former occurring at low T_0 and the latter at high T_0, whereas at intermediate T_0 unsteady combustion was observed.

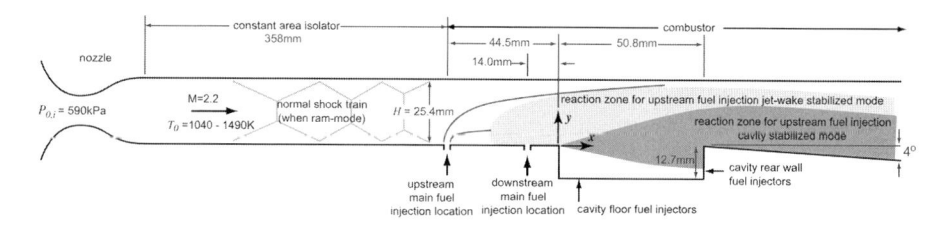

Figure 5.16 Schematic of the University of Michigan supersonic combustion facility, from [119].

LES for different operating conditions of the supersonic combustor have been performed by Zettervall and Fureby [121], using three different reaction mechanisms: the skeletal D7 [111], the comprehensive J20 [122], and the newly developed comprehensive Z22 [121]. The need for an alternative comprehensive H$_2$–air mechanism was identified in [121] based on comparisons between reaction mechanism predictions of the laminar flame speed, S_L; flame temperature and ignition delay time τ_{ign}; and experimental data, revealing that most global and skeletal reaction mechanisms underpredict τ_{ign} at low temperatures whereas the most detailed reaction mechanisms overpredict τ_{ign}. Figure 5.17 presents instantaneous volumetric renderings of temperature for the four cases using selected reaction mechanisms. For Case A, all LES predicts that T increases somewhat along the isolator due to the breakup of the shock train in the nozzle. For the D7 mechanism ignition occurs early in the shear-layer around the jet being discharged porthole injector, resulting in a jet-wake stabilized flame that is not consistent with the experimental observations, suggesting that the flame is stabilized by the cavity. For the J20 mechanism, no ignition of the injected H$_2$ takes place due to the too long τ_{ign} mentioned earlier. For the Z22 mechanism, unsteady combustion is observed to occur in or around the cavity shear-layer, suggesting *cavity-stabilized combustion* as observed experimentally. The LES results for Case B using the D7 mechanism behaves very similar to that of Case A, with too early ignition and jet-wake stabilized combustion instead of a mix between cavity and jet-wake stabilized combustion as observed in the experiment. Sporadic ignition events can be observed to occur in the cavity and just downstream of the cavity for the J20 mechanism, but these are not strong or large enough to ignite the H_2 plume, and hence no combustion occurs. The Z22 mechanism is observed to predict an unsteady mix of cavity and jet-wake stabilized combustion, as the experiments suggest. All three reaction mechanism predicts early ignition for Cases D and G, and combustion occurs in the shear layers surrounding the H_2 jets resulting in *jet-wake stabilized combustion* as also observed experimentally.

Figure 5.17 The Driscoll and Micka combustor. Instantaneous volumetric rendering of the temperature for Case A using the D7, J20, and Z22 mechanisms, Case B using the Z22 mechanism, Case D using the Z22 mechanism and Case G using the Z22 mechanism. The temperature ranges from $0.98T_{peak}$ to $0.02T_{peak}$.

Figure 5.18 summarizes the various behaviors shown by the Driscoll and Micka combustor as the stagnation temperature, T_0, is varied. The computational results shown are for the Z22 reaction mechanism. The central line plot in Fig. 5.18 shows the fraction of time the combustion spent in cavity stabilized, intermediate oscillatory, or jet-wake stabilized combustion modes as determined from the experimental high-speed images [119]. The dashed line corresponds to an approximate curve fit based on the experimental data (symbols). From the experiments, T_0 appears to be the dominant variable in determining the combustion stabilization mode and at high T_0 (> 1350 K) combustion was almost always stabilized by the jet-wake. For low T_0 (< 1150 K) combustion was stabilized primarily by the cavity shear layer, whereas for intermediate T_0 combustion was found to oscillate between the cavity and jet-wake stabilization modes. The instantaneous and time-averaged heat-release LES results using Z22 for Case A ($T_0 = 1130$ K), Case B ($T_0 = 1240$ K), Case D ($T_0 = 1370$ K), and Case G ($T_0 = 1500$ K) are added to this graph as well as the mean experimental combustion luminosity images. The fraction of time combustion is spent in the jet-wake stabilized mode is extracted from the LES and superimposed on the line plot. Very good

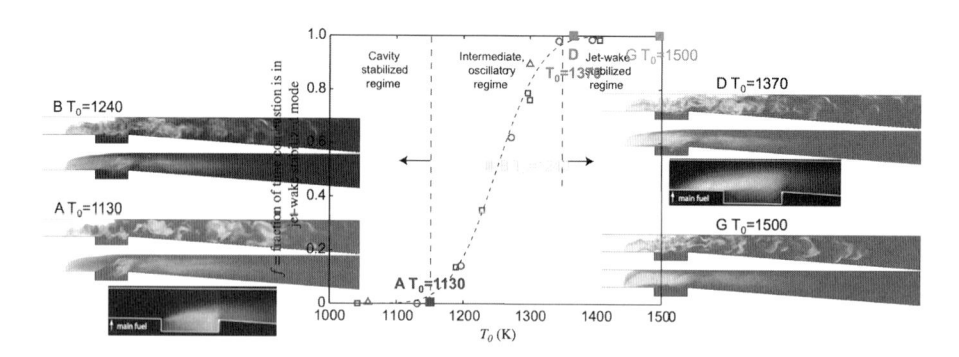

Figure 5.18 The Driscoll and Micka combustor, Combustion stabilization mode versus T_0 from the experiments and from the LES using the Z22 reaction mechanism.

agreement between the experimental and LES results is observed. As described above, this agreement is a direct consequence of using the Z22 reaction mechanism, as the D7 and J22 reaction mechanisms burn either too quickly or too slowly.

5.6 Future Directions

As computational capacity increases, CFD simulations including finite-rate chemistry are becoming increasingly important in research and development of combustion system. As we move toward new fuels with various chemical structures, alcohols, esters, ethers, and hydrocarbon chains with various chain lengths, branching, and degrees of saturation, the demand on development of new mechanism is high. Mechanism development to a large extent relies on the fundamental understanding of the combustion characteristics, ignition delay time, laminar burning velocity, and extinction strain rate, and there is a constant demand on extensive and high-quality experimental studies. Experimental challenges include reaching high pressures of relevance for the applications and obtaining accurate data at low temperatures. Lean, low-temperature combustion is increasingly important as a NOx mitigation strategy.

Hydrocarbon fuels representative of the fossil fuels of use today can to a large extent be accurately modeled using the best available mechanisms. For many of the oxygenated biofuels there are very few or no reliable mechanisms, neither detailed nor reduced. The difference in chemical structure compared to the alkane fuels results in new subsets of reactions are becoming increasingly important. This means that the understanding gained from alkane fuels cannot be directly transferred to the oxygenated fuels, and each new group of fuels presents a new dimension of the chemical kinetics. An important part of the development of reduced kinetic mechanisms is therefore to identify the important reaction subsets for different types of fuels.

An important part in future development is the ability to use CFD to predict pollutant formation, also for minor pollutants. An example is acetaldehyde production from ethanol combustion. The aldehyde is a minor product but highly toxic and needs to be accurately assessed.

References

[1] C. K. Law, "Combustion at a crossroads: Status and prospects," *Proc. Combust. Inst.*, **31**, 1–29, 2007.

[2] R. Hilbert, F. Tap, H. El-Rabii, and D. Thevenin, "Impact of detailed chemistry and transport models on turbulent combustion simulations," *Prog. Energy Combust. Sci.*, **30**, 61–117, 2004.

[3] S. Hochgreb, "Mind the gap: Turbulent combustion model validation and future needs," *Proc. Combust. Inst.*," **37**, 2091–2107, 2018.

[4] K. Kohse-Höinghaus, "Clean combustion: Chemistry and diagnostics for a systems approach in transportation and energy conversion," *Prog. Energy Combust. Sci.*, **65**, 1–5, 2018.

[5] E. Fedina, C. Fureby, G. Bulat, and W. Meier, "Assessment of finite rate chemistry large eddy simulation combustion models," *Flow Turbul. Combust.*, **99**, 385–409, 2017.

[6] C. Fureby, *The Volvo Validation Rig – A Comparative Study of Large Eddy Simulation Combustion Models at Different Operating Conditions*, ser. AIAA SciTech Forum. American Institute of Aeronautics and Astronautics, 2018.

[7] S. B. Pope, "Small scales, many species and the manifold challenges of turbulent combustion," *Proc. Combust. Inst.*, **34**, 1–31, 2013.

[8] A. M. Fiore, V. Naik, D. V. Spracklen, et al., "Global air quality and climate," *Chem. Soc. Rev.*, **41**, 6663–6683, 2012.

[9] M. Pelucchi, C. Cavallotti, E. Ranzi, A. Frassoldati, and T. Faravelli, "Relative reactivity of oxygenated fuels: Alcohols, aldehydes, ketones, and methyl esters," *Energy & Fuels*, **30**, 8665–8679, 2016.

[10] S. Chu and A. Majumdar, "Opportunities and challenges for a sustainable energy future," *Nature*, **488**, 294–303, 2012.

[11] R. E. Dunmore, L. K. Whalley, T. Sherwen, et al., "Atmospheric ethanol in london and the potential impacts of future fuel formulations," *Faraday Discussions*, **189**, 105–120, 2016.

[12] M. Z. Jacobson, "Effects of ethanol (e85) versus gasoline vehicles on cancer and mortality in the united states," *Environ. Sci. Technol.*, **41**, 4150–4157, 2007.

[13] B. Franzelli, E. Riber, and B. Cuenot, "Impact of the chemical description on a large eddy simulation of a lean partially premixed swirled flame," *Comptes Rendus Mecanique*, **341**, 247–256, 2013.

[14] T. Jaravel, E. Ribert, B. Cuenot, and G. Bulat, "Large eddy simulation of an industrial gas turbine combustor using reduced chemistry with accurate pollutant prediction," *Proc. Combust. Inst.*, **36**, 3817–3825, 2017.

[15] J. Zador, C. A. Taatjes, and R. X. Fernandes, "Kinetics of elementary reactions in low-temperature autoignition chemistry," *Prog. Energy Combust. Sci.*, **37**, 371–421, 2011.

[16] A. A. Konnov, A. Mohammad, V. R. Kishore, N. I. Kim, C. Prathap, and S. Kumar, "A comprehensive review of measurements and data analysis of laminar burning velocities for various fuel+air mixtures," *Prog. Energy Combust. Sci.*, **68**, 197–267, 2018.

[17] A. T. Holley, X. Q. You, E. Dames, H. Wang, and F. N. Egolfopoulos, "Sensitivity of propagation and extinction of large hydrocarbon flames to fuel diffusion," *Proc. Combust. Inst.*, **32**, 1157–1163, 2009.

[18] T. Turanyi and A. S. Tomlin, *Analysis of kinetic reaction mechanisms*. Berlin/Heidelberg: Springer-Verlag, 2014.

[19] S. J. Klippenstein, "From theoretical reaction dynamics to chemical modeling of combustion," *Proc. Combust. Inst.*, **36**, 77–111, 2017.

[20] J. A. Miller, M. J. Pilling, and E. Troe, "Unravelling combustion mechanisms through a quantitative understanding of elementary reactions," *Proc. Combust. Inst.*, **30**, 43–88, 2005.

[21] W. Tsang, "Chemical kinetic database for combustion chemistry 2. methanol," *J. Phys. Chem. Ref. Data*, **16**, 471–508, 1987.

[22] W. Tsang, "Chemical kinetic data-base for combustion chemistry 3. propane," *J. Phys. Chem. Ref. Data*, **17**, 887–952, 1988.

[23] W. Tsang, "Chemical kinetic data-base for combustion chemistry 4. isobutane," *J. Phys. Chem. Ref. Data*, **19**, 1–68, 1990.

[24] W. Tsang and R. F. Hampson, "Chemical kinetic database for combustion chemistry .1. methane and related-compounds," *J. Phys. Chem. Ref. Data*, **15**, 1087–1279, 1986.

[25] D. L. Baulch, C. T. Bowman, C. J. Cobos, R. A. Cox, T. Just, J. A. Kerr, M. J. Pilling, D. Stocker, J. Troe, W. Tsang, R. W. Walker, and J. Warnatz, "Evaluated kinetic data for combustion modeling: Supplement ii," *J. Phys. Chem. Ref. Data*, **34**, 757–1397, 2005.

[26] S. L. Peukert and J. V. Michael, "High-temperature shock tube and modeling studies on the reactions of methanol with d-atoms and ch3-radicals," *J. Phys. Chem. A*, **117**, 10186–10195, 2013.

[27] M. Cathonnet, J. C. Boettner, and H. James, "Study of methanol oxidation and self ignition in the temperature-range 500-600-degrees-c," *Journal De Chimie Physique Et De Physico-Chimie Biologique*, **79**, 475–478, 1982.

[28] J. T. Jodkowski, M. T. Rayez, J. C. Rayez, T. Berces, and S. Dobe, "Theoretical study of the kinetics of the hydrogen abstraction from methanol. 3. reaction of methanol with hydrogen atom, methyl, and hydroxyl radicals," *J. Phys. Chem. A*, **103**, 3750–3765, 1999.

[29] S. J. Klippenstein, L. B. Harding, M. J. Davis, A. S. Tomlin, and R. T. Skodje, "Uncertainty driven theoretical kinetics studies for CH3OH ignition: Ho2+ch3oh and o-2+ch3oh," *Proc. Combust. Inst.*, **33**, 351–357, 2011.

[30] I. M. Alecu and D. G. Truhlar, "Computational study of the reactions of methanol with the hydroperoxyl and methyl radicals. 2. accurate thermal rate constants," *J. Phys. Chem. A*, **115**, 14599–14611, 2011.

[31] J. Warnatz, "Rate coefficients in the C/H/O system," in *Combustion Chemistry*, W. C. Gardiner Jr., Ed., 1984, 197–360. Berlin/Heidelberg: Springer-Verlag.

[32] J. C. Prince and F. A. Williams, "Short chemical-kinetic mechanisms for low-temperature ignition of propane and ethane," *Combust. Flame*, **159**, 2336–2344, 2012.

[33] J. C. Prince, F. A. Williams, and G. E. Ovando, "A short mechanism for the low-temperature ignition of n-heptane at high pressures," *Fuel*, **149**, 138–142, 2015.

[34] D. F. Davidson and R. K. Hanson, "Interpreting shock tube ignition data," *Int. J. Chem. Kint.*, **36**, 510–523, 2004.

[35] C.-J. Sung and H. J. Curran, "Using rapid compression machines for chemical kinetics studies," *Prog. Energy Combust. Sci.*, **44**, 1–18, 2014.

[36] S. S. Goldsborough, S. Hochgreb, G. Vanhove, M. S. Wooldridge, H. J. Curran, and C. J. Sung, "Advances in rapid compression machine studies of low- and intermediate-temperature autoignition phenomena," *Prog. Energy Combust. Sci.*, **63**, 1–78, 2017.

[37] W. Liu, C. K. Law, and T. F. Lu, "Multiple criticality and staged ignition of methane in the counterflow," *Int. J. Chem. Kinetics*, **41**, 764–776, 2009.

[38] A. Ansari and F. N. Egolfopoulos, "Flame ignition in the counterflow configuration: Reassessing the experimental assumptions," *Combust. Flame*, **174**, 37–49, 2016.

[39] T. F. Lu and C. K. Law, "Toward accommodating realistic fuel chemistry in large-scale computations," *Prog. Energy Combust. Sci.*, **35**, 192–215, 2009.

[40] F. N. Egolfopoulos, N. Hansen, Y. Ju, K. Kohse-Hoinghaus, C. K. Law, and F. Qi, "Advances and challenges in laminar flame experiments and implications for combustion chemistry," *Prog. Energy Combust. Sci.*, **43**, 36–67, 2014.

[41] V. A. Alekseev, J. D. Naucler, M. Christensen, E. J. K. Nilsson, E. N. Volkov, L. P. H. de Goey, and A. A. Konnov, "Experimental uncertainties of the heat flux method for measuring burning velocities," *Combust. Sci. Technol.*, **188**, 853–894, 2016.

[42] M. Goswami, S. C. R. Derks, K. Coumans, et al., "The effect of elevated pressures on the laminar burning velocity of methane plus air mixtures," *Combust. Flame*, **160**, 1627–1635, 2013.

[43] Z. Chen, "On the accuracy of laminar flame speeds measured from outwardly propagating spherical flames: Methane/air at normal temperature and pressure," *Combust. Flame*, **162**, 2442–2453, 2015.

[44] E. Ranzi, A. Frassoldati, R. Grana, A. Cuoci, T. Faravelli, A. P. Kelley, and C. K. Law, "Hierarchical and comparative kinetic modeling of laminar flame speeds of hydrocarbon and oxygenated fuels," *Prog. Energy Combust. Sci.*, **38**, 468–501, 2012.

[45] A. T. Holley, Y. Dong, M. G. Andac, and F. N. Egolfopoulos, "Extinction of premixed flames of practical liquid fuels: Experiments and simulations," *Combust. Flame*, **144**, 448–460, 2006.

[46] G. Esposito and H. K. Chelliah, "Skeletal reaction models based on principal component analysis: Application to ethylene-air ignition, propagation, and extinction phenomena," *Combust. Flame*, **158**, 477–489, 2011.

[47] B. Fiorina, D. Veynante, and S. Candel, "Modeling combustion chemistry in large eddy simulation of turbulent flames," *Flow, Turbul. Combust.*, **94**, 3–42, 2015.

[48] N. Zettervall, C. Fureby, and E. J. K. Nilsson, "Small skeletal kinetic mechanism for kerosene combustion," *Energy & Fuels*, **30**, 9801–9813, 2016.

[49] H. Wang, R. Xu, K. Wang, C. T. Bowman, R. K. Hanson, D. F. Davidson, K. Brezinsky, and F. N. Egolfopoulos, "A physics-based approach to modeling real-fuel combustion chemistry - I. evidence from experiments, and thermodynamic, chemical kinetic and statistical considerations," *Combust. Flame*, **193**, 502–519, 2018.

[50] E. J. K. Nilsson and A. A. Konnov, "Role of hoco chemistry in syngas combustion," *Energy & Fuels*, **30**, 2443–2457, 2016.

[51] A. Fomin, T. Zavlev, V. A. Alekseev, I. Rahinov, S. Cheskis, and A. A. Konnov, "Experimental and modelling study of (CH2)-C1 in premixed very rich methane flames," *Combust. Flame*, **171**, 198–210, 2016.

[52] M. Christensen, E. J. K. Nilsson, and A. A. Konnov, "A systematically updated detailed kinetic model for CH_2O and CH_3OH combustion," *Energy & Fuels*, **30**, 6709–6726, 2016.

[53] V. A. Alekseev, M. Christensen, and A. A. Konnov, "The effect of temperature on the adiabatic burning velocities of diluted hydrogen flames: A kinetic study using an updated mechanism," *Combust. Flame*, **162**, 1884–1898, 2015.

[54] W. K. Metcalfe, S. M. Burke, S. S. Ahmed, and H. J. Curran, "A hierarchical and comparative kinetic modeling study of C1-C2 hydrocarbon and oxygenated fuels," *Int. J. Chem. Kinet.*, **45**, 638–675, 2013.

[55] C. V. Naik, K. V. Puduppakkam, and E. Meeks, "An improved core reaction mechanism for saturated C0-C4 fuels," *J. Eng. Gas Turb. Power*, **134**, 2012.

[56] Y. Xin, D. A. Sheen, H. Wang, and C. K. Law, "Skeletal reaction model generation, uncertainty quantification and minimization: Combustion of butane," *Combust. Flame*, **161**, 3031–3039, 2014.

[57] T. F. Lu and C. K. Law, "A directed relation graph method for mechanism reduction," *Proc. Combust. Inst.*, **30**, 1333–1341, 2005.

[58] Y. Qiu, L. Yu, L. Xu, Y. Mao, and X. Lu, "Workbench for the reduction of detailed chemical kinetic mechanisms based on directed relation graph and its deduced methods: Methodology and n-cetane as an example," *Energy & Fuels*, **32**, 7169–7178, 2018.

[59] T. Nagy and T. Turnyi, "Reduction of very large reaction mechanisms using methods based on simulation error minimization," *Combust. Flame*, **156**, 417–428, 2009.

[60] T. Lvas, D. Nilsson, and F. Mauss, "Automatic reduction procedure for chemical mechanisms applied to premixed methane/air flames," *Proc. Combust. Inst.*, **28**, 1809–1815, 2000.

[61] S. H. Lam and D. A. Goussis, "The CSP method for simplifying kinetics," *Int. J. Chem. Kinetics*, **26**, 461–486, 1994.

[62] S. H. Lam and D. A. Goussis, *Basic theory and demonstration of computational singular perturbation for stiff equations*, Modelling and Simulation of Systems, **3**. Amsterdam: Baltzer Scientific Publishing Co., 1989.

[63] U. Maas and S. B. Pope, "Simplifying chemical kinetics: Intrinsic low-dimensional manifolds in composition space," *Combust. Flame*, **88**, 239–264, 1992.

[64] Y. Gao, R. Q. Shan, S. Lyra, C. Li, H. Wang, J. H. Chen, and T. F. Lu, "On lumped-reduced reaction model for combustion of liquid fuels," *Combust. Flame*, **163**, 437–446, 2016.

[65] G. Goldin, Z. Y. Ren, Y. Gao, T. F. Lu, H. Wang, and R. Xu, "HEEDS opimized hychem mechanism," *Proc. the ASME Turbo Expo: Turbomachinery Technical Conference and Exposition*, GT2017–64 407, 2017.

[66] H. Wang, X. You, A. V. Joshi, S. G. Davis, A. Laskin, F. Egolfopoulos, and C. K. Law, *USC Mech Version II. High-Temperature Combustion Reaction Model of H2/CO/C1-C4 Compounds*, 2007.

[67] A. Larsson, N. Zettervall, T. Hurtig, E. J. K. Nilsson, A. Ehn, P. Petersson, M. Alden, J. Larfeldt, and C. Fureby, "Skeletal methane-air reaction mechanism for large eddy simulation of turbulent microwave-assisted combustion," *Energy & Fuels*, **31**, 1904–1926, 2017.

[68] N. Zettervall, C. Fureby, and E. J. K. Nilsson, "Small skeletal kinetic reaction mechanism for ethylene-air combustion," *Energy & Fuels*, **31**, 14138–14149, 2017.

[69] N. Zettervall, K. Nordin-Bates, E. J. K. Nilsson, and C. Fureby, "Large eddy simulation of a premixed bluff body stabilized flame using global and skeletal reaction mechanisms," *Combust. Flame*, **179**, 1–22, 2017.

[70] M. D. Smooke and V. Giovangigli, *Formulation of the Premixed and Nonpremixed Test Problems*. New York: Springer-Verlag, 1991, p. 384.

[71] G. P. Smith, D. M. Golden, F. Frenklach et al., "Gri-mech 3.0, 1999."

[72] E. Sher and S. Refael, "A simplified reaction scheme for the combustion of hydrogen enriched methane air flame," *Combust. Sci. Technol.*, **59**, 371–389, 1988.

[73] S. Refael and E. Sher, "Reaction-kinetics of hydrogen-enriched methane air and propane air flames," *Combust. Flame*, **78**, 326–338, 1989.

[74] A. Ehn, P. Petersson, J. J. Zhu, Z. S. Li, M. Alden, E. J. K. Nilsson, J. Larfeldt, A. Larsson, T. Hurtig, N. Zettervall, and C. Fureby, "Investigations of microwave stimulation of a turbulent low-swirl flame," *Proc. Combust. Inst.*, **36**, 4121–4128, 2017.

[75] W. P. Jones and R. P. Lindstedt, "Global reaction schemes for hydrocarbon combustion," *Combust. Flame*, **73**, 233–249, 1988.

[76] C. K. Westbrook and F. L. Dryer, "Chemical kinetic modeling of hydrocarbon combustion," *Prog. Energy Combust. Sci.*, **10**, 1–57, 1984.

[77] C. K. Westbrook and F. L. Dryer, "Simplified reaction-mechanisms for the oxidation of hydrocarbon fuels in flames," *Combust. Sci. Technol.*, **27**, 31–43, 1981.

[78] E. J. Hu, X. T. Li, X. Meng, Y. Z. Chen, Y. Cheng, Y. L. Xie, and Z. H. Huang, "Laminar flame speeds and ignition delay times of methane-air mixtures at elevated temperatures and pressures," *Fuel*, **158**, 1–10, 2015.

[79] J. M. Bergthorson and M. J. Thomson, "A review of the combustion and emissions properties of advanced transportation biofuels and their impact on existing and future engines," *Renewable & Sustainable Energy Rev.*, **42**, 1393–1417, 2015.

[80] W. Liu, R. Sivaramakrishnan, M. J. Davis, S. Som, D. E. Longman, and T. F. Lu, "Development of a reduced biodiesel surrogate model for compression ignition engine modeling," *Proc. Combust. Inst.*, **34**, 401–409, 2013.

[81] O. Herbinet, W. J. Pitz, and C. K. Westbrook, "Detailed chemical kinetic mechanism for the oxidation of biodiesel fuels blend surrogate," *Combust. Flame*, **157**, 893–908, 2010.

[82] M. Mehl, W. J. Pitz, C. K. Westbrook, and H. J. Curran, "Kinetic modeling of gasoline surrogate components and mixtures under engine conditions," *Proc. Combust. Inst.*, **33**, 193–200, 2011.

[83] M. Mehl, J. Y. Chen, W. J. Pitz, S. M. Sarathy, and C. K. Westbrook, "An approach for formulating surrogates for gasoline with application toward a reduced surrogate mechanism for cfd engine modeling," *Energy & Fuels*, **25**, 5215–5223, 2011.

[84] R. Xu, K. Wang, S. Banerjee, et al., "A physics-based approach to modeling real-fuel combustion chemistry - ii. reaction kinetic models of jet and rocket fuels," *Combust. Flame*, **193**, 520–537, 2018.

[85] A. Felden, L. Esclapez, E. Riber, B. Cuenot, and H. Wang, "Including real fuel chemistry in les of turbulent spray combustion," *Combust. Flame*, **193**, 397–416, 2018.

[86] C. K. Westbrook, C. V. Naik, O. Herbinet, et al., "Detailed chemical kinetic reaction mechanisms for soy and rapeseed biodiesel fuels," *Combust. Flame*, **158**, 742–755, 2011.

[87] Y. Pei, M. Mehl, W. Liu, T. Lu, W. J. Pitz, and S. Som, "A multi-component blend as a diesel fuel surrogate for compression ignition engine applications," *J. Eng. Gas Turb. Power*, **GTP-15-1057**, 2015.

[88] W. J. Pitz and C. J. Mueller, "Recent progress in the development of diesel surrogate fuels," *Prog. Energy Combust. Sci.*, **37**, 330–350, 2011.

[89] K. Kundu, P. Penko, and S. Yang, *Simplified Jet-A/air combustion mechanisms for calculation of NO(x) emissions*, ser. Joint Propulsion Conferences. American Institute of Aeronautics and Astronautics, 1998.

[90] S. M. Sarathy, P. Osswald, N. Hansen, and K. Kohse-Hoinghaus, "Alcohol combustion chemistry," *Prog. Energy Combust. Sci.*, **44**, 40–102, 2014.

[91] N. M. Marinov, "A detailed chemical kinetic model for high temperature ethanol oxidation," *Int. J. Chem. Kinet.*, **31**, 183–220, 1999.

[92] F. C. Minuzzi, C. Bublitz, and A. L. De Bortoli, "Development of a reduced mechanism for ethanol using directed relation graph and sensitivity analysis," *J. Math. Chem.*, **55**, 1342–1359, 2017.

[93] W.-W. Kim and S. Menon, *A new dynamic one-equation subgrid-scale model for large eddy simulations*, ser. Aerospace Sciences Meetings. American Institute of Aeronautics and Astronautics, 1995.

[94] V. Sabelnikov and C. Fureby, "Les combustion modeling for high re flames using a multi-phase analogy," *Combust. Flame*, **160**, 83–96, 2013.

[95] W. P. Jones, A. J. Marquis, and F. Wang, "Large eddy simulation of a premixed propane turbulent bluff body flame using the eulerian stochastic field method," *Fuel*, **140**, 514–525, 2015.

[96] E. Giacomazzi, F. R. Picchia, and N. Arcidiacono, "On the distribution of lewis and schmidt numbers in turbulent flames," in *30th Italian Combustion Section Meeting*, 2007.

[97] A. J. Aspden, N. Zettervall, and C. Fureby, "An a priori analysis of a dns database of turbulent lean premixed methane flames for les with finite-rate chemistry," *Proc. Combust. Inst.*, **37**, 2601–2609, 2019.

[98] U. Stopper, M. Aigner, W. Meier, R. Sadanandan, M. Sthr, and I. S. Kim, "Flow field and combustion characterization of premixed gas turbine flames by planar laser techniques," *J. Eng. Gas Turb. and Power*, **131**, 021 504–021 504–8, 2008.

[99] U. Stopper, M. Aigner, H. Ax, W. Meier, R. Sadanandan, M. Stohr, and A. Bonaldo, "Piv, 2d-lif and 1d-raman measurements of flow field, composition and temperature in premixed gas turbine flames," *Exp. Therm. Fluid Sci.*, **34**, 396–403, 2010.

[100] U. Stopper, W. Meier, R. Sadanandan, M. Stohr, M. Aigner, and G. Bulat, "Experimental study of industrial gas turbine flames including quantification of pressure influence on flow field, fuel/air premixing and flame shape," *Combust. Flame*, **160**, 2103–2118, 2013.

[101] S. K. Sadasivuni, G. Bulat, V. Sanderson, and N. Swaminathan, "Application of scalar dissipation rate model to siemens dle combustors," *Proc. the ASME Turbo Expo 2012: Power for Land, Sea and Air*, GT2012–68 483, 2012.

[102] G. Bulat, W. P. Jones, and A. J. Marquis, "Large eddy simulation of an industrial gas-turbine combustion chamber using the sub-grid pdf method," *Proc. Combust. Inst.*, **34**, 3155–3164, 2013.

[103] G. Bulat, E. Fedina, C. Fureby, W. Meier, and U. Stopper, "Reacting flow in an industrial gas turbine combustor: Les and experimental analysis," *Proc. Combust. Inst.*, **35**, 3175–3183, 2015.

[104] A. Filosa, B. Noll, M. Di Domenico, and M. Aigner, *Numerical Investigations of a Low Emission Gas Turbine Combustor using Detailed Chemistry*, ser. AIAA Propulsion and Energy Forum. American Institute of Aeronautics and Astronautics, 2014.

[105] I. Langella, Z. X. Chen, N. Swaminathan, and S. K. Sadasivuni, "Large-eddy simulation of reacting flows in industrial gas turbine combustor," *J. Propul. Power*, **34**, 1269–1284, 2018.

[106] A. Sjunnesson, P. Henrikson, and C. Lofstrom, *CARS measurements and visualization of reacting flows in a bluff body stabilized flame*, ser. Joint Propulsion Conferences. American Institute of Aeronautics and Astronautics, 1992.

[107] A. Sjunnesson, C. Nelson, and E. Max, "Lda measurements of velocities and turbulence in a bluff body stabilized flame," Volvo Aero, Tech. Rep., 1991.

[108] A. Sjunnesson, S. Olovsson, and B. Sjöblom, "Validation rig - a tool for flame studies," Volvo Aero, Tech. Rep., 1991.

[109] S. I. Moller, E. Lundgren, and C. Fureby, "Large eddy simulation of unsteady combustion," *Symp. (International) Combust.*, **26**, 241–248, 1996.

[110] C. Fureby, "A computational study of combustion instabilities due to vortex shedding," *Proc. Combust. Inst.*, **28**, 783–791, 2000.

[111] D. Davidenko, I. Gökalp, E. Dufour, and P. Magre, *Numerical Simulation of Hydrogen Supersonic Combustion and Validation of Computational Approach*, International Space Planes and Hypersonic Systems and Technologies Conferences. American Institute of Aeronautics and Astronautics, 2003.

[112] J. Kim and S. B. Pope, "Effects of combined dimension reduction and tabulation on the simulations of a turbulent premixed flame using a large-eddy simulation/probability density function method," *Combust. Theory Model.*, **18**, 388–413, 2014.

[113] A. Ghani, T. Poinsot, L. Gicquel, and G. Staffelbach, "LES of longitudinal and transverse self-excited combustion instabilities in a bluff-body stabilized turbulent premixed flame," *Combust. Flame*, **162**, 4075–4083, 2015.

[114] C. Fureby, *A Comparative Study of Large Eddy Simulation (LES) Combustion Models applied to the Volvo Validation Rig*. AIAA SciTech Forum. American Institute of Aeronautics and Astronautics, 2017.

[115] B. Rochette, O. Vermorel, L. Y. Gicquel, T. Poinsot, and D. Veynante, *ARC versus two-step chemistry and third-order versus second-order numeric scheme for Large Eddy Simulation of the Volvo burner*, ser. AIAA SciTech Forum. American Institute of Aeronautics and Astronautics, 2018.

[116] A. Comer, M. Ihme, Li, C., S. Menon, J. Oefelein, B. Rankin, V. Sankaran, and V. Sankaran 2017. *Proceedings of the Second Model Validation for Propulsion (MVP2) Workshop* (2018) *AIAA Science and Technology Forum and Exposition*, Florida; *Proceedings of the Third Model Validation for Propulsion (MVP3) Workshop* (2019) *AIAA Science and Technology Forum and Exposition*, California.

[117] D. C. Haworth, R. J. Blint, B. Cuenot, and T. J. Poinsot, "Numerical simulation of turbulent propane-air combustion with nonhomogeneous reactants," *Combust. Flame*, **121**, 395–417, 2000.

[118] D. Micka and J. Driscoll, *Dual-Mode Combustion of a Jet in Cross-Flow with Cavity Flameholder*, ser. Aerospace Sciences Meetings. American Institute of Aeronautics and Astronautics, 2008.

[119] D. J. Micka and J. F. Driscoll, "Combustion characteristics of a dual-mode scramjet combustor with cavity flameholder," *Proc. Combust. Inst.*, **32**, 2397–2404, 2009.

[120] D. J. Micka, "Combustion stabilization, structure and spreading in a laboratory duel-mode scramjet combustor," Ph.D. dissertation," 2010.

[121] N. Zettervall and C. Fureby, *A Computational Study of Ramjet, Scramjet and Dual-mode Ramjet Combustion in Combustor with a Cavity Flameholder*, AIAA SciTech Forum. American Institute of Aeronautics and Astronautics, 2018.

[122] C. J. Jachimowski, "An analytical study of the hydrogen-air reaction mechanism with application to scramjet combustion," NASA Langley Research Center, Tech. Rep., 1988.

6 MILD Combustion

Y. Minamoto, N. A. K. Doan, and N. Swaminathan

6.1 Introduction

Alternative combustion concepts and technologies are explored constantly to design and develop efficient and cleaner combustion systems in order to mitigate the environmental impact of conventional combustion. Fuel lean premixed combustion is known to be a potential method to meet these demands [1, 2], but it is highly susceptible to thermo-acoustic instability, which is undesirable for smooth and stable operation of practical combustors. One method to avoid the instabilities is to preheat the reactant mixture by using the heat recovered from the exhaust. This would also enhance the thermal efficiency while improving the combustion stability. However, preheating increases the flame temperature, leading to an increase in thermal nitrogen oxides (NOx) formation in conventional combustion. This adverse effect of preheating using the recovered heat limits its use for systems employing conventional combustion techniques.

Combustion involving both preheating and dilution of reactants with burnt products (through their recirculation) helps us to achieve efficient, cleaner, and "silent" combustion [3–5]. The recirculation of products can be either internal or external to the combustion chamber and this dilution reduces the oxygen available for combustion, which limits the temperature rise and thus the NOx formation. This technology is used mainly for furnaces [6, 7] with internal recirculation strategies for maintaining combustion stability. This led to a particular combustion mode called MILD (moderate, intense, or low dilution) combustion with no visible flames on photographs, which prompted the terminology "flameless combustion" for this mode. This flameless attribute observed in experiments and numerical simulations is shown in Fig. 6.1. Photographs of a burner operated under conventional and MILD combustion conditions [8] are shown in Fig. 6.1a and b respectively. In contrast to the premixed case, neither a visible flame nor high gradients of light intensity are observed in the MILD combustion case. This is highlighted further in Figs. 6.1c–e showing volume rendered images of the temperature field obtained from direct numerical simulations (DNSs) of a conventional premixed flame and MILD combustion in premixed and non-premixed modes [9, 10]. The absence of a visible flame under MILD combustion conditions is clear.

Generally, MILD combustion occurs when the reactants are preheated to a temperature, T_r, larger than the reference autoignition temperature, T_{ign}, of a given fuel and

MILD Combustion 241

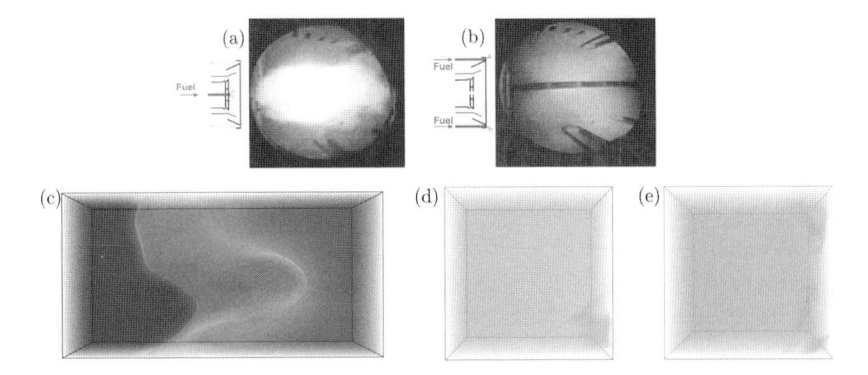

Figure 6.1 Comparison of direct photographs of a burner operating under conventional (a) and MILD (b) combustion conditions. (Reprinted from [8] with permission). Parts (c), (d), and (e) respectively show the volume-rendered image of the temperature field obtained from direct numerical simulation (DNS) of premixed conventional combustion and MILD combustion under premixed and non-premixed conditions.

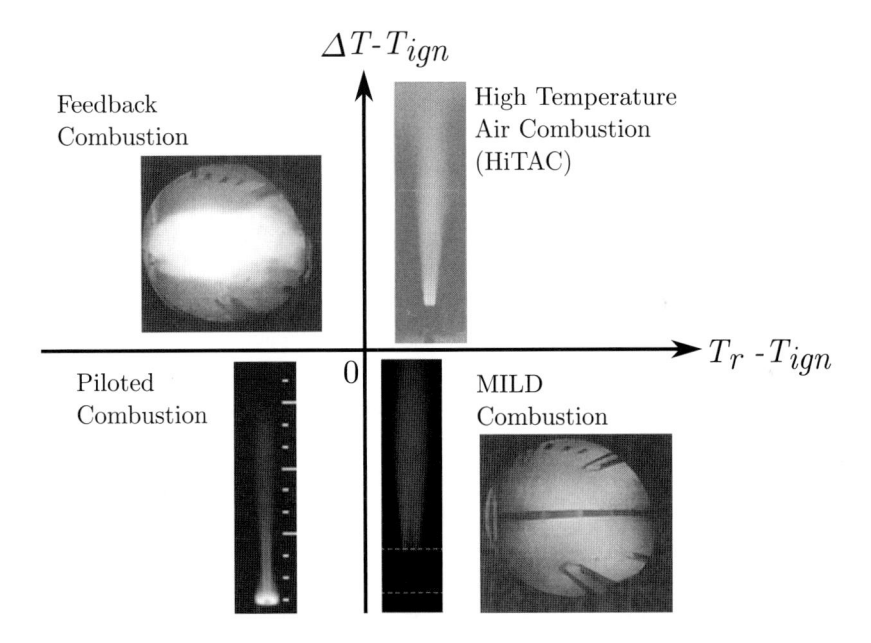

Figure 6.2 Combustion type diagram [5] showing feedback and MILD combustion [8], HiTAC [11], and piloted combustion [12].

the maximum temperature rise, $\Delta T = T_p - T_r$, is smaller than T_{ign}, where T_p is the product temperature [5]. The small temperature rise results from the large amount of dilution with products leading to low oxygen level available for combustion. The hot products also contain radical species, which help to achieve stable combustion.

A diagram such as that shown in Fig. 6.2 can be used to distinguish MILD combustion from other modes such as conventional, pilot assisted, and high-temperature air combustion (HiTAC). The temperature rise is larger than T_{ign} for the conventional (also known as feedback) and HiTAC. The piloted assisted combustion is located in the third quadrant because of the relatively smaller temperature rise and the reactants are typically at ambient temperature. The MILD combustion is in the fourth quadrant for obvious reasons and the flameless like appearance is the result of homogeneous combustion releasing heat uniformly across a larger region compared to conventional combustion. This physical situation combined with averaging (over the camera shutter opening time) effects alludes to flameless-like appearance as observed in past experimental studies [8, 13, 14]. If one looks at this combustion at the right time scale using laser diagnostics with fast camera [15] or DNS data [16], then the existence of flames, autoignition, and their interactions can be observed. Also, high preheating involved makes the combustion stability and dynamics less sensitive to the inflow conditions of MILD mixtures compared to those for conventional combustion [17–22]. More theoretical insights are given in Section 6.2.

These unique characteristics of MILD combustion offer a number of advantages. First, the heat recovery from the exhaust improves the thermal efficiency [3–5]. Second, the low combustion temperature and dilution significantly reduces the thermal NOx formation [3–5, 14, 23, 24]. Typically, it takes few seconds to produce a substantial level of thermal NOx at about 1900 K and this decreases to a few milliseconds when the temperature is about 2300 K [3]. The maximum flame temperature is typically lower than 1900 K in MILD combustion, leading to very low NOx formation rates [3, 5], which has been observed since the early days of MILD combustion research as shown in Fig. 6.3 [3, 17]. This figure shows the NOx production decreases sharply when the combustion mode is switched from conventional to MILD (at $t = 1100$ s) for a burner investigated in [17]. The NOx production in various industrial burners with capacities ranging from 6 to 250 kW operating at various temperatures is shown in Fig. 6.3b [3], and this result also highlights that NOx production can decrease by an order of magnitude for MILD combustion. Third, combustion noise decreased substantially when conventional lifted flame operation was changed to MILD mode [3]. Fourth, combustion instabilities are suppressed because of the small unsteady temperature rise even when the recirculation rate exceeds 30%, which is an upper limit for stable combustion using normal ambient air [3, 4]. The recirculation rate is defined as the ratio of the recirculated exhaust gas mass flow rate to fresh reactants mass flow rate. Thus, MILD combustion allows (1) enhancement of system thermal efficiency, (2) reduction of pollutants emission, and (3) avoidance of undesirable combustion instabilities, leading to quieter combustion.

The high preheating temperature also allows maintenance of stable combustion even in a high-velocity jet without a need for internal recirculation zones [3, 5, 25]. Thus, the design of a MILD combustor would no longer be constrained by the requirements of recirculation zones or flame holders, which is advantageous for high-speed combustion. Furthermore, MILD conditions are achieved quite easily in practical devices using conventional techniques such as exhaust or flue gas recirculation (EGR or FGR) or staged fuel injection [3, 5, 26] or the HiPerMix concept [21]. These multiple advantages have renewed interest in the use of MILD combustion as one of the "green" technologies for thermal power generation [5, 19, 21, 27]. The progress achieved so far on this topic is reviewed by Perpignan et al. [28], specifically with a view for its application to gas turbines. There are various challenges that need to be addressed and understood for better use of MILD combustion technologies for combustion applications with lower carbon footprints.

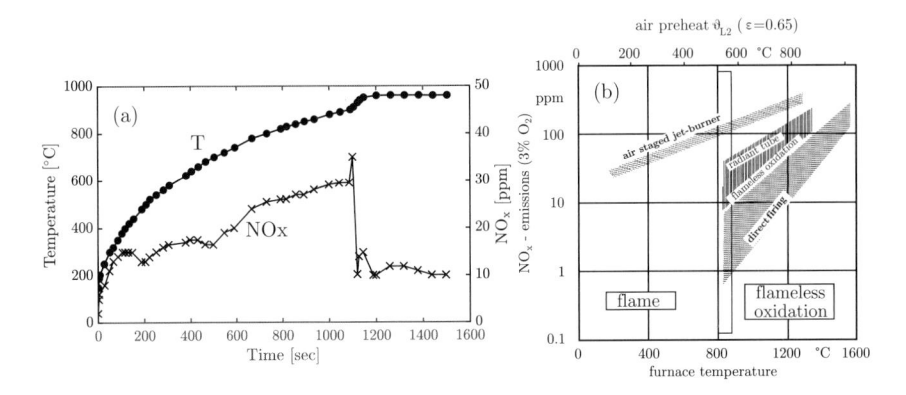

Figure 6.3 (a) NOx production and temperature evolution in a combustor switching from a stable flame to MILD combustion mode. (Reprinted from [17]). (b) NOx level corrected to 3% O_2 as a function of furnace temperature. (Reprinted from [3]).

Cavaliere and de Joannon [5] reviewed the past studies on this topic conducted up to early 2000 and it is summarized briefly here. The MILD combustion was influenced by the physical conditions such as combustor design and flow conditions, and thermodynamic and chemical parameters of the system. The thermochemical state of the mixture covered the whole range from the frozen to fully reacted (equilibrium) conditions. The dilution using hot gases extended the pressure–temperature region where oxidation is suppressed under conventional combustion conditions, implying that the inflammable mixture under conventional situation could become flammable. The preheated reactant had enough enthalpy for self-ignition and thus the recirculated gases played a dominant role in reducing the oxygen level rather than stabilizing MILD combustion. The lower heat release rates in MILD combustion compared to conventional cases resulted from a lower mass flux across the stoichiometric isosurface in MILD cases, which was verified by analysing a double mixing layer between diluted oxidizer and fuel. Based on all of these observations, they suggested that the chemical states in MILD combustion might evolve as in a well-stirred reactor (WSR).

In the following, we shall give an overview of the progress made on this topic in the past 15 years and identify outstanding issues for further investigation. Definitions of MILD combustion are discussed in Section 6.2. Insights obtained using zero- and one-dimensional model reactors are discussed in Section 6.3. Flow configurations employed for MILD combustion experiments and applications are discussed in Section 6.4.1. Section 6.5 discusses the results on turbulent MILD combustion. The future outlook along with potential applications are discussed in the final section.

6.2 Definitions of MILD Combustion

Several definitions of MILD combustion were proposed in past studies [3, 5, 29] and the original proposition in [29] related MILD combustion processes to the behavior of a stochastic homogeneous reactor. This definition was motivated by the homogeneous temperature fields observed in the initial experimental studies. Under this condition, the ignition and extinction points no longer exist in the classical S-shaped curve and there is a monotonic shift from unburned to burned conditions. This situation was noted to be MILD combustion [29].

This was shown formally by Oberlack et al. [29] using the conservation equations for fuel and temperature in a perfectly stirred reactor (PSR) evolving in time. These equations are

$$m \frac{dY_F}{dt} = \dot{m}(Y_{F,r} - Y_F) - \dot{\omega}_F W_F V \tag{6.1}$$

$$mC_p \frac{dT}{dt} = \dot{m} C_p (T_r - T) + \dot{\omega}_F W_F Q V, \tag{6.2}$$

where m is the mass inside the PSR with volume V and $\dot{m}$ is the mass flow rate through the reactor. The mass fraction of the fuel in the incoming stream at a temperature T_r is $Y_{F,r}$ and its molecular weight is W_F. The heating value of the fuel is Q. The reaction rate is expressed using a single-step reaction with an activation energy E as

$$\dot{\omega}_F = B \left(\frac{\rho Y_F}{W_F} \right) \exp \left(\frac{-E}{RT} \right) \tag{6.3}$$

for fuel-lean premixed combustion. The universal gas constant is R. The residence time through the reactor is $t_r = m/\dot{m}$, which can be used to normalize the time, t. If one normalizes Y_F using $Y_{F,r}$, T by T_r and recognizing that $\rho = m/V$, Eqs. (6.1) and (6.2) can be written as

$$\frac{d\widehat{Y}}{d\tau} = 1 - \widehat{Y} - \text{Da}\widehat{Y} \exp\left(-\frac{\widehat{E}}{\widehat{T}}\right) \tag{6.4}$$

$$\frac{d\widehat{T}}{d\tau} = 1 - \widehat{T} + \widehat{Q}\text{Da}\widehat{Y} \exp\left(-\frac{\widehat{E}}{\widehat{T}}\right), \tag{6.5}$$

where $\text{Da} = Bt_r$, $\widehat{Q} = QY_{F,r}/(C_p T_r)$, and $\widehat{E} = E/(RT_r)$. It can be shown that $\widehat{Y} = (1 + \widehat{Q} - \widehat{T})/\widehat{Q}$ by multiplying the $d\widehat{Y}/d\tau$ equation by $\widehat{Q}$ and adding it to the $d\widehat{T}/d\tau$ equation and recognizing the steady-state solution of the resulting equation is $(1 + \widehat{Q}) = \widehat{T} + \widehat{Q}\widehat{Y}$. By using this result, Eq. (6.5) can be written as

$$\frac{d\widehat{T}}{d\tau} = 1 - \widehat{T} + \text{Da}(1 + \widehat{Q} - \widehat{T}) \exp\left(-\frac{\widehat{E}}{\widehat{T}}\right), \tag{6.6}$$

which gives

$$1 - \widehat{T}_s + \text{Da}(1 + \widehat{Q} - \widehat{T}_s) \exp\left(-\frac{\widehat{E}}{\widehat{T}_s}\right) = 0 \tag{6.7}$$

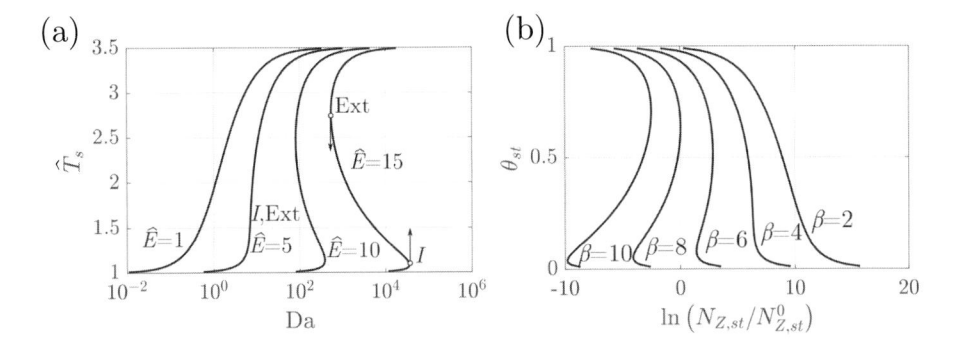

Figure 6.4 S-shaped curve illustrating the condition of MILD combustion for the (a) premixed conditions and (b) non-premixed.

for steady state. The variation of the steady-state temperature $\widehat{T}_s$ with Da is shown in Fig. 6.4a for $\widehat{Q} = 2.5$, for various values of $\widehat{E}$. The ignition point is denoted as I and the extinction point is "Ext." For a given value of $\widehat{Q}$, the ignition and extinction (turning) points move toward each other as $\widehat{E}$ decreases and for a particular value of $\widehat{E}$ the distinction disappears and there is a monotonic variation from unburned to burned state, which is seen for $\widehat{E} = 5$ in Fig. 6.4a. The condition for the absence of the turning points can be deduced from Eq. (6.7) using $d\,\mathrm{Da}/d\,\widehat{T}_s = 0$ and this condition is given by

$$\widehat{E} \leq 4 \left(\frac{1+\widehat{Q}}{\widehat{Q}} \right). \tag{6.8}$$

This can be deduced to be

$$\widehat{E} \leq 4 \left(1 + \frac{C_p T_r}{Q Y_{F,r}} \right) = 4 \left(1 + \frac{T_r}{\Delta T} \right). \tag{6.9}$$

The second part is obtained by approximating $(QY_{F,r})/(C_p T_r)$ by $(T - T_r)/T_r = \Delta T/T_r$.

Another definition proposed in [5] suggests that MILD combustion occurs when $T_r > T_{ign}$ and $\Delta T < T_{ign}$. This definition is commonly used as it clearly differentiates MILD combustion from the three other combustion types as shown in Fig. 6.2. In particular, MILD combustion differs from *feedback* (conventional) combustion, as the former cannot be sustained without reactants preheating, and it differs from *high temperature* combustion, as it achieves a relatively low temperature rise.

It should be noted that the aforementioned two definitions could be applied for premixed combustion because of the use of the residence time, t_r, inside the reactor. The mixing time denoted by the scalar dissipation rate N_Z of the mixture fraction plays an important role in non-premixed situation. Hence, the analysis of Oberlack et al. [29] was reformulated in [30] using unsteady flamelet equations and that formulation gave

$$\frac{d\theta_{st}}{d\tau} + \frac{N_{Z,st}}{N^0_{Z,st}} \theta_{st} - \dot{\omega}(\theta_{st}) = 0, \tag{6.10}$$

where τ is time normalized using $N_{Z,st}^0$ and $\theta_{st} = (T_{st} - T_{st,u})/\Delta T_{st}$. The reaction rate is

$$\dot{\omega}(\theta_{st}) = \text{Da}\exp(\beta_{ref} - \beta)\frac{(1-\alpha)(1-\theta_{st})^2}{[1-\alpha(1-\theta_{st})]\theta_{st}}\exp\left(-\frac{\alpha\beta(1-\theta_{st})}{1-\alpha(1-\theta_{st})}\right). \quad (6.11)$$

It is to be noted that Eq. (6.10) is written for stoichiometric condition, i.e., at $Z = Z_{st}$ in the mixture fraction space. The various symbols are defined as follows: $\beta = E/(RT_{st,b}) = T_a/T_{st,b}$, $\alpha = (T_{st,b} - T_{st,u})/T_{st,b} = \Delta T_{st}/T_{st,b}$, $\text{Da} = B/N_{Z,st}^0$. The subscripts u and b denote the unburnt and burnt conditions respectively. Under steady state condition Eq. (6.10) gives

$$\frac{N_{Z,st}}{N_{Z,st}^0}\theta_{st} = \dot{\omega}(\theta_{st}). \quad (6.12)$$

The variation of θ_{st} with $\ln(N_{Z,st}/N_{Z,st}^0)$ is shown in Fig. 6.4b for $\text{Da} = 100$. As for the premixed case, here one also observes the S-shaped curve showing ignition and extinction (turning) points, which disappears for particular combination of β and α. The turning points are obtained by using $dN_{Z,st}/d\theta_{st} = 0$ for Eq. (6.12), which gives

$$(\beta^2 + 6\beta + 1)\alpha^2 - (6\beta - 2)\alpha + 1 < 0 \quad (6.13)$$

for monotonic variation from unburned to burned state (MILD combustion). This equation yields

$$\frac{\Delta T}{T_r} < (1+Q)f(\beta) \quad ; \quad f(\beta) = \frac{(6\beta - 2) \pm \sqrt{32\beta^2 - 48\beta}}{2(\beta^2 + 6\beta + 1)} \quad (6.14)$$

for the MILD combustion.

The conditions given by Eqs. (6.9) and (6.14) and $\Delta T \leq T_{ign}$ are plotted in Fig. 6.5 for $T_{ign} = 1000$ K and $E = 40$ kcal/mol. The region enclosed by these borders satisfies all of the three conditions for MILD combustion for the given parameters. Although this is useful information, one must be mindful of the various approximations used in the above analysis. Just to spell them out, (1) a single-step reaction was assumed, (2) the combustion was taken to be fuel-lean in the analysis of the premixed system, (3) adiabatic conditions were assumed, and (4) the Lewis number was taken to be unity. The Lewis number could be seen as a second-order effect for MILD combustion. The one-step reaction used includes the temperature sensitivity and finite rate chemistry effects but does not account for the presence of radicals and intermediates, which play a key role in thermochemical explosions required to stabilize MILD combustion as highlighted in [3]. The analyses may be extended to include chain reactions [29] but the analytical treatment is likely to be intractable. This would also be the case for nonadiabatic conditions. However, the numerical solutions with chain reactions and also for nonadiabatic conditions give similar conclusions. More importantly, practical MILD combustion involves inhomogeneous mixtures containing unburned and partially and fully burned mixtures undergoing simultaneous mixing and chemical reactions and thus it may not fit into one of the idealized cases investigated in past studies. Indeed, the DNS analysis discussed in [31] showed that

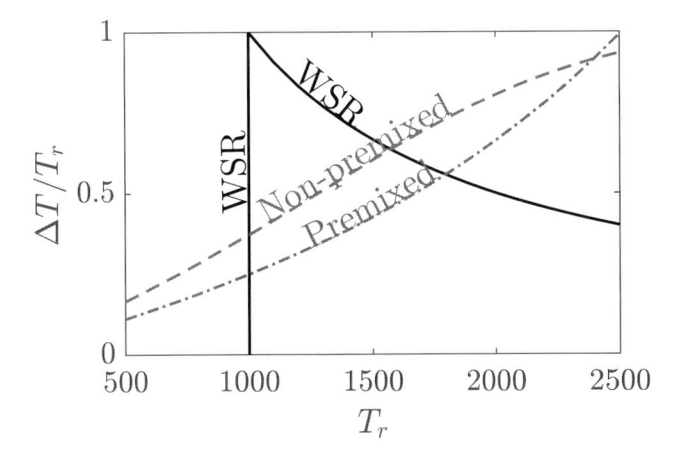

Figure 6.5 Combustion regime map for MILD combustion according to the definitions of Cavaliere and de Joannon [5] (marked as WSR), Oberlack et al. [29] (premixed), and Evans et al. [30] (non-premixed).

θ_{st} increased with increase in $N_{Z,st}$ during the inception phase (lower branch of the S-curve), which is contrary to the behavior seen in Fig. 6.4. The increase in θ_{st} was shown to be related to the presence of radicals in the flow. Thus further investigations are required.

6.3 Investigations Using Zero- and One-Dimensional Model Reactors

Many experimental and numerical studies are devoted to investigating fundamental aspects of MILD combustion using canonical laminar configurations such as the well-stirred reactor (WSR), counterflow flames, and plug flow reactor (PFR). These studies focused on the combustion behavior, influence of dilution levels, autoignition processes, and chemical kinetics. This section presents a summary of these studies.

6.3.1 Well-Stirred Reactor

The homogeneous temperature fields with lower gradients of species observed in the initial experiments on MILD combustion motivated the use of WSR to gain further understanding. The compatibility of flameless features of MILD combustion with a two-stage oxidation process from chemical pathway analysis was confirmed for a methane–air mixture using WSR calculations in [8]. The first of these two stages corresponds to an oxidation in rich diluted conditions (oxygen-starved environment), which is followed by further mixing of radicals with oxygen to complete the oxidation. The so-called flameless feature of MILD combustion would come from the relatively low production of chemiluminescent species from this particular two-stage oxidation mechanism. The transient behavior of this combustion mode was studied further using experiments and simulations in [32, 33]. The dynamic combustion modes with temperature oscillations arise because of the competition between the oxidation and recombination reactions at temperatures and dilution levels that are typical in MILD combustion. These oscillations, along with a map of combustion behavior, are shown in Fig. 6.6. Adding hydrogen to the MILD mixture inhibits these oscillations and thereby increases the range of inlet temperature and dilution levels for stable operation in MILD mode even under nonadiabatic conditions [34]. Hence, the temperature oscillations observed could have been amplified by the influences of heat losses in addition to the competing kinetics mechanisms noted earlier. However, a similar dynamical behavior was observed for adiabatic conditions also [35] suggesting the dynamic behavior ensues predominantly from the competing kinetics mechanisms. This highlights the important role of chemical kinetics in MILD combustion. Thus, the results obtained in past studies using a single-step reaction must be treated cautiously.

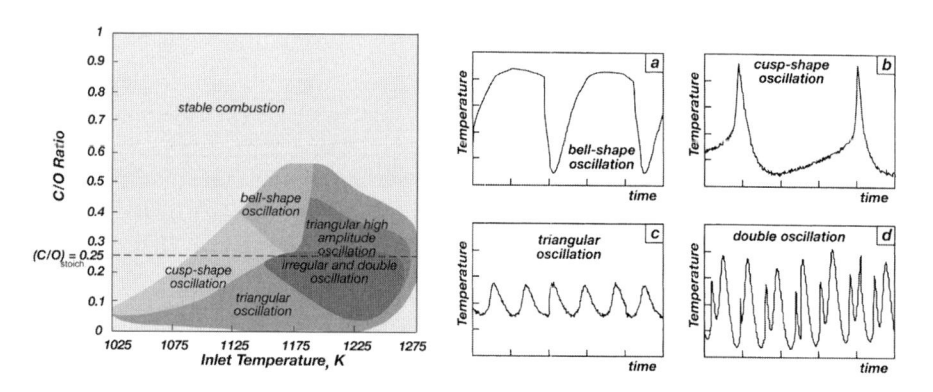

Figure 6.6 Map of combustion behavior for methane oxidation in diluted conditions (90% in volume of N_2 and 10% of mixture of CH_4/O_2) in a WSR (left) and typical temperature oscillations observed in the reactor (right). Reprinted from [32].

6.3.2 Counterflow Flame

The counterflow configuration is also used to investigate MILD combustion [15, 36–42]. These various studies have explored the influence of preheating and dilution of either fuel or oxidizer on MILD combustion. This configuration involves two opposing streams whose composition and temperature can be varied independently to be representative for MILD combustion, allowing one to develop maps of heat release rate variation with the dilution and temperature of the two streams. Typical examples of these are shown in Fig. 6.7.

The mixture of CH_4–O_2 at various equivalence ratios impinging against a flow of hot inert (N_2) was considered in [37]. In addition to classical deflagrative regimes of combustion, associated to conventional mixture inside the flammability limits of the CH_4–O_2 mixture, the existence of a mode called homogeneous charged diffusion ignition (HCDI) was observed. This mode was appearing for high inert temperature with low temperature rise, ascribing it to the MILD combustion regime. Also, this mode showed characteristics of MILD combustion, namely a distribution of reactions

MILD Combustion 249

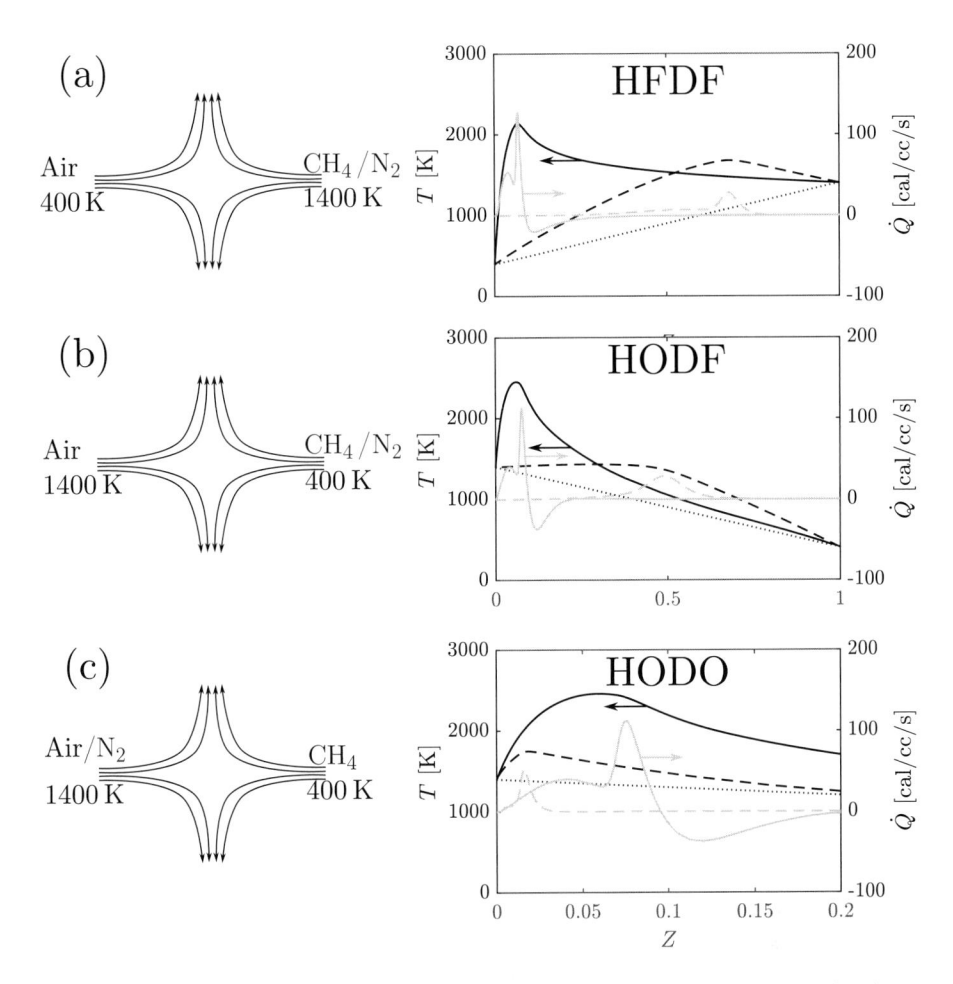

Figure 6.7 Evolution of temperature (black) and heat release rate (gray) with mixture fraction for an undiluted or conventional case (full lines) and a MILD case (5% of fuel or air) (dashed lines) in the corresponding counterflow configurations shown on the left.

over a wide range of mixture fractions (and thus also over a large physical space) and a smooth transition from HCDI to deflagrative regimes. The same configuration, but with a stream of hot air on one side and methane diluted with nitrogen at ambient temperature on the other side (hot oxidizer diluted fuel [HODE] conditions), was considered in [38] and the results showed a larger broadening of the reaction zone (in mixture fraction space and thus also in physical space) compared to a conventional case without dilution. Similar behaviors were observed for an ambient air stream opposing a stream of hot and diluted methane (hot fuel diluted fuel [HFDF] condition) [39]. The hot oxidant diluted oxidant (HODO) configuration involved a stream of methane at ambient temperature opposing a hot and diluted oxidant stream [40]. The oxidation in all these configurations could occur only if the temperature of the preheated stream was high enough to promote autoignition. Also, the peak heat release rate locations were uncorrelated with the stoichiometric mixture fraction under those autoigniting conditions. Figure 6.7 shows typical temperature profile (black) and heat release rate (gray) variations with mixture fraction, Z, for the three, HFDF, HODF, and HODO, configurations depicting both MILD (dashed lines) and conventional (full lines) combustion. In particular, the double peak for heat release rate observed for the conventional high-temperature combustion disappears for MILD conditions. The absence of a pyrolytic regions can also be observed at high dilution.

The transient evolutions in HODF and HFDF configurations were further studied using 2D laminar simulation using FLUENT in [41]. Compared to the steady analysis performed in earlier studies, the range of conditions (in terms of dilution and preheat temperature) yielding MILD combustion was found to be smaller, implying that higher temperatures were needed to obtain autoignition. A stream of gaseous kerosene impinging against an oxygen stream preheated and diluted with hot combustion product was investigated in [42]. Contrary to other works discussed earlier, no large broadening of the reaction zone was observed and the existence of clearly defined peaks for heat release rate and OH, comparable to a conventional flame, was noted.

6.3.3 Plug Flow Reactor

The plug flow reactor has also been used in past experimental and numerical studies [43–46]. These studies focused on the ignition delay time for mixture under MILD conditions and analysis of the chemical pathways involved. A mixture of oxygen with CH_4 [43] or C_3H_8 [44] preheated and diluted with N_2 showed a significant slowdown of the kinetic pathways during the ignition process. Also, it was suggested that existing chemical mechanisms may need to be revised to appropriately capture the competition between the various chemical pathways (oxidation/recombination) under MILD conditions. The use of CO_2 or H_2O instead of N_2 for the dilution was examined in [45] and these diluents decreased the system reactivity when compared to N_2. Furthermore, the effects of these product species on the autoignition time was found to be influenced by the temperature. This study suggested that there were possibilities for CO_2 and H_2O to act as (1) reactants for elementary reactions, (2) collision partners with larger third-body coefficients compared to N_2 for three-body recombination reactions, and (3) they could decompose either producing or consuming radical species. When the temperature is lower than the autoignition temperature (low T) the third-body effect (2) prevails, promoting ignition. This effect gradually decreases with increasing temperature (similar to or larger than the autoignition temperature) and it shifts toward the kinetic effect (3), which inhibits ignition through endothermic reactions. These analyses for C_1-C_2 fuel blend were performed in [46] showing similar behaviors but the presence of C_2 species decreased the autoignition time. However, all of these studies focusing on ignition delay time and chemical pathways showed discrepancies between the calculated and measured ignition delay times, especially for fuel-rich regions. This suggests that there is a strong need to develop appropriate chemical mechanism for MILD conditions. The work in [36] considered the PFR but the reactant mixture was diluted using either major species or all the species present in the product of a laminar flame. The presence of intermediate radicals was observed to decrease the ignition delay.

6.3.4 Insights from Laminar Calculations

All of the aforementioned studies have highlighted some fundamental features of MILD combustion and have in particular emphasized the importance of the chemical kinetics in MILD combustion. It was observed both in the WSR and PFR configurations that there is a strong competition between oxidation and recombination reaction pathways, which could lead to dynamic behavior (temperature oscillations discussed earlier in Fig. 6.6) under MILD combustion. Furthermore, the counterflow flame results showed that the heat release rate variations for MILD conditions were observed to be significantly different from those in conventional high-temperature combustion. Indeed, neither double peak nor pyrolysis is observed in MILD combustion. Finally, as discussed in Section 6.2, MILD combustion in practice involves a highly inhomogeneous mixture containing fuel, oxidizer, intermediate species, and fully burnt and unburnt gases, and thus these laminar configurations invoke large simplifications of the actual physics of MILD combustion, despite the useful insights obtained. Hence, some care is required in extending these results to turbulent MILD combustion situations.

6.4 Past Experimental Explorations

6.4.1 Typical Configurations to Achieve MILD Combustion

Various experimental configurations used in past studies to achieve MILD combustion are shown in Fig. 6.8. A common feature to all of these configurations is the presence of diluted hot oxidizer either produced through combustor aerodynamics (see Figs. 6.8a and 6.8b) or supplied intentionally as in Figs. 6.8c and 6.8d.

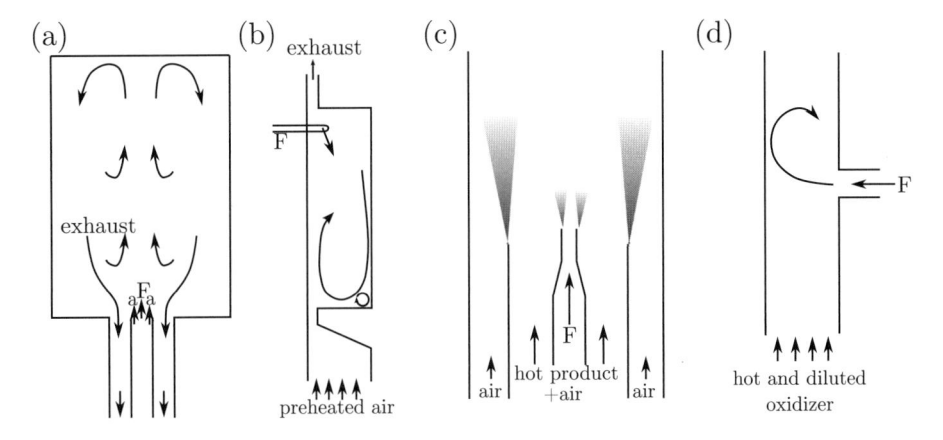

Figure 6.8 Schematic representation of MILD combustion burners.

6.4.1.1 Internal Recirculation Configuration

An *Internal recirculation configuration* is created aerodynamically either by arranging the inlets and outlets appropriately as in Fig. 6.8a or by integrating bluff bodies in the design as in Fig. 6.8b. The choice depends on the application and other factors such as pressure drop, cooling requirements, and so forth. The first choice is commonly used in furnace-like applications [13, 17, 47–49] involving a fuel jet in the center surrounded by air jets, which may be preheated. The large recirculation region created by these jets entrains the background mixture containing O_2 and hot gases. These recirculation regions also involve strong shear layers, producing turbulence to enhance the mixing of fuel with hot diluted oxidizer stream.

The furnace must be run under conventional combustion conditions first in order to preheat it and to set up the required background fluid having hot combustion products mixed with air into which fuel is injected. Another configuration [4] to produce an internal recirculation zone involves a bluff body and fuel that is injected into this zone, as shown schematically in Fig. 6.8b. This burner also starts with conventional combustion and moves into the MILD mode subsequently. However, the cooling requirement for this configuration is larger compared to those involving jets shown in Fig. 6.8a. Either gaseous, liquid, or possibly solid fuels can be used in both of these configurations. The time required to transition from conventional combustion to MILD mode is of great importance from a practical perspective and this transition must be as quick as possible.

6.4.1.2 Jet in Hot and Diluted Coflow

Jet in hot and diluted coflow (JHC) has been one of the most widely used configurations for MILD combustion studies since 2000. This configuration is used in [25, 50–57]. Its schematic is shown in Fig. 6.8c, which involves a high-momentum central jet issuing fuel into a coflow of combustion products mixed with air. The combustion products are supplied by an upstream secondary burner. The dilution level and temperature of the coflow can be accurately controlled. Another cooler coflowing air stream can be used to cool the combustor walls [50]. The shear and mixing layers develop between the streams of air, vitiated air (combustion gases + air), and fuel as shown by the shaded triangular regions in Fig. 6.8c. The mixing of vitiated air and fuel streams leads to ignition at a downstream location. There is a possibility for entrainment of the cooler air into the reacting region further downstream. This can lead to local extinction.

6.4.1.3 Jet in Hot Crossflow

A *Jet in hot crossflow* as (JICF) shown in Fig. 6.8d is often used to achieve high levels of fuel–air mixing over a shorter (residence) time compared to characteristic flame time, and hence it is used for many engineering applications, including gas turbines. The JICF involves complex three-dimensional flow structures made of inhomogeneous shear layer vortices, wake vortices, horseshoe vortices, and large-scale counterrotating vortex pairs (CVPs), as shown in Fig. 6.9, spanning a broad range of length and time scales [58]. While these flow structures play pivotal roles in enhancing the mixing and influencing flame stabilization, they are challenging to investigate and model. Combustion in this configuration generally involves lifted flames. The distance from jet exit to the flame leading edge (lift-off length) is an important parameter from a combustor design viewpoint. However, the lift-off length depends on the balance between large-scale convection, mixing, and chemical reactions that are influenced by the complex flow structures. Thus, there are significant research efforts to understand the mixing and flame stabilization mechanisms for JICF configurations. Passive scalar transport in this configuration is studied using experiments [59] and numerical simulations [60], and these studies show that the mixing problem cannot be treated as a direct analogue of the flow problem as usually done for self-similar jets. The heat release coupled with organized turbulent motion in the JICF influence the scalar transport, which is different from the passive scalar transport. Also, preferential diffusion of species, with mass diffusivities varying by as much as fivefold, can modify the scalar mixing characteristics [61, 62]. Therefore, flame stabilization depends on the balance of all of these complex transport phenomena.

The flame anchoring and stabilization in the near field of a fuel jet in crossflow was studied using DNS [63, 64]. The effects of fuel injection angle with respect to the cross flow direction [65] and fuel composition [66] were also investigated. The flame stabilization in the near field of the fuel jet through partially premixed flame propagation mechanism was noted in experimental [67, 68] and DNS [63, 64] studies. Indeed, the fluid velocity in the local flame normal direction at the flame base was reported to be strongly correlated to the laminar flame speed of the stoichiometric mixture [68]. Combustion of fuel jet in a hot vitiated cross flow has also been investigated in past experiments [15, 69] and direct numerical simulations [69]. The hot transverse coflow to the fuel jet adds complexity to the flame stabilization mechanism, since the propagating flame leading edge alone is no longer dominant in the presence

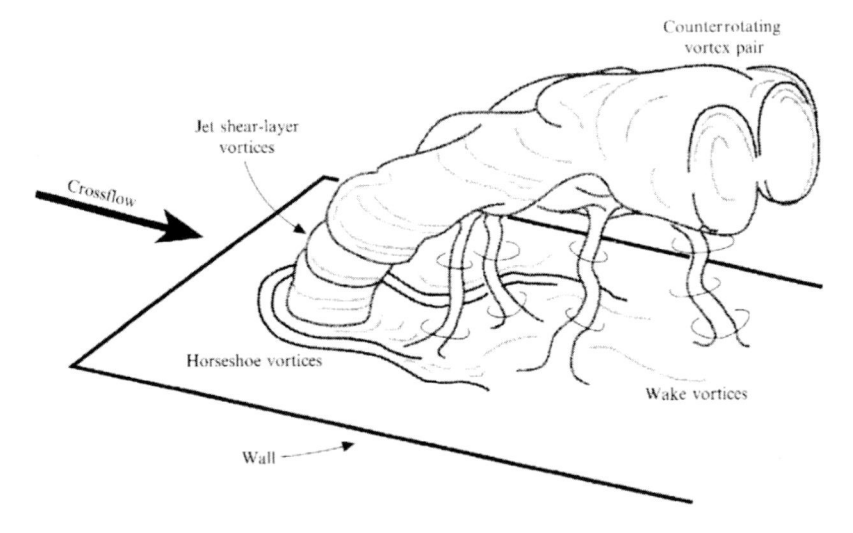

Figure 6.9 Typical flow features in a JICF [58].

of autoignition caused by the highly reactive mixtures coming from the hot vitiated air. More targeted investigations are required to closely understand the stabilization and combustion mechanisms under conditions relevant for MILD combustion.

6.4.2 Structure and Identification of Reaction Zones

Early MILD combustion experiments predominantly used configurations involving internal exhaust gas recirculation like in furnaces [3, 4, 13, 17], demonstrating the NOx reduction and other advantages of MILD combustion along with various physical features involved in this combustion. In particular, the absence of visible and audible flame was highlighted in [3], which led the authors to call MILD combustion as *flameless* combustion. Further investigations on this topic used laboratory-scale combustors involving separate streams of fuel at ambient temperature and hot and diluted oxidizer [17, 25, 49, 56], shown schematically in Figs. 6.8 and 6.9. The fuel stream is also diluted sometimes instead of oxidizer. The former arrangement is called hot oxidant-diluted oxidant (HODO), and the latter is called hot fuel-diluted oxidant (HFDO) [5]. MILD combustors operating under partially premixed and fully premixed conditions have also been investigated in past studies [24, 70].

Advanced laser diagnostics have been used to shed light on the unique characteristics of "invisible" flames by measuring instantaneous temperature and species fields [13, 17, 25, 49, 71]. Reaction zones distributed in space are observed under MILD conditions when compared to the conventional combustion using laser thermometry images. Rayleigh temperature and OH-PLIF measurements showed homogeneous temperature fields with lower OH intensity compared to that from conventional combustion [13, 17]. Also, the gradients in the OH field was observed to be lower. Additional laser Doppler velocimetry highlighted the existence of recirculation regions, their features and their role in mixing flue gases with fuel–air mixtures to sustain the reactions. As a result of this recirculation, the reaction kinetics was said to be slowed down, leading to reduced burning intensity. This combination led the authors to suggest that chemical kinetics time scales had become similar to typical flow time scales.

The MILD combustion in JHC of Adelaide [50] or Delft [55] has been investigated using laser diagnostics and numerical simulations to gather closer insights. Both of these burner designs are similar to what is described in Section 6.4.1 and the Delft JHC burner allows having a coflow coming from a partially premixed flame and not just fully premixed flame like in the Adelaide burner. The following paragraphs describe the physical observations obtained from these various experiments and how those observations extend or not to the other configurations.

As discussed earlier, MILD combustion does not have visible flames (reaction zones) and thus advanced laser diagnostics were used to identify and characterize these reaction zones. Medwell et al. [25] performed CH_2O–PLIF in a JHC configuration with CH_4–H_2 as fuel. Visual inspections of these PLIF images suggested the presence of spatially distributed reaction zones. Similar observations were made in another experimental study by [71]. These PLIF images presented in Fig. 6.10 show

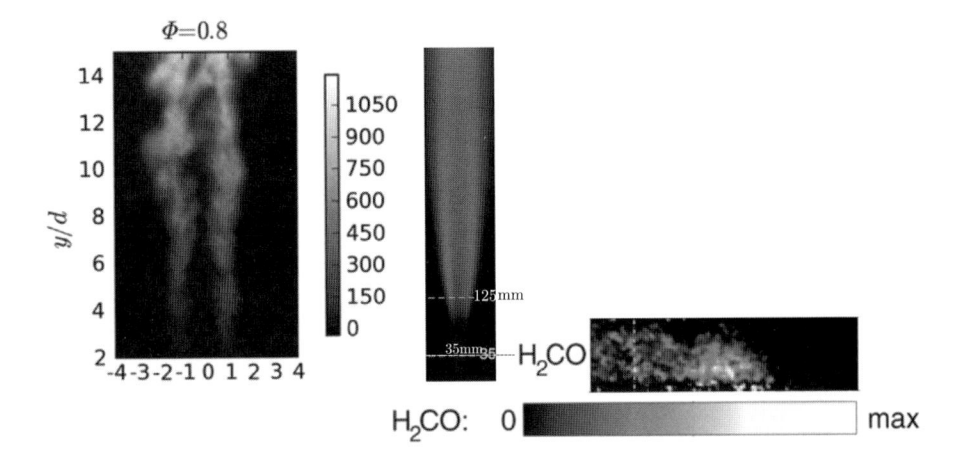

Figure 6.10 CH$_2$O–PLIF measurements of [25] (left) and [71] (right) in JHC burners.

the spatially distributed nature of the reaction zones. However, OH-PLIF images from these two experimental works showed locations of peak OH with clear gradients, suggesting the existence of thin reaction zones. The intensities of the OH–PLIF were, however, found to be much smaller and the general thickness of OH structure was larger than in conventional flames, as can be seen in Fig. 6.11.

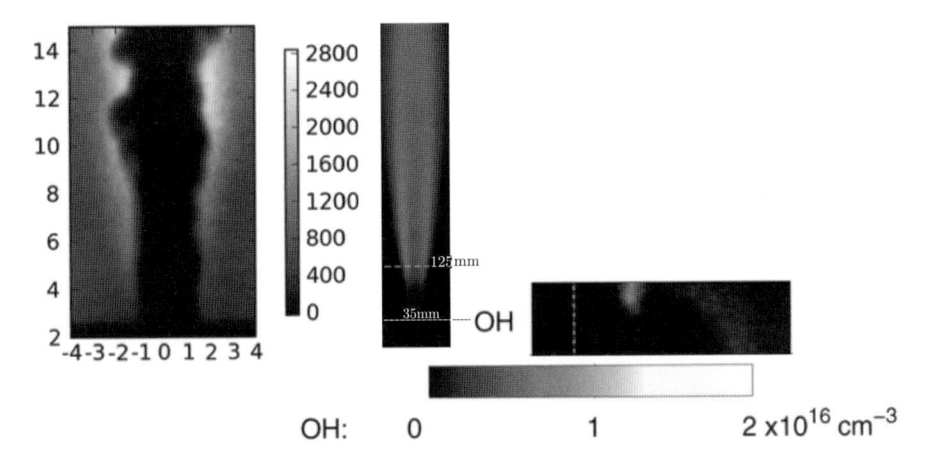

Figure 6.11 OH–PLIF measurements of [71] (left) and [25] (right).

The OH–PLIF measurements performed by [49] in a furnace-like configuration showed OH and temperature to be spatially distributed with a patchy appearance, but a bimodal PDF reported for the temperature suggested thin reaction zones. So, the nature of the reaction zones in MILD combustion is unclear from these laser diagnostics. In addition to these ambivalent aspects of reaction zones, their identification is challenging for the following reason. The use of only OH–PLIF to identify heat-releasing (reaction) zones in MILD combustion is insufficient because the locally produced OH level may be smaller or comparable to the OH present in the recirculating gases and so differentiating them could become difficult. However, the chemiluminescence imaging using OH* showed that the heat release occurred over a large region of the MILD combustor investigated in [48] and this insight is quite contrary to that suggested by OH–PLIF [25] (i.e., thin regions with large heat release). A comparison of the OH-PLIF and OH* chemiluminescent images from a JICF setup showed substantial level of OH in regions with no or very small heat release rate as shown in Fig. 6.12 depicting averaged chemiluminescent and PLIF images [15]. The instantaneous OH–PLIF images showed the presence of thin OH fronts as observed in DNS studies [10, 16]. The mismatch between the OH* and OH was also observed in the calculations of laminar flames under MILD conditions [36]. This particular observation was also noted in [57], suggesting that OH may not cover the entire region where chemical reactions occur in MILD combustion. Indeed, it was suggested that fuel decomposition could occur in fuel-rich regions where OH was not an adequate marker. All of these suggest that one must be cautious in interpreting OH–PLIF images of MILD combustion, and another marker, such as CH or simultaneous OH and CH_2O markers, could prove to be more useful.

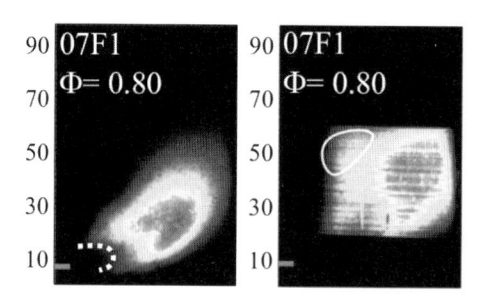

Figure 6.12 OH* chemiluminescent measurements (left) and OH–PLIF measurements (right) from [15]. The dashed while line marks the region with no heat release and the solid white line marks the region with substantial OH but with no heat release.

The presence of OH gradients in thin regions in MILD combustion [25, 36, 71] signifies that there are flames present. However, autoignition could also be of importance given the high reactant temperature [5]. The heated coflow was shown to play a key role in sustaining the combustion through ignition of mixture that was locally extinguished by the entrained surrounding cold air. This was shown using CH_2O-PLIF in [25, 53]. Also, UV-luminescence recordings were used to study the evolution of the ignition kernels in an experimental setup similar to a JHC and it was observed that the ignition kernels were produced constantly and they grew in size as they were convected downstream [55]. Similar behaviors were also observed for a JICF setup [15]. The importance of autoignition was also stressed in [51] showing that the ignition kernels at the flame base were appearing at the most reactive mixture fraction Z_{MR}. These kernels were then progressing toward the stoichiometric and richer mixtures as they were convected downstream.

As discussed, MILD combustion involves intense mixing between the fuel and oxidizer streams achieved using recirculating flows. Also, the chemical time scale is relatively large due to the intense dilution. As a result, the interaction between turbulence and reaction zones is generally significant compared to conventional combustion. The effects of turbulence and strain on the MILD reaction zones is investigated in [54], suggesting that the convolutions and stretching of the reaction zones are caused by the turbulent vortices, which leads to thinning of the reaction zone as pictured in Fig. 6.13. The authors of [54] also suggested that this thinning process caused OH "weakening" and also an increase in the CH_2O–PLIF signal. Furthermore, an increase in the strain rate combined with an accompanying reduction of the reaction rate contributes to an additional transport of O_2 across the reaction zones, which leads to a partial premixing. This behavior is thought to lead to a stabilization of MILD combustion reaction zones by increasing the formation of intermediate radicals.

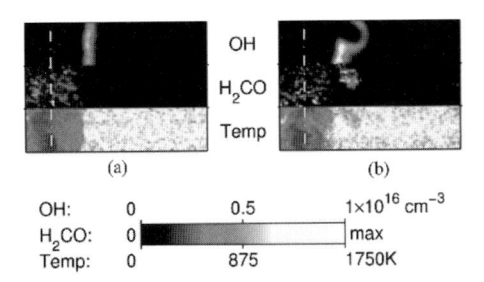

Figure 6.13 Illustration of the OH-thinning from [54].

The turbulence–flame interaction in a JHC configuration was investigated in [56, 57] using simultaneous PIV and OH–PLIF measurements. It was reported that the flow could be divided into a turbulent jet-influenced region and a quasi-laminar coflow-influenced region. The separation between these two regions was sharp and similar to a viscous superlayer in a turbulent boundary layer. It was also observed that most reactions were happening in the quasi-laminar region. A further correlation between region of low vorticity/strain and strong OH signals was observed. It was suggested that this behavior was due to the strong role of mixture fraction on the chemistry. Indeed, for the chosen mixture, autoignition phenomena and most chemical reactions occurred at a relatively small mixture fraction, which thus yielded a flame in the lean part of the mixing layer, which was in the quasi-laminar coflow region.

6.5 DNS of MILD Combustion

6.5.1 Reaction Zone Shape and Structure

Several investigations performed in past studies using advanced laser diagnostics described in Section 6.4 suggested that combustion under MILD conditions may belong to a "distributed" combustion regime in the turbulent combustion regime diagram (see Figs. 2.1 and 2.2). Further investigations are required to understand various other aspects, such as reaction zone dynamics and their role in the distributed combubstion or "flameless" aspects of MILD combustion. It is not easy to gather this information using just PLIF techniques because of their various limitations, specifically for MILD combustion applications [72]. Also, these measurements rely on an indirect measure for heat release rate based on OH and CH$_2$O PLIF signals, since a direct measurement of the heat release rate is difficult with the current measurement techniques. On the other hand, direct numerical simulations can provide complementary information required to improve our understanding of MILD combustion although the scale for DNS is generally small (of the order of few cubic centimeters), which is because of the computational resource requirements for DNS; see Section 1.6 of Chapter 1 and Section 2.2 of Chapter 2. As discussed in those sections, DNS solves all the conservation laws and transport equations for mass, momentum, energy, and species concentrations, without using closure models except for the chemical kinetics, which is represented using the Arrhenius rate expression. Thus, all relevant quantities are available for analyses and the turbulence–flame interaction phenomena are captured inherently. Currently, only two groups have performed DNS of MILD combustion to investigate different aspects of this combustion, although there are several other DNS of turbulent combustion [69, 73, 74] with diluted or highly preheated mixtures showing features similar to MILD combustion, and therefore these simulations would be useful for MILD combustion research.

The ignition in a temporally evolving turbulent mixing layer between fuel and a hot coflow is investigated using 2D [76] and 3D [75] DNS. The objective is to accurately simulate the experiments in the JHC [50] but in a smaller domain amenable for DNS. The numerical configuration consists of a rectangular domain of 20×20 mm^2 for the 2D and $10 \times 20 \times 5$ mm^3 for the 3D cases. It is noted that, given the reported Reynolds number based on Taylor microscale and Kolmogorov length scale, the turbulent integral length scale can be estimated to be about 8 mm, suggesting that the domain lengths (10 mm and 5 mm) in the periodic directions are insufficient to capture the entire turbulent kinetic energy cascade in the 3D case. These domains contain fuel in the middle surrounded by hot and diluted oxidizer. These fuel and oxidizer layers flow in opposite streamwise directions to mimic the shear and mixing layers between the fuel and coflowing streams in the experiments of Dally et al. [50]. The results showed autoignition occurring along the most reactive mixture fraction (Z_{MR}) isosurface with delays depending on the local temperature and scalar dissipation rate on the isosurface. Ignition kernels developing into propagating flames were not observed. Molecular diffusion played an important role, and its effect on the overall ignition process was assessed by comparing various one-dimensional laminar mixing layers computed without initial fluid velocity but with different Lewis numbers. The results showed that a constant non-unity Lewis number approach predicted ignition delay times close to those given by methods with detailed treatments for diffusion velocities, but unity Lewis number approximations yielded 10 times longer ignition delay times. These results signify the importance of thermal and mass diffusion processes on the ignition of MILD mixtures. The 3D turbulence encompasses small-scale structures that induce more variations in the species distribution as shown in Fig. 6.14. These

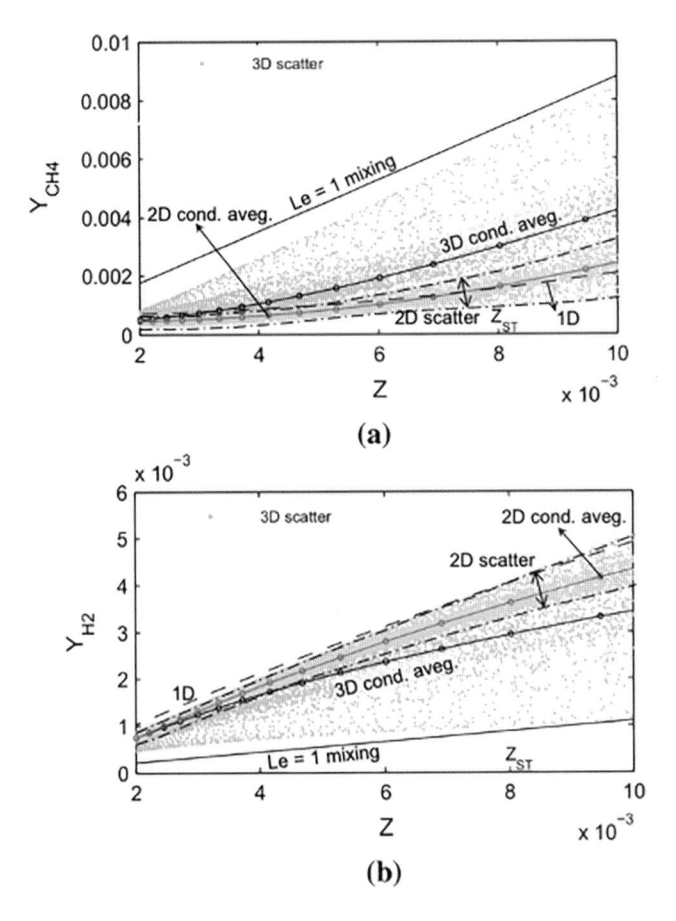

Figure 6.14 Distribution of methane (a) and hydrogen (b) around a stoichiometric mixture fraction. Dots represent scatters and lines with symbols represent averages conditioned on mixture fraction. These DNS results are shown for 0.8 ms from 3D (black) and 2D (red) cases. The mixing layer solutions for unity Lewis number (solid lines) and constant nonunity Lewis numbers (dashed lines) from [75] are also shown.

variations cause differences in the ignition delay times, which can lead to subsequent variations in the turbulence–flame interaction in 3D compared to 2D cases. Thus, the DNS of MILD combustion should be performed with 3D turbulence.

The discussion in Section 6.2 suggests that the initial and boundary conditions for simulating MILD combustion are not quite well determined and this becomes more challenging for non-premixed MILD combustion. The DNS studies [75, 76] discussed earlier mimic the initial conditions of past experiments [50], but it is unclear where these conditions would lie in the combustion mode diagram shown in Fig. 6.2. The spatial variations of temperature in the 3D DNS [75] are shown in Fig. 6.15 along with a direct photograph of the respective, HM1, experiment. Sharp gradients, denoting thin fronts, can be seen in both images, which could be because of the initial and boundary conditions used for both experiments and simulations. Indeed, the appearance of thin sheet-like reaction zones in these images contradicts the views gathered from laser thermometry measurements in [13, 17, 25, 49, 71] showing distributed/patchy temperature fields; see the discussion in Section 6.4.2.

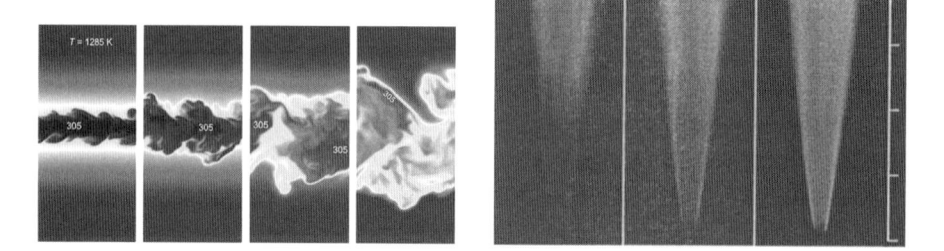

Figure 6.15 The four frames on the left show the spatial temperature variation from the 3D MILD combustion DNS of the flame HM1 in a mixing layer configuration respectively for $t = 1, 3, 8$, and 12 ms. Temperature ranges from 305 K (dark blue) to 1285 K (red) [75]. Direct photograph of flames, HM1, HM2, and HM3 (from left to right) from experiments of [50] are shown in the right half.

The comparison of the DNS results with the corresponding experiment HM1 in Fig. 6.15 reveals an interesting observation. The lower part closer to the jet nozzle in the experimental flame is invisible, suggesting "flameless" combustion. This is because the oxygen concentration is low in those regions. In contrast, the DNS results show a large temperature gradient in those regions (earlier stage). This difference could be due to (1) the difference in Reynolds number and (2) the mixing of oxidizer and fuel generally takes longer time in a slot jet (used in the DNS) than round jet (used in the experiments). Hence, the turbulent mixing prior to ignition could change mixture conditions (see Fig. 6.2) and thus the ignition locations. So, the comparison shown in Fig. 6.15 suggests that the balance between turbulent mixing and flame or ignition time scales is important to achieve MILD reaction zones observed in experiments, even if the mixture's thermochemical conditions are matched.

The importance of the ratio between the two, turbulent mixing and thermochemical, time scales is recognized in the other DNS studies [9, 16, 77–80] by constructing the initial and boundary conditions very carefully. This series of studies considered premixed MILD combustion with internal recirculation of exhaust gases as in a furnace-like configuration [13]. Since the process of exhaust gas recirculation (EGR) has a relatively large time scale compared to that plausible for typical DNS, these direct simulations are split into two stages. The first stage considers the initial mixing between fresh premixed and hot exhaust gases, where time scales for mixing and ignition are taken into account, and the second stage involves the combustion of the diluted and preheated reactants. The DNS results are analyzed in great detail and many important insights obtained on the MILD combustion physics are available in [9, 16, 77–80].

A visual examination [16] suggested that there are regions with relatively strong chemical activities giving thin and sheet-like reaction zones similar to those of

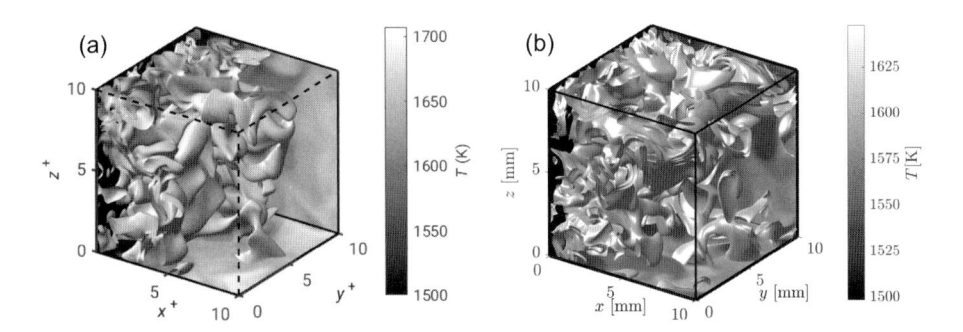

Figure 6.16 Isosurface of normalized heat release rate (gray) and temperature field shown in the bottom and side surfaces from the DNS of (a) premixed [77] and (b) non-premixed [10] MILD combustion. The heat release rate is normalized using $\rho_r S_L c_p \Delta T / \delta_{th}$, where the S_L and δ_{th} are propagation speed and thermal thickness of an MIFE (MILD flame element) having an equivalence ratio of 0.8, which corresponds to the global equivalence ratios used in the DNSs.

the classical premixed flamelets discussed in Chapter 3. However, compared to conventional turbulent premixed flames these reaction zones are highly convoluted and distributed over a much larger portion of the computation domain as shown in Fig. 6.16a. These reaction zones interact frequently with each other, leading to an apparent thickening and volumetric reaction zones. One gets the so-called flameless combustion when these reaction zones are averaged over an appropriate time scale as shown in Fig. 6.1. Furthermore, these interactions increase in frequency with an increase in the turbulence or dilution levels. Also, the scalar gradients in the tangential directions of these reaction zones are observed to be comparable to the normal gradient [16], which further confirms non-flamelet like behavior for MILD reaction zones. As discussed in Chapter 3, the tangential gradients must be much smaller than the normal gradient for flamelet combustion. Furthermore, the PDF of reaction progress variable based on temperature is observed to be non-bimodal, which again confirms the non-flamelet behavior on an average sense for MILD combustion [9, 78].

The non-flamelet behavior is because of strong interplay among flame-like structures, autoignition regions, and frequent interactions of flames. All of these lead to reactions distributed over a region rather than on a surface, as for the conventional combustion. This feature showing pancake-like reaction zones are demonstrated clearly using a novel analysis involving Minkowski functionals (MFs), which is described briefly below and full details can be found in [78]. A given three-dimensional object with an enclosed volume $\mathcal{V}$ bounded by Surface S with surface area S will have 4 MFs given by [81]

$$\mathcal{F}_0 = \mathcal{V}, \quad \mathcal{F}_1 = \frac{S}{6}, \quad \mathcal{F}_2 = \frac{1}{3\pi} \int_S \frac{\kappa_1 + \kappa_2}{2} dS, \quad \mathcal{F}_3 = \frac{1}{2\pi} \int_S \kappa_1 \kappa_2 \, dS,$$

where κ_1 and κ_2 ($\kappa_1 \geq \kappa_2$) are the two principal curvatures at a given point on S and these functionals are Galilean invariant morphological properties of the object. The object's topological invariant is related to the fourth MF $\mathcal{F}_3$, which is the Euler characteristics, χ, of the object. Hence $\mathcal{F}_3 = \chi$ and this functional is related to the three-dimensional genus $G = 1 - 0.5\chi$, which is defined as the number of cuts that can be made along a simple curve on the object without splitting it. For an object made by a single closed surface $G = 0$ and thus $\mathcal{F}_3 = 2$. If there is a hole on the object, say for example a torus, then $G = 1$ and thus $\mathcal{F}_3 = 0$. For a pretzel $G = 2$ and thus $\mathcal{F}_3 = -2$. Hence, $\mathcal{F}_3$ is related to the number of holes, N_{hole}, on the object through $\mathcal{F}_3 \approx 1 - N_{\text{hole}}$. Here, the reaction zones educed from the DNS data are the objects. Now, three characteristic length scales for the reaction zones can be defined as

$$\text{Thickness, } T \equiv \frac{\mathcal{F}_0}{2\mathcal{F}_1}; \quad \text{Width, } W \equiv \frac{2\mathcal{F}_1}{\pi \mathcal{F}_2}; \quad \text{Length, } L \equiv \frac{3\mathcal{F}_3}{4(G+1)}$$

and these quantities, ordered as $T < W < L$, are only representative scales and do not give exact dimensions in the three directions except for a spherical object. For the special case of a sphere with radius r one can verify that $T = W = L = r$ (note that $G = 0$ for a sphere made of a single closed surface). Two dimensionless *shape finders* can be defined using the above three length scales as [82, 83]

$$P = \frac{W - T}{W + T}, \quad \text{and} \quad F = \frac{L - W}{L + W},$$

which are called as *planarity* and *filamentarity*. A large value of P with small F suggests sheet-like objects whereas the opposite denotes a long tube-like objects. Small values for both P and F denote blobs. Hence, a map of shape finder can be constructed using P and F and this map will be discussed latter. These shape finders are used in various areas of physics, as discussed by Leung et al. for noncombusting [84] and by Minamoto et al. for combusting flows [9].

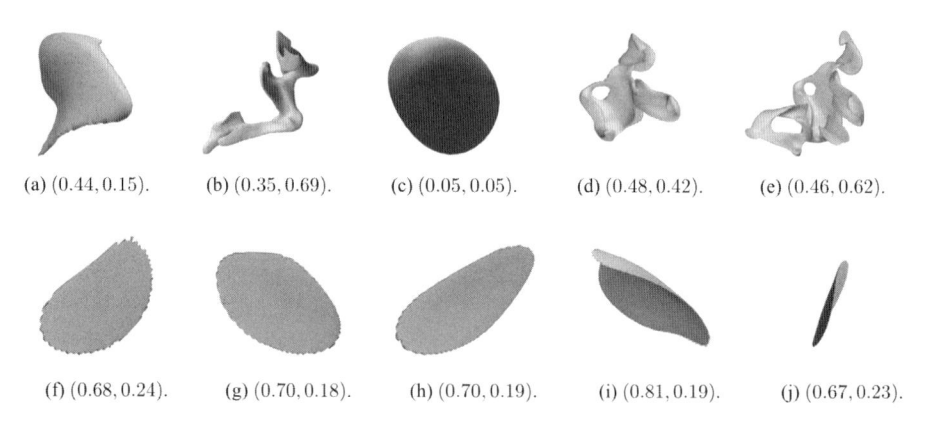

(a) (0.44, 0.15).   (b) (0.35, 0.69).   (c) (0.05, 0.05).   (d) (0.48, 0.42).   (e) (0.46, 0.62).

(f) (0.68, 0.24).   (g) (0.70, 0.18).   (h) (0.70, 0.19).   (i) (0.81, 0.19).   (j) (0.67, 0.23).

Figure 6.17 Typical reaction zones extracted from turbulent premixed MILD (top row) and conventional (bottom row) combustion cases directly simulated by Minamoto [85]. The corresponding values of P and F are also shown.

Typical reaction zones extracted from the DNSs of Minamoto [85] are shown in Fig. 6.17 along with the corresponding values of P and F. These structures are extracted using thresholds as explained in [9] for turbulent premixed combustion under both MILD (top row) and conventional (bottom row) conditions. A statistical view can be gathered by plotting P and F together for all the reaction zones in the

DNS domain, which is shown in Fig. 6.18 along with a map of the shape finder. The MILD combustion results are shown for a reference case (○), another case (△) with only turbulence level (u'/S_L) lowered by about 40%, and a third case (□) which is exactly the same as the reference case but with the mean oxygen mole fraction reduced to 2.5% from 3.5%. The conventional premixed combustion results show that the reaction zones are sheet-like while the MILD reaction zones have varied morphology ranging from blobs to pancakes, which is unique to MILD combustion. Further elaborate detail can be found in [9, 85]. Similar behavior is also observed for turbulent MILD combustion under non-premixed conditions [86].

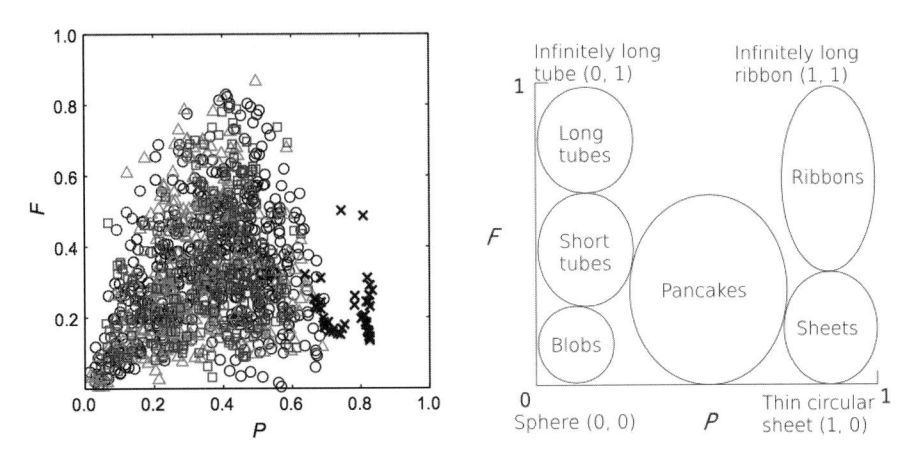

Figure 6.18 Scatterplots of P and F for reaction zones in turbulent premixed combustion under MILD (○, □, and △) and conventional (×) conditions directly simulated by Minamoto [85]. A map of the shape finder is shown on the right.

6.5.2 Are There Flames?

The MF analysis helps us to recognize that the most probable shape of the MILD reaction zones is pancake-like and not sheet-like as in the conventional combustion case, but it does not allow one to see if these reaction zones are flames or autoignition or a combination of these two phenomena. Some insights can be gathered using the balance of various terms appearing in the transport equation for c given by

$$\underbrace{\frac{\partial \rho c}{\partial t} + \frac{\partial \rho u_i c}{\partial x_i}}_{\mathscr{C}:\text{ convection}} = \underbrace{\frac{\partial}{\partial x_j}\left(\rho D_c \frac{\partial c}{\partial x_j}\right)}_{\mathscr{D}:\text{ diffusion}} + \underbrace{\dot{\omega}_c}_{\mathscr{R}:\text{ reaction}}. \quad (6.15)$$

The balance among the convective, diffusive, and reactive processes may be assessed by studying $\mathscr{B}$:

$$\mathscr{B} \equiv |\mathscr{C} - \mathscr{D}| - |\mathscr{R}|. \quad (6.16)$$

Hence $\mathscr{B} < 0$ signifies a reaction-dominated region, $\mathscr{B} > 0$ implies that the region is dominated by convection–diffusion balance, and $\mathscr{B} = 0$ suggests that there is a balance among convection, diffusion, and reaction like in a steady flame. The local dominance of flame or ignition (reaction) is studied using the spatial variation of $\mathscr{B}$ along with normalized reaction rate of c shown in Fig. 6.19 for the most diluted case investigated in [78, 85]. The shape finders (morphologies) of these reaction zones in this case are shown in Fig. 6.18 using squares. The isolines of $\mathscr{B}^+ = 1.5$ (the superscript + denotes that $\mathscr{B}$ is normalized using $\rho_r S_L / \delta_{th}$) and -1.5 depict the spatial variation of $\mathscr{B}^+$. The reaction-dominated regions having $\mathscr{B}^+ < 0$ are marked using black lines and one of these regions is marked using a black box with solid lines, which is enlarged for clarity on the side (see the bottom right frame in Fig. 6.19). The propagating-flame dominated regions with $\mathscr{B}^+ > 0$ are marked using white lines and one such region is marked by a white box, which is also enlarged for clear exposition of $\mathscr{B}^+ > 1.5$, with intense reactions signifying the presence of propagating flames. There are many reaction zones having positive $\mathscr{B}^+$ for some parts and negative $\mathscr{B}^+$ for the rest. Four such regions are marked using black boxes with dashed lines. Also, there are regions with entangled attributes of ignition and propagating flame. The central region enlarged on the side is a good example of such a region with both ignition and propagating-flame phenomena occurring close by. Hence, the ignition and flames coexist in turbulent premixed MILD combustion.

The $\mathscr{B}$ analysis for non-premixed MILD combustion was also conducted in an earlier study using a species mass fraction transport equation [87]. Lagrangian tracking of fluid elements along with this $\mathscr{B}$ analysis for non-premixed MILD combustion showed that a lean mixture ignites first in the upstream and then develops

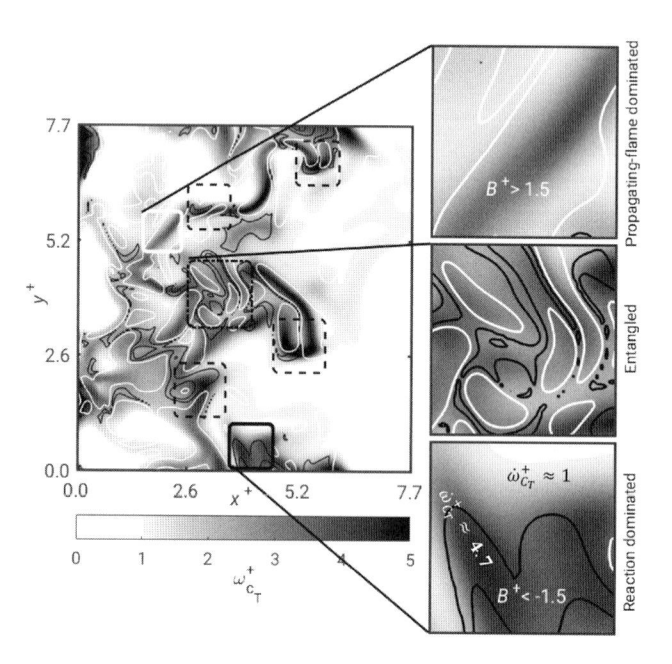

Figure 6.19 Typical contours of normalized reaction rate, $\dot{\omega}_c^+ = \rho_r S_L \dot{\omega}_c / \delta_{th}$, is shown using a pseudo-color map along with $\mathscr{B}^+ = 1.5$ (white lines) and $\mathscr{B}^+ = -1.5$ (black lines) in the mid x–y plane of the DNS domain shown in Fig. 6.16a. The results are taken from [78, 85].

into flame-like structures subsequently in downstream locations. The favored mixture fraction for ignition is possibly related to the so-called "most reactive mixture fraction" [88]. The rich mixture with larger ignition delay time burns with a flame-like structure because of the reaction fronts of lean mixtures that have ignited already and the ignition fronts propagating toward the richer mixtures. The length scales of the incoming mixture fraction and progress variable fields play an important role. A smaller length scale for the mixture fraction field, implying a higher stratification, yields a higher probability to have sequential ignition rather than an evolution toward a propagating flame. Hence, the combustion under MILD conditions is not just an autoignition process as has been commonly perceived and involves autoignition, both rich and lean premixed flames, interacting flames, and a close (shorter than the induction time or length) existence of ignition and flame phenomena. Further elaborate details can be found in [87].

6.5.3 Comments on Markers

Laser diagnostics of MILD combustion discussed in Section 6.4.2 predominantly employed OH–PLIF, which led to differing views on the nature of MILD reaction zones. For example, the OH–PLIF images discussed in Fig. 6.11 showed thin regions of OH with a clear peak and strong gradients, which were also observed in [13, 17, 49]. These regions marked using the OH–PLIF did not correspond to regions with large heat release rate identified using OH* and thus other markers may be required for MILD combustion as elaborated in Section 6.4.2. Minamoto and Swaminathan [16] concluded by analyzing PLIF images synthesized using DNS data of premixed MILD combustion that markers composed of at least two scalars, for example $OH \times CH_2O$, are required for MILD combustion and OH alone is inadequate. This is also confirmed in another DNS study [89].

In the classical flames, OH is generally large near the reaction zones and thus it serves as a good marker. Also, the gradient of OH is relatively a better marker for reaction zones rather than the OH level itself. All of this becomes less clear for MILD combustion because it includes vitiated (recirculated) gases containing a substantial level of minor species such as OH. Furthermore, the local production of OH in MILD reaction zones is likely to be small because of a small temperature rise (few tens of Kelvin) across these zones. Thus, one must be cautious in using OH-PLIF for MILD combustion studies. This is investigated in Fig. 6.20 using DNS data of non-premixed MILD combustion of Doan et al. [10]. This figure shows the PDF of $\Delta Y_{OH} = Y_{OH}^R - Y_{OH}^{CD}$, where Y_{OH}^R is the local OH mass fraction and Y_{OH}^{CD} is its value when there is no reaction (implying that there is only convection and diffusion) and thus it represents the local OH value that would have come from the incoming vitiated (recirculated) gases. Hence, ΔY_{OH} represents the effect of the MILD reaction on the local OH mass fraction. If $\Delta Y_{OH} > 0$ then there is local production of OH by MILD reaction and vice versa. The 3D field of $\Delta Y_{OH} > 0$ can be used to construct a PDF conditioned on the heat release rate, $\dot{Q}$ as shown in Fig. 6.20. This result shows that the most probable value of ΔY_{OH} is about 1×10^{-4}, which is nearly 1/10th of the most

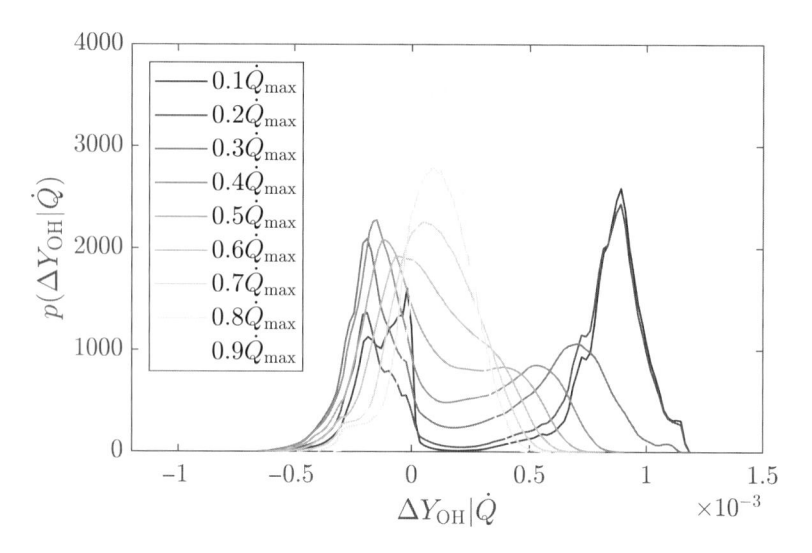

Figure 6.20 PDF of ΔY_{OH} constructed using DNS data of non-premixed MILD combustion [10]. The time of the DNS snapshot chosen for this analysis is arbitrary.

probable largest value observed in the low heat releasing regions. The right peak for the low heat release rate corresponds to the burnt mixtures while the left peak is for the mixtures beginning to react. This behavior is similar for other cases with more dilution investigated in [10]. These results point out that LIF imaging for MILD combustion needs more care and closer attention than that required for conventional combustion. The chemiluminescent species such as OH* is observed to work well to mark MILD reaction zones [72] but fine (small scale) features of the MILD reaction zone are hard to discern using chemiluminescence because it uses a line of sight technique (the information is averaged along the line of sight). Further careful exploration is required to identify good markers for heat release rate in MILD combustion.

6.6 Modeling of MILD Combustion

Turbulent combustion, in general, involves highly nonlinear turbulence–chemistry interactions (TCIs), which introduce challenges for modeling the mean or filtered reaction rates. These challenges and various approaches developed to overcome them are discussed in Chapters 3 and 4. Many of those approaches have also been used for simulations of MILD combustion using Reynolds-averaged Navier–Stokes (RANS) and large-eddy simulation (LES) paradigms and those studies are reviewed briefly below. It is worth remembering the challenging aspects of MILD combustion physics highlighted in previous sections of this chapter and this would be helpful to appreciate the limitations of some of the modeling approaches despite showing a good comparison with measurements in past studies.

6.6.1 RANS Calculations

The averaged conservation equations for mass, momentum, energy, and one or a few key scalars to represent combustion are solved as discussed in Chapter 1. For non-premixed combustion, a Favre-averaged mixture fraction and its variance are commonly used and premixed systems require the Favre-averaged reaction progress variable and its variance. All of these four scalars and also the covariance, $\widetilde{Z''c''}$, are required for MILD combustion because chemical reactions and mixing of fuel with vitiated/diluted air occur together. Nevertheless, it is helpful to have some perspective of past studies and so they are discussed briefly below based on the modeling approach used to obtain the mean reaction rate.

6.6.1.1 Flamelet-Based Approach

The flamelets approach discussed elaborately in Chapter 3 assumes that the chemical time scales are shorter than the Kolmogorov time scale, which also implies that reactions occur in thin reaction layers typically smaller than the Kolmogorov length scale. Thus, the turbulence does not affect the structure of the laminar reaction zones embedded in the flow. This approach for MILD combustion is generally questionable as both experiments [50] and laminar calculations [39] have shown a broadening of the reaction zones in MILD combustion, further suggesting that reaction zones could have typical length scales similar to those of the turbulence field. The past numerical studies [18, 49, 90–92] did not show reasonable agreement between measured and computed temperature and species variations; specifically the peak temperature and pollutant mass fractions were noted to be overestimated. This results provide some support to question the applicability of the flamelets to MILD combustion. Also, the various insights on the MILD reaction zones shape and structure discussed in Section 6.5 raise many questions for this approach. However, the a priori tests conducted using DNS data in [79, 85] suggest that the flamelet-based approach could work well if the MILD flame elements (MIFEs) are constructed carefully. Major species mass fractions and temperature computed using MIFEs compare quite well with DNS results but the minor species are predicted poorly. It is also suggested that the interaction of reaction zones, which is observed often in MILD combustion, needs to be included in the MIFE approach for its improved performance, and the methodology to include the interaction is yet to be explored.

6.6.1.2 Eddy Dissipation Concept Model

This model assumes that reactions occur in regions of turbulent kinetic energy dissipation. Since this dissipation is through a viscous mechanism it is implied that the reactions occur in small scales, which are called fine structures [93]. These regions are then taken to be perfectly stirred reactors (PSRs) and the residence time in these reactors and their volume fractions (of the fine structures) are estimated using a turbulent kinetic energy cascade model.

The EDC model was shown to be an improvement over flamelets [91, 92] and used in many past MILD combustion calculations [18, 94–100]. A reasonable agreement for major species and temperature was observed, which was attributed to the ability of this model to include detailed chemistry [22, 96]. Indeed, MILD combustion chemistry is complex, involving competing chemical pathways as discussed in Section 6.3, and thus the model should be able to accommodate these for MILD combustion [97]. Furthermore, the EDC model yielded improved results if the residence times in the fine structures or their volume fractions were increased by tuning the model parameters [94, 95]. This approach was explored further through a parametric study to find optimum values for the model parameters [98, 101] or by relating these parameters to the local Reynolds and Damköhler numbers [102]. These studies showed good improvement for major species and temperature but the minor species such as OH were overestimated. Large discrepancies with measurements were observed when the dilution level is large. This suggests that RANS-EDC works if the model parameters are selected carefully on a case-to-case basis and hence it is not robust. Further improvements to this model were proposed through using the partially stirred reactor (PaSR) or plug flow reactor (PFR), instead of PSR, equations [103] to obtain reaction rates in the fine structure or by introducing an ad hoc approach involving the mixing time scale in the fine structures to capture local extinction [104] despite the possibility for local extinction is remote for MILD combustion involving hot gases with radicals (diluents). Both of these approaches showed some improvements.

6.6.1.3 Conditional Moment Closure

This approach assumes that the fluctuations of reactive scalars are controlled by fluctuations of a key quantity: the mixture fraction in non-premixed systems or progress variable in premixed systems. Thus, it uses conditional moments of reactive scalars and does not assume whether the chemical scales are smaller or larger than the relevant turbulence scales. Thus, it can be used for a wide range of combustion regimes without parameter tuning [105]. Elaborate details of this approach are discussed in Section 3.4 of Chapter 3.

The mass fractions and temperature obtained using this approach [106] for non-premixed jet flames in diluted hot coflow [50] compared well with the measurements, but the NOx concentration was underestimated. The conditional source term estimation (CSE) approach shares many features with CMC but the conditional averages are obtained by inverting an integral equation in CSE while they are transported in CMC. The CSE approach was also used to simulate MILD combustion [107] in the Delft-JHC burner [55, 56] and a small nonadiabatic furnace operating under MILD conditions [108]. By treating the D-JHC burner as a three-stream mixing problem, two conditioning variables, the standard mixture fraction and another one based on oxygen elemental mass fraction, were used. Some satisfactory comparisons between computed and measured statistics were shown but there is a large room for improvement.

6.6.1.4 Transported PDF Approach

The transport equation for the composition joint PDF is solved along with some closure models for the small scale (micro-) mixing [109]. The reaction rate is closed as discussed earlier in Chapter 3. The mean of any thermochemical quantities can be deduced using the joint PDF. This model was used in various studies of MILD combustion [98, 110–114]. Generally, good agreements with measurements were obtained for mean temperature and major species mass fractions. This approach was shown to yield better results compared to the EDC model [110]. Also, the standard micromixing models, the Eulerian minimum spanning tree (EMST) and interaction-by-exchange-with-mean (IEM), were shown to be adequate for MILD combustion [112]. However, the peak temperature and CO were overestimated for the lean regions [114]. Hence, further development and improvement of mixing models are imperative.

6.6.1.5 Tabulated Chemistry Approach

The PSR providing the reaction rate in fine structures of the EDC model can be used to build a look-up table containing mean reaction rate required for the RANS calculation. This approach has resulted from the a priori analysis in [79, 80], demonstrating that the reaction rate variations in premixed MILD combustion are represented quite well by a PSR whose state is governed by the local Favre-averaged progress variable, $\tilde{c}$, and its variance, σ_c^2. Thus, the look-up table will have these two variables as the controlling variable. Zhen et al. [115] extended that approach to simulate the burner in [50], with the look-up table having $\tilde{c}$, σ_c^2, $\tilde{Z}$, σ_Z^2 and the covariance, σ_{cZ}, as the controlling variable. Now, the mean reaction rate is given by

$$\overline{\dot{\omega}_c} = \int_0^1 \int_0^1 \dot{\omega}_c(\xi, \zeta) \, P(\xi, \zeta) \, d\xi \, d\zeta,$$

where ξ and ζ are the sample space variables for the mixture fraction and progress variable. The joint PDF is modeled through the copula method using the marginal PDFs of c and Z, and the covariance. This approach is based on the FlaRe framework described in Section 3.5 of Chapter 3. The PSR calculations supply $\dot{\omega}_c(\xi, \zeta)$ in the above equation. The results of this approach are similar to those from the transported PDF approach [114] but with a substantially lower computational cost. The above approach is also extended to include a heat loss effect [116] in the confined cyclonic experimental burner investigated in [117]. Computed results compared well with measurements. Further details on this approach and the comparisons can be found in [115] and [116].

Göktolga et al. [118] proposed a tabulated chemistry approach called Multistage FGM to include autoignition and flame propagation by using two reaction progress variables: one for the autoignition precursor reaction and another one for the overall oxidation. A priori assessment of this model results using 2D mixing layer DNS results shows reasonable behavior for heat release and species production. However, its performance in comparison to measurements is yet to be evaluated.

6.6.1.6 Overall Comparison

From the various works cited earlier, it becomes clear that the classical flamelets model is inadequate for MILD combustion simulations since the basic assumption of this model is not obeyed in MILD combustion. Furthermore, the EDC model requires adjustments to reflect the effects of reactions distributed over larger volumes under MILD conditions in order to get good comparisons with measurements [91]. Both CMC and transported PDF methods can provide reasonable agreements but at a much larger computational cost. The tabulated chemistry approach employing PSR [115] seems to have good potential for MILD combustion, as it provides good accuracy (agreements with measurements) at a lower computational cost.

6.6.2 LES

The background of LES is described in Section 1.7. As discussed there, this approach resolves large dynamic scales of turbulence along with models to represent the effects of subgrid scale processes. Combustion typically occurs in subgrid scales and thus it needs to be modeled. Generally, the LES is expected to yield improved results. There are only a few LES computations of MILD combustion and they have used various SGS combustion closures. The experimental cases used to validate these simulations are either D-JHC [55, 56] or A-JHC [50] burners. The tabulated chemistry approach with a homogeneous reactor and Eulerian stochastic fields method is used by Kulkarni and Polifke [119] in their LES of one condition of the D-JHC burner and observed that lift-off height is captured quite satisfactorily but the temperature is overestimated. The LES with transported PDF model, as SGS combustion closure yielded early ignition leading to temperature overprediction for the D-JHC cases; however, the trend of decreasing lift-off height, L_f, as the jet Reynolds number drops was captured quite well [120]. A severe underprediction of L_f with overpredictions of both mean and rms temperature was reported when CSE was used as SGS combustion closure to simulate D-JHC cases [121]. Furthermore, the CSE approach predicted a trend for L_f variation with the Reynolds number opposite to that observed in experiments. The influences of preferential diffusion effects in the D-JHC burner were also investigated using the flame-generated manifold (FGM) approach but no comparisons of measured and computed statistics were shown [122]. From the above discussion, it is quite clear that these SGS models are not quite satisfactory for MILD combustion and further development is required. However, good comparisons of temperature and species mass fraction statistics were shown for A-JHC cases using the flamelet progress variable (FPV) approach by treating the A-JHC burner as a three-stream mixing and reacting problem [123, 124]. The FPV approach is a tabulated chemistry approach using a counterflow non-premixed flame as the canonical configuration. This approach, similar to FlaRe described in Section 3.5, uses one mixture fraction and a reaction progress variable, and its extension to a three-stream problem involves an additional mixture fraction as is used for CSE approach also [121].

The spray combustion under MILD conditions in the D-JHC burner was also investigated using experiments [125] and LES [126, 127]. The LES used FGM based on an autoigniting mixing layer as SGS closure, and reasonable agreements for the droplet distribution and temperature variations were obtained. However, a significant underestimate of L_f was observed, for example, an underestimate of the lift-off height by about 2 mm (9.5 D_o, where the liquid injector orifice diameter is D_o) in the LES observed for $(T_{cf}, \overline{X_{O2,cf}}\%) = (1600, 6.89)$ changes to about 33 D_o for (1350, 9.3). The hot coflow temperature and its oxygen mole fraction are denoted using T_{cf} and $\overline{X_{O2,cf}}$. While these LES studies have provided some insights into the performance of the various SGS closures, the physical insights gathered from the DNS discussed in Section 6.5 highlight some important questions on the applicability of these models to MILD combustion simulations. Further studies are required.

6.7 Potential Applications and Future Outlook

Highly preheated and diluted combustion modes were tested in small-scale industrial burner setups before they were named "MILD" or "flameless" in the late 1990s. These combustion modes were applied first in small-scale furnaces in Japan, the USA, and Europe, and those are reviewed in [4].

The application of MILD combustion to gas turbine engines is more complex. The issues and challenges are discussed in [28] along with a conceptual design of an MILD combustor for gas turbine applications. A staged or secondary injection of fuel into vitiated hot gas coming from an upstream burner is used in many low NOx ground-based gas turbine engines [128, 129]. Some gas turbine applications using an MILD combustion mode partially are reviewed in [5]. Hayashi and Mizobuchi reviewed the techniques that could be employed to control combustion and emissions by utilizing hot burned gases [130]. Further experimental and numerical investigations to explore the physical and chemical aspects of turbulent combustion under MILD conditions in practical configurations are required for full exploitation of this combustion mode.

The Cross-ministerial Strategic Innovation Promotion Program (SIP) in Japan involving both automobile industries and universities is leading R&D activities for internal combustion (IC) engines utilizing thermochemical conditions characterized by high compression ratio, low equivalence ratio, and high EGR rate for compression and spark ignition engines, to the limit of homogeneous charge compression ignition conditions [131]. Understanding the physics and chemistry of MILD combustion thoroughly is key for developing robust predictive CFD models and tools for these applications and it is one of the main objectives of the SIP activities.

Ignition of diluted mixtures is difficult to predict and concerted research activities using DNS and advanced laser diagnostics are needed to gain further understanding. The conditions for many of the current combustion systems will shift from their conventional regimes toward MILD regimes to meet the ever-increasing environmental and pollutant emission regulations while supplying energy required for the future society [2]. This shift will bring in additional scientific and technological challenges; for example, the turbulent combustion is likely to be distributed in space and time with comparable chemical and turbulence time scales. Indeed, turbulent MILD combustion is a scientifically challenging and rich topic to explore and is also an important topic to study from the future energy perspective. The outstanding scientific challenges include:

1. Finding a robust and universal definition of MILD combustion. Indeed, earlier definitions of MILD combustion rely on simplified laminar canonical configurations such as PSR or counterflow flame and do not consider the important role played by radicals in the inception stage of MILD combustion. DNS results and their analysis showed the crucial role of radicals.
2. Understanding the role of initial/inflow mixture conditions and their subsequent mixing on MILD combustion. Specifically, would MILD combustion observed in industrial applications such as gas turbine engines with staged combustors be represented appropriately by JHC burners or DNS configurations investigated in past studies? Indeed, experiments using a JHC configuration do not show some of the features (homogeneous temperature field with small temperature or species gradients) as observed in MILD combustion studies using furnaces or DNS with EGR. Thus, there is some ambiguity as to whether the JHC is fully representative of MILD combustion configurations unless the conditions are controlled closely as has been done for some of the cases investigated using a D-JHC burner [55, 56].
3. Understanding and selecting appropriate experimental diagnostics, which are reliable to identify reaction or heat-releasing regions under MILD conditions. The MILD combustion involves large amount of exhaust gases containing many of the scalars used in laser diagnostics and so the use of conventional marker for it appears to be inadequate based on the DNS data analysis which showed markers consisting of two scalars are more appropriate for MILD combustion.
4. Finding an appropriate canonical candidate (a reactor or flame) for subgrid scale modeling of MILD combustion. DNS data showed that PSR could be used suitably [79, 80], which is also demonstrated through RANS [115, 116], though further careful investigations are needed. Previous modeling of MILD combustion has shown their limitations as described in Section 6.6.
5. Understanding the physical and chemical mechanisms of liquid fuel combustion under MILD conditions. Most of the practical fuels are liquids and when these fuels are injected into MILD mixtures there could be flash boiling. What are the physical and chemical consequences of it on the "greener" aspects of MILD combustion? How can one realistically capture these effects in a simpler modeling framework that is attractive for industry use?
6. On the practical side, what are appropriate flow arrangements with small pressure loss so that MILD combustion conditions could be achieved with minimal startup transients? This is important for gas turbine applications. The additional weight gain that may come from flow path arrangements would be an important factor to consider for aero applications. Some of the developments utilizing hot burned gases for gas turbine applications are discussed by Hayashi and Mizobuchi [130].

References

[1] D. Dunn-Rankin and P. Therkelsen, *Lean Combustion: Technology and Control*. San Diego: Academic Press, 2016.

[2] K. N. C. Bray and N. Swaminathan, "Fundamentals and challenges," in N. Swaminathan and K. N. C. Bray, Eds., *Turbulent Premixed Flames*. Cambridge: Cambridge University Press, 2011, pp. 1–40.

[3] J. A. Wünning and J. G. Wünning, "Flameless oxidation to reduce thermal no-formation," *Prog. Energy Combust. Sci.*, **23**, 81–94, 1997.

[4] M. Katsuki and T. Hasegawa, "The science and technology of combustion in highly preheated air," *Symp. Combust.*, **27**, 3135–3146, 1998.

[5] A. Cavaliere and M. de Joannon, "Mild combustion," *Prog. Energy Combust. Sci.*, **30**, 329–366, 2004.

[6] J. A. Wünning, "Flameless oxidation with highly preheated air," *Chemie Ing. Tech.*, **63**, 1243–1245, 1991.

[7] G. Woelk and J. Wünning, "Controlled combustion by flameless oxidation," in *Jt. Meet. Br. Ger. Sect. Combust. Institute, Cambridge*, 1993.

[8] M. de Joannon, A. Saponaro, and A. Cavaliere, "Zero-dimensional analysis of diluted oxidation of methane in rich conditions," *Proc. Combust. Inst.*, **28**, 1639–1646, 2000.

[9] Y. Minamoto, N. Swaminathan, R. S. Cant, and T. Leung, "Reaction zones and their structure in MILD combustion," *Combust. Sci. Technol.*, **186**, 1075–1096, 2014.

[10] N. A. K. Doan, N. Swaminathan, and Y. Minamoto, "DNS of MILD combustion with mixture fraction variations," *Combust. Flame*, **189**, 173–189, 2018.

[11] T. Fujimori, D. Riechelmann, and J. Sato, "Effect of liftoff on NOx emission of turbulent jet flame in high-temperature coflowing air," *27th Symp. Combust.*, 1149–1155, 1998.

[12] M. J. Dunn, A. R. Masri, R. W. Bilger, and R. S. Barlow, "Finite rate chemistry effects in highly sheared turbulent premixed flames," *Flow, Turbul. Combust.*, **85**, 621–648, 2010.

[13] I. B. Ozdemir and N. Peters, "Characteristics of the reaction zone in a combustor operating at mild combustion," *Exp. Fluids*, **30**, 683–695, 2001.

[14] N. Krishnamurthy, P. J. Paul, and W. Blasiak, "Studies on low-intensity oxy-fuel burner," *Proc. Combust. Inst.*, **32**, 3139–3146, 2009.

[15] J. A. M. Sidey and E. Mastorakos, "Visualization of MILD combustion from jets in cross-flow," *Proc. Combust. Inst.*, **35**, 3537–3545, 2015.

[16] Y. Minamoto and N. Swaminathan, "Scalar gradient behaviour in MILD combustion," *Combust. Flame*, **161**, 1063–1075, 2014.

[17] T. Plessing, N. Peters, and W. J. G., "Laseroptical investigation of highly preheated combustion with strong exhaust gas recirculation," *Symp. Combust.*, **27**, 3197–3204, 1998.

[18] F. Christo and B. B. Dally, "Modeling turbulent reacting jets issuing into a hot and diluted coflow," *Combust. Flame*, **142**, 117–129, 2005.

[19] I. O. Awosope, N. H. Kandamby, and F. C. Lockwood, "Flameless oxidation modelling: on application to gas turbine combustors," *J. Energy Inst.*, **79**, 75–83, 2006.

[20] C. Galletti, A. Parente, and L. Tognotti, "Numerical and experimental investigation of a mild combustion burner," *Combust. Flame*, **151**, 649–664, 2007.

[21] O. Lammel, H. Schutz, G. Schmitz, R. Luckerath, M. Stohr, B. Noll, M. Aigner, M. Hase, and W. Krebs, "FLOX® combustion at high power density and high flame temperatures," *J. Eng. Gas Turbines Power*, **132**, p. 121503, 2010.

[22] A. Parente, J. C. Sutherland, B. B. Dally, L. Tognotti, and P. Smith, "Investigation of the MILD combustion regime via principal component analysis," *Proc. Combust. Inst.*, **33**, 3333–3341, 2011.

[23] H. Mohamed, H. Bentîcha, and S. Mohamed, "Numerical modeling of the effects of fuel dilution and strain rate on reaction zone structure and NOx formation in flameless combustion," *Combust. Sci. Technol.*, **181**, 1078–1091, 2009.

[24] P. Li and J. Mi, "Influence of inlet dilution of reactants on premixed combustion in a recuperative furnace," *Flow, Turbul. Combust.*, **87**, 617–638, 2011.

[25] P. R. Medwell, P. A. M. Kalt, and B. B. Dally, "Simultaneous imaging of OH, formaldehyde, and temperature of turbulent nonpremixed jet flames in a heated and diluted coflow," *Combust. Flame*, **148**, 48–61, 2007.

[26] S. Hayashi and Y. Mizobuchi, "Utilization of hot burnt gas for better control of combustion and emissions," in *Turbulent Premixed Flames*, N. Swaminathan and K. N. C. Bray, Eds., Cambridge: Cambridge University Press, 2011, pp. 365–378.

[27] R. Lückerath, W. Meier, and M. Aigner, "FLOX combustion at high pressure with different fuel compositions," *J. Eng. Gas Turbine Power*, **130**, p. 011505, 2008.

[28] A. A. V. Perpignan, A. G. Rao, and D. J. E. M. Roekaerts, "Flameless combustion and its potential towards gas turbines," *Prog. Energy Combust. Sci.*, **69**, 28–62, 2018.

[29] M. Oberlack, R. Arlitt, and N. Peters, "On stochastic Damkohler number variations in a homogeneous flow reactor," *Combust. Theory Model.*, **4**, 495–509, 2000.

[30] M. J. Evans, P. R. Medwell, H. Wu, A. Stagni, and M. Ihme, "Classification of Non-Premixed MILD and Autoignitive Flames," *Proc. Combust. Inst.*, **36**, 4297–4304, 2017.

[31] N. A. K. Doan and N. Swaminathan, "Role of radicals on MILD combustion inception," *Proc. Combust. Inst.*, doi.org/10.1016/j.proci.2018.07.038, 2019.

[32] M. de Joannon, P. Sabia, A. Tregrossi, and A. Cavaliere, "Dynamic Behavior of Methane Oxidation in Premixed Flow Reactor," *Combust. Sci. Technol.*, **176**, 769–783, 2004.

[33] M. de Joannon, A. Cavaliere, T. Faravelli, E. Ranzi, P. Sabia, and A. Tregrossi, "Analysis of process parameters for steady operations in methane mild combustion technology," *Proc. Combust. Inst.*, **30**, 2605–2612, 2005.

[34] P. Sabia, M. de Joannon, S. Fierro, A. Tregrossi, and A. Cavaliere, "Hydrogen-enriched methane mild combustion in a well stirred reactor," *Exp. Therm. Fluid Sci.*, **31**, 469–475, 2007.

[35] P. Sabia, G. Sorrentino, A. Chinnici, A. Cavaliere, and R. Ragucci, "Dynamic behaviors in methane MILD and oxy-fuel combustion. Chemical effect of CO_2," *Energy & Fuels*, **29**, 1978–1986, 2015.

[36] J. A. M. Sidey, E. Mastorakos, and R. L. Gordon, "Simulations of autoignition and laminar premixed flames in methane/air mixtures diluted with hot products," *Combust. Sci. Technol.*, **186**, 453–465, 2014.

[37] M. de Joannon, A. Matarazzo, P. Sabia, and A. Cavaliere, "Mild combustion in homogeneous charge diffusion ignition (HCDI) regime," *Proc. Combust. Inst.*, **31**, 3409–3416, 2007.

[38] M. de Joannon, P. Sabia, G. Sorrentino, and A. Cavaliere, "Numerical study of mild combustion in hot diluted diffusion ignition (HDDI) regime," *Proc. Combust. Inst.*, **32**, 3147–3154, 2009.

[39] M. de Joannon, G. Sorrentino, and A. Cavaliere, "MILD combustion in diffusion-controlled regimes of hot diluted fuel," *Combust. Flame*, **159**, 1832–1839, 2012.

[40] M. de Joannon, P. Sabia, G. Cozzolino, G. Sorrentino, and A. Cavaliere, "Pyrolitic and oxidative structures in hot oxidant diluted oxidant (HODO) MILD combustion," *Combust. Sci. Technol.*, **184**, 1207–1218, 2012.

[41] G. Sorrentino, D. Scarpa, and A. Cavaliere, "Transient inception of MILD combustion in hot diluted diffusion ignition (HDDI) regime: A numerical study," *Proc. Combust. Inst.*, **34**, 3239–3247, 2013.

[42] J. A. M. Sidey and E. Mastorakos, "Simulations of laminar non-premixed flames of kerosene with hot combustion products as oxidiser," *Combust. Theory Model.*, **20**, 958–973, 2016.

[43] P. Sabia, M. de Joannon, A. Picarelli, and R. Ragucci, "Methane auto-ignition delay times and oxidation regimes in MILD combustion at atmospheric pressure," *Combust. Flame*, **160**, 47–55, 2013.

[44] P. Sabia, M. de Joannon, M. Lubrano Lavadera, P. Giudicianni, and R. Ragucci, "Autoignition delay times of propane mixtures under MILD conditions at atmospheric pressure," *Combust. Flame*, **161**, 3022–3030, 2014.

[45] P. Sabia, M. Lubrano Lavadera, P. Giudicianni, G. Sorrentino, R. Ragucci, and M. de Joannon, "CO_2 and H_2O effect on propane auto-ignition delay times under mild combustion operative conditions," *Combust. Flame*, **162**, 533–543, 2015.

[46] P. Sabia, M. Lubrano Lavadera, G. Sorrentino, P. Giudicianni, R. Ragucci, and M. de Joannon, "H_2O and CO_2 dilution in MILD combustion of simple hydrocarbons," *Flow, Turbul. Combust.*, **96**, 433–448, 2016.

[47] G. Szego, B. B. Dally, and G. Nathan, "Scaling of NOx emissions from a laboratory-scale mild combustion furnace," *Combust. Flame*, **154**, 281–295, 2008.

[48] A. S. Veríssimo, A. M. A. Rocha, and M. Costa, "Operational, combustion, and emission characteristics of a small-scale combustor," *Energy & Fuels*, **25**, 2469–2480, 2011.

[49] B. B. Dally, E. Riesmeier, and N. Peters, "Effect of fuel mixture on moderate and intense low oxygen dilution combustion," *Combust. Flame*, **137**, 418–431, 2004.

[50] B. B. Dally, A. N. Karpetis, and R. S. Barlow, "Structure of turbulent non-premixed jet flames in a diluted hot coflow," *Proc. Combust. Inst.*, **29**, 1147–1154, 2002.

[51] E. Abtahizadeh, A. Sepman, F. Hernández-Pérez, J. A. van Oijen, A. Mokhov, L. P. H. de Goey, and H. Levinsky, "Numerical and experimental investigations on the influence of preheating and dilution on transition of laminar coflow diffusion flames to mild combustion regime," *Combust. Flame*, **160**, 2359–2374, 2013.

[52] R. L. Gordon, A. R. Masri, and E. Mastorakos, "Simultaneous Rayleigh temperature, OH- and CH_2O-LIF imaging of methane jets in a vitiated coflow," *Combust. Flame*, **155**, 181–195, 2008.

[53] P. R. Medwell, P. A. M. Kalt, and B. B. Dally, "Imaging of diluted turbulent ethylene flames stabilized on a Jet in Hot Coflow (JHC) burner," *Combust. Flame*, **152**, 100–113, 2008.

[54] P. R. Medwell, P. A. M. Kalt, and B. B. Dally, "Reaction zone weakening effects under hot and diluted oxidant stream conditions," *Combust. Sci. Technol.*, **181**, 937–953, 2009.

[55] E. Oldenhof, M. J. Tummers, E. H. van Veen, and D. J. E. M. Roekaerts, "Ignition kernel formation and lift-off behaviour of jet-in-hot-coflow flames," *Combust. Flame*, **157**, 1167–1178, 2010.

[56] E. Oldenhof, M. J. Tummers, E. H. van Veen, and D. J. E. M. Roekaerts, "Role of entrainment in the stabilisation of jet-in-hot-coflow flames," *Combust. Flame*, **158**, 1553–1563, 2011.

[57] E. Oldenhof, M. J. Tummers, E. H. van Veen, and D. J. E. M. Roekaerts, "Conditional flow field statistics of jet-in-hot-coflow flames," *Combust. Flame*, **160**, 1428–1440, 2013.

[58] T. F. Fric and A. Roshko, "Vortical structure in the wake of a transverse jet," *J. Fluid Mech.*, **279**, 1–47, 1994.

[59] L. K. Su and M. G. Mungal, "Simultaneous measurements of scalar and velocity field evolution in turbulent crossflowing jets," *J. Fluid Mech.*, **513**, 1–45, 2004.

[60] S. Muppidi and K. Mahesh, "Direct numerical simulation of passive scalar transport in transverse jets," *J. Fluid Mech.*, **598**, 335–360, 2008.

[61] L. L. Smith, R. W. Dibble, L. Talbot, R. S. Barlow, and C. D. Carter, "Laser Raman scattering measurements of differential molecular diffusion in nonreacting turbulent jets of H_2/CO_2 mixing with air," *Phys. Fluids*, **7**, p. 1455, 1995.

[62] L. Smith, R. Dibble, L. Talbot, R. Barlow, and C. Carter, "Laser Raman scattering measurements of differential molecular diffusion in turbulent nonpremixed jet flames of H_2/CO_2 fuel," *Combust. Flame*, **100**, 153–160, 1995.

[63] R. W. Grout, A. Gruber, C. S. Yoo, and J. H. Chen, "Direct numerical simulation of flame stabilization downstream of a transverse fuel jet in cross-flow," *Proc. Combust. Inst.*, **33**, 1629–1637, 2011.

[64] R. W. Grout, A. Gruber, H. Kolla, P. T. Bremer, J. C. Bennett, A. Gyulassy, and J. H. Chen, "A direct numerical simulation study of turbulence and flame structure in transverse jets analysed in jet-trajectory based coordinates," *J. Fluid Mech.*, **706**, 351–383, 2012.

[65] H. Kolla, R. W. Grout, A. Gruber, and J. H. Chen, "Mechanisms of flame stabilization and blowout in a reacting turbulent hydrogen jet in cross-flow," *Combust. Flame*, **159**, 2755–2766, 2012.

[66] Y. Minamoto, H. Kolla, R. W. Grout, A. Gruber, and J. H. Chen, "Effect of fuel composition and differential diffusion on flame stabilization in reacting syngas jets in turbulent cross-flow," *Combust. Flame*, **162**, 3569–3579, 2015.

[67] E. Hasselbrink and M. G. Mungal, "Non-premixed methane transverse jet flames," *27th Symp. Combust.*, 1167–1173, 1998.

[68] D. Han and M. Mungal, "Stabilization in turbulent lifted deflected-jet flames," *Proc. Combust. Inst.*, **29**, 1889–1895, 2002.

[69] S. Lyra, B. Wilde, H. Kolla, J. M. Seitzman, T. C. Lieuwen, and J. H. Chen, "Structure of hydrogen-rich transverse jets in a vitiated turbulent flow," *Combust. Flame*, **162**, 1234–1248, 2015.

[70] J. Mi, P. Li, B. B. Dally, and R. A. Craig, "Importance of initial momentum rate and air-fuel premixing on moderate or intense low oxygen dilution (MILD) combustion in a recuperative furnace," *Energy & Fuels*, **23**, 5349–5356, 2009.

[71] C. Duwig, B. Li, Z. Li, and M. Aldén, "High resolution imaging of flameless and distributed turbulent combustion," *Combust. Flame*, **159**, 306–316, 2012.

[72] N. A. K. Doan and N. Swaminathan, "Analysis of markers for combustion mode and heat release in MILD combustion using DNS data," *Combust. Sci. Technol.*, **191**, 1059–1078, 2019.

[73] C. S. Yoo, E. S. Richardson, R. Sankaran, and J. H. Chen, "A DNS study on the stabilization mechanism of a turbulent lifted ethylene jet flame in highly-heated coflow," *Proc. Combust. Inst.*, **33**, 1619–1627, 2011.

[74] N. Fukushima, M. Katayama, Y. Naka, T. Oobayashi, M. Shimura, Y. Nada, M. Tanahashi, and T. Miyauchi, "Combustion regime classification of HCCI/PCCI combustion using Lagrangian fluid particle tracking," *Proc. Combust. Inst.*, **35**, 3009–3017, 2015.

[75] M. U. Göktolga, J. A. van Oijen, and L. P. H. de Goey, "3D DNS of MILD combustion: A detailed analysis of heat loss effects, preferential diffusion, and flame formation mechanisms," *Fuel*, **159**, 784–795, 2015.

[76] J. A. van Oijen, "Direct numerical simulation of autoigniting mixing layers in MILD combustion," *Proc. Combust. Inst.*, **34**, 1163–1171, 2013.

[77] Y. Minamoto, T. D. Dunstan, N. Swaminathan, and R. S. Cant, "DNS of EGR-type turbulent flame in MILD condition," *Proc. Combust. Inst.*, **34**, 3231–3238, 2013.

[78] Y. Minamoto, N. Swaminathan, R. S. Cant, and T. Leung, "Morphological and statistical features of reaction zones in MILD and premixed combustion," *Combust. Flame*, **161**, 2801–2814, 2014.

[79] Y. Minamoto and N. Swaminathan, "Modelling paradigms for MILD combustion," *Int. J. Adv. Eng. Sci. Appl. Math.*, **6**, 65–75, 2014.

[80] Y. Minamoto and N. Swaminathan, "Subgrid scale modelling for MILD combustion," *Proc. Combust. Inst.*, **35**, 3529–3536, 2015.

[81] H. Minkowski, "Volumen und oberfläche," *Mathematische Annalen*, **57**, 447–495, 1903.

[82] V. Sahni, B. S. Sathyaprakash, and S. F. Shandarin, "Shape nders: A new shape diagnostic for large-scale structure," *Astrophys. J.*, **495(1)**, L5–L8, 1998.

[83] J. V. Sheth and V. Sahni, "Exploring the geometry, topology and morphology of large scale structure using Minkowski functionals," *Curr. Sci.*, **88**, 1101–1116, 2005.

[84] T. Leung, N. Swaminathan, and P. A. Davidson, "Geometry and interaction of structures in homogeneous isotropic turbulence," *J. Fluid Mech.*, **710**, 453–481, 2012.

[85] Y. Minamoto, "Physical Aspects and Modelling of Turbulent MILD Combustion," PhD thesis, Cambridge University, Department of Engineering, 2013.

[86] N. A. K. Doan, "Physical Insights of Non-Premixed MILD Combustion Using DNS," PhD thesis, Cambridge University, Department of Engineering, 2018.

[87] N. A. K. Doan and N. Swaminathan, "Autoignition and flame propagation in non-premixed MILD combustion," *Combust. Flame*, **201**, 234–243, 2019.

[88] E. Mastorakos, "Ignition of turbulent non-premixed flames," *Prog. Energy Combust. Sci.*, **35**, 57–97, 2009.

[89] Z. M. Nikolaou and N. Swaminathan, "Heat release rate markers for premixed combustion," *Combust. Flame*, **161**, 3073–3084, 2014.

[90] P. J. Coelho and N. Peters, "Numerical simulation of a mild combustion burner," *Combust. Flame*, **124**, 503–518, 2001.

[91] S. R. Shabanian, P. R. Medwell, M. Rahimi, A. Frassoldati, and A. Cuoci, "Kinetic and fluid dynamic modeling of ethylene jet flames in diluted and heated oxidant stream combustion conditions," *Appl. Therm. Eng.*, **52**, 538–554, 2013.

[92] A. Rebola, P. J. Coelho, and M. Costa, "Assessment of the performance of several turbulence and combustion models in the numerical simulation of a flameless combustor," *Combust. Sci. Technol.*, **185**, 600–626, 2013.

[93] B. F. Magnussen, "On the structure of turbulence and a generalized eddy dissipation concept for chemical reaction in turbulent flow," in *19th Am. Inst. Aeronaut. Astronaut. Aerosp. Sci. Meet.*, 1981.

[94] J. Aminian, C. Galletti, S. Shahhosseini, and L. Tognotti, "Key modeling issues in prediction of minor species in diluted-preheated combustion conditions," *Appl. Therm. Eng.*, **31**, 3287–3300, 2011.

[95] A. De, E. Oldenhof, P. Sathiah, and D. Roekaerts, "Numerical simulation of Delft-Jet-in-Hot-Coflow (DJHC) flames using the eddy dissipation concept model for turbulence-chemistry interaction," *Flow, Turbul. Combust.*, **87**, 537–567, 2011.

[96] D. Lupant and P. Lybaert, "Assessment of the EDC combustion model in MILD conditions with in-furnace experimental data," *Appl. Therm. Eng.*, **75**, 93–102, 2015.

[97] A. Mardani, S. Tabejamaat, and S. Hassanpour, "Numerical study of CO and CO_2 formation in CH4/H2 blended flame under MILD condition," *Combust. Flame*, **160**, 1636–1649, 2013.

[98] M. J. Evans, P. R. Medwell, and Z. F. Tian, "Modeling lifted jet flames in a heated coflow using an optimized eddy dissipation concept model," *Combust. Sci. Technol.*, **187**, 1093–1109, 2015.

[99] A. Frassoldati, P. Sharma, A. Cuoci, T. Faravelli, and E. Ranzi, "Kinetic and fluid dynamics modeling of methane/hydrogen jet flames in diluted coflow," *Appl. Therm. Eng.*, **30**, 376–383, 2010.

[100] C. Galletti, a. Parente, M. Derudi, R. Rota, and L. Tognotti, "Numerical and experimental analysis of NO emissions from a lab-scale burner fed with hydrogen-enriched fuels and operating in MILD combustion," *Int. J. Hydrogen Energy*, **34**, 8339–8351, 2009.

[101] A. Mardani, "Optimization of the eddy dissipation concept (EDC) model for turbulence-chemistry interactions under hot diluted combustion of CH_4/H_2," *Fuel*, **191**, 114–129, 2017.

[102] A. Parente, M. R. Malik, F. Contino, A. Cuoci, and B. B. Dally, "Extension of the eddy dissipation concept for turbulence/chemistry interactions to MILD combustion," *Fuel*, **163**, 98–111, 2016.

[103] Z. Li, A. Cuoci, A. Sadiki, and A. Parente, "Comprehensive numerical study of the Adelaide Jet in Hot-Coflow burner by means of RANS and detailed chemistry," *Energy*, **139**, 555–570, 2017.

[104] J. Aminian, C. Galletti, and L. Tognotti, "Extended EDC local extinction model accounting finite-rate chemistry for MILD combustion," *Fuel*, **165**, 123–133, 2016.

[105] A. Y. Klimenko and R. W. Bilger, "Conditional moment closure for turbulent combustion," *Prog. Energy Combust. Sci.*, **25**, 595–687, 1999.

[106] S. H. Kim, K. Y. Huh, and B. B. Dally, "Conditional moment closure modeling of turbulent nonpremixed combustion in diluted hot coflow," *Proc. Combust. Inst.*, **30 I**, 751–757, 2005.

[107] J. Labahn, D. Dovizio, and C. Devaud, "Numerical simulation of the Delft-Jet-in-Hot-Coflow (DJHC) flame using conditional source-term estimation," *Proc. Combust. Inst.*, **35**, 3547–3555, 2015.

[108] J. W. Labahn and C. Devaud, "Species and temperature predictions in a semi-industrial MILD furnace using a non-adiabatic conditional source-term estimation formulation," *Combust. Theory Model.*, **21**, 466–486, 2017.

[109] D. C. Haworth, "Progress in probability density function methods for turbulent reacting flows," *Prog. Energy Combust. Sci.*, **36**, 168–259, 2010.

[110] F. Christo and B. B. Dally, "Application of transport PDF approach for modelling MILD combustion," in *15th Australas. Fluid Mech. Conf.*, December, 2004, 1–4.

[111] A. De, A. Dongre, and R. Yadav, "Numerical investigation of Delft-Jet-In-Coflow (DHJC) burner using probability density function (PDF) transport modeling," in *ASME Turbo Expo 2013 Turbine Tech. Conf. Expo.*, 2013.

[112] A. De and A. Dongre, "Assessment of turbulence-chemistry interaction models in MILD combustion regime," *Flow, Turbul. Combust.*, **94**, 439–478, 2015.

[113] R. L. Gordon, a. R. Masri, S. B. Pope, and G. M. Goldin, "A numerical study of auto-ignition in turbulent lifted flames issuing into a vitiated co-flow," *Combust. Theory Model.*, **11**, 351–376, 2007.

[114] J. Lee, S. Jeon, and Y. Kim, "Multi-environment probability density function approach for turbulent CH4/H2 flames under the MILD combustion condition," *Combust. Flame*, **162**, 1464–1476, 2015.

[115] Z. Chen, V. M. Reddy, S. Ruan, N. A. K. Doan, W. L. Roberts, and N. Swaminathan, "Simulation of MILD combustion using perfectly stirred reactor model," *Proc. Combust. Inst.*, **36**, 4279–4286, 2017.

[116] Z. X. Chen, N. A. K. Doan, X. J. Lv, N. Swaminathan, G. Ceriello, G. Sorrentino, and A. Cavaliere, "Numerical study of a cyclonic combustor under moderate or intense low-oxygen dilution conditions using non-adiabatic tabulated chemistry," *Energy & Fuels*," 2018.

[117] G. Sorrentino, P. Sabia, M. de Joannon, A. Cavaliere, and R. Ragucci, "The effect of diluent on the sustainability of MILD combustion in a cyclonic burner," *Flow, Turbul. Combust.*, **96**, 449–468, 2016.

[118] M. U. Göktolga, J. A. van Oijen, and L. H. de Goey, "Modeling MILD combustion using a novel multistage FGM method," *Proc. Combust. Inst.*, **36**, 4269–4277, 2017.

[119] R. M. Kulkarni and W. Polifke, "LES of Delft-Jet-In-Hot-Co fl ow (DJHC) with tabulated chemistry and stochastic fields combustion model," *Fuel Process. Technol.*, **107**, 138–146, 2013.

[120] R. Bhaya, A. De, and R. Yadav, "Large eddy simulation of mild combustion using PDF-based turbulence-chemistry interaction models," *Combust. Sci. Technol.*, **186**, 1138–1165, 2014.

[121] J. W. Labahn and C. B. Devaud, "Large eddy simulations (LES) including conditional source-term estimation (CSE) applied to two Delft-Jet-in-Hot-Coflow (DJHC) flames," *Combust. Flame*, **164**, 68–84, 2016.

[122] E. Abtahizadeh, P. de Goey, and J. van Oijen, "LES of Delft Jet-in-Hot Coflow burner to investigate the effect of preferential diffusion on autoignition of CH4/H2 flames," *Fuel*, **191**, 36–45, 2017.

[123] M. Ihme and Y. C. See, "LES flamelet modeling of a three-stream MILD combustor: Analysis of flame sensitivity to scalar inflow conditions," *Proc. Combust. Inst.*, **33**, 1309–1317, 2011.

[124] M. Ihme, J. Zhang, G. He, and B. B. Dally, "Large-eddy simulation of a jet-in-hot-coflow burner operating in the oxygen-diluted combustion regime," *Flow, Turbul. Combust.*, **89**, 449–464, 2012.

[125] H. Correia Rodrigues, M. J. Tummers, E. H. van Veen, and D. Roekaerts, "Spray flame structure in conventional and hot-diluted combustion regime," *Combust. Flame*, **162**, 759–773, 2014.

[126] L. Ma and D. Roekaerts, "Modeling of spray jet flame under MILD condition with non-adiabatic FGM and a new conditional droplet injection model," *Combust. Flame*, **165**, 402–423, 2016.

[127] L. Ma and D. Roekaerts, "Structure of spray in hot-diluted coflow flames under different coflow conditions: A numerical study," *Combust. Flame*, **172**, 20–37, 2016.

[128] F. Guethe, R. Lachner, B. Schuermans, F. Biagioli, W. Geng, A. Inauen, S. Schenker, R. Bombach, and W. Hubschmid, "Flame imaging on the ALSTOM EV-burner: thermo acoustic pulsations and CFD-validation," in *44th AIAA Aerospace Sciences Meeting and Exhibit, Aerospace Sciences Meetings, AIAA2006-437*, 2006.

[129] F. Guethe, D. Stankovic, F. Genin, K. Syed, and D. Winkler, "Flue gas recirculation of the alstom sequential gas turbine combustor tested at high pressure," in *Proceedings of ASME Turbo Expo 2011, GT2011-45379*, 2011.

[130] S. Hayashi and Y. Mizobuchi, "Utilization of hot burnt gas for better control of combustion and emissions," in N. Swaminathan and K. N. C. Bray, Eds., *Turbulent Premixed Flames*. Cambridge: Cambridge University Press, 2011, pp. 365–378.

[131] SIP "Pioneering the future: Japanese science, technology and innovation," www8.cao.go.jp/cstp/panhu/sip_english/sip_en.html, Accessed July 12, 2017.

7 Supersonic Combustion

A. Mura and V. Sabelnikov

7.1 Introduction and Background

As emphasized in the previous chapters, by far most combustion studies are devoted to low-speed flows, i.e., flows that are typified by a characteristic M that remains rather small compared with unity: $M < 1.0$. In this respect, it should be recalled that the Mach number $M = u/a$, with u as the norm of the velocity and a as the speed of sound, can be related to the ratio of kinetic energy and thermal (or sensible) energy. Its squared value, M^2, is indeed a measure of this ratio of kinetic energy, associated to the flow at the macroscopic scale, and internal energy, associated with the random and disordered motion of molecules. Its squared value is also used to evaluate normalized density variations $\Delta\rho/\rho$ but it should be underlined that there are many other possible sources for density variation, which can occur whatever the value of the Mach number. They are associated, for instance, (1) to thermal expansion (e.g., heat release by chemical reactions), (2) to multicomponent flow mixing with chemical species featuring strong density differences, or (3) volume variations associated to boundary changes, as operated in reciprocating engines [1]. Thus, in low Mach number flows, the variations of density mainly result from either (i) composition changes, which are associated to mixing processes taking place in the multicomponent flow, or (2) heat release induced by the exothermicity of chemical reactions. This is in sharp contrast with high-speed flows such as those considered in the present chapter, since local compression–expansion and viscous heating effects may also induce nonnegligible density variations.

At first sight, the most well known exception to low Mach number combustion is provided by detonations, i.e., combustion waves propagating at supersonic speeds. According to the classical Zeldovich–von Neumann–Döring (ZND) model, it can be pictured as follows: a leading shock wave propagates into the mixture of (fully) premixed reactants with strong compression and high pressure levels reached at the von Neumann state. This favors the initiation of chemical reactions. Heat is released by the corresponding exothermal chemical reactions, thus leading to an increased velocity of the shock wave. As deduced from the solution, sonic velocity conditions are reached behind the reaction zone (in combustion products at thermodynamic equilibrium) and this ensures some protection of the reactive wave from possible perturbations. Typical values of the Chapman–Jouguet velocity are of the order of kilometers per second. The real structure of the detonation waves may, however, display some significant differences with the above picture [2]. The reality is indeed by far more complex: the ZND model is unstable. As shown by Campbell and Woodhead [3], detonation is an oscillating phenomenon. It has a multidimensional nature and is shown to be a system made up of incident waves, Mach waves, and transverse waves. Its structure may be characterized from triple points evolution and associated cellular structure (detonation cell). The corresponding velocity of propagation may differ from the Chapmann–Jouguet detonation velocity. Moreover, it is noteworthy that their formation, as a result of either direct initiation or deflagration to detonation transition, still concentrates important research efforts. In this respect, the main differences between deflagration and detonation waves will be emphasized (in the third section of this chapter) and the transitions between them discussed.

From a general point of view, high-speed reactive flows are found in various situations relevant to security and safety issues, hypersonic propulsion systems, rocket engines and igniters [4], atmospheric reentry flows, and astrophysics. As far as the last point, i.e., astrophysics, is concerned, it should be emphasized that exothermic processes associated with nuclear fusion and fission reactions are of fundamental importance. For instance, violent explosive burning takes place in astrophysical transients. Unexpectedly, burning in astrophysical plasmas and chemical systems on Earth indeed display some similarities during such explosive burning [5–7]. In both situations, flow dynamics can be described using continuum Navier–Stokes equations provided that suitable energy source terms and transport representations are used. Explosive thermonuclear burning often occurs in the form of flames and detonations, with the overall structure and dynamics similar to that of their terrestrial counterparts. Such a conclusion holds in thermonuclear (type Ia) supernovae. Some remarkable differences do, however, exist between chemical and astrophysical combustion. They are associated, for instance, to the equation of state (EoS) and to the transport processes. In particular, in the core of a white dwarf, plasma becomes degenerate (electrons occupy all possible quantum states below a certain energy level), thus leading its EoS to be modified in such a manner that the pressure p becomes virtually independent of temperature [8]. The corresponding thermonuclear flames thus feature a moderate density jump. This is in sharp contrast with standard flames (i.e., terrestrial flames) in atmospheric conditions, which are characterized by larger density ratios.

As regards the applications to propulsion, a typical example of high-speed combustion is the rotating detonation engine (RDE). The basic concept of an RDE lies on a detonation wave that travels around an annulus, which is a circular channel. Fuel and oxidizer are continuously injected in this channel and detonation is initiated in the corresponding annular spacing between the two cylinders [9]. Detonation products are expelled in a nozzle. Theoretically, detonation should be more efficient than the conventional mode operating at an almost constant pressure p (Brayton or Joule cycle). If a sufficient gain in efficiency can be practically obtained, this would lead to a possible reduction of fuel consumption and combustion emission.

Finally, one of the most typical examples of supersonic combustion applied to propulsion is the scramjet engine [10–13]. For flight Mach number values larger than 5 (hypersonic flight), attention was focused in the late 1950s on the possibility of performing the combustion at supersonic speed in order to avoid the detrimental effects of dissociation. In ramjets, such dissociation processes can indeed be induced by the high temperature increase due to strong deceleration. The main elements of such a scramjet are (1) an inlet that compresses the air scooped from the atmosphere, (2) an isolator that protects the inlet operation from flow disturbances in the lower hypersonic regime associated to flight Mach number values ranging between 4 and 8 (dual-mode combustion), (3) a combustion chamber within which fuel is injected and burned with the main air stream, and (4) a nozzle that expands the resulting combustion products (after recombination) to free stream pressure. In the dual-mode combustion regime disturbances may be induced by the heat release taking place in the combustor. These pressure disturbances may propagate upstream of fuel injection and alter the inlet operation. This difficulty is circumvented by the installation of a short duct between the inlet and the combustor, which is known as the isolator. It avoids shock trains to disrupt the inlet operation and maintains a stable airflow to the combustion chamber. It aims at (1) maximizing the static pressure and the stagnation pressure recovery and (2) minimizing the flow distortion. The flow structure inside such an isolator can be described as follows: In the absence of boundary layers, there would be no shock train; the presence of the incoming boundary layers, however, produces a series of oblique and normal shocks that spread the pressure rise along the duct [14]. At some locations inside the isolator the flow reattaches to the wall and the shock train weakens, tending to disappear and improving the homogeneity of the flow inside the combustion chamber, in such a manner that efficient mixing and burning conditions are reached. As the combustion back pressure increases, the shock train length increases, thus leading to a sufficient rise of the combustor inlet pressure. However, if the shock train reaches the inlet, the mass of air captured will decrease, increasing the pressure and thermal loads, increasing drag and decreasing thrust. The integration of the scramjet into the propulsive system requires the aircraft forebody to compress the suctioned air in order to achieve adequate pressure and temperature for combustion and to maximize the pressure recovery. The inlet must be able to capture the amount of air required by the engine and decelerate it to the required speed without significant pressure loss and producing the least amount of drag. Also, the air inlet must be large enough so as to provide the necessary thrust even at high altitudes. Its design is thus crucial to the successful operation of the engine, since it is the only provider of airflow and compression. Finally, the combustion chamber appears as one of the most critical components of the scramjet engine, since it entails complicated phenomena such as compressibility effects, heat release, residence time, shock interaction, and turbulent mixing. The main challenges on its development are related to the improvement of mixing and burning efficiencies within a short residence time (of the order of 1 millisecond) in order to stabilize the flame while minimizing the total pressure loss. Over the years, several fuel injection techniques have been proposed, such as transverse fuel injection from a wall orifice, backward facing step with transverse injection, and angled injection. Combined with cavity flame holders, all of them aim at favoring ignition and stabilization of the combustion processes. The scramjet combustion chamber will be kept as a relevant example of combustion in supersonic flows throughout the whole chapter.

The objective of this chapter is twofold: it indeed aims at providing (1) the practitioners with the theoretical concepts underpinning the basic rules of supersonic combustion engine design, and (2) the researcher with a general overview of the current state of the art in the field of combustion and turbulent combustion in high-speed flows.

This chapter is organized as follows. The basics related to high-speed combustion in homogeneous one-dimensional flows will be first recalled. Special emphasis will be placed on their possible transitions and thermal chocking conditions. It is a common belief that the matter gathered in the corresponding section corresponds to rather well-known basic knowledge but the experience of the authors is that today many people are computing reactive high-speed flows without having this required background on supersonic combustion. The resulting section should help to fill, at least partly, the corresponding gap. Then, some specific features of compressible turbulence and turbulent mixing in nonreactive and reactive flows will be discussed. Since in most applications reactants are initially unmixed, some emphasis will be placed on scalar mixing between fuel and oxidizer as a prerequisite before combustion takes place. Examples of turbulent combustion in high-speed flows will be subsequently provided and discussed. Finally, the discussion will close with current challenges and some suggestions for future research.

7.2 Supersonic Reactive Flows: Governing Equations

In this section, the presentation starts with a few basics on high-speed flow combustion. First the Navier–Stokes equations governing the dynamics and thermodynamics of multicomponent reactive high-speed flows are presented. Thus, coming back to the basic set of conservation equations discussed in Chapter 1, we consider their steady one-dimensional form and analyze the corresponding possible solutions. In comparison to their low-speed counterparts, reactive high-speed flows raise some specific issues as an outcome of the special couplings between flow and chemical reactions that are triggered in these conditions. First, because of the large Reynolds number values that are considered, the role of convective processes increases and, as a consequence, molecular transport effects can be neglected outside of (1) the shock waves internal structure, which can be treated as a discontinuity; (2) boundary layers; and (3) shear layers featuring large gradients of velocity and composition (temperature and species mass fractions). Therefore, the inviscid Euler equations are suited to describe all physical phenomena taking place outside of these various regions. Second, there is a strong coupling between chemical reactions and compressibility effects, i.e., compression waves before their evolution into shocks, and rarefaction (i.e., expansion) waves (reversible effects). This is because the gas temperature is sensitive to compression/rarefaction waves. Put in different words, there is a permanent exchange between internal energy (molecular scale) and kinetic energy (macroscopic scale). In this respect, it is noteworthy that the dissipation of the kinetic energy inside the shock waves also results in an increase of the internal energy (irreversible effects). In fact, it does not seem useless to remind here that the Mach number M is nothing but the squared root of the ratio between the kinetic energy (of the flow) and thermal energy or static enthalpy, i.e., the average kinetic energy of molecules. In the vicinity of sonic conditions (M = 1), the kinetic energy is the same order of magnitude as the static enthalpy per unit mass and it becomes the leading-order contribution as the Mach number value is increased beyond unity. Therefore, since the kinetic energy dissipation induced by molecular viscosity effects, in boundary layers, shear layers, or shock waves, is proportional to the squared Mach number M^2, moderate changes in velocity result in significant changes in static temperature and pressure. Considering the sensitivity of chemical reactions to temperature and pressure, this induces a coupling between the velocity field and chemical reactions. Another remarkable point is that the heat release due to combustion will add relatively smaller amounts of energy as the Mach number is increased beyond unity: The influence of combustion on the flowfield may thus be expected to be less important as the Mach number value is increased. It is worth noting that, in low Mach number combustion, the aforementioned coupling between the velocity field and the chemistry can be neglected. In such low Mach number flows, dilatation is mainly due to the heat release by chemical reaction. Moreover, it should be emphasized that the thermodynamic pressure component can be considered as constant or at least homogeneous in space when it varies in time. Only the gradients of the dynamic pressure component will be kept in the momentum conservation equation and the corresponding spatial variations remain unable to alter the chemical reaction rates.

The set of Navier–Stokes equations applies provided that the system is at thermodynamic equilibrium with an equirepartition of energy at the molecular level: molecules have the same average energy associated with each independent degree of freedom [15]. This hypothesis assumes that (1) the characteristic scale of the flowfield variations ℓ remains sufficiently large with respect to the mean free path ℓ_{pm} – the value of ℓ_{pm} is of the order of 0.1 μm in standard conditions of pressure and temperature – and (2) the characteristic time of the flow evolution is much larger than the average time between two successive collisions τ_m, the value of which is around 10^{-10} s in standard conditions. Simple estimates of ℓ_{pm} and τ_m may be found in the literature [15]. These hypotheses are fulfilled in low Mach number flows. However, as far as high-speed flows are concerned, it is necessary to be more cautious; in particular, such a hypothesis is no longer satisfied inside shock waves, the thickness of which is of the same order of magnitude as the mean free path ℓ_{pm}. Finally, it does not hold for some hypersonic flows at low pressure levels or at very high temperatures.

Neglecting body forces, which are negligible for high-speed flows, and assuming summations over repeated indices, the Navier–Stokes equations for compressible multicomponent reactive flows read as follows [16, 17]. These equations, which were presented and discussed in Chapter 1, are rewritten for the sake of convenience in a form suitable for supersonic flows:

- Mass conservation equation

$$\frac{\partial \rho}{\partial t} + \frac{\partial}{\partial x_i}(\rho u_i) = 0, \quad (7.1)$$

where u_i is the velocity component in direction x_i.
- Species conservation equation

$$\frac{\partial \rho Y_\alpha}{\partial t} + \frac{\partial}{\partial x_i}\left(\rho \left(u_i + \mathcal{U}_{i,\alpha}\right) Y_\alpha\right) = \dot{\omega}_\alpha, \quad (7.2)$$

where Y_α is the mass fraction of chemical species α. The quantity $\mathcal{U}_{i,\alpha}$ denotes its diffusion velocity component in direction x_i and $\dot{\omega}_\alpha$ is the chemical production rate of mass of species α per unit volume and per unit time.
- Momentum conservation equation

$$\frac{\partial \rho u_i}{\partial t} + \frac{\partial}{\partial x_j}\left(\rho u_j u_i + p\delta_{ij} - \tau_{ij}\right) = 0, \quad (7.3)$$

where δ_{ij} is the Kronecker delta (i.e., permutation or substitution tensor) and τ_{ij} denotes the viscous tensor

$$\tau_{ij} = 2\mu S_{ij} + \left(\kappa - \frac{2}{3}\mu\right) S_{kk}\delta_{ij}, \quad (7.4)$$

with $S_{ij} = (A_{ij} + A_{ji})/2$ the symmetric part of the velocity gradient tensor $A_{ij} = \partial u_i/\partial x_j$, μ is the shear viscosity (or first viscosity coefficient), while κ is the bulk viscosity. At this level, the second viscosity coefficient $\lambda = \kappa - 2\mu/3$ is often introduced. The Stokes hypothesis, which consists in assuming a negligible bulk viscosity (i.e., $\lambda = -2\mu/3$), is often invoked but could be questioned for highly compressible flows such as those considered herein [18, 19]. The neglection of bulk viscosity is strictly valid for monoatomic gases. For polyatomic gases, the scattered available experimental and numerical data show that it is certainly not zero and actually often far from negligible. The ratio κ/μ can display significant variations. For dihydrogen, which, as we will see later on, is a reference fuel for high-speed combustion applications, this ratio can exceed 30. The shear viscosity may be calculated from the well-known Sutherland formula [20].
- Total energy conservation equation

$$\frac{\partial \rho e_t}{\partial t} + \frac{\partial}{\partial x_i}\left(\rho u_i e_t + p u_i - \tau_{ij} u_j + q_i\right) = 0, \quad (7.5)$$

where $e_t = e + (u_i u_i)/2$ is the total energy and q_i is the i-component of the molecular heat flux vector

$$q_i = \rho \sum_{\alpha=1}^{\alpha=N_s} Y_\alpha \mathcal{U}_{i,\alpha} h_\alpha - \lambda \frac{\partial T}{\partial x_i}, \quad (7.6)$$

where radiative heat transfer has been neglected. In Eq. (7.6), the quantity λ denotes the thermal conductivity. It is noteworthy that, to be consistent with the notations of

Chapter 1, thermal conductivity is denoted by λ but the same notation is also used to refer to the second viscosity coefficient discussed earlier.

- Kinetic energy conservation equation

$$\frac{\partial}{\partial t}\left(\rho\frac{u_k u_k}{2}\right) + \frac{\partial}{\partial x_i}\left(\rho u_i \frac{u_k u_k}{2} + p u_i - \tau_{ij} u_j\right) = p\frac{\partial u_i}{\partial x_i} - \tau_{ij}\frac{\partial u_i}{\partial x_j}, \quad (7.7)$$

where the last term corresponds to the viscous dissipation of kinetic energy $\varepsilon = \tau_{ij}\partial u_i/\partial x_j$, which is also often termed dissipation function. This contribution can be rewritten in terms of S_{ij}:

$$\varepsilon = 2\mu\left(S_{ij} - \delta_{ij} S_{kk}/3\right)\left(S_{ij} - \delta_{ij} S_{kk}/3\right) + \kappa S_{kk} S_{kk}, \quad (7.8)$$

and it becomes noteworthy that, since it is a sum of squared terms, this contribution is a sink term in the kinetic energy transport equation. An order of magnitude analysis shows that this term scales as the squared Mach number, i.e., it is $\mathcal{O}(M^2)$. It is therefore nonnegligible in high Mach number flows.

It should be emphasized that the energy conservation equation may be presented in various other forms, which can be based on the total enthalpy, $h_t = h + (u_i u_i)/2$,

$$\frac{\partial \rho h_t}{\partial t} + \frac{\partial}{\partial x_i}\left(\rho u_i h_t - \tau_{ij} u_j + q_i\right) = \frac{\partial p}{\partial t}; \quad (7.9)$$

on the enthalpy,

$$\frac{\partial \rho h}{\partial t} + \frac{\partial}{\partial x_i}\left(\rho u_i h + q_i\right) = \frac{\partial p}{\partial t} + u_i\frac{\partial p}{\partial x_i} + \tau_{ij}\frac{\partial u_i}{\partial x_j}; \quad (7.10)$$

or on the internal energy,

$$\frac{\partial \rho e}{\partial t} + \frac{\partial}{\partial x_i}\left(\rho u_i e + q_i\right) = -p\frac{\partial u_i}{\partial x_i} + \tau_{ij}\frac{\partial u_i}{\partial x_j}. \quad (7.11)$$

The above set of conservation equations has to be supplemented with an equation of state (EoS). In the case of a thermodynamically perfect gas, i.e., ideal gas, it reads

$$p = \rho \frac{R_u}{\mathcal{W}} T = \rho R T, \quad (7.12)$$

where R_u denotes the universal gas constant and $\mathcal{W}$ is the mean molecular weight of the mixture:

$$\frac{1}{\mathcal{W}} = \sum_{\alpha=1}^{\alpha=N_s} \frac{Y_\alpha}{\mathcal{W}_\alpha}, \quad (7.13)$$

with $\mathcal{W}_\alpha$ the molar mass of the αth species.

It is also possible to derive a conservation equation for the entropy

$$\frac{\partial \rho s}{\partial t} + \frac{\partial}{\partial x_i}\left(\rho u_i s + \frac{q_i}{T}\right) = -\frac{q_i}{T^2}\frac{\partial T}{\partial x_i} + \frac{\tau_{ij}}{T}\frac{\partial u_i}{\partial x_j} + \frac{1}{T}\sum_{\alpha=1}^{\alpha=N_s}\frac{\mu_\alpha^c}{\mathcal{W}_\alpha}\dot{\omega}_\alpha, \quad (7.14)$$

where the last term does involve the chemical potential of species α, i.e., μ_α^c. The corresponding contribution becomes zero in the two limiting situations of either chemical equilibrium or frozen flows (i.e., $\dot{\omega}_\alpha = 0$ for all chemical species).

In the above equations, the enthalpy is given by

$$h = \sum_{\alpha=1}^{\alpha=N_s} Y_\alpha h_\alpha = \sum_{\alpha=1}^{\alpha=N_s} Y_\alpha \left(h_{s,\alpha} + h_{f,\alpha}^0 \right) = h_s + \sum_{\alpha=1}^{\alpha=N_s} Y_\alpha h_{f,\alpha}^0, \quad (7.15)$$

where the enthalpy of chemical species α is obtained from

$$h_\alpha = h_\alpha(T) = h_{s,\alpha} + h_{f,\alpha}^0 = \int_{T_0}^{T} c_{p,\alpha} dT + h_{f,\alpha}^0, \quad (7.16)$$

where $h_{f,\alpha}^0$ stands for the enthalpy of formation of species α taken at a reference temperature T_0. The quantity $c_{p,\alpha}$ is the specific heat capacity of species α at constant pressure and $h_{s,\alpha}$ denotes the sensible enthalpy. The sensible enthalpy h_s of the multicomponent mixture is

$$h_s = \sum_{\alpha=1}^{\alpha=N_s} Y_\alpha h_{s,\alpha} = \sum_{\alpha=1}^{\alpha=N_s} \left(Y_\alpha \int_{T_0}^{T} c_{p,\alpha} dT \right) = \int_{T_0}^{T} c_p dT \text{ with } c_p = \sum_{\alpha=1}^{\alpha=N_s} Y_\alpha c_{p,\alpha}.$$

If one takes a closer look at the above budgets, the dissipation function appears as a production term in the enthalpy and energy equations. This is an irreversible conversion of kinetic energy into thermal energy due to viscous friction, and the corresponding exchange is nothing but an entropy production.

The most detailed mathematical description of molecular diffusion processes in multicomponent flows may be found elsewhere [21] and the corresponding solution is often replaced by simpler approximations. In most of the numerical studies, first, the barodiffusion (Dufour effect) and the thermodiffusion (Sorêt effect) are neglected, and, second, Fick's law is used. The latter should be a good approximation for combustion in air, since the mass fraction of fuel is usually small with respect to the mass fraction of nitrogen, which can be considered as a diluting species. However, it should be recognized that the corresponding effects may sometimes have a substantial influence in the case of H_2–air mixtures [22, 23] and, except for the sake of simplicity, it remains therefore difficult to justify the use of these approximations in the general case.

The simpler mixture-averaged formulation of the molecular diffusion flux is based on a modified version of the Hirschfelder and Curtiss approximation [24]:

$$\rho Y_\alpha \mathcal{U}_{i,\alpha} = -\rho D_\alpha^m \frac{W_\alpha}{W} \frac{\partial X_\alpha}{\partial x_i} - \rho Y_\alpha V_i^c \quad (7.17)$$

$$= -\rho D_\alpha^m \frac{W_\alpha}{W} \frac{\partial X_\alpha}{\partial x_i} - \rho Y_\alpha \sum_{\beta=1}^{\beta=N_s} D_\beta^m \frac{W_\beta}{W} \frac{\partial X_\beta}{\partial x_i},$$

where the mixture-averaged diffusion coefficient D_α^m of species α is evaluated from

$$D_\alpha^m = \frac{1 - Y_\alpha}{\sum_{\beta=1, \beta \neq \alpha}^{\beta=N_s} X_\beta / D_{\beta\alpha}}, \tag{7.18}$$

where $X_\beta = (\rho_\beta/\rho) \cdot (W/W_\beta)$ denotes the molar fraction of chemical species β.

It is noteworthy that the second (i.e., the last) term of the right-hand-side (RHS) of Eq. (7.17) is a corrective term that ensures the total mass conservation. At each time step, we evaluate each component of the correction velocity $V_i^c = \sum_{\beta=1}^{\beta=N_s} D_\beta^m (W_\beta/W) \partial X_\beta / \partial x_i$, which are added to the velocity component so as to enforce the compatibility between the discrete forms of species mass fractions and total mass conservation equations.

The description of fuel oxidation does involve a large number of elementary reaction steps. In practice, reduced kinetic schemes are considered instead. They include only the most important reaction steps, with some approximations applied on the reaction rates so as to provide a satisfactory level of accuracy within given ranges of pressure, temperature, and mixture composition. It is assumed that the reactive flow corresponds to a chemical system of N_s species reacting through N_r elementary reaction steps. The lth reaction step can be written as follows:

$$\sum_{\alpha=1}^{\alpha=N_s} v_{\alpha,l}^f A_\alpha \rightleftarrows \sum_{\alpha=1}^{\alpha=N_s} v_{\alpha,l}^r A_\alpha, \tag{7.19}$$

where A_α is the chemical symbol of species α, $v_{\alpha,l}^f$ and $v_{\alpha,l}^r$ are the molar stoichiometric coefficients of species α for the forward and reverse rate of the lth reaction.

The mass reaction rate that appears in the RHS of Eq. (7.2) reads

$$\dot{\omega} = W_\alpha \sum_{l=1}^{l=N_r} v_{\alpha,l} \varpi_l, \tag{7.20}$$

where $\varpi_l = \varpi_{f,l} - \varpi_{r,l}$ is the rate of progress of the lth reaction step; the rates of forward and reverse reactions $\varpi_{f,l}$ and $\varpi_{r,l}$ are written as

$$\varpi_{f,l} = K_{f,l}(T) \prod_{\alpha=1}^{\alpha=N_s} X_\alpha^{v_{\alpha,l}^f} \quad \text{and} \quad \varpi_{r,l} = K_{r,l}(T) \prod_{\alpha=1}^{\alpha=N_s} X_\alpha^{v_{\alpha,l}^r}, \tag{7.21}$$

where the rate constants $K_{f,l}$ and $K_{r,l}$ are modeled using the Arrhenius law e.g., the expression of $K_{f,l}$ reads

$$K_{f,l} = A_{f,l} T^{n_l} \exp\left(-\frac{E_l}{R_u T}\right). \tag{7.22}$$

The preexponential constant $A_{f,l}$, the temperature exponent n_l, and the activation energy E_l of the lth reaction step are deduced from experiments and, only in some cases, they can be determined from theoretical calculations. The reverse rates $K_{r,l}$ may be evaluated from their forward counterparts using the definition of equilibrium constants.

7.3 Steady Premixed Combustion Waves in High-Speed Flows

We consider one-dimensional flows, and three different types of compressible premixed combustion waves will be described. In the first situation, the heat release induced by exothermic chemical reactions takes place in a hot high-speed subsonic flow of reactants. It is usually referred to as a deflagration wave. In the second case, the heat release occurs in a supersonic flow of reactants. Finally, the last situation corresponds to a chemically frozen leading shock wave followed by a deflagration [15, 25–27].

Let us begin with compressible combustion waves in high-speed (subsonic or supersonic) one-dimensional plane flows of premixed reactants. This model is commonly used for the analysis and description of fast flame propagation [28, 29]. It is assumed that the combustion wave travels from the right to the left, i.e., in the negative direction. Thus, the incoming flow velocity (on the left) is positive in a frame of reference attached to the wave. The exothermic chemical reactions produce heat release, hereafter denoted by Q, the level of which can be evaluated at each location x from the expression

$$Q(x) = \sum_{\alpha=1}^{\alpha=N_s} (Y_{\alpha,in} - Y_\alpha(x)) h^0_{f,\alpha}, \qquad (7.23)$$

where $Y_{\alpha,in}$ denotes the mass fraction of chemical species α at the inlet. Provided that the combustion products reach the equilibrium state, the heat release reaches its maximum possible value

$$Q = \sum_{\alpha=1}^{\alpha=N_s} (Y_{\alpha,in} - Y_{\alpha,eq}) h^0_{f,\alpha}. \qquad (7.24)$$

Because of the thermal choking phenomena, two distinct classes of steady[1] and unsteady configurations must be considered separately [15, 25]. Thermal choking is indeed a distinctive feature of heat addition to high-speed flows. It means the following: for any steady one-dimensional flow with an inlet Mach number M_{in}, there is indeed a maximum amount of heat Q_c that can be released at a given Mach number [10, 30].

As a result of the thermal choking, the flow velocity reaches the sonic speed at some plane x^* (i.e., $u = a^*$ and $M^* = 1$) and the value of Q_c can be deduced from the total enthalpy budget

$$Q_c = \int_{T_{in}}^{T_{eq}} c_p dT - \frac{u_{in}^2}{2} + \frac{a^{*2}}{2}, \qquad (7.25)$$

where a^* is the speed of sound at location x^*. Under the simplifying assumption of constant specific heats and molecular weights, it reduces to

---

[1] It should be recalled that the frame of reference is attached to the combustion wave.

$$Q_c = c_p T_{in} \left(1 - \frac{1}{M_{in}^2}\right)^2 \frac{M_{in}^2}{2(1+\gamma)}, \quad (7.26)$$

where γ denotes the ratio of specific heat capacities and $M = u/a$ with $a = \sqrt{\gamma p/\rho} = \sqrt{\gamma R T}$ with $R = R_u/W$, R_u being the universal gas constant.

Any attempt to add more energy than Q_c to a flow of a given Mach number M_{in} will result in an unsteady flow featuring a chemically frozen shock followed by a deflagration [15, 25]. The restriction imposed by Eq. (7.26) may be reformulated as follows: For a given energy addition Q, there exists a critical value of the inlet Mach number

$$M_{in,c} = \sqrt{1 + \frac{1+\gamma}{2} \frac{Q}{c_p T_{in}}} \pm \sqrt{\frac{1+\gamma}{2} \frac{Q}{c_p T_{in}}}, \quad (7.27)$$

where the plus or minus signs in front of the second term of the RHS refer to supersonic and subsonic inlet flows, respectively. The supersonic inlet flows characterized by $M_{in} > M_{in,c} > 1$ and subsonic inlet flows characterized by $M_{in} < M_{in,c} < 1$ are not affected by thermal choking.

The case associated to $Q = Q_c$ corresponds to the classical Zeldovich–Neumann–Döring detonation with finite-rate chemistry, i.e., to the configuration featuring a chemically frozen leading shock wave followed by a deflagration with combustion products which, approaching the equilibrium state, accelerate up to the sonic velocity. The detonation thus propagates at the Chapman–Jouguet (CJ) speed. The corresponding Mach number value M_{CJ} is readily deduced from Eq. (7.27) with the positive sign and $Q = Q_c$. At this level, it seems worth emphasizing that the detonation wave propagating at $M = M_{CJ}$ is the sole self-sustaining steady compressible combustion wave, which may, however, be unstable.

7.3.1 Situations with Released Energy Below the Critical Value

Let us first consider the case of a supersonic inlet flow. For convenience, we introduce the subscript − and + to denote gas flow quantities taken upstream and downstream of the leading shock, if any. There exist two possible situations [15, 25]: (1) supersonic inlet flow of cold reactants featuring low static temperature (e.g., normal room temperature), where the chemical reaction rates may be neglected at the inlet flow conditions, i.e., no self-ignition; and (2) supersonic inlet flow of hot reactants, where the chemical reaction rates are not negligible and self-ignition processes may take place. In the first case, we may obtain the combination of a chemically frozen shock followed by a deflagration wave provided that the reactants' post-shock temperature is sufficiently high. The so-called cold-boundary difficulty [17, 29] is avoided, since the changes in cold inlet flow proceed so slowly that they can be ignored during the travel time of the shock and combustion wave system across the domain under consideration. With this reservation, case (1) is applicable to unbounded space. The initial conditions for the reacting subsonic flow correspond to the post-shock state. In the second case, there are two possible scenarios. The first is similar to the one just described, with a chemically frozen leading shock wave followed by a deflagration, while the second scenario is similar to combustion in the supersonic inlet flow. The combustion proceeds from the inlet condition to the fully burned gas state along the supersonic branch of the Rayleigh line, with molecular effects remaining negligible. It should be emphasized that the subcase (2) is applicable to the half-space $0 < x < +\infty$ [15, 25]. The corresponding set of steady conservation equations, written in the reference frame attached to the wave, follows from the Navier–Stokes set of Eqs. (7.2), (7.1), (7.2), (7.3), and (7.5):

$$\frac{d\rho u}{dx} = 0, \qquad \rho u = \dot{m} = \text{cst} \tag{7.28}$$

$$\frac{d}{dx}(p + \rho u^2) = \frac{4}{3}\frac{d}{dx}\left(\mu \frac{du}{dx}\right) \tag{7.29}$$

$$\dot{m}\frac{dY_\alpha}{dx} = -\frac{d\rho \mathcal{U}_{x,\alpha} Y_\alpha}{dx} + \dot{\omega}_\alpha \tag{7.30}$$

$$\dot{m}\frac{dh_t}{dx} = \frac{dq_x}{dx}, \tag{7.31}$$

where $\rho u^2 + p = \dot{m}u + p$ is the impulse function (or dynalpy) and q_x and $\mathcal{U}_{x,\alpha}$ denote the component, in direction x, of the molecular heat flux and molecular diffusion velocity, respectively. It should be noticed that the bulk viscosity has been neglected in the expression of the viscous tensor. Initial conditions for the subcase (1) correspond to subsonic flow behind the chemically frozen shock wave (frozen Rankine–Hugoniot relation). The mass flow rate per unit surface $\dot{m} = \rho u$, dynalpy $\dot{m}u + p$, total enthalpy $h_t = h + u^2/2$, and composition remain constant across the shock wave; therefore we have

$$\dot{m} = \rho_+ u_+ = \rho_- u_- \tag{7.32}$$

$$\dot{m}u_+ + p_+ = \dot{m}u_- + p_- \tag{7.33}$$

$$h_+ + u_+^2/2 = h_- + u_-^2/2 \tag{7.34}$$

$$Y_{\alpha,+} = Y_{\alpha,-} = Y_{\alpha,in}. \tag{7.35}$$

For flows with constant specific heat capacity and molecular weight, the above set of equations yields the relations between quantities taken upstream and downstream of the leading shock wave:

$$M_+^2 = \frac{1 + (\gamma - 1)M_-^2/2}{\gamma M_-^2 - (\gamma - 1)/2} \tag{7.36}$$

$$\frac{p_+}{p_-} = \frac{\gamma M_-^2 - (\gamma - 1)/2}{(\gamma + 1)/2} \tag{7.37}$$

$$\frac{T_+}{T_-} = \frac{(1+(\gamma-1)M_-^2/2)(2\gamma M_-^2/(\gamma-1)-1)}{(\gamma+1)^2 M_-^2/(2(\gamma-1))} \qquad (7.38)$$

$$\frac{\rho_+}{\rho_-} = \frac{u_-}{u_+} = \frac{1+(\gamma-1)M_-^2/2}{(\gamma+1)M_-^2/2}. \qquad (7.39)$$

The application of the Navier–Stokes equation to the frozen shock (i.e., formally $-\infty < x < +\infty$), yields the relations between (1) the jumps of both the kinetic energy and internal energy across the frozen shock and (2) the profiles of the velocity, pressure, and dissipation function inside the shock wave internal structure

$$\dot{m}\left(\left(\frac{u_k u_k}{2}\right)_{+\infty} - \left(\frac{u_k u_k}{2}\right)_{-\infty}\right) = -\int_{-\infty}^{+\infty} u\frac{dp}{dx}\,dx - \int_{-\infty}^{+\infty} \rho\varepsilon\,dx < 0 \qquad (7.40)$$

$$\dot{m}(h_{+\infty} - h_{-\infty}) = \int_{-\infty}^{+\infty} u\frac{dp}{dx}\,dx + \int_{-\infty}^{+\infty} \rho\varepsilon\,dx > 0. \qquad (7.41)$$

Since $u\,dp/dx$ is strictly positive, the second expression is positive-definite, while the first is negative-definite.

The total enthalpy jump is equal to zero, i.e.,

$$h_{t,+\infty} - h_{t,-\infty} = 0, \qquad (7.42)$$

while the enthalpy jump reads

$$h_{+\infty} - h_{-\infty} = \int_{T_{-\infty}}^{T_{+\infty}} c_p\,dT \quad \text{with} \quad c_p = \sum_{\alpha=1}^{N_s} Y_\alpha c_{p,\alpha}. \qquad (7.43)$$

From Eqs. (7.42) and (7.43), one may obtain the kinetic energy jump

$$\dot{m}\left(\left(\frac{u_k u_k}{2}\right)_{+\infty} - \left(\frac{u_k u_k}{2}\right)_{-\infty}\right) = -\int_{T_{-\infty}}^{T_{+\infty}} c_p\,dT < 0. \qquad (7.44)$$

The following conclusions follow from Eqs. (7.40), (7.41), (7.43), and (7.43): First, the kinetic energy and enthalpy jumps are due to two mechanisms: (1) compressibility effects associated to the integral of $u dp/dx$ and (2) dissipation function associated to the integral of $\rho\varepsilon$. The determination of these two integrals requires knowledge of the internal shock structure. In this respect, it should be stressed that the integral $\int_{-\infty}^{+\infty} \rho\varepsilon\,dx$ remains finite in the limit of vanishingly small value of the molecular viscosity $\nu \to \epsilon$. Indeed, the shock thickness δ_S is inversely proportional to the kinematic viscosity ν. For moderate Mach number values, it scales as $\nu/(c \cdot (M-1))$. Therefore, the norm of the velocity gradient scales as $|u_+ - u_-|/\nu$. The dissipation may thus be approximated from $|u_+ - u_-|^2/\nu$, in such a manner that $\rho\varepsilon$, once integrated over the characteristic shock thickness δ_S, i.e., $\int_{-\infty}^{+\infty} \rho\varepsilon\,dx$, scales as $\rho_m |u_+ - u_-|^2$ with ρ_m an intermediate value of the density that satisfies $\rho_- < \rho_m < \rho_+$. Second, the transformation of the kinetic energy within the shock results not only from the dissipation of kinetic energy (irreversible process), but also from compressibility effects, as it is for a compression wave without any shocks (reversible process). Third, the last set of relations shows that the enthalpy and kinetic energy jumps do not depend on the internal structure of the shock. Similarly, from the entropy conservation equation

$$\frac{d}{dx}\left(\dot{m} s + \frac{q_x}{T}\right) = -\frac{q_x}{T^2}\frac{dT}{dx} + \frac{\rho\varepsilon}{T} \qquad (7.45)$$

follows the expression that relates the entropy jump to the internal shock structure:

$$\dot{m}(s_{+\infty} - s_{-\infty}) = \int_{-\infty}^{+\infty}\left(-\frac{q_x}{T^2}\frac{dT}{dx} + \frac{\rho\varepsilon}{T}\right)dx > 0. \qquad (7.46)$$

As it was done above for the kinetic energy jump, it can be shown that the integral present in the RHS of Eq. (7.46) remains finite. It is noteworthy that the entropy jump across the shock is due to molecular transport only. This is in constrast to Eqs. (7.40) and (7.41), obtained for the kinetic energy and enthalpy jumps, which do include an additional contribution associated to compression. It is noteworthy that the entropy jump may also be obtained directly from the entropy definition and does not depend on the values of molecular transport coefficients. Considering the expression of the heat flux,

$$q_x = -\lambda\frac{dT}{dx} = -\rho c_p \alpha\frac{dT}{dx}, \qquad (7.47)$$

Eq. (7.46) reduces to

$$\dot{m}(s_{+\infty} - s_{-\infty}) = \int_{-\infty}^{+\infty}\left(\frac{\rho c_p}{T^2}\alpha\frac{dT}{dx}\frac{dT}{dx} + \frac{\rho\varepsilon}{T}\right)dx > 0, \qquad (7.48)$$

where the quantity

$$N_T = \alpha\left(\frac{dT}{dx}\frac{dT}{dx}\right) \qquad (7.49)$$

is the temperature SDR[2]. For flows with constant mean molecular weight and specific heat [30], the entropy jump may be determined from

$$s_{+\infty} - s_{-\infty} = -r\ln\left(\frac{p_{t,+\infty}}{p_{t,-\infty}}\right) > 0, \qquad (7.50)$$

with the total pressure ratio given by

$$\frac{p_{t,+\infty}}{p_{t,-\infty}} = \left[\frac{(\gamma+1)M_-^2/2}{1+(\gamma-1)M_-^2/2}\right]^{\frac{\gamma}{\gamma-1}} \cdot \left[\frac{2\gamma}{\gamma+1}M_-^2 - \frac{\gamma-1}{\gamma+1}\right]^{-\frac{1}{\gamma-1}} < 1. \qquad (7.51)$$

Let us now consider the case of a deflagration in hot subsonic inlet flow. This is the condition discussed at the beginning of Section 7.3. Such a deflagration wave is described by Eqs. (7.1)–(7.5) supplemented with initial conditions [25]. It should be underlined that, in contrast to Chapman–Jouguet detonations, steady compressible combustion waves with $Q < Q_c$, whatever they correspond to either (1) supersonic

---

[2] SDR is the acronym for scalar dissipation rate.

cold or (2) subsonic hot flows of reactants, are not self-sustaining steady compressible combustion waves because the flow behind the deflagration wave remains subsonic. Thus, it may be affected by any disturbances coming from the post-deflagration region. The steadiness of compressible combustion waves with $Q < Q_c$ may, however, be maintained by some external process. In particular, the propagation speed of these waves is fixed by the initial conditions [25].

Some additional insights may be gained from a preliminary order of magnitude analysis. As stated earlier, the conservation Eqs. (7.1)–(7.5) have been written under the assumption of the local thermodynamic equilibrium. However, this assumption is fulfilled only for flows in which the characteristic length scales relevant to the gradients of any quantity (composition, velocity, etc.) remain much larger than the mean free path of molecules, i.e.,

$$\ell \gg \ell_{pm}, \tag{7.52}$$

with the order of magnitude of the molecular transport coefficients being related the mean free path as follows [24]:

$$\mathcal{O}(\alpha) = \mathcal{O}\left(\frac{\lambda}{\rho c_p}\right) = \mathcal{O}(D_\alpha) = \mathcal{O}(\nu) = a\,\ell_{pm}, \tag{7.53}$$

The above condition may also be written

$$\frac{u}{\nu}\ell \gg \frac{u}{\nu}\ell_{pm}, \tag{7.54}$$

which, using Eq. 7.53, becomes

$$\frac{u}{\nu}\ell \gg \frac{u}{a\ell_{pm}}\ell_{pm}, \quad \text{i.e.,} \quad \text{Re}_\ell \gg M, \tag{7.55}$$

where $\text{Re}_\ell = u \gg \ell/\nu$.

The above expression may also be written in the following form:

$$\frac{u/\ell}{D_\alpha/\ell^2} \gg 1, \tag{7.56}$$

which implies that the molecular diffusion flux in the flow direction remains negligible with respect to its convective counterpart.

The left-hand side of Eq. (7.55) is nothing but the order of magnitude of the ratio between the convective and diffusive terms. Therefore, when the Mach number value approaches (or is greater than) unity, the hypothesis of local equilibrium remains satisfied only if the diffusive terms in the flow direction are negligible with respect to their convective counterparts. In fact, the violation of this condition signs the birth of a shock wave inside which the NSE no longer apply. A refined analysis of the corresponding region would require a Boltzmann's description but this difficulty is usually circumvented considering jump relationships across the shock wave. At this level, it should be noted that, in the case of shear layer flows, transverse diffusive transport could remain the same order of magnitude as the convection terms in the main flow direction [20].

Neglecting molecular transport terms, the momentum, species mass fractions, and energy conservation equations, i.e., Eqs. (7.29)–(7.31), reduce to algebraic budgets [25, 30]:

$$\frac{dI}{dx} = \frac{d}{dx}(p + \dot{m}u) = 0, \quad p + \dot{m}u = p_- + \dot{m}u_- \qquad (7.57)$$

$$\dot{m}\frac{dY_\alpha}{dx} = \dot{\omega}_\alpha \qquad (7.58)$$

$$\dot{m}\frac{dh_t}{dx} = 0, \quad h_t = h + \frac{u^2}{2} = h_- + \frac{u^2_-}{2}. \qquad (7.59)$$

Together with the EoS, mass conservation $\dot{m} = \rho u = \rho_- u_-$, and given the inlet state, the above set of algebraic equations determines the primary variables: density ρ, velocity u, species mass fractions Y_α, temperature T, and pressure p. It is noteworthy that the final state, corresponding to the chemical equilibrium of combustion products for which $\dot{\omega}_\alpha = 0$, can even be determined without any resolution of Eq. (7.58).

At this level, it is commonly assumed that the progress of chemical reactions can be characterized thanks to a single scalar variable, namely a progress variable. Used in conjunction with a constant specific heat and molecular weight approximation, this assumption allows to express the total enthalpy as follows:

$$h_t = c_p T + \frac{u^2}{2} + q\, c, \qquad (7.60)$$

where q is the amount of heat released per unit mass of reactants and c denotes the progress variable, which is defined to be zero in the fresh reactants and unity in the burned products at thermodynamic equilibrium. Such a progress variable has been introduced within the framework of single-step (i.e., global) chemistry.

The use of high-energy asymptotics and single-step chemistry reveals that, for the steady configuration with a supersonic inlet flow $M_{in} > 1$ and with $Q < Q_c$, a frozen shock is followed by a deflagration at a fixed distance downstream of the shock. The reaction zone behind the shock wave consists of a weakly reacting induction domain followed by a thin reaction zone, within which chemical reaction proceeds very rapidly toward the burned gases state. If the leading shock is too weak, i.e., when $1 < M_+^2 < (3\gamma - 1)/\gamma(3 - \gamma) > 1$ (and consequently $1/\gamma < M_- < 1$), this rapidly reacting region does not appear and chemical energy is preferentially converted into kinetic energy whilst temperature actually decreases [25–27]. Therefore, the condition $M_+^2 > (3\gamma - 1)/\gamma(3 - \gamma) > 1$ has to be satisfied for ignition to take place within the deflagration attached to steady shock. It is worth noting that $(3\gamma - 1)/\gamma(3 - \gamma)$ is approximately 1.2 for $\gamma = 7/5$. Finally, it is noteworthy that the asymptotic analysis of the subsonic inlet flow, $M_{in} < 1$, for the situation of hot reactants flows in half-space $0 < x < \infty$ has been also conducted by Clarke [25–27].

7.3.2 Situations with Released Energy Above the Critical Value

As stated earlier, there are no steady-state flow conditions for situations associated to $Q > Q_c$. In such cases, the flow is essentially unsteady and features a leading shock followed by a deflagration. The resort to asymptotics within the framework of single-step chemistry shows that, as time elapses, an unsteady region develops within the shock wave–reaction zone complex. The relevant characteristic time scale of this process is the chemical induction time at the Von Neumann state, just behind the leading shock wave. It has been established that there is a time period over which the evolution of the complex is such that the deflagration drives the leading shock on its own, without any assistance from other shock-driving mechanisms. This may provide an explanation for the phenomena observed in the early stages of detonation initiation. The reader is referred to the work of Clark [25–27] for further details.

7.3.3 Fast Compressible Flames and Their Transition to Detonation

Deflagration-to-detonation transition (DDT) is a typical and important example of the evolution of an unsteady fast compressible and reactive wave toward detonation. The first stage of DDT is the acceleration of the subsonic deflagration, which culminates in the form of a choked flame [28, 31–33]. The second stage is the subsequent transition of this chocked flame into a detonation. As emphasized by Oran and Gamezo [32], the DDT is a basic combustion problem that remains one of the major unsolved problems in theoretical combustion theory. The history, background, and state-of-the-art of fast compressible flames and their transition to Chapman–Jouguet deflagrations and detonations may be found in several references and, for the sake of conciseness, the present discussion is only restricted to a few recent ones and references therein. The corresponding set of references concerns mainly DDT in a confined environment. The possibility of DDT in unconfined environment is a much more involved problem [32, 34]. Recent experimental studies have been conducted to analyze DDT for five different hydrocarbon fuels, namely methane, ethane, ethylene, acetylene, and propane with oxygen as the oxidizer [33]. DDT with a single obstacle featuring different geometries and with multiple orice plates has been also the subject of recent experimental investigations [35]. Significant numerical and theoretical efforts have also been spent in this direction [32, 36–38]. The accurate numerical simulation of DDT requires the resolution of the smallest chemical and fluid mechanical scales. Put in other words, the physical processes relevant to the flame thickness, induction delay time and associated length scales with a representative multistep chemistry, as well as the Kolmogorov time and length scales, have to be resolved. Several physical mechanisms are responsible for creating the conditions in which the DDT can occur. These are (1) flame-shock couplings; (2) turbulence due to flame–shock interaction; (3) flame–turbulence interaction; (4) birth of hot spots, which can trigger the Zeldovich gradient mechanism of DDT [28, 29]; (5) shocks and flames interaction with boundary layers and obstacles; (6) flame–flame interactions; (7) turbulent mixing of hot products with unburned reactants; and (8) direct shock ignition. To alleviate the restrictions associated to three-dimensional direct numerical simulation (DNS) limitations, a common strategy consists in using large-eddy simulation (LES) with different subgrid models. In this respect, the compressible linear-eddy model (CLEM) has been used in some studies [36–38], while the artificial thickened flame model was retained in some other recent studies [39].

7.4 Influence of Wall Friction and Heat Transfer

The preliminary estimation of propulsive system characteristics (thrust, specific impulse, etc.) is often carried out with the aid of modified one-dimensional equations suited to account for the changes in the cross-stream section area and relevant to the real geometry of the propulsion system. The total pressure losses – due to wall shear stress, fuel injection devices, and so forth – and the heat transfer at the walls can be taken into account using phenomenological or empirical relationships. Such one-dimensional computations are cheap and allow determination of the possible ranges of variation of input parameters as well as the overall geometry of the propulsion system so as to obtain a net positive thrust. Then, the relevance of these one-dimensional computations may be checked on the basis of more advanced CFD simulations.

In contrast to Section 7.3, the analysis is now extended to the general case of non-premixed and partially premixed reactants. The distribution of the heat release $Q(x)$ along the combustor is prescribed by using a function $\Lambda(x)$ obtained from either experiments or empirical correlations, $Q(x) = (1 - \Lambda(x))q$, with $0 \leq \Lambda(x) \leq 1$ and $d\Lambda(x)/dx \leq 0$. The values $\Lambda = 0$ and $\Lambda = 1$ refer to complete combustion, i.e., combustion products, usually assumed to be at thermodynamic equilibrium, and to reactants at the combustor inlet x_{in}, respectively. It is worth noting that $\Lambda(x)$ may also depend on some flow parameters such as the Mach number M. Assuming (1) constant mean molecular weight and specific heats, and (2) no gas injection across the combustor wall at abscissa $x > x_{in}$, the one-dimensional conservation equations read [30]

$$d(\dot{m}A) = 0, \qquad \dot{m}A = \rho u A \qquad (7.61)$$

$$dI = pdA_w - \tau_w \Pi dx, \qquad I = (p + \rho u^2)A = p(1 + \gamma M^2)A \qquad (7.62)$$

$$\dot{m}Adh_t = Q_w \Pi dx, \qquad h_t = c_p T + \left(\frac{u^2}{2}\right) + \Lambda q = c_p T_t + \Lambda q \qquad (7.63)$$

$$\dot{m}Ac_p dT_t = \dot{m}AdQ + Q_w \Pi dx, \qquad dQ = -qd\Lambda > 0, \qquad (7.64)$$

where τ_w denotes the wall shear stress and Q_w is related to the heat transfer at the wall per unit wetted area (in $W \cdot m^{-2}$). The quantity dA_w is the differential of the wetted area, i.e., the differential of the area that is in contact with the gas flow, over which the friction and heat transfer exert. The channel perimeter is Π, and finally τ_w and Q_w are defined by

$$\tau_w = C_f \left(\frac{\rho u^2}{2} \right) \tag{7.65}$$

and

$$Q_w = \rho u c_p \mathrm{St}(T_w - T_{aw}) = \dot{m} c_p \mathrm{St}(T_w - T_{aw}), \tag{7.66}$$

where C_f and St denote the coefficient of friction and the Stanton number, respectively. The Reynolds analogy allows writing $C_f = 2\,\mathrm{St}$ [30]. The quantities T_w and T_{aw} denote the actual wall temperature and the adiabatic wall temperature, respectively. The latter is defined by

$$T_{aw} = T\left(1 + r\frac{\gamma-1}{2}M^2\right), \tag{7.67}$$

where r denotes the recovery factor.

Equations (7.61), (7.63), and (7.64) reduce to

$$\frac{dI}{I} = \frac{1}{1+\gamma M^2}\frac{dA}{A} - \frac{\gamma M^2}{2(1+\gamma M^2)}\left(\frac{4C_f}{D}\right)dx \tag{7.68}$$

$$\frac{dT_t}{T_t} = \left(\frac{dT_t}{T_t}\right)_Q + \left(\frac{dT_t}{T_t}\right)_w, \tag{7.69}$$

where

$$\left(\frac{dT_t}{T_t}\right)_Q = \frac{dQ}{c_p T_t} \tag{7.70}$$

$$\left(\frac{dT_t}{T_t}\right)_w = \frac{(T_w - T_{aw})}{T_t}\left(\frac{4\,\mathrm{St}}{D}\right)dx, \tag{7.71}$$

with $(dT_t/T_t)_Q$ and $(dT_t/T_t)_w$ referring to the combustion heat release and the wall heat transfer effects, respectively. The quantity $D = 4A/\Pi$ denotes the hydraulic diameter of the channel.

The differential equations for M, u, T, ρ, p, p_t, and the entropy s are derived using the perfect gas EoS, the definition of Mach number $M = u/a$, the definitions of the stagnation quantities: temperature T_t, pressure p_t, and density ρ_t:

$$T_t = T\left(1 + \frac{\gamma-1}{2}M^2\right) \tag{7.72}$$

$$p_t = p\left(1 + \frac{\gamma-1}{2}M^2\right)^{\gamma/(\gamma-1)} \tag{7.73}$$

$$\rho_t = \rho\left(1 + \frac{\gamma-1}{2}M^2\right)^{1/(\gamma-1)} \tag{7.74}$$

and the Gibbs equation [17, 29]

$$Tds = c_p dT - \frac{1}{\rho}dp = c_p dT_t - \frac{1}{\rho_t}dp_t. \qquad (7.75)$$

Rearrangement of the various terms yields equations for differential changes:

$$\frac{dM}{M} = -\frac{1+(\gamma-1)M^2/2}{1-M^2}\frac{dA}{A} + \frac{(1+\gamma M^2)(1+(\gamma-1)M^2/2)}{2(1-M^2)}\frac{dT_t}{T_t}$$
$$+ \frac{\gamma M^2(1+(\gamma-1)M^2/2)}{2(1-M^2)}\frac{4C_f}{D}dx$$

$$\frac{du}{u} = \frac{-1}{1-M^2}\frac{dA}{A} + \frac{1+(\gamma-1)M^2/2}{1-M^2}\frac{dT_t}{T_t} + \frac{\gamma M^2}{2(1-M^2)}\frac{4C_f}{D}dx$$

$$\frac{dT}{T} = \frac{(\gamma-1)M^2}{1-M^2}\frac{dA}{A} + \frac{(1-\gamma M^2)(1+(\gamma-1)M^2/2)}{1-M^2}\frac{dT_t}{T_t} - \frac{\gamma(\gamma-1)M^4}{2(1-M^2)}\frac{4C_f}{D}dx$$

$$\frac{d\rho}{\rho} = \frac{M^2}{1-M^2}\frac{dA}{A} + \frac{(1+(\gamma-1)M^2/2)}{1-M^2}\frac{dT_t}{T_t} - \frac{\gamma M^2}{2(1-M^2)}\frac{4C_f}{D}dx$$

$$\frac{dp}{p} = \frac{\gamma M^2}{1-M^2}\frac{dA}{A} + \frac{\gamma M^2(1+(\gamma-1)M^2/2)}{1-M^2}\frac{dT_t}{T_t}$$
$$- \frac{\gamma M^2(1+(\gamma-1)M^2/2)}{2(1-M^2)}\frac{4C_f}{D}dx$$

$$\frac{dp_t}{p_t} = -\frac{\gamma M^2}{2}\frac{dT_t}{T_t} - \frac{\gamma M^2}{2}\frac{4C_f}{D}dx$$

$$\frac{ds}{R} = \frac{\gamma}{\gamma-1}(1+(\gamma-1)M^2/2)\frac{dT_t}{T_t} + \frac{\gamma M^2}{2}\frac{4C_f}{D}dx. \qquad (7.76)$$

It is noteworthy that the one-dimensional budgets written above in the case of variable cross section area involve variables that are actually the averages obtained along the cross section area of the channel.

The above set of equations (7.76) describes the continuous evolution of the gas flow quantities along the combustor. If the leading shock appears, the jump relationships across the shock follow from Eqs. (7.61)–(7.64), and they are fully consistent with Eqs. (7.32)–(7.34). It is worth noting that the conservation equation for the impulse function, i.e., Eqs. (7.62) and (7.68), do not contain terms proportional to dT_t associated to the heat release and wall heat transfer; these factors alter the impulse function implicitly through the Mach number.

The set of equations (7.76) shows that (1) addition (or loss) of heat plays a role similar to a contraction (respectively a divergence) of the flow at least for p, ρ, u, and M; (2) an increase (respectively a decrease) in the flow velocity corresponds to a decrease (respectively increase) in the pressure level. These two behaviors are verified regardless of the Mach number of the flow. On the other hand, it is clear that the influence exerted on the flow parameters by the addition of energy or by a variation of section may differ depending on whether it is subsonic or supersonic [30].

7.5 Thermal Choking in Constant Cross Section Area Channel

The present section is focused on thermal choking conditions for a constant cross section area channel flow without wall friction and heat transfer. Differential Eqs. (7.61)–(7.64) with $C_f = \text{St} = 0$ and $dA = 0$ reduce to algebraic equations:

$$\rho u = \rho_{in} u_{in} = \dot{m} \tag{7.77}$$

$$p + \dot{m}u = p + \dot{m}u_{in} \tag{7.78}$$

$$c_p T + \frac{u^2}{2} + \Lambda q = c_p T_{in} + \frac{u_{in}^2}{2} + Q(x). \tag{7.79}$$

As already stated in this section, the solution to the above equations exists if, at the given M_{in}, the heat release after complete combustion does not exceed the critical value Q_c. The value Q_c is given by Eq. (7.26). The thermal choking happens if $Q = Q_c \leq q$. At at given section of the combustor $x = x^*$, the Mach number becomes equal to unity $M(x^*) = M^* = 1$. It should be noted that the distributions of the heat release $Q(x)$ along the combustor depends on the supersonic ($M_{in} > 1$) or subsonic ($M_{in} < 1$) inlet flow. In this respect, as emphasized in the forthcoming section of this chapter, the residence time and mixing efficiency may be significantly altered in supersonic flows. Therefore, the characteristic length on which the heat release takes place is greater for supersonic inlet flows compared to their subsonic counterparts and, consequently, the values of x^* are different for supersonic and subsonic inlet flows. It is larger in supersonic flows, i.e., $x^*_{sup} > x^*_{sub}$. In the absence of any possible ambiguity, the subindexes will be sometimes omitted below. It should be emphasized that the sonic velocity point $x = x^*$ is a singular point for the set of differential equations (7.76). If the condition $Q = Q_c \leq q$ is met, i.e., if the thermal choking occurs before complete combustion takes place, the steady solution cannot be maintained in the region $x > x^*$, and it will develop some unsteadiness. At the sonic point $M^* = 1$, algebraic Eqs. (7.77)–(7.79) yield

$$\frac{u^*}{u_{in}} = \frac{\rho_{in}}{\rho^*} = \frac{\gamma M_{in}^2 + 1}{(1 + \gamma)M_{in}^2} \tag{7.80}$$

$$\frac{u^*}{u_{in}} = \frac{\rho_{in}}{\rho^*} = \frac{\gamma M_{in}^2 + 1}{(1 + \gamma)} \tag{7.81}$$

$$\frac{u^*}{u_{in}} = \frac{\rho_{in}}{\rho^*} = \frac{\left(\gamma M_{in}^2 + 1\right)^2}{(1 + \gamma)^2 M_{in}^2}. \tag{7.82}$$

If the condition $Q = Q_c = q$ is met, i.e., thermal choking happens simultaneously with complete combustion, the solution can be extended to the region $x > x^*$;

this is the classical Chapman–Jouguet solution. There are two possibilities: either (i) $d\Lambda/dx \neq 0$ at $x = x^*$ or (2) $d\Lambda/dx = 0$ at $x = x^*$. In case (2), x^* may be equal to infinity but it should be recalled that the values of C_f and St have been assumed equal to zero.

The analysis of the solutions of Eqs. (7.77)–(7.79) may be found in several textbooks [30]. When the heat release is less than the critical value Q_c, i.e., $Q < Q_c < q$, there are two classes of possible solutions. We limit ourselves to a schematic presentation of the solutions in the Clapeyron's coordinates $(p, v = 1/\rho)$, as depicted in Fig. 7.1.

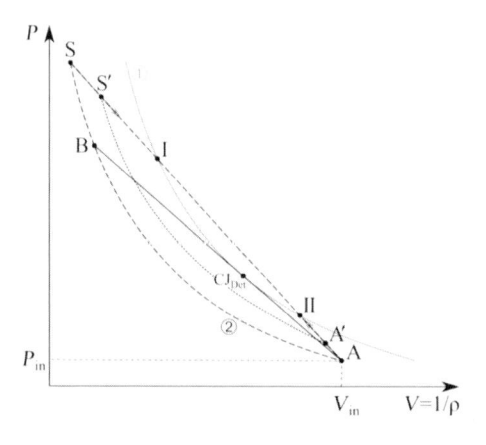

Figure 7.1 Schematic representation in the plane $(p, v = 1/\rho)$ of the possible trajectories corresponding to steady one-dimensional flow in a constant cross section area. The curve marked as ① is the Hugoniot curve with heat release and the curve ② is the Hugoniot curve associated to the nonreactive shock wave.

The first class of solutions, hereafter denoted class (a), is depicted by ASI trajectory in Fig. 7.1. This is the classic solution associated to a strong (overdriven) detonation for which the leading chemically frozen shock appears in the inlet flow (point S in Fig. 7.1). Combustion takes place downstream of the shock in the deflagration wave (in the subsonic flow), i.e., between the points S and I in Fig. 7.1. It is worth noting that this solution can be generalized. The leading shock can be replaced by a region where the flow is no longer one-dimensional. Indeed, the transition from a supersonic flow to a subsonic flow can occur through an extended region of the three-dimensional flow including complex shock waves and wave interactions (oblique shock waves, Mach waves, recirculation zones), often referred to as shock train or "pseudo-shock" in the literature [14, 30, 40].

The second class of solution, hereafter denoted class (b), is depicted by the trajectory of type AII (i.e., A to II) in Fig. 7.1. The flow is supersonic. It may be reached provided that the static temperature in the inlet supersonic flow is sufficiently high to trigger self-ignition.

In principle, nothing precludes the possible existence of a composite solution between (a) and (b). A typical example of such a composite solution AA'S'I is depicted in Fig. 7.1. It could start upstream in the form of a continuous supersonic flow, M > 1 (with chemical reactions taking place from A to A'), then undergo the frozen shock wave (from A' to S') and finally, downstream of the shock wave, evolve as a subsonic flow, M < 1 (with chemical reactions taking place from S' to I).

The solutions of classes (a) and (b) reach the thermal choking at $Q = Q_c < q$, trajectories ABCJ$_{Det}$ and ACJ$_{Det}$ in Fig. 7.1. As already emphasized earlier, these solutions cannot be continued to the region $x > x^*$ since, as follows from Eq. (7.76), $|dM/dx| = \infty$ at $x = x^*$, and the solutions become unsteady. If the equality $Q = Q_c = q$ holds (i.e., $\Lambda(x^*) = 0$), the trajectory ABCJ$_{Det}$ corresponds to the Chapman–Jouguet (C-J) detonation with the point CJ$_{Det}$ named the C-J point [28–31].

7.6 Stability Analysis in the Vicinity of Thermal Choking Conditions

The stability of steady one-dimensional flow can be studied in the vicinity of thermal choking conditions. It is assumed here that the thermal choking happens at the section $x = x^*$ when the combustion is almost complete, i.e., $Q(x^*) = Q_c$ is approximately equal to q and $\Lambda(x^*)$ is almost zero. Experimental observations show that for a steady one-dimensional supersonic flow with an amount of heat addition close to thermal choking, i.e., for sufficiently small but finite values of the ratio $(Q - Q_c)/Q_c$, only the discontinuous flow presenting a combination a frozen shock (or shock train/pseudo-shock) followed by subsonic flow with heat release is realized [14, 40, 42, 43]. This experimental observation may be explained using the quasi-static approach and (nonlinear) stability analysis of the steady flow with heat addition close to thermal choking state.

7.6.1 Constant Cross Section with Wall Friction and Heat Exchange

To illustrate the idea of the quasi-static approach, let us first consider the constant cross section area flow, taking into account the loss of impulse function due to wall friction, i.e., Eqs. (7.62) and (7.68), with $dA = 0$.

Let us consider two steady inlet flows, supersonic and subsonic, that achieve thermal choking conditions, $M^* = 1$, at section $x = x^*$. As underlined earlier, the abscissa values x^* are different for supersonic and subsonic inlet flows ($x_{sup} > x_{sub}$). It is assumed that $Q(x^*) = Q_c$ approximately equal to q corresponds to almost complete combustion for both inlet flows. The point $x = x^*$, as seen from the set of equations (7.76), is singular, and the solutions cannot be continued beyond this point to the region $x > x^*$. However, the set of equations (7.76) can be solved analytically in the vicinity of the singular point $x = x^*$ for $x^* - x \to \epsilon$, with ϵ any arbitrary small positive number. In particular, in the vicinity of x^*, the first equation reduces to

$$\frac{dM^2}{dx} = \frac{K}{D(1 - M^2)}, \quad x^* - x \to \epsilon. \qquad (7.83)$$

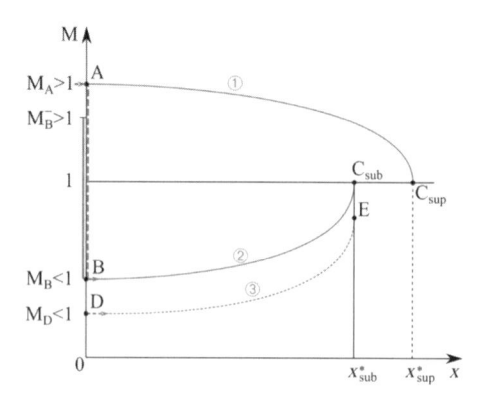

Figure 7.2 Schematic illustration of the trajectories relevant to one-dimensional flows close to thermal choking state in the plane (M, x). Steady flows in a constant cross section area with wall friction and heat transfer.

where K denotes a dimensionless parameter

$$K = \frac{2D}{c_p T^*}\left(\frac{dQ}{dx}\right)_{x=x^*} + \frac{1+\gamma}{2}\left[(1+\gamma)\frac{T_w - T_{aw}}{T_t}\text{St} + \gamma C_f\right]. \qquad (7.84)$$

Since $Q(x^*) = Q_c$ is almost equal to q, the first term present in the RHS of this expression is small compared to the second contribution. At this level, it should be recalled that Eq. (7.83) is valid whether the flow is supersonic or subsonic at the inlet.

Thus, we obtain

$$M \to 1 \pm \sqrt{K(x^* - x)/4D}, \qquad \left|\frac{dM}{dx}\right| \to \frac{1}{4}\sqrt{\frac{K}{(x^* - x)}D}, \qquad x^* - x \to \epsilon, \qquad (7.85)$$

where the signs + and − refer to supersonic and subsonic continuous flows, respectively. Using Eq. (7.85), and similar equations for other quantities, the whole set of equations can be solved in the reverse direction to find the inlet conditions. In Fig. 7.2, a schematic illustration of these two continuous trajectories is depicted in the plane (M,x): supersonic ($M_{in,sup} = M_A > 1$) AC_{sup} and subsonic ($M_{in,sup} = M_B < 1$) BC_{sub} trajectories with thermal choking at x^*_{sup} and x^*_{sub}, respectively. In addition to this, the discontinuous trajectory ADE – assuming the same conditions as for the two continuous flows heat addition Q_c – with the frozen leading shock (AD) followed by heat release in subsonic flow (DE) is also depicted in Fig. 7.2. The Mach numbers on the two sides of the leading shock are denoted by $M_A = M_{in,sup}$ and M_D, respectively. Because of the viscous effects, and in contrast to the case without viscous effects, the two inlet Mach numbers, supersonic $M_A = M_{in,sup}$ and subsonic M_B, are no longer interrelated by the Rankine–Hugoniot relationship, i.e., M_B is not equal to the Mach number value M_D obtained downstream of the frozen shock.

Let us show that $M_B > M_D$. To this end, we compare, first, the inlet impulse functions for the trajectories corresponding to supersonic and subsonic flows. Integrating Eq. (7.62) with $dA = 0$ over the range $x_{in} \leq x \leq x*$ yields

$$I(x_{in}) = I^* + \Pi \int_{x_{in}}^{x^*} \tau_w dx = I^* + \frac{\Pi}{2} \int_{x_{in}}^{x^*} C_f u dx \qquad (7.86)$$

$$I^* = I(x^*) = \frac{\gamma+1}{2\gamma} \dot{m} a^*, \quad a^* = \sqrt{\gamma R T^*}, \quad T^* = \frac{2}{\gamma+1} T_t^*, \qquad (7.87)$$

where I^* denotes the impulse function at M = 1, and a^* denotes the speed of sound at M = 1.

Considering that the velocity along the supersonic flow trajectory is larger than along the subsonic flow trajectory and assuming, for the sake of simplicity but without loss of generality, that (1) C_f is a constant parameter and (2) $T^*_{t,sup} = T^*_{t,sub}$ (i.e., $I^*_{sup} = I^*_{sub}$), we conclude from these two equations that the inlet impulse function for the supersonic flow I_A is always greater than the inlet impulse function for the subsonic flow I_B.

Second, we remind that the impulse function of the discontinuous flow is conserved across the leading shock, i.e., $I_D = I_A$. Therefore, we conclude that the impulse functions of two subsonic flows, associated to the trajectories DE and BC, are such that $I_D > I_B$. Third, from this inequality, it follows that the inlet Mach numbers of two subsonic flows, depicted in Fig. 7.2, satisfy the inequality $M_D < M_B$, which is a direct consequence of the relationship that exists between the impulse function and the Mach number [30]:

$$\frac{I}{I^*} = f(M), \quad f(M) = \frac{1+\gamma M^2}{M\sqrt{2(\gamma+1)[1+(\gamma-1)M]}}, \quad f(M=1) = 1. \qquad (7.88)$$

The dimensionless impulse function, $f(M)$, displays a minimum at M = 1 and monotonically increases with M for M < 1 ($f(M) \to \infty, M \to 0$) and for M > 1 ($f(M) \to \gamma/\sqrt{(\gamma^2-1)}, M \to \infty$), i.e., (1) for subsonic flows, if $I_2 > I_1$, then $M(I_2) < M(I_1)$ and vice versa, and (2) for supersonic flows, if $I_2 > I_1$, then $M(I_2) > M(I_1)$ and vice versa. Therefore, the subsonic flow, depicted by the trajectory DE in Fig. 7.2, is not altered by thermal choking at the same level of heat addition Q_c as the subsonic flow depicted by BC_{sub} in Fig. 7.2.

Let M_B^- be the supersonic inlet Mach number linked with M_B through the Rankine–Hugoniot relationship. Then, as it can be seen from Fig. 7.2, the discontinuous flows with the supersonic inlet Mach numbers in the interval $M_B^- < M_{in} < M_A$ are not affected by the thermal choking. Indeed, in these discontinuous flows, the subsonic Mach numbers behind of leading shocks will remain bounded in such a manner that $M_D < M < M_B$. This explains why, for a supersonic flow, the conditions of which are close to those associated to thermal choking, only the solution preceded by a shock wave or a pseudo-shock zone is observed in experiments. Finally, it should be noticed that the flow with $M_{in} > M_A$ – at the same Q_c as for flows with M_A and M_B – is not affected by the thermal choking, as previously emphasized after Eq. (7.27), which defines the minimum value $M_{in,c}$.

7.6.2 Variable Cross Section: Divergent Channel

The present discussion is limited to the case of a divergent channel, i.e., $dA/dx > 0$, because this is the geometry retained for dual mode scramjet combustors. Channel divergence and heat release are assumed to be chosen in such a manner that the Mach number of the supersonic inlet flow decreases due to the combined impact of combustor divergence and heat release, i.e., the first equation of (7.76) may be rewritten as

$$\frac{d\mathrm{M}^2}{dx} = \frac{G(x)}{D(1-\mathrm{M}^2)}, \quad x \leq x^*, \tag{7.89}$$

where the dimensionless function $G(x)$ is positive. Then, the analysis remains approximately the same as the one already conducted for the constant cross section area case. The thermal choking (sonic) point $x = x^*$ is singular, and the supersonic and subsonic solutions to the set of equations (7.76) cannot be continued beyond this point in the region $x > x^*$. The Mach number M approaches unity as

$$\mathrm{M} \to 1 \pm \sqrt{G(x^*)(x^* - x)/4D}, \quad x^* - x \to \epsilon, \tag{7.90}$$

where the signs $+$ and $-$ refer to the supersonic and subsonic continuous flows, respectively.

An explicit expression for Q_c cannot be obtained. The evaluation of Q_c requires seeking the solution of the system of equations (7.76) for a given set of functions $A(x)$ and $Q(x)$. Supersonic ($\mathrm{M}_{in,sup} = \mathrm{M}_A > 1$) branch AC_{sup} and subsonic ($\mathrm{M}_{in,sub} = \mathrm{M}_B < 1$) branch BC_{sub} with thermal choking at x^*_{sup} and x^*_{sub}, respectively, and the discontinuous trajectory ADE with the frozen leading shock (AD) followed by heat release in subsonic flow (DE) are similar to those depicted in Fig. 7.2, and they are therefore not reproduced in an additional separate figure. The two inlet Mach numbers, supersonic $\mathrm{M}_{in,sup} = \mathrm{M}_A$, and subsonic $\mathrm{M}_{in,sub} = \mathrm{M}_B$, are not interrelated by the Rankine–Hugoniot relationship, i.e., M_B is not equal to Mach number M_D behind the frozen shock. Similar to the case with the constant cross section area, these values are such that $\mathrm{M}_B > \mathrm{M}_D$. Indeed, integrating Eq. (7.62) for the impulse function I over the interval $x_{in} \leq x \leq x^*$ yields

$$I(x_{in}) = I^* - \int_{x_{in}}^{x^*} p\frac{dA}{dx}\,dx + 2\dot{m}A \int_{x_{in}}^{x^*} C_f \frac{u}{D}\,dx, \tag{7.91}$$

where $\dot{m}A = \rho u A = cst$. The second term in the RHS of this equation, i.e., the input from the cross section area variation, is negative for a divergent channel both for the supersonic AC_{sup} and subsonic DE trajectories. The pressure along the trajectory corresponding to supersonic flow is smaller with respect to the subsonic flow. Therefore, preserving the assumption C_f in the last term, it is concluded from this equation that the inlet impulse function for the supersonic flow I_A is always greater than the inlet impulse function for the subsonic flow I_B, i.e., $I_A > I_B$. The impulse function of the discontinuous flow is conserved across the leading shock, i.e., $I_D = I_A$, and finally the inequality $I_D > I_B$ is recovered.

The situation here is similar to that of the constant cross section area flow with viscous friction: The discontinuous flows with the supersonic inlet Mach numbers from the interval $M_B^- < M_{in} < M_A$ are not altered by thermal choking, with M_B^- the supersonic inlet Mach number linked with M_B through the Rankine–Hugoniot relationship.

The analysis reported in the above sections of this chapter allowed delineation of different possible high-speed combustion modes, to discuss some transitions between them as well as the possible birth of thermal chocking and stability issues in the vicinity of the corresponding conditions. This analysis, however, does provide no detail about the dynamics, no detail about the unsteady intermediate states, and no detail about the way the transitions may proceed between them. Thus, several questions remain open: What is the route toward these steady states? Are they always reached? What is the relevance of a simplified quasi-one-dimensional representation for more realistic high-speed flow conditions featuring separated injection of fuel and oxidizer, complex geometry effects, turbulence, and so forth? Therefore, the next sections of this chapter aim at enlarging the scope of the presentation by providing a brief survey on the description of turbulent mixing and combustion in high-speed flows.

7.7 Turbulent Mixing in Compressible Flows

In contrast to the simplified picture of one-dimensional steady-state flows of fully premixed reactants discussed in the previous sections, the reality of reactive supersonic flows in practical devices is often associated with unsteady and three-dimensional turbulent flows of non-premixed reactants. Since scalar mixing at large and small scales (i.e., molecular micro-mixing) is a prerequisite before combustion may occur, the description of turbulent mixing in compressible flows does appear as one of the principal issues associated with high-speed combustion. Such a description offers a challenging task for the modelers since the mixing process is influenced by variable density effects and associated pressure influence. Indeed, as emphasized in the introduction, these density variations may result not only from composition changes – including temperature changes – but also from viscous heating and local compression – expansion effects. These density changes, whatever their origins, can be characterized with the dilatation $P_A \equiv \partial u_k / \partial x_k$, i.e., the first invariant of the velocity gradient tensor $A_{ij} \equiv \partial u_i / \partial x_j$. In this respect, such density variations have been previously shown to be a source of counter-gradient diffusion effects, thus influencing the turbulent transport at the largest scales of the flow [44, 45]. Their potential influence on the scalar mixing at the smallest scales, i.e., scalar dissipation rate and micro-mixing, has been also studied in relevant situations including planar shock–turbulence interaction [46] and mixing in highly under-expanded jets [23, 47]. It should be underlined that, just for the sake of conciseness, only a few preliminary notions will be provided before embarking on the description of recent research results. It has been indeed chosen to place more emphasis on the discussion of some very recent studies devoted to turbulent mixing in supersonic flows. The readers are therefore referred to reference textbooks of the literature for a more exhaustive introduction to the basic knowledge of compressible flows [1, 48, 49].

One geometry of reference is the coaxial jet and, among others, well-documented experimental databases of supersonic coflowing jets of helium with air and argon with air have been documented by Cutler and coworkers [50–52]. Let us begin by considering the planar spatially developing mixing layer as a first canonical example of turbulent mixing in compressible flows. The indexes 1 and 2 refer to the highest and lowest speed streams, respectively, with U_1 and U_2 the corresponding velocities. Then, it is standard to introduce the convective Mach numbers $M_{c_1} = (U_1 - U_c)/a_1$ and $M_{c_2} = (U_2 - U_c)/a_2$, where the quantity a denotes the speed of sound and U_c is the convective velocity. The expression of the latter can be simplified as follows: $U_c = (a_1 U_2 + a_2 U_1)/(a_1 + a_2)$, in such a manner that the convective Mach number reduces to

$$M_c = M_{c_1} = M_{c_2} = \frac{U_1 - U_2}{a_1 + a_2}. \tag{7.92}$$

This convective Mach number has been used to evaluate the possible influence of compressibility effects on the mixing layer development [53, 54], which is generally characterized by the vorticity thickness

$$\delta = \delta_\omega = \frac{U_1 - U_2}{|\partial \widetilde{u}_1/\partial x_2|} \tag{7.93}$$

and by the associated normalized growth rate

$$\frac{\dot{\delta}}{\dot{\delta}_i} \text{ with } \dot{\delta} = \frac{d\delta}{dx_1}, \tag{7.94}$$

where $\dot{\delta}_i$ refers to incompressible conditions of reference.

This mixing layer normalized growth rate has been found to decay with the convective Mach number M_c [53–55]. When M_c increases, the normalized pressure fluctuations decrease and thus reduce the communication across the large-scale structures of the mixing layer: here is a *lack of pressure communication*. The stabilizing effect of compressibility is explained by a dampened growth rate of instability modes and an inhibited influence of pressure–strain correlations [56]. The corresponding evolution of pressure–strain terms for increasing values of M_c results from the decrease of both pressure and strain-rate fluctuations. In this respect, recent analyses indeed revealed that the decay of pressure fluctuations is not the sole reason for the reduced pressure–strain correlation, since the strain-rate fluctuations themselves decrease in a similar amount [57].

Thus, there is a strong dependency of the mixing layer normalized growth rate on the convective Mach number and it is therefore rather tempting to use M_c as a similarity parameter, with two distinct mixing layers being expected to display similar compressibility effects provided that their convective Mach number value remains the

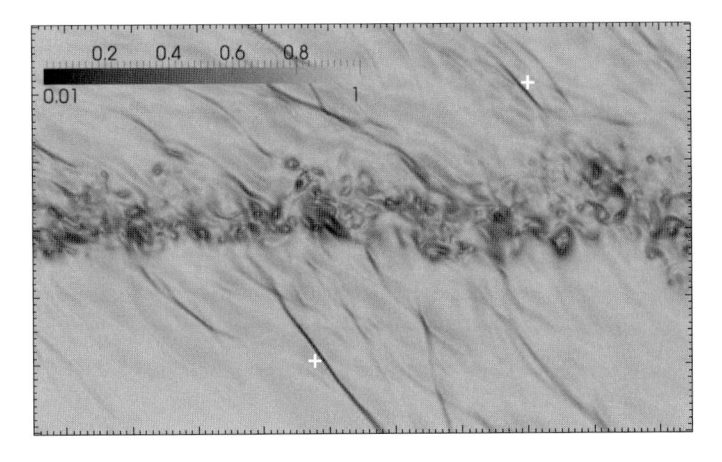

Figure 7.3 Numerical Schlieren imaging of a nonreactive spatially developing mixing layer at $M_c = 1.1$ (Courtesy of P. J. Martinez-Ferrer)

same. A generalized form, featuring a dependency to the density ratio $s = \rho_2/\rho_1$, is generally adopted for the normalized growth rate:

$$\frac{\dot{\delta}}{\dot{\delta}_{inc}} = \frac{U_1 - U_2}{2U_c} f_c(M_c, s) \Phi(M_c), \qquad (7.95)$$

where $\Phi(M_c)$ is a function that can be determined from experimental data and $f_c(M_c, s)$ is a correction factor, close to unity. The analysis of various experimental data conducted by Barre et al. [58] indeed shows that, even reduced, some departures of available data from a unique curve $\Phi(M_c)$ are still present.

It is also noteworthy that some shocklet structures may appear for sufficiently large values of this convective Mach number; see the vicinity of the white crosses in Fig. 7.3. It has been indeed verified that Rankine–Hugoniot relations hold across the discontinuity at these locations [59]. It should be specified that this figure displays the numerical schlieren imaging of a nonreactive spatially developing mixing layer at $M_c = 1.1$ with simulation details provided in two companion papers [57, 60].

On the other hand, the direct influence of through interactions of mixing layers with shock waves has also been studied both numerically and experimentally [61–63]. Such a canonical flowfield is typical of the shock-mixing layer interactions that take place in compressible flows of practical interest [64]. For instance, supersonic jets at high nozzle-pressure ratio (NPR) give rise to complex compressible structures, where shocks and expansions waves interact with the turbulent outer shear layer [47]. It is also relevant to scramjet intakes and combustors, where shock waves interact with the shear layers issued from the injection systems. On the one hand, it is clear that the occurrence of shock waves in supersonic combustors induces pressure losses that cannot be avoided but, on the other hand, the resulting shock interactions with mixing layers contribute to scalar dissipation (i.e., mixing) rate enhancement and may favor combustion stabilization in high-speed flows.

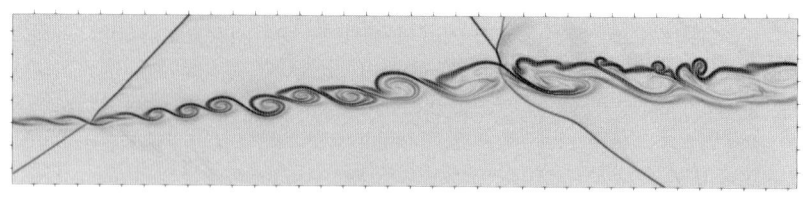

(a) with consideration of the bulk viscosity

(b) with a bulk viscosity set to zero

Figure 7.4 Instantaneous numerical Schlieren images obtained at normalized time $t\,(U_1 - U_2)/\delta_{\omega,0} = 70.0$ during a shock mixing layer interaction problem. (Courtesy of R. Boukharfane)

A typical example of a mixing layer impacted by a shock wave is presented in Fig. 7.4. The upper stream of the mixing layer corresponds to the fuel inlet, i.e., a mixture containing hydrogen, and the bottom inlet stream to vitiated air. The impinging shock wave is issued from the left bottom side, i.e., oxidizer inlet stream., with an angle of 33 degrees. The corresponding computational setup has been described elsewhere [19]. Figure 7.4 is typical of the topology of the shock wave refraction in supersonic planar mixing layers. First, the interaction between the incident shock and the mixing layer enhances the vorticity growth. Further downstream, the shock wave reflects on the top boundary condition, and the resulting reflected shock wave interacts with the organized vortices. Such instantaneous flow visualizations are very revealing of some local features of the shear layer, which are filtered out once fields or cross-stream profiles of averaged quantities are considered instead. For instance, quantitative comparisons of the onset of the streamwise vortices formation can be obtained from the instantaneous fields of the dimensionless magnitude of the density gradient (i.e., numerical schlieren), reported in Fig. 7.4. In this figure, it is remarkable that, in the absence of bulk viscosity, the normalized abscissa $x_1/\delta_{\omega,0}$ at which the vortex roll-up processes take place, is significantly smaller than the one associated with the situation featuring nonzero value of the bulk viscosity ($\delta_{\omega,0}$ denotes the initial vorticity thickness at the inlet of the computational domain). It should be emphasized here that only one typical set of instantaneous snapshots is reported in this figure just for the sake of conciseness, but the corresponding vortex roll-up process is systematically observed to take place early in the absence of volume viscosity.

A detailed and quantitative analysis of the bulk viscosity influence on the development of such a shock-impinged mixing layer has been recently provided in the literature [19]. The observed differences occur because of the nonnegligible value of the ratio of the bulk to shear viscosity, which can reach levels larger than 11 in the conditions associated with Fig. 7.4. In fact, the physical meaning of the bulk viscosity is not obvious but can be illustrated as follows. In either an expansion or a contraction of the mixture, the work done by the pressure modifies immediately the translational energy of the molecules, while a certain time lag is needed for the translational and internal energy to reequilibrate through inelastic collisions. The bulk viscosity is associated with this relaxation phenomenon and is evaluated from this internal energy relaxation time lag.

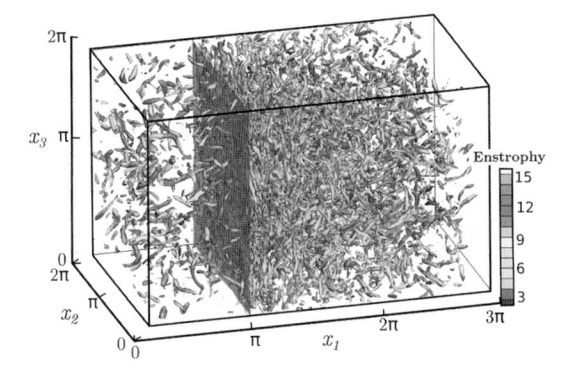

Figure 7.5 Instantaneous isosurface of the λ_2-criterion colored by the enstrophy magnitude. The shock wave is visualized by an isosurface of negative dilatation colored by pressure. (Courtesy of R. Boukharfane)

From the computational point of view, a more standard configuration to study shock–turbulence interaction is the one depicted in Fig. 7.5, where you have an homogeneous isotropic turbulence (HIT) interacting with a planar shock wave. The flow is coming from the left and the intensification of turbulent activity is clearly visible just downstream of the shock. A wealth of studies has been devoted to the analysis of such shock–turbulence interactions over the past 60 years. It has been previously treated analytically for small-amplitude disturbances using linearized Euler equations in conjunction with Rankine–Hugoniot jump conditions. Such conditions are met when there is a large-scale separation between the smallest turbulence length scale, i.e., the Kolmogorov length scale and the shock thickness. Under this hypothesis, the solutions behind the shock are obtained as a sum of plane wave solutions, assuming that each plane wave component of turbulence independently interacts with the shock. This approach is referred to as the linear interaction analysis [65–68]. As far as experimental studies are concerned, the majority of results concerns the shock–boundary layer interaction but the complexity of this flowfield makes the fundamental physics of the interaction difficult to sort out in comparison to the interaction of homogeneous turbulence with a planar shock wave. In Fig. 7.5, one can notice that the flow is strongly anisotropic in the post-shock region: the orientations of the turbulent structures (as deduced from the λ_2-criterion) are indeed parallel to the shock surface. The features of the corresponding post-shock turbulence have been analyzed in detail; for instance, the characteristics of the anisotropy tensor b_{ij} have been plotted in the Lumley triangle [69]. Upstream of the shock wave, the properties of the inlet turbulence – which is a HIT – are recovered, while downstream of the shock wave the characteristics are those of a cigar-shaped turbulence [46]. In a second step of the analysis, the structure of the turbulent flow is analyzed on the basis of the second and third invariants of the anisotropic (i.e., traceless) part of the velocity gradient tensor, $A_{ij}^* = A_{ij} - \delta_{ij} P_A$, where P_A denotes the first invariant of A_{ij}. Upstream of the shock wave, the joint probability analysis of the corresponding invariants, i.e., Q_{A^*} and R_{A^*}, displays the well-known teardrop shape [70]. Most of the collected data indeed follow either a stable-focus/stretching (SFS) or unstable-node/saddle/saddle (UNSS) local flow topology. Such a shape has been previously put into evidence by scrutinizing homogeneous isotropic turbulence computations, and its relevance has been also confirmed experimentally. However, significant changes of local flow dynamics and topology occur across the shock wave [46, 71]. They are associated with the enhancement of both (1) stable-node/saddle/saddle (SNSS) and (2) stable-focus compression (SFC). In this respect, it should be noted that, as far as experiments are concerned, there have been only a few attempts to study the interaction of a shock wave with homogeneous turbulence [72, 73]. As emphasized earlier, most of the experimental data have been indeed gathered in inhomogeneous turbulence conditions, e.g., shock-boundary layer interaction or shock mixing layer interaction geometry.

The preceding discussion shows that it has been a long time since the influence of shock waves on the turbulence dynamics concentrated important research efforts. This is in sharp contrast with the analysis of scalar mixing in such conditions, which received considerably less interest over the years. However, a significant step in this direction has recently been made [46]. In this study a detailed computational investigation of spatially decaying homogeneous scalar turbulence interacting with a planar shock wave was conducted. Comparisons were carried out with reference computations performed without shock waves and three distinct values of the shock Mach number M_s considered. With a more homogeneous scalar field and significantly reduced scalar gradient or SDR, there is a clear scalar mixing enhancement observed downstream of the shock wave. This is quite remarkable on the longitudinal evolutions of the variance and SDR normalized by their values taken just upstream of the shock wave. There is indeed a significant increase of the mixing rate just downstream of the shock wave and a stronger relaxation of composition (scalar) fluctuations compared to the unshocked case. In the evolution of the SDR, one of the leading-order source terms is the turbulence–scalar interaction (TSI) term, which renders the couplings between the velocity gradient components and the anisotropy of scalar dissipation rate tensor. Once expressed in the eigenframe of the strain-rate tensor, one can notice that it depends on its eigenvalues and on the orientation between the eigenvectors and the unit vector normal to the scalar isosurfaces. If an alignment of the scalar with the principal straining direction is privileged, the SDR tends to decrease, while if an alignment with the principal compression direction is priviliged, it tends to increase the SDR. It is therefore relevant to proceed with the analysis of the PDFs of the cosine angles between the normal direction and the eigenvectors. Its detailed inspection shows that there is a privileged alignment with the compression direction, which tends to be enhanced by the shock wave. This tendency increases with the Mach number: the shock wave is found to intensify not only the turbulent kinetic energy but also the scalar mixing. This is in line with early experimental studies from which it was concluded that shock waves may favor mixing and combustion stabilization [74].

The preceding discussion has been focused on the description of turbulent mixing in simple flow geometries, e.g., two-dimensional sheared flows. From a more practical viewpoint, there exist various methods to proceed with fuel injection within the airflow issued from the supersonic inlet and subsequent mixing enhancement: (1) transverse or crossflow fuel injection through small orifices [75]; (2) recirculating flows in cavities [76]; (3) hypermixer pylons featuring fuel injection systems, such as those used in ONERA-JAXA experiments [77]; or (4) strut injectors [78, 79] are among the most standard strategies. In practice, the corresponding strategies are often combined to favor ignition and stabilization of the combustion processes; see for instance [80]. As far as numerical simulations are concerned, the relevance of the retained strategy may be checked through a mixing efficiency analysis. The corresponding fuel mixing degree is indeed a critical parameter to evaluate the supersonic combustor performance as a whole. In this respect, there exist many possible definitions of the mixing efficiency in the literature; see for instance [75]. This quantity, i.e., the fuel mixing degree, can be defined as the mass flow rate ratio of reactants that would react to the total mass flow rate of reactants. For instance, considering hydrogen as the fuel, it may be written as follows [81, 82]:

$$\eta_m(x_1) = \frac{\int_{\mathcal{A}(x_1)} \overline{\rho}\,\widetilde{u}_1\, Y_{H_2,r}\, dA}{\int_{\mathcal{A}(x_1)} \overline{\rho}\,\widetilde{u}_1\, \widetilde{Y}_{H_2}\, dA}, \qquad (7.96)$$

where, following [81], the mass fraction of reactants that would be involved in the reaction $Y_{H_2,r}$ can be approximated by assuming a complete oxidation of the available amount of hydrogen:

$$Y_{H_2,r} = \begin{cases} \widetilde{Y}_{H_2} & \widetilde{Y}_{H_2} \leq Y_{H_2,st} \\ Y_{H_2,st}(1-\widetilde{Y}_{H_2})/(1-Y_{H_2,st}) & \widetilde{Y}_{H_2} \geq Y_{H_2,st}, \end{cases} \qquad (7.97)$$

with $Y_{H_2,st}$ the fuel mass fraction at stoichiometry. In Eq. (7.96), the elementary transverse surface element dA is evaluated from the product $\Delta x_2 \cdot \Delta x_3$ and, at each location x_1 along the combustor length, the integral is performed over the whole transverse section $\mathcal{A}(x_1)$.

Nonreactive flow simulations, such as those discussed earlier, can be used to get some preliminary insights on the reactive flow developments. For instance, from the nonreactive flow simulation, the combustion stabilization location can be roughly estimated thanks to tabulated self-ignition delays. Ignition probability density functions can also be approximated from the resolved composition PDF [83]. However, it should be fairly acknowledged that such estimates, which do ignore the possible influence of the combustion and associated heat release on the compressible flow development, provide only crude approximations that are highly questionable. In particular, the development of reactive compressible flow in confined conditions relevant to scramjet operation has been shown to be highly sensitive to the mixture reactivity – inlet temperature and equivalence ratio – and chemical reaction development [84]. It should also be mentioned that the above estimate of combustion stabilization location is relevant only for sufficiently high combustor inlet temperatures, when stabilization really proceeds by a self-ignition mechanism. At lower combustor inlet temperatures, when the self-ignition is no longer possible, forced ignition is required with the subsequent stabilization resulting from flame propagation often assisted by the recirculation of hot products in combustor cavities. In the general case, stabilization results from the interplay of both mechanisms, i.e., flame propagation and self-ignition [80], which makes the use of the above estimates questionable. It is quite clear that more insights should be gained from reactive flow simulations such as those discussed in the next section.

7.8 Turbulent Combustion in High-Speed Flows

In the first part of this discussion, we will recall the most essential features of the physics of high speed reactive flows, the physical processes at play, the corresponding characteristic time scales, and, if possible (considering our current and limited state-of-knowledge of this kind of flows), the reactive flow regimes.

Supersonic reactive flows are associated with quite extreme combustion conditions and one of the first questions that may arise indeed concerns the combustion regimes relevant to such conditions. This specific – and still open question – has been central to several studies since the work of Balakrishnan and Williams [83, 85–88].

The local conditions encountered in such supersonic reactive flows are highly variable from one location to another and it is therefore not possible to rely on the single knowledge of the inlet flow conditions to determine a priori the relevant turbulent combustion regimes. One can argue that this is not a limitation specific to supersonic flows and that this restriction also holds for low Mach number reactive flows, which is indeed the case. However, it must be emphasized that this variability is dramatically exacerbated in supersonic flow conditions, since not only the local flow characteristics but also the local thermodynamic conditions, in terms of pressure and temperature, display significant variations, which may be induced by (1) viscous dissipation of kinetic energy in shear and boundary layers, (2) compression induced by shock waves, and (3) rarefaction waves. The corresponding variations of thermodynamic conditions will significantly influence the chemical kinetics and, thereby, the nature of the turbulence-chemistry interaction (TCI) at the local level.

Moreover, from a practical viewpoint, the construction of turbulent combustion diagrams requires the choice of relevant time and length scales and this raises additional questions. For instance, as underlined in some of the above references [83, 88], as regards the choice of the chemical time scale, depending on the local conditions, either self-ignition or flame characteristic time scales may be relevant. The construction of any turbulent combustion diagrams may therefore require three or even more characteristic axes because of the multiple characteristic chemical times. This is consistent with the conclusions recently made by Driscoll, who analyzed a large set of experimental results obtained from highly turbulent methane-air flames [89–91]. He indeed proposed adding third and fourth axes to the Borghi diagram [92] to take into account residence time, mixing time, and ignition time (in the case of preheated reactants). From this analysis, he also concluded that intense turbulence does not create large and uniform reaction zones, but instead produces pockets of distributed reactions that are interconnected to each other by flamelets, a situation that can be referred to as the partially distributed reaction (PDR) regime. Moreover, he noted that PDR regimes correspond to local blobs of distributed reactions that are connected by thin flamelets. These blobs were found to be associated with either the merging or the breaking of flamelets. This type of combustion is also referred to as the distributed flamelets regime. Moreover, based on this set of experimental data, the same research group gained some new insights on the physics of autoignition for highly preheated reactants. The aforementioned concepts of partially distributed combustion regimes and autoignition physics for sufficiently preheated streams of reactants appear to be basically close to Shchetinkov's concept of microvolume combustion regime and to Shchetinkov's analysis of the combustion mechanism in a scramjet combustor, respectively [93, 94]. The interested reader may refer to Section 2.1 of this book entitled *Historical and physical perspective of turbulent combustion* for further details. At this level, it should also be underlined that spotty combustion has been indeed recently put into evidence in high-speed flow conditions [60] and the corresponding conditions are expected to be akin to MILD conditions [95]. In the general case, the construction of such a diagram also requires discriminating between premixed and non-premixed combustion modes [87] and, in this respect, the limitations of available flame index [96] are well known. The corresponding indices are indeed based on a one-dimensional flame picture with Fick diffusion model and a single-step chemistry.[3] Moreover, as suggested in the work of Balakrishnan and Williams [85], it is likely that the sole consideration of characteristic time and length scales of flow variation, in addition to chemical time and length scales, may no longer be sufficient. Some additional inputs associated to compressibility effects and exchanges between kinetic energy and temperature may also be relevant. Despite these limitations, it should be emphasized that the conclusions drawn from the most recent analyses [88, 98], conducted on the basis of highly resolved numerical simulations, performed in conditions relevant to scramjet engine operations, put into evidence important finite-rate chemistry effects and Damköhler number values that remain smaller (or of the order) of unity. Such a conclusion provides some support to the use of reactor-like description (e.g., PSR and PaSR) of turbulent combustion in high-speed flows. However, it should be acknowledged that many other stategies have been used, with some success, to tackle the problem of turbulent combustion modeling in supersonic flow conditions.

In a second part of the presentation, we proceed with a survey of turbulent combustion modeling applied to supersonic flows. In this respect, it should be stated that, over the past 30 years, almost all standard turbulent combustion modeling frameworks have been considered and tested to simulate turbulent combustion in high-speed flows.

---

[3] It is noteworthy that the last two issues have been addressed in some recent references [4, 97].

As far as the probability density function (PDF) methodology is concerned, if one except the work of Eiffer and Kollmann [99] and the one of Mobus et al. [100], there were only few efforts directed toward the full resolution of the PDF transport equation in supersonic flow conditions. The most recent steps in this direction have been made using the Eulerian Monte Carlo (EMC) formalism [101, 102]. This limited resort to the transported PDF method is in sharp contrast to the presumed PDF strategy, which has been retained as a possible modeling framework in many computational studies of turbulent combustion in high-speed flows, especially in the research group of Hassan, Baurle, and coworkers [103–105]. Flamelet models have been also tested [106–108], sometimes with specific extensions so as to incorporate the possible exchanges between kinetic energy and sensible enthalpy [109, 110] but it must be acknowledged that such a generalization still require further work [60, 111]. From a general point of view, the large input associated to flow kinetic energy term in energy conservation equation as well as the important pressure variation inside the flowfield introduce some dependences on these additional parameters, which deserves to be taken into account in the flamelet tabulations and this specific feature significantly complicates the flamelet library construction. LES and RANS computations of turbulent combustion in supersonic flows have been also performed with some success on the basis of either partially stirred reactors, i.e., PaSR [78, 79, 112], thus accounting for the turbulence–chemistry interactions (TCI), or under the perfectly stirred reactor (PSR) or quasi-laminar chemistry assumption [113].

Finally, the present section closes with a nonexhaustive presentation of some experimental test cases of reference for supersonic combustion computations. Indeed, the relevance of the various modeling strategies discussed earlier has been tested against data obtained in experimental setup that are considered as high-speed combustion benchmarks in the scientific community. Several experimental test rigs have been indeed designed to investigate turbulent combustion in supersonic flows over the past 50 years and, in many of them, the fuel – often hydrogen, methane, and ethylene (or a mixture between them) – is injected into a main flow of hot air that can be vitiated or not. In the corresponding high-temperature air flows, self-ignition may be triggered within the mixing layer that develops downstream of the fuel injection. For instance, such conditions are met in the coflowing jets experiments conducted by Beach [114] and by Cheng et al. [115], which consist of a supersonic burner that provides a choked main jet of hydrogen surrounded by an annular axisymmetric hot coflowing jet of vitiated air at Mach 2. Such open, i.e., unconfined, coflowing jet geometries have been often retained as reference test cases for the validation of supersonic combustion computations. For instance, the non-premixed supersonic jet flame studied experimentally by Cheng et al. [115] has been widely used as a pertaining test case for the computational modelling of turbulent combustion in supersonic flows [100, 104, 112, 116–119]. At this level, it should be emphasized that the important role played by pressure waves issued from the highly compressible external mixing layer with the ambient air has been recently put into evidence [119]. Indeed, in the corresponding set of experiments, ignition is supported by the corresponding pressure waves. This is firmly confirmed thank to an index that reflects the correlation between the heat release and pressure waves. Its consideration indeed allows to emphasize the crucial role played by the shock-wave structure on the stabilization of this supersonic flame. Another open geometry of reference is the JISCF studied at Stanford University [120]. It has been simulated first within the flamelet framework [108] and more recently with combustion represented by a well-stirred reactor that is given an initial condition and integrated in time until it reaches equilibrium [121]. The development and stabilization of combustion downstream of a wedge or strut injector model, such as the one studied by Waidmann et al. [122], has been also retained as a relevant geometry to assess the ability of computational models to describe the interaction between shock waves and hydrogen injection, which is of key importance for combustion stabilization. In this respect, the resulting well-documented experimental database has been used in several numerical investigations [78, 106, 107, 123].

All the experimental databases mentioned above have been widely used for turbulent combustion modeling validation but it should be fairly acknowledged that the operative conditions associated to these various Laboratory setups remain often rather far from those operated in scramjet prototypes. On the one hand, the total temperature levels reached in the wedge injector experiments conducted by Waidmann et al. [122] remain too moderate to trigger some typical effects of supersonic combustion conditions and, on the other hand, the development of free coflowing jets in the atmosphere is not representative of the large-scale mixing phenomena taking place in the more realistic geometry of wall injection into a confined high-enthalpy air stream. Other experiments were therefore developed to study high-speed mixing and combustion in a simple scramjet-like engine environment. For instance, a direct-connect supersonic combustor model, often referred to as the SCHOLAR experiment, has been developed at the NASA Langley Research Center. The model consists of a copper section (approximately 550 mm in length) followed by a longer carbon steel section attached to the aft end of the copper section. The copper section contains an inclined single injector that delivers gaseous hydrogen into a vitiated air stream [124]. As far as numerical simulations are concerned, the computation of this geometry has also concentrated some research efforts. For instance, it has been computed within the RANS framework using presumed PDF in conjunction with a detailed chemistry description [125]. Another investigation [126] considered both (1) the infinitely fast chemistry approximation and (2) the quasi-laminar approximation together with a single-step finite-rate chemistry so as to discriminate the influence of chemical kinetics in the flowfield thermochemical behavior. Finally, LES of the same experimental device were also performed [127].

Experiments covering scramjet-like engine environment have been also performed on the LAPCAT-II geometry, which is operated in the LAERTE facility at the ONERA Palaiseau Research Center. The LAPCAT-II combustor is fed with a hot vitiated air stream at Mach 2 issued from a de Laval nozzle. Total temperature levels can be as high as 1800–1900 K while the total pressure levels can reach 1.0–1.2 MPa. The corresponding conditions are expected to be relevant to real scramjet engines operations and concentrate some recent and ongoing computational studies [84, 128]. The experimental data include spontaneous flame emission images, wall-pressure measurements, and planar laser-induced fluorescence (PLIF) images of hydroxyl (OH) radicals, which are useful to validate numerical simulations. The LAERTE facility has been also used to operate other devices, such as the NAL supersonic combustor equipped with two conventional two-stage injection struts and an alternating-wedge injection strut. Large-eddy simulations of the corresponding geometry have been performed and successfully compared to measurements and flow visualizations [79, 113], with the resulting set of computational data subsequently used so as to complement available experimental data.

Other reference datasets have been obtained within the framework of the HyShot research program. The objective was to demonstrate the feasibility of igniting and maintaining supersonic combustion under realistic flight conditions. On the one hand, flight tests have been performed. The Hyshot scramjet was accelerated up to Mach 8 with a rocket and, between the altitudes of 23 km and 35 km, gaseous hydrogen was injected into the scramjet and pressure was sampled. A flight Mach number between 7 and 8 was recorded. On the other hand, ground tests have been carried out at different conditions in the T4 shock tunnel at the University of Queensland [75, 129] and in the high enthalpy shock tunnel of the DLR German Aerospace Center in Göttingen [130] in order to obtain correlations with flight-based tests. The HyshotII databases, which have been gathered within the framework of this research program, are now considered as a relevant and challenging test case for computational studies. In this respect, numerical simulations of this geometry have been performed with the PaSR model [131, 132], with the EDC model [133], and with the flamelet model within either the RANS [134] or the LES framework [135].

Relevant ground-based experiments have also been conducted on the HIFiRE Direct Connect Rig (HDCR) at NASA Langley Research Center (LaRC) in support of the HIFiRE flight test campaign [136]. It corresponds to a cavity-based hydrocarbon-fueled scramjet combustor, which was tested in a direct-connect fashion in the NASA LaRC Arc Heated Scramjet Test Facility. The model included a constant-area cross section isolator duct attached to a combustor containing five stages of fuel injectors. During the experiment, only the primary and secondary injectors located upstream of the cavity and downstream of the cavity trailing edge, respectively, were fueled. As already emphasized in the previous section, cavity flameholders are used as a way to (1) increase the fuel-residence times and (2) favor fuel–air mixing. A detailed description of various cavity flameholder designs and their performances has been provided by Ben-Yakar and Hanson [76]. The corresponding geometry has been retained as a pertaining test case for both Reynolds-averaged Navier–Stokes [87] and large-eddy simulations [108]. Finally, there exist many other recent experimental databases that have been gathered on geometries featuring wall-mounted cavities and the corresponding set of data served as experimental benchmarks to evaluate the capabilities of hybrid and dynamic RANS/LES computational models [137, 138]. In this respect, the experiments performed by Micka and Driscoll offers a very challenging test case [80]. This study was conducted on a dual-mode scramjet combustor with normal fuel injection upstream of a cavity flame-holder. Depending on the inlet vitiated air stream temperature, two stabilization modes are recovered: cavity-stabilized regime and jet-wake stabilized regime. At lower combustor inlet temperatures the stabilization is reached by the flame propagation mechanism (cavity-stabilized regime). At larger combustor inlet temperatures the stabilization is obtained through the self-ignition processes (jet-wake stabilized regime). Thus, combustion is either anchored at the cavity leading edge and spreads into the main flow at an approximately constant angle (low stagnation temperature levels) or stabilizes a short distance downstream of the hydrogen injection, in its wake, and features a curved leading edge (high stagnation temperature levels). For intermediate values of the operative temperature, some oscillations between the two modes are expected, and reproducing the behavior observed for these intermediate operative temperatures so as to understand how these oscillations may happen and how long each stabilization mode is active offers an interesting and challenging perspective for future computational studies.

7.9 Current Challenges and Future Research Needs

The use of reactor-like descriptions of SGS turbulence–chemistry interactions is continuously growing. This includes the EDC, PSR, and PaSR closures as well as the E-PaSR and U-PaSR models [83, 119, 139], which correspond to more recent developements of the PaSR approach. E-PaSR and U-PaSR stand for extented PaSR and unsteady PaSR, respectively. Thus, as far as the description of combustion chemistry is concerned, the development of reduced chemical schemes is needed. Some remarkable progress has been accomplished in this last direction for the hydrogen-air system [140, 141] but further efforts are still necessary for other fuels including ethylene, methane, and their mixtures, which are commonly used in supersonic combustion benchmarks. From a more general point view and in regard to the modeling of turbulent combustion in high-speed flows, additional works also remain needed to develop closures that can handle both self-ignition and flame propagation and their interplays.

The description and modeling of variable-density and compressible turbulence still offer open roads for future research. Indeed, even in the absence of shock waves or at rather moderate compressibility levels [57, 142], the characterization of variable-density turbulence still presents important difficulties that require further work. In this respect, it remains unclear if the Favre-averaging (or Favre-filtering) framework is better suited to variable density flows than the standard Reynolds averaging. Further efforts are required to address this specific issue that may also answer another question related to the existence of possible counter-gradient diffusion in reactive flows. The influence of shock waves [46, 143, 144] as well as possible couplings with variable-density effects [145] also raise fundamental and still unanswered questions.

Finally, from a general viewpoint, the characterization and the understanding of unsteady combustion phenomena and combustion oscillations development in scramjet engine combustors also appear as crucial practical issues that are receiving considerable interest [80, 146–148].

To conclude, there are other issues that have not been presented in detail in this chapter. This was just for the sake of conciseness and interested readers may find additional information as well as other viewpoints in two recently published review papers [13, 149].

Acknowledgments

It would have not been possible to present some of the content reported in this chapter without the experience gained from discussions shared with some colleagues and former PhD students. In this respect, the authors would like to thank Roland Borghi, Radouan Boukharfane, Romain Buttay, Luis Fernando Figueira da Silva, Christer Fureby, Laurent Gomet, Guillaume Lehnasch, Pedro José Martínez Ferrer, Yann Moule, Fábio Henrique Eugênio Ribeiro, and Anthony Techer for several stimulating interactions. The financial support of ONERA and Ministry of Education and Science of the Russian Federation (Contract No. 14.G39.31.0001 of 13.02.2017) is gratefully acknowledged.

References

[1] P. Chassaing, R. A. Antonia, F. Anselmet, and L. Joly, *Variable Density Turbulence*, R. Moreau, Ed. New York: Kluwer Academic, 2002.

[2] D. Desbordes and H. N. Presles, "Multi-scaled cellular detonation," in *Shock Waves Science and Technology Library: Detonation Dynamics*, F. Zhang, Ed. Berlin/Heidelberg: Springer, 2012, 281–338.

[3] C. Campbell and D. W. Woodhead, "Striated photographic records of explosion-waves," *J. Chem. Soc.*, 1572–1578, 1927.

[4] R. Buttay, L. Gomet, G. Lehnasch, and A. Mura, "Highly resolved numerical simulation of combustion downstream of a rocket engine igniter," *Shock Waves*, **27**, 655–674, 2017.

[5] A. M. Gamezo, V. N. Khokhlov, E. S. Oran, A. Y. Chtchelkanova, and R. O. Rosenberg, "Thermonuclear supernovae: simulations of the delagration stage and their implications," *Science*, **299**, 77–80, 2003.

[6] E. Oran, "Astrophysical combustion," *Proc. Combust. Inst.*, **30**, 1823–1840, 2005.

[7] E. P. Hicks, "Rayleigh-Taylor unstable flames – fast or faster?" *Astrophys. J.*, **803**, p. 72, 2015.

[8] A. Poludnenko, "Astrophysical combustion: from a laboratory flame to a thermonuclear supernova," in *Proceedings of the Twenty-fifth International Colloquium on the Dynamics of Explosions and Reactive Systems*, 2015.

[9] P. Wolanski, "Detonative propulsion," *Proc. Combust. Inst.*, **34**, 125–158, 2013.

[10] W. H. Heiser and D. T. Pratt, *Hypersonic Airbreathing Propulsion*. AIAA Education Series, 1994.

[11] M. K. Smart, "Scramjets," RTO-EN-AVT-150-09, Tech. Rep., NATO, Science and Technology Organisation. DOI: 10.14339/RTO-EN-AVT-150, 2008.

[12] C. Segal, *The Scramjet Engine: Processes and Characteristics*. Cambridge: Cambridge University Press, 2009.

[13] J. Urzay, "Supersonic combustion in air-breathing propulsion systems for hypersonic flight," *Ann. Rev. Fluid Mech.*, **50**, 593–627, 2018.

[14] K. Matsuo, Y. Miyazato, and H. D. Kim, "Shock train and pseudo-shock phenomena in internal gas flows," *Prog. Aero. Sci.*, **35**, 33–100, 1999.

[15] J. F. Clarke and M. McChesney, *Dynamics of Relaxing Gases*. Oxford: Butterworths, 1976.

[16] K. K. Kuo, *Principles of Combustion*, New York: John Wiley & Sons, Wiley-Interscience publication. 1986.

[17] F. A. Williams, *Combustion Theory*. Boca Raton, FL: CRC Press, 2018.

[18] G. Billet, V. Giovangigli, and G. De Gassowski, "Impact of volume viscosity on a shock–hydrogen-bubble interaction," *Combust. Theory Model.*, **12**, 221–248, 2008.

[19] R. Boukharfane, P. J. Martínez Ferrer, A. Mura, and V. Giovangigli, "On the role of bulk viscosity in compressible reactive shear layer developments," *Euro. J. Mech. - B/Fluids*, **77**, 32–47, 2019.

[20] H. Schlichting, *Boundary-Layer Theory*, McGraw-Hill series in mechanical engineering. New York: McGraw-Hill, 1979.

[21] S. Chapman and T. G. Cowling, *The Mathematical Theory of Non-uniform Gases*. Cambridge University Press,, 1970.

[22] L. F. Figueira da Silva and B. Deshaies, "The influence of equivalence ratio and Sorêt effect on the ignition of hydrogen-air mixtures in supersonic boundary layers," *Symp. (International) Combust.*, **25**, 29–36, 1994.

[23] R. Buttay, G. Lehnasch, and A. Mura, "Turbulent mixing and molecular transport in highly under-expanded hydrogen jets," *Int. J. Hydrogen Energy*, **43**, 8488–8505, 2018.

[24] J. O. Hirschfelder and C. F. Curtiss, *Molecular Theory of Gases and Liquids*. New York: John Wiley & Sons, 1969.

[25] J. F. Clarke, *Finite Amplitude Waves in Combustible Gases*. Philadelphia: SIAM, 1985, 183–245.

[26] J. F. Clarke, *Combustion and Compressibility in Gases*, C. S.-L. C.M. Brauner, Ed. NATO ASI Series (Series E: Applied Sciences) Berlin/Heidelberg: Springer, 1988.

[27] J. F. Clarke, "Fast flames, waves and detonation," *Prog. Energy Combust. Sci.*, **15**, 241–271, 1989.

[28] J. H. S. Lee, *The Detonation Phenomenon*. Cambridge: Cambridge University Press, 2008.

[29] Y. B. Zeldovich, G. I. Barenblatt, V. B. Librovich, and G. M. Makhviladze, *The Mathematical Theory of Combustion and Explosions*. Moscow: Akademiiânauk SSSR, Consultant Bureau, 1985.

[30] A. H. Shapiro, *The Dynamics and Thermodynamics of Compressible Fluid Flow*. New York: John Wiley & Sons, 1953.

[31] K. I. Shchelkin and Y. K. Troshin, *Gasdynamics of Combustion*. Mono Book Corporation, 1965.

[32] E. Oran and V. Gamezo, "Origins of the deflagration-to-detonation transition in gas-phase combustion," *Combust. Flame*, **148**, 4–47, 2007.

[33] M. Saif, W. Wang, A. Pekalski, M. Levin, and M. I. Radulescu, "Chapman-Jouguet deflagrations and their transition to detonation," *Proc. Combust. Inst.*, **36**, 2771–2779, 2017.

[34] A. Y. Poludnenko, T. A. Gardiner, and E. S. Oran, "Spontaneous transition of turbulent flames to detonations in unconfined media," *Phys. Rev. Let.*, **107**, p. 054501, 2011.

[35] J. Gray, "Reduction in the run-up distance for the deflagration to detonation transition and applications to pulse detonation combustion," Ph.D. dissertation, TU Berlin, 2018.

[36] B. M. Maxwell, "Turbulent combustion modelling of fast flames and detonations using compressible LEM-LES," Ph.D. dissertation, Carleton-Ottawa Institute for Mechanical and Aerospace Engineering, University of Ottawa (Canada), 2016.

[37] B. Maxwell, R. R. Bhattacharjee, S. S. M. Lau-Chapdelaine, S. A. E. G. Falle, G. J. Sharpe, and M. I. Radulescu, "Influence of turbulent fluctuations on detonation propagation," *J. Fluid Mech.*, **818**, 646–696, 2017.

[38] B. Maxwell, A. Pekalski, and M. Radulescu, "Modelling of the transition of a turbulent shock-flame complex to detonation using the linear eddy model," *Combust. Flame*, **192**, 340–357, 2018.

[39] S. Yu and S. Navarro-Martinez, "Modelling of deflagration to detonation transition using flame thickening," *Proc. Combust. Inst.*, **35**, 1955–1961, 2015.

[40] L. Crocco, *One-Dimensional Treatment of Steady Gas Dynamics*. Princeton: Princeton University Press, 1958, **3**, 64–348.

[41] F. A. Williams, *Combustion Theory*. San Francisco: Benjamin/Cummings, 1985.

[42] C. Segal, M. G. Owens, and S. Mullagilli, "Thermal chocking investigation in a supersonic combustor," *ISABE paper 99-7053*," 1999.

[43] S. Mashio, K. Kurashina, T. Bamba, S. Okimoto, and T. Kaji, "Unstart phenomenon due to thermal choke in scramjet module," *AIAA paper 2001-1887*," 2001.

[44] K. H. Luo and K. N. C. Bray, "Combustion-induced pressure effects in supersonic diffusion flames," *Symp. (International) Combust.*, **27**, 2165–2171, 1998.

[45] S. Serra, V. Robin, A. Mura, and M. Champion, "Density variations effects in turbulent diffusion flames: modeling of unresolved fluxes," *Combust. Sci. Technol.*, **186**, 1370–1391, 2014.

[46] R. Boukharfane, Z. Bouali, and A. Mura, "Evolution of scalar and velocity dynamics in planar shock-turbulence interaction," *Shock Waves*, **28**, 1117–1141, 2018.

[47] R. Buttay, G. Lehnasch, and A. Mura, "Analysis of small-scale scalar mixing processes in highly under-expanded jets," *Shock Waves*, **26**, 193–212, 2016.

[48] A. Smits and J. P. Dussauge, *Turbulent Shear Layers in Supersonic Flow*. Berlin/Heidelberg: Springer, 2006.

[49] T. B. Gatski and J. P. Bonnet, *Compressibility, Turbulence and High Speed Flow*. Philadelphia: Elsevier, 2013.

[50] A. A. Carty and A. D. Cutler, "Development and validation of a supersonic helium-air coannular jet facility," NASA report CR-1999-209717, Tech. Rep., 1999.

[51] A. Cutler, G. S. Diskin, J. P. Drummond, and J. A. White, "Supersonic coaxial jet experiment for computational fluid dynamics code validation," *AIAA J.*, **44**, 585–592, 2006.

[52] C. W. Cfifton and A. D. Cutler, "A supersonic argon/air coaxial jet experiment for computational fluid dynamics code validation," NASA report CR-2007-214866, Tech. Rep., 2007.

[53] D. W. Bogdanoff, "Compressibility effects in turbulent shear layers," *AIAA J.*, **21**, 926–927, 1983.

[54] D. Papamoschou and A. Roshko, "The compressible turbulent shear layer: an experimental study," *J. Fluid Mech.*, **197**, 453–477, 1988.

[55] C. Pantano and S. Sarkar, "A study of compressibility effects in the high-speed turbulent shear layer using direct simulation," *J. Fluid Mech.*, **451**, 329–371, 2002.

[56] S. Sarkar, "The stabilizing effect of compressibility in turbulent shear flow," *J. Fluid Mech.*, **182**, 163–186, 1995.

[57] P. J. Martínez Ferrer, G. Lehnasch, and A. Mura, "Compressibility and heat release effects in high-speed reactive mixing layers: growth rates and turbulence characteristics," *Combust. Flame*, **180**, 284–303, 2017.

[58] S. Barre, C. Quine, and J. P. Dussauge, "Compressibility effects on the structure of supersonic mixing layers: experimental results," *J. Fluid Mech.*, **259**, 47–78, 1994.

[59] P. J. Martínez Ferrer, "Etude par simulation numérique de l'auto-allumage en écoulement turbulent cisaillé supersonique," Ph.D. dissertation, Ecole Nationale Supérieure de Mécanique et d'Aérotechnique, 2013.

[60] P. J. Martínez Ferrer, G. Lehnasch, and A. Mura, "Compressibility and heat release effects in high-speed reactive mixing layers: structure of the stabilization zone and modeling issues relevant to turbulent combustion in supersonic flows," *Combust. Flame*, **180**, 304–320, 2017.

[61] J. R. Nuding, "Interaction of compressible shear layers with shock waves: an experimental study," *AIAA paper 1996-4515*," 1996.

[62] J. C. Hermanson and B. M. Cetegen, "Shock-induced mixing of nonhomogeneous density turbulent jets," *Phys. Fluids*, **12**, 1210–1225, 2000.

[63] S. M. V. Rao, S. Asano, I. Imani, and T. Saito, "Effect of shock interactions on mixing layer between co-flowing supersonic flows in a confined duct," *Shock Waves*, **28**, 1–17, 2017.

[64] C. Huete, A. L. Sánchez, and F. A. Williams, "Diffusion-flame ignition by shock-wave impingement on a hydrogen/air supersonic mixing layer," *J. Propul. Power*, **33**, 256–263, 2017.

[65] F. K. Moore, "Unsteady oblique interaction of a shock wave with a plane disturbance," NACA Report TR-1165, Tech. Rep., 1954.

[66] H. S. Ribner, "Shock-turbulence interaction and the generation of noise," NACA Report TR-1233, Tech. Rep., 1954.

[67] H. S. Ribner, "Convection of a pattern of vorticity through a shock wave," NACA Report TR-1164, Tech. Rep., 1954.

[68] H. S. Ribner, "Spectra of noise and amplified turbulence emanating from shock-turbulence interaction," *AIAA J.*, **25**, 436–442, 1987.

[69] J. L. Lumley and G. R. Newman, "The return to isotropy of homogeneous turbulence," *J. Fluid Mech.*, **82**, 161–178, 1977.

[70] A. Ooi, J. Martin, J. Soria, and M. S. Chong, "A study of the evolution and characteristics of the invariants of the velocity-gradient tensor in isotropic turbulence," *J. Fluid Mech.*, **381**, 141–174, 1999.

[71] J. Ryu and D. Livescu, "Turbulence structure behind the shock in canonical shock-vortical turbulence interaction," *J. Fluid Mech.*, **756**, R1-1–R1-12, 2014.

[72] L. Jacquin, C. Cambon, and E. Blin, "Turbulence amplification by a shock wave and rapid distorsion theory," *Phys. Fluids A*, **5**, 2539–2550, 1993.

[73] S. Barre, D. Alem, and J. P. Bonnet, "Experimental study of normal shock/homogeneous turbulence interaction," *AIAA J.*, **34**, 968–974, 1994.

[74] H. Huh and J. F. Driscoll, "Diffusion-flame ignition by shock-wave impingement on a hydrogen/air supersonic mixing layer," *Symp. (International) Combust.*, **26**, 2933–2939, 1996.

[75] Y. Moule, V. A. Sabelnikov, A. Mura, and M. Smart, "Computational fluid dynamics investigation of a Mach 12 scramjet engine," *J. Propul. Power*, **30**, 461–473, 2014.

[76] A. Ben-Yakar and R. K. Hanson, "Cavity flame-holders for ignition and flame stabilization in scramjets: an overview," *J. Propul. Power*, **17**, 869–877, 2001.

[77] T. Sunami, P. Magre, A. Bresson, F. Grisch, M. Orain, and M. Kodera, "Experimental study of strut injectors in a supersonic combustor using OH-PLIF," *AIAA Paper 2005-3304*," 2005.

[78] A. Mura and J. F. Izard, "Numerical simulation of supersonic nonpremixed turbulent combustion in a scramjet combustor model," *J. Propul. Power*, **26**, 858–868, 2010.

[79] C. Fureby, K. Nordin-Bates, K. Petterson, A. Bresson, and V. A. Sabelnikov, "A computational study of supersonic combustion in strut injector and hypermixer flow fields," *Proc. Combust. Inst.*, **35**, 2127–2135, 2015.

[80] D. J. Micka and J. F. Driscoll, "Combustion characteristics of a dual-mode scramjet combustor with cavity flameholder," *Proc. Combust. Inst.*, **32**, 2397–2404, 2009.

[81] C. Liu, Y. Zhao, Z. Wang, H. Wang, and M. Sun, "Dynamics and mixing mechanism of transverse jet injection into a supersonic combustor with cavity flameholder," *Acta Astronautica*, **136**, 90–100, 2017.

[82] Z. Cai, X. Liu, C. Gong, M. Sun, Z. Wang, and X. S. Bai, "Large-eddy simulation of the fuel transport and mixing process in a scramjet combustor with rearwall-expansion cavity," *Acta Astronautica*, **126**, 375–381, 2016.

[83] F. H. E. Ribeiro, R. Boukharfane, and A. Mura, "Highly-resolved large-eddy simulations of combustion stabilization in a scramjet engine model with cavity flameholder," *Computers & Fluids*," 2019.

[84] G. Pelletier, M. Ferrier, A. Vincent-Randonnier, V. A. Sabelnikov, and A. Mura, "Wall roughness effects on combustion development in confined supersonic flow," https://doi.org/10.2514/1.B37842 (in press).

[85] G. Balakrishnan and F. A. Williams, "Turbulent combustion regimes for hypersonic propulsion employing hydrogen-air diffusion flames," *J. Propul. Power*, **10**, 434–437, 1994.

[86] A. Ingenito and C. Bruno, "Physics and regimes of supersonic combustion," *AIAA J.*, **48**, 515–525, 2010.

[87] J. Quinlan, J. C. McDaniel, T. G. Drozda, G. Lacaze, and J. C. Oefelein, "A priori analysis of flamelet-based modeling for a dual-mode scramjet combustor," *AIAA 2014-3743*," 2014.

[88] A. Techer, "Simulation aux grandes échelles implicites et explicites de la combustion supersonique," Ph.D. dissertation, ENSMA Poitiers, 2017.

[89] J. F. Driscoll, "Premixed turbulent combustion regimes of thickened and distributed reactions," in *Proceedings of the ninth Mediterranean Combustion Symposium*, 2015.

[90] J. F. Driscoll, "Premixed turbulent combustion in high Reynolds number regimes of thickened flamelets and distributed reactions," Air Force Research Laboratory Report AFRL-AFOSR-VA-TR-2016-0136, Tech. Rep., 2016.

[91] A. Skiba, T. M. Wabel, C. D. Carter, S. D. Hammack, J. E. Temme, and J. F. Driscoll, "Premixed flames subjected to extreme levels of turbulence: Flame structure and a new measured regime diagram," *Combust. Flame*, **189**, 407–432, 2018.

[92] R. Borghi, "On the structure and morphology of turbulent premixed flames," in *Recent Advances in the Aerospace Science*, C. Casci, Ed. Plenum Publishing, 1985, 117–138.

[93] E. S. Shchetinkov, "Calculation of flame velocity in turbulent stream," *Symp. (International) Combust.*, **7**, 583–589, 1958.

[94] E. S. Shchetinkov, *Physics of Gases Combustion*. Nauka (Science), Moscow (in Russian), 1965.

[95] N. Swaminathan, "Physical insights on MILD combustion from DNS," *Front. Mech. Eng.*, **5**, 2019.

[96] H. Yamashita, M. Shimada, and T. Takeno, "A numerical study on flame stability at the transition point of jet diffusion flames," *Symp. (International) Combust.*, **26**, 27–34, 1996.

[97] S. Zhao, Z. Bouali, and A. Mura, "Computational investigation of weakly turbulent flame kernel growths in iso-octane droplet clouds in CVC conditions," *Flow Turbul. Combust.*," 2019.

[98] A. Techer, Simulation aux grandes échelles implicite et explicite de la combustion supersonique, PhD Thesis, University of Poitiers, 2017.

[99] P. Eiffer and W. Kollmann, "PDF predictions of supersonic hydrogen flames," *AIAA paper 1993-0448*," 1993.

[100] H. Mobus, P. Gerlinger, and D. Bruggemann, "Scalar and joint scalar-velocity-frequency Monte Carlo PDF simulation of supersonic combustion," *Combust. Flame*, **132**, 3–24, 2003.

[101] C. Gong, M. Jangi, X. S. Bai, J. H. Liang, and M. Sun, "Large-eddy simulation of hydrogen combustion in supersonic flows using an Eulerian stochastic fields method," *Int. J. Hydrogen Energy*, **42**, 1264–1275, 2017.

[102] Y. Paixao de Almeida and S. Navarro-Martinez, "Large eddy simulation of a supersonic lifted flame using the Eulerian stochastic fields method," *Proc. Combust. Inst.*, **37**, 3693–3701, 2019.

[103] R. A. Baurle, G. A. Alexopoulos, and H. A. Hassan, "An assumed joint PDF approach for supersonic turbulent combustion," *J. Propul. Power*, **10**, 473–485, 1994.

[104] R. A. Baurle and S. S. Girimaji, "Assumed PDF turbulence-chemistry closure with temperature-composition correlations," *Combust. Flame*, **134**, 131–148, 2003.

[105] X. Xiao, H. A. Hassan, and R. A. Baurle, "Modeling scramjet flows with variable turbulent Prandtl and Schmidt numbers," *AIAA J.*, **45**, 1415–1423, 2007.

[106] M. Oevermann, "Numerical investigation of turbulent hydrogen combustion in a scramjet using flamelet modeling," *Aero. Sci. Technol.*, **4**, 463–480, 2000.

[107] M. Berglund and C. Fureby, "LES of supersonic combustion in a scramjet engine model," *Proc. Combust. Inst.*, **31**, 2497–2504, 2007.

[108] A. Saghafian, L. Shunn, D. A. Philips, and F. Ham, "Large eddy simulations of the HIFiRE scramjet using a compressible flamelet/progress variable approach," *Proc. Combust. Inst.*, **35**, 2163–2172, 2015.

[109] L. L. Zheng and K. N. C. Bray, "The application of new combustion and turbulence models to H_2-air nonpremixed supersonic combustion," *Combust. Flame*, **99**, 440–448, 1994.

[110] V. A. Sabelnikov, B. Deshaies, and L. F. Figueira da Silva, "Revisited flamelet model for nonpremixed combustion in supersonic turbulent flows," *Combust. Flame*, **114**, 577–584, 1998.

[111] Z. Gao, J. Wang, C. Jiang, and C. Lee, "Application and theoretical analysis of the flamelet model for supersonic turbulent combustion flows in the scramjet engine," *Combust. Theory Model.*, **18**, 652–691, 2014.

[112] J. F. Izard, G. Lehnasch, and A. Mura, "A Lagrangian model of combustion in high-speed flows: application to scramjet conditions," *Combust. Sci. Technol.*, **181**, 1372–1396, 2009.

[113] M. Berglund, E. Fedina, C. Fureby, J. Tegner, and V. A. Sabelnikov, "Finite rate chemistry large-eddy simulation of self-ignition in supersonic combustion ramjet," *AIAA J.*, **48**, 540–550, 2010.

[114] H. L. Beach, "Supersonic mixing and combustion of a hydrogen jet in a coaxial high-temperature test case," *AIAA Paper 72-1179*," 1972.

[115] T. S. Cheng, J. A. Wehrmeyer, R. W. Pitz, O. Jarrett, and G. B. Northam, "Raman measurement of mixing and finite-rate chemistry in a supersonic hydrogen-air diffusion flame," *Combust. Flame*, **99**, 157–173, 1994.

[116] R. A. Baurle, A. T. Hsu, and H. A. Hassan, "Assumed and evolution probability density functions in supersonic turbulent combustion calculations," *J. Propul. Power*, **11**, 1132–1138, 1995.

[117] L. Gomet, V. Robin, and A. Mura, "Influence of residence and scalar mixing time scales in non-premixed combustion in supersonic turbulent flows," *Combust. Sci. Technol.*, **184**, 1471–1501, 2012.

[118] P. Boivin, A. Dauptain, C. Jiménez, and B. Cuenot, "Simulation of a supersonic hydrogen-air autoignition-stabilized flame using reduced chemistry," *Combust. Flame*, **159**, 1779–1790, 2012.

[119] Y. Moule, V. A. Sabelnikov, and A. Mura, "Highly resolved numerical simulation of combustion in supersonic hydrogen-air coflowing jets," *Combust. Flame*, **161**, 2647–2668, 2014.

[120] M. Gamba and G. M. Mungal, "Ignition, flame structure and near-wall burning in transverse hydrogen jets in supersonic crossflow," *J. Fluid Mech.*, **780**, 226–273, 2015.

[121] G. V. Candler, N. Cymbalist, and P. E. Dimotakis, "Wall-modeled large-eddy simulation of autoignition-dominated supersonic combustion," *AIAA J.*, **55**, 2410–2423, 2017.

[122] W. Waidmann, F. Alff, M. Bohm, U. Brummund, M. Clauss, and M. Oschwald, "Supersonic combustion of hydrogen/air in a scramjet combustion chamber," *Space Technol.*, **15**, 421–429, 1995.

[123] U. Wepler and W. W. Koschel, "Numerical investigation of turbulent reacting flows in a scramjet combustor model," *AIAA Paper 2002- 3572*," 2002.

[124] A. Cutler, P. Danehy, R. Springer, and D. DeLoach, "CARS thermometry in a supersonic combustor for CFD code validation," *AIAA Paper 2002-0743*," 2002.

[125] P. Keistler, X. Xiao, H. Hassan, and A. Cutler, "Simulation of the SCHOLAR supersonic combustion experiments," *AIAA Paper 2007-835*," 2007.

[126] M. S. R. Chandra Murty and D. Chakraborty, "Numerical simulation of angular injection of hydrogen fuel in scramjet combustor," *Proc. Inst. Mech. Eng., Part G: J. Aero. Eng.*, **226**, 861–872, 2012.

[127] A. Ingenito and C. Bruno, "LES of a supersonic combustor with variable turbulent Prandtl and Schmidt numbers," *AIAA Paper 2008-515*," 2008.

[128] A. Vincent-Randonnier, V. A. Sabelnikov, A. Ristori, N. Zettervall, and C. Fureby, "An experimental and computational study of hydrogen-air combustion in the LAPCAT II supersonic combustor," *Proc. Combust. Inst.*, **37**, 3703–3711, 2019.

[129] M. K. Smart, N. E. Hass, and A. Paull, "Flight data analysis of the HyShot 2 scramjet flight experiment," *AIAA J.*, **44**, 2366–2375, 2006.

[130] K. Hannemann, S. Karl, J. M. Schramm, and J. Steelant, "Methodology of a combined ground based testing and numerical modelling analysis of supersonic combustion flow paths," *Shock Waves*, **20**, 353–366, 2010.

[131] C. Fureby, M. Chapuis, E. Fedina, and S. Karl, "CFD analysis of the HyShot II scramjet combustor," *Proc. Combust. Inst.*, **33**, 2399–2405, 2011.

[132] M. Chapuis, E. Fedina, C. Fureby, K. Hannemann, S. Karl, and J. Martinez Schramm, "A computational study of the HyShot II combustor performance," *Proc. Combust. Inst.*, **34**, 2101–2109, 2013.

[133] A. Ingenito, D. Cecere, and E. Giacomazzi, "Large eddy simulation of turbulent hydrogen-fuelled supersonic combustion in an air cross-flow," *Shock Waves*, **23**, 481–494, 2013.

[134] R. Pecnik, V. E. Terrapon, F. Ham, G. Iaccarino, and H. Pitsch, "Reynolds-averaged Navier-Stokes simulations of the HyShot II scramjet," *AIAA J.*, **50**, 1717–1732, 2012.

[135] J. Larsson, S. Laurence, I. Bermejo-Moreno, J. Bodart, S. Karl, and R. Vicquelin, "Incipient thermal choking and stable shock-train formation in the heat-release region of a scramjet combustor," *Combust. Flame*, **162**, 907–920, 2015.

[136] K. Jackson, M. Gruber, and S. Buccellato, "HIFiRE flight 2 project overview and status update 2011," *AIAA Paper 2011-2202*," 2011.

[137] A. S. Potturi and J. R. Edwards, "Large-eddy/Reynolds-averaged Navier-Stokes simulation of cavity-stabilized ethylene combustion," *Combust. Flame*, **162**, 1176–1192, 2015.

[138] E. A. Hassan, D. M. Peterson, K. Walters, and E. A. Luke, "Dynamic hybrid RANS/LES computations of a supersonic cavity," *AIAA Paper 2016-1118*," 2016.

[139] V. A. Sabelnikov and C. Fureby, "LES combustion modeling for high Re flames using a multi-phase analogy," *Combust. Flame*, **160**, 83–96, 2013.

[140] P. Boivin, A. L. Sánchez, and F. A. Williams, "Four-step and three-step systematically reduced chemistry for wide-range H_2-air combustion problems," *Combust. Flame*, **160**, 76–82, 2013.

[141] A. L. Sánchez and F. A. Williams, "Recent advances in understanding of flammability characteristics of hydrogen," *Prog. Energy Combust. Sci.*, **41**, 1–55, 2014.

[142] G. S. Sidharth and G. V. Candler, "Subgrid-scale effects in compressible variable-density decaying turbulence," *J. Fluid Mech.*, **846**, 428–459, 2018.

[143] Y. P. M. Sethuraman, K. Sinha, and J. Larsson, "Density variations effects in turbulent diffusion flames: modeling of unresolved fluxes," *Theo. Comput. Fluid Dyn.*, **32**, 629–654, 2018.

[144] C. H. Chen and D. A. Donzis, "Shock-turbulence interactions at high turbulence intensities," *J. Fluid Mech.*, **870**, p. 813–847, 2019.

[145] Y. Tian, F. A. Jaberi, and D. Livescu, "Density effects on the post-shock turbulence structure and dynamics," *J. Fluid Mech.*, **880**, 935–968, 2019.

[146] K. C. Lin, K. Jackson, R. Behdadnia, T. A. Jackson, F. Ma, and V. Yang, "Acoustic characterization of an ethylene-fueled scramjet combustor with a cavity flameholder," *J. Propul. Power*, **26**, 1161–1170, 2010.

[147] M. L. Fotia and J. F. Driscoll, "Ram-scram transition and flame/shock-train interactions in a model scramjet experiment," *J. Propul. Power*, **29**, 261–273, 2013.

[148] H. Ouyang, W. Liu, and M. Sun, "The influence of cavity parameters on the combustion oscillation in a single-side expansion scramjet combustor," *Acta Astronautica*, **137**, 52–59, 2017.

[149] E. D. Gonzalez-Juez, A. R. Kerstein, R. Ranjan, and S. Menon, "Advances and challenges in modeling high-speed turbulent combustion in propulsion systems," *Prog. Energy Combust. Sci.*, **60**, 26–67, 2017.

8 Liquid Fuel Combustion

Z. Bouali, A. Mura, and J. Reveillon

Multiphase flows are ubiquitous in environmental flows (e.g., river flows), thermal energy systems, and reactive flows relevant to propulsion or energy production. Therefore, their analysis has attracted significant research efforts. This chapter deals with combustion of liquid–gas mixtures, and the solid–gas mixture combustion is discussed in the next chapter. Regardless of potential applications, characterization of the interface that separates the different phases often constitutes one of the main challenge, since the interface evolution controls to a large extent the transfer of mass, momentum, and energy. Since interfaces are presently considered in fluid media, the corresponding heterogeneous media can flow, at least at some locations. Therefore, one specific target is to characterize the corresponding flowfield not only in terms of momentum transfer but also mass exchanges, mixing, and possible reactions among various components. Determining the space–time evolution of the fluid–gas interface in liquid fuel combustion is challenging. Also, the description of heat and mass transfer in two-phase flows, specifically involving liquid–gas mixtures, poses further challenges discussed in this chapter. This includes phase transitions in the form of vaporization processes, i.e., evaporation or boiling. Similar processes, such as devolatilization, char combustion, ash formation, etc., are involved in solid–gas combustion, which are discussed in the next chapter. The determination of the evolution of the joint statistics of velocity and characteristic length scales (i.e., granulometry) in two-phase flows is a key point for analyzing liquid fuel combustion. These statistics are governed mainly by the interaction of turbulence with the interface. In this respect, liquid atomization and turbulence belong to a more general class of fragmentation processes that operate at rather high Reynolds and Weber numbers. In such flow regimes, inertia dominates over viscous and surface tension forces, except at the smallest scales where molecular effects become effective. In single-phase flows, the smallest length scales of interest are the Kolmogorov scale for the velocity field and the Batchelor scale for the scalar field. As emphasized later on, for fluid media featuring interfaces, the reference scalar field is often an indicator function associated with strong variations of density, viscosity, etc., and the classical theory of turbulence does not incorporate feedback from the scalar field onto the velocity field. However, this scalar field indeed acts on the velocity field, for instance, through pressure terms and there are some questions and specific issues regarding cross-scale (i.e., backscatter) energy transfer processes. This short introduction clearly emphasizes the multiscale characteristics of multiphase reactive flows and the present chapter aims to provide a brief description of the state-of-the-art in this field.

As a first step toward studying liquid fuel combustion, it is worth emphasizing that *multiphase* systems should be understood as composite systems featuring at least two flowing *materials*, either liquid and gas (in this chapter) or solid particles in gas (in the next chapter). Indeed, there is still some ambiguity regarding the use of this terminology versus *multicomponent* systems and the reader is referred to the textbook of Drew and Passman [1] for further details. As a consequence, in the present framework, the density difference between the two phases, i.e., the liquid (or solid) and the gas, is expected to be significant. It exceeds 100 in many applications, and this raises specific modeling issues for turbulent two-phase flows [2, 3]. Such a vast density difference is absent in multicomponent systems.

Two-phase flow combustion is encountered in many practical applications ranging from pool fires to diesel engines. It is also relevant to liquid propellant rockets. From a practical point of view, the design of liquid-fueled combustion systems is subject to a large number of constraints: it is necessary to atomize the liquid, to vaporize the resulting droplets, to mix the fuel vapor with surrounding air, and to ignite and burn the mixture within the available residence time in a practical system. In regard to the ignition of turbulent sprays, readers are referred to the recent study of Neophytou et al. [4] and to the recent review of Mastorakos [5].

Depending on the application, various multiphase flow topologies can be identified. They are often classified as either separated or disperse flows, which consist of finite particles, drops, or bubbles distributed in a connected volume of the continuous phase, while the separated flows consist of continuous streams of different fluids separated by interfaces. In this chapter, attention is focused on situations that involve the liquid as the *disperse* phase and the gas as the *continuous* phase. However, in the forthcoming sections, atomization processes will be discussed and one may locally have the liquid as the continuous phase and the gas as the disperse phase, e.g., cavitation processes in high-pressure injection systems. The above choice is especially relevant for two-phase flow combustion applications, since mass transfer (i.e., evaporation) should be optimized. In this respect, the main focus of this chapter is on developed sprays and only a brief overview of phenomena involved in atomization processes is provided. The presentation indeed focuses on two-phase flow combustion developing in regions where the dispersed phase (liquid) volume fraction does not exceed 10^{-3}. Such a condition is required to enhance evaporation and mixing of the fuel vapor with the oxidizer. This is in contrast with the dense regions where atomization, droplet collision, and coalescence are the dominant processes. Coalescence is the process by which several droplets interact and merge with each other. In the dilute regime, the dispersed phase may have a significant influence on the continuous (gaseous) phase turbulence. The presence of liquid droplets even with small volume fractions of the order of 10^{-6} can modify the carrier fluid turbulence spectrum; this is the so-called turbulence modulation [6, 7].

This chapter is organized as follows. In the next section, the formation of sprays through atomization processes together with their statistical description will be discussed. This will be followed by a short discussion of the available frameworks for two-phase flow description and associated computational fluid dynamics methods. The modeling issues related to evaporation and combustion are discussed in Section 8.3. Two-phase flow turbulent combustion regimes are described afterwards in Section 8.4, and the chapter closes with a short discussion on the current challenges in this topic.

8.1 Two-Phase Flow Topology and Spray Statistics

As emphasized in the introduction, there exists a wide spectrum of multiphase flows that are characterized by the simultaneous presence of at least two distinct phases such as gas–solid, liquid–solid, or gas–liquid, including gas–liquid–solid. However, this chapter focuses on two-phase flows of liquid–gas mixtures and combustion engines are considered as relevant and typical examples.

In engines, the injection system generates a spray with desired droplet size distribution at a given location. The topology of the liquid–gas flow that emerges from the injector changes progressively from the separated liquid jet flowing in the air to a spray of small droplets dispersed in the surrounding gas (Fig. 8.1). This transition is governed by several factors including the internal flow in the injector and the external liquid–gas and liquid–liquid interactions. Depending on the injection technology and nozzle geometry, the liquid leaves the nozzle with either a little or significant level of turbulence intensity. Some vapor bubbles may also be produced when the local pressure in the liquid phase falls below the saturation pressure, which is known as cavitation.

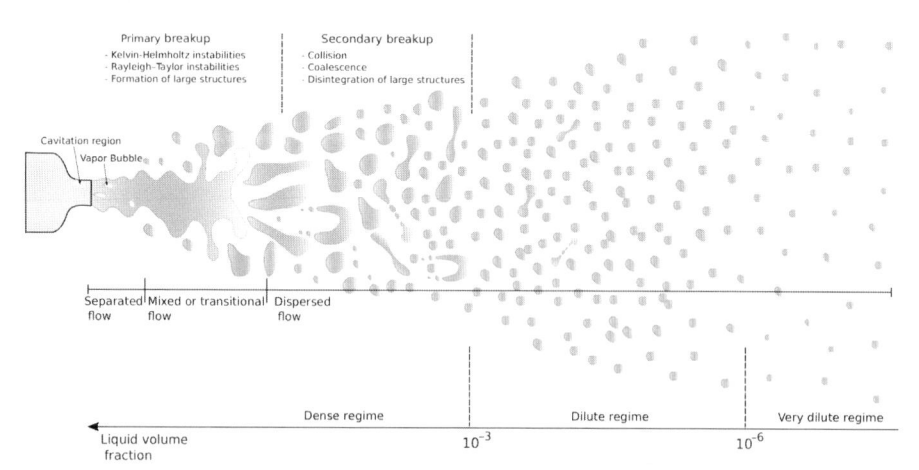

Figure 8.1 Schematic representation of atomization processes and various flow regimes in classical sprays.

Once the liquid penetrates into the gaseous environment, Kelvin–Helmholtz instabilities develop and lead to the birth of initial ripples at the liquid–gas interface. Then, the liquid–gas boundary undergoes an acceleration, which is induced by the density jump through the interface (Rayleigh–Taylor instability), thus producing liquid ligaments that are further stretched in the air stream and break into droplets [8]. These initial steps of liquid disintegration into ligaments and large droplet formation are referred to as the primary breakup process.

The aerodynamic forces that are induced by the velocity difference between the two phases is a source of instability and results in a secondary atomization (i.e., secondary breakup) of the ligaments and liquid blobs into even smaller droplets. Based on experimental observations, this secondary atomization step has been divided into four main regimes [9]: bag breakup, multimode breakup, shear/stripping breakup, and catastrophic breakup. These main regimes may themselves include other subregimes [10].

The secondary atomization regime depends mainly on two dimensionless numbers: (1) the Weber number We $= \rho U a / \sigma$, where ρ is the gas density, U the gas-droplet relative velocity, a the droplet initial diameter, and σ the surface tension; and (2) the liquid Ohnesorge number, which is defined as Oh $= \mu_l / \sqrt{(\rho_l a \sigma)}$, where μ_l and ρ_l are the dynamic viscosity and density of the liquid respectively.

According to Hsiang and Faeth [9, 11], the value of We required for regime transitions, as illustrated in Fig. 8.2, remains approximately constant for Oh values smaller than 0.1 (i.e., for liquid with relative low viscosity). For Oh $>$ 0.1, this value progressively increases with Oh. As a consequence, the critical value We$_c$ required to initiate the atomization processes is around 13 for Oh $<$ 0.1, while it is 30 for Oh equal to unity. It is noteworthy that, for Weber numbers smaller than We$_c$, atomization does not occur, and only droplets are formed. It is also noteworthy that, in automotive applications [12], it is often claimed that the determination of the secondary breakup of gasoline and diesel sprays depends exclusively on the Weber number (cf. Fig. 8.2).

Finally, the fact that the breakup process is not instantaneous is another important feature of secondary atomization. The breakup process indeed requires some time to take place and it is considered that this time has been reached once the droplet and/or all resulting liquid fragments do no longer undergo further breakups [13].

The liquid–gas flow in the region very close to the nozzle can be considered as a separated (or transitional) flow and the rest of the flowfield can be considered as a dispersed flow. Depending on liquid volume fraction, it can be classified either as a dense or a dilute spray. The dense region, which is dominated by atomization phenomena, refers to the zone where the liquid volume fraction is above 10^{-3}. In this region, droplet collision and coalescence occur frequently and can significantly influence the topology and dynamics of the disperse phase.

In the dilute regime, the liquid volume fraction lies within the range between 10^{-6} and 10^{-3}. In this regime, the probability to have liquid–liquid interactions remains quite negligible. However, the influence of the dispersed phase on the turbulent continuous phase (two-way interactions) may be significant. Finally, in the highly dilute regime, where the influence of the dispersed liquid phase on the turbulence in the gas becomes negligible, the liquid volume fractions is less than 10^{-6}. The corresponding liquid droplets act as tracers that do not alter the gas flow.

Let us now proceed with a statistical description of spray using distribution functions. In dispersed sprays, each droplet can be described by its diameter, position, velocity, temperature, and composition. Thus, a spray distribution function can be introduced to define the probable number of droplets having, at time t, (1) a position between $\mathbf{x} - d\mathbf{x}/2$ and $\mathbf{x} + d\mathbf{x}/2$, (2) a diameter within the range between $D - dD/2$ and $D + dD/2$, (3) a velocity ranging from $\mathbf{u} - d\mathbf{u}/2$ to $\mathbf{u} + d\mathbf{u}/2$, (4) a temperature between $T - dT/2$ and $T + dT/2$, and (5) a composition ranging from $\alpha - d\alpha/2$ to $h\alpha + d\alpha/2$.

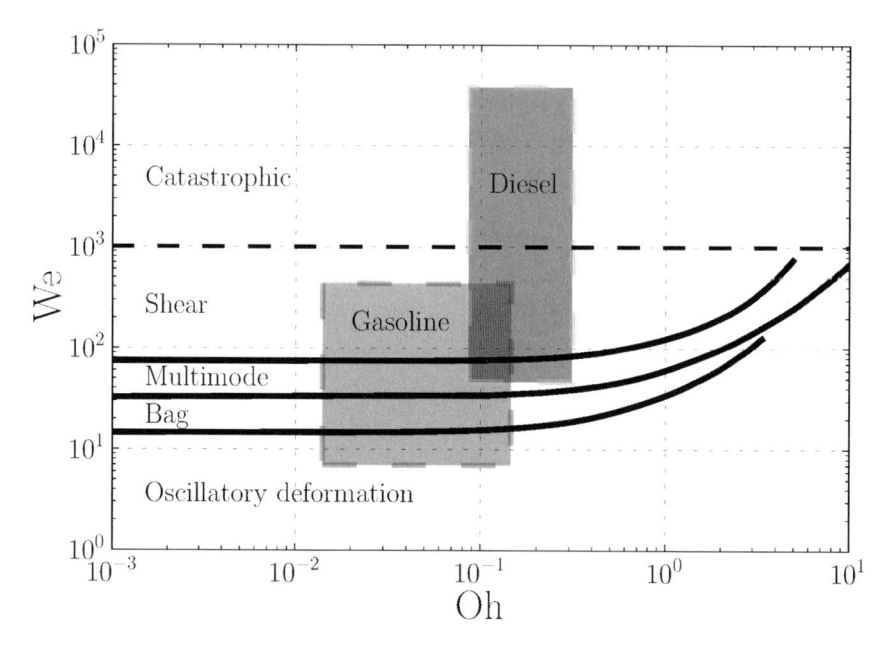

Figure 8.2 Map of secondary atomization regimes as functions of Weber and Ohnesorge numbers. Adapted from Hsiang and Faeth (1993) and Chryssakis and Assanis (2011).

For a homogeneous composition of the liquid and without any special consideration of the distribution dependence on velocity and temperature, the standard size distribution function $f(D)$ or droplet size distribution of the spray can be obtained by integrating the distribution function, defined earlier, over time and space. This is the fraction of droplet having a diameter around D. From a practical viewpoint, and at the discrete level, this function is readily obtained by classifying liquid elements into groups or classes, the diameters of which lie between $D - \Delta D/2$ and $D + \Delta D/2$. The histogram of droplet sizes may be then obtained by evaluating the number of elements present in each group or class. At the continuous level, the counterpart of this discrete histogram corresponds to the number distribution $f_0(D)$, which is a probability density function (PDF) that characterizes the spray. The quantity $f_0(D)$ gives the probability of finding a droplet with a diameter lying in the range from $D - \Delta D/2$ to $D + \Delta D/2$, divided by ΔD. It has therefore the dimension of the inverse of a length. Other PDFs characterizing the spray are the length $f_1(D)$, area $f_2(D)$, and volume

Table 8.1 Mean diameters

a	b	Name	Field of application
1	0	Linear	Comparisons, evaporation
2	0	Surface	Absorption
3	0	Volume	Hydrology
2	1	Surface diameter	Adsorption
3	1	Volume diameter	Evaporation, molecular diffusion
3	2	Sauter	Combustion, mass transfer, and efficiency studies
4	3	De Brouckere	Combustion equilibrium

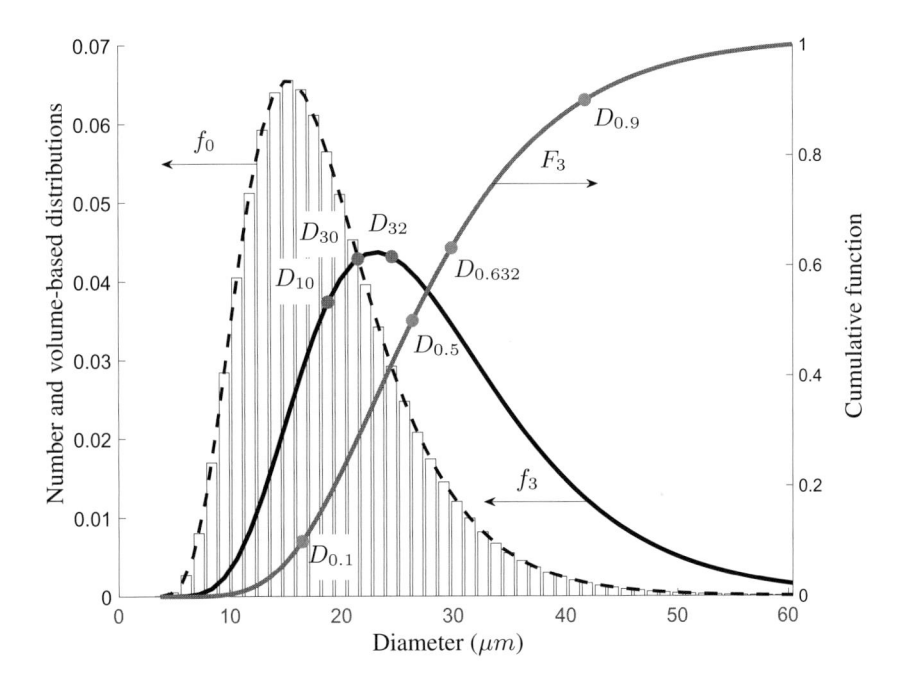

Figure 8.3 Typical droplet size distribution.

$f_3(D)$ distributions with the last one (i.e., $f_3(D)$) being by far the most popular among them. Actual sprays contain droplets having finite maximum and minimum sizes. For the sake of convenience, it is commonly assumed that $0 < D < \infty$. Common sense, indeed, tells us that the probability of finding a droplet in each class must be positive, while the probability of finding a droplet in any of the classes must reach 100%:

$$f(D) \geq 0 \quad \text{and} \quad \int_0^\infty f(D)d(D) = 1. \tag{8.1}$$

The most popular drop size distributions are built on the basis of spray experimental data (empirical fitting approach) and they are formulated for the number-based distribution (log-normal, Nukiyama–Tanasawa, log-hyperbolic) or for the volume-based distribution (root-normal, normal or modified Rosin-Rammler). One drawback is that these functions do not guarantee a satisfactory extrapolation beyond the range of experimental data for which they were fitted. Several approaches were therefore proposed to overcome this limitation: the maximum entropy formalism (MEF); the discrete probability function (DPF); and one numerical approach, the stochastic breakup model.

The distribution of the spray can be characterized with a reduced number of parameters. Indeed, for practical purposes, it is convenient to work only with average diameters such as those defined below instead of considering the whole droplet size distribution. From a general viewpoint, these diameters are defined as follows:

$$(D_{ab})^{a-b} = \frac{\int_0^\infty f_0(D) D^a \, dD}{\int_0^\infty f_0(D) D^b \, dD}, \qquad (8.2)$$

where a and b are two positive integers, the sum of which is called the order of the mean diameter. The values of a and b for some of the commonly used mean diameters are given in Table 8.1. These mean diameters can also be evaluated from a discrete expression:

$$(D_{ab})^{a-b} = \left[\sum_{i=1}^{i=N_c} N_i D_i^a \bigg/ \sum_{i=1}^{i=N_c} N_i D_i^b \right], \qquad (8.3)$$

where N_i is the number of elements (i.e., droplets) in the ith class and N_c denotes the total number of classes. In the spray literature, the droplet size distribution is generally characterized using one single mean droplet diameter, which somehow leads to ignore the possible effects of the actual polydispersity of the spray.

Mean droplet diameters were standardized by Mugele and Evans [14] and it noteworthy that they can be evaluated from any type of distribution (number, length, area, or volume):

$$(D_{ab})^{a-b} = \frac{\int_0^\infty f_n(D) D^{a-n} \, dD}{\int_0^\infty f_n(D) D^{b-n} \, dD}, \qquad (8.4)$$

where n is an integer for the distribution type. It can be readily checked that D_{ab} indeed has the units of a length. Commonly used mean diameters include the arithmetic mean diameter D_{10}, the volume mean diameter D_{30}, and the Sauter mean diameter D_{32}. It is also possible to define some representative diameters, listed in Table 8.2, using the cumulative function

$$F_n(D_\eta) = \int_0^{D_\eta} f_n(D) \, dD = \eta. \qquad (8.5)$$

The cumulative function F_3 is sketched in Fig. 8.3 with some of the representative diameters, listed in Table 8.2, marked. Various statistical moments for the droplet diameter can also be defined as given in Table 8.3.

Table 8.2 Representative diameters

	Name	Value of the cumulative
$D_{0.1}$	—	10% of volume in droplets smaller than $D_{0.1}$
$D_{0.5}$	MMD[a]	50%
$D_{0.9}$	—	90%
$D_{0.999}$	Maximum diameter	99.9%
D_{peak}	Peak diameter	—

[a] MMD, mass median diameter.

Table 8.3 Definition of some characteristic moments

	Number-based distribution	Volume-based distribution
Mean	D_{10}	D_{43}
Variance	$D_{20}^2 - D_{10}^2$	$D_{53}^2 - D_{43}^2$
Skewness	$\dfrac{D_{30}^3 - 3D_{20}^2 D_{10} + 2D_{10}^3}{(D_{20}^2 - D_{10}^2)^{3/2}}$	$\dfrac{D_{63}^3 - 3D_{53}^2 D_{43} + 2D_{43}^3}{(D_{53}^2 - D_{43}^2)^{3/2}}$
Kurtosis	$\dfrac{D_{40}^4 - 4D_{30}^3 D_{10} + 6D_{20}^2 D_{10}^2 - 3D_{10}^4}{(D_{20}^2 - D_{10}^2)^2}$	$\dfrac{D_{73}^4 - 4D_{63}^3 D_{43} + 6D_{53}^2 D_{43}^2 - 3D_{43}^4}{(D_{53}^2 - D_{43}^2)^2}$

To describe droplet size dispersion, Tate [15] proposed considering the following uniformity index:

$$\text{DUI} = \frac{1}{D_{0.5}} \sum_{i=1}^{i=N_c} V_i \left(D_{0.5} - D_i \right), \tag{8.6}$$

where $D_{0.5}$ denotes the mass median characteristic diameter. The dispersion of the distribution is also often characterized from the relative span factor $\Delta = (D_{0.9} - D_{0.1})/D_{0.5}$ or from a dispersion boundary factor $\Delta_B = (D_{0.999} - D_{0.5})/D_{0.5}$. The dispersion index δ is defined as

$$\delta = \int_0^{D_{max}} D f_3(D) \, dD. \tag{8.7}$$

In this section, the standard definitions required for a statistical description of the spray are introduced. Special emphasis has been placed on the droplet size distribution function, which allows characterization of the spray and its polydispersity. In the next subsection, the discussion is first extended to the mathematical description of two-phase flows in general and then focuses on the transport of the spray joint probability density function.

8.2 Mathematical Framework and Description of Two-Phase Flows

The formulation of a mathematical model for two-phase flows requires appropriate field equations and constitutive relationships to be settled. However, the derivation of such equations is significantly more complicated than for strictly continuous and homogeneous media associated to single-phase flows. The difficulties in deriving the corresponding set of balance equations for such heterogeneous media are associated with the presence of discontinuities. Indeed, the local form of the conservation equations is deduced from integral balances of mass, momentum, and energy but this requires the characteristic variables (e.g., density, velocity, temperature, pressure, etc.) to be continuously differentiable in the region (i.e., volume) of integration. Thus, the presence of interfaces featuring flow variable discontinuities introduces some difficulties in the mathematical description of two-phase flows. From a general point of view, a two-phase flow can be considered as a field that is subdivided into single-phase regions with moving boundaries between them. The standard differential balance of mass, momentum, and energy holds for each subregion but it does not, however, for the set of these sub-regions without violating the conditions of continuity. The mathematical description of the fluid medium must therefore be handled in the sense of generalized functions or distributions. The preceding discussion is at the basis of the introduction for the *two-fluid* framework, which is also often referred to as the Eulerian–Eulerian description of two-phase flows. Each phase is described separately, and the first step consists in delineating (i.e., identifying) the surface that effectively distinguishes them. It is treated as a moving boundary, the position of which can be described through an implicit equation $f(\mathbf{x},t) = 0$. One can define the interface as a constant property surface or isolevel surface of the characteristic property, Φ i.e., $f(\mathbf{x},t) = \Phi - \Phi^f$ where Φ^f denotes the peculiar iso-value of interest; the superscript f denotes that this is indeed the corresponding value at the front location. The above equation is sometimes referred to as a topological equation but it is more standardly known as the local instant formulation [16, 17]. For instance, regions with the liquid (resp. gaseous) phase are identified by $f(\mathbf{x},t) > 0$ (resp. $f(\mathbf{x},t) < 0$). In this manner, the phase function or phase indicator function (PIF) defined by $\chi_l(\mathbf{x},t) = 1$ in the liquid (and $\chi_l(\mathbf{x},t) = 0$ elsewhere) can be evaluated from $\chi_l(\mathbf{x},t) = 1 - \chi_g(\mathbf{x},t) = H(f(\mathbf{x},t))$, where $H(X)$ is the Heaviside generalized function, which is zero for $X \leq 0$ and unity for $X > 0$. The moving interface is described using the following equation:

$$\frac{\partial f}{\partial t} + \mathbf{w}_i \cdot \nabla f = 0, \tag{8.8}$$

with $\mathbf{w}_i$ the velocity field at the interface.

In each phase or in regions where $f(\mathbf{x},t) \neq 0$, the standard (i.e., single-phase) conservation must be satisfied. For example, the continuity or mass conservation is

$$\frac{\partial \rho_k}{\partial t} + \nabla \cdot (\rho_k \mathbf{u}_k) = 0, \tag{8.9}$$

where the index k stands for either the liquid phase (index l) or the gaseous phase (index g). In regions where $f(\mathbf{x}, t) = 0$, the conservation equations are replaced by local jump conditions. For instance, the mass conservation equation becomes

$$\sum_{k=1}^{k=2} \rho_k (\mathbf{u}_{ki} - \mathbf{w}_i) \cdot \mathbf{n}_{ki} = 0, \qquad (8.10)$$

where $\mathbf{n}_{ki}$ is the outward unit normal vector for each of the two phases, i.e., $\mathbf{n}_{li} = -\nabla f / \|\nabla f\|$ and $\mathbf{n}_{gi} = \nabla f / \|\nabla f\|$, while $\mathbf{u}_{ki}$ denotes the i-component of the flow velocity at the interface in phase k. Equation (8.10) expresses nothing but the mass balance at the interface and it is often referred to as the local instant mass jump condition. Other jump relations can be derived for momentum, energy, and chemical species mass fractions if required, i.e., for multicomponent flows.

The local formulation, which is briefly presented above, is essential for two-fluid models or Eulerian–Eulerian models. These models are formulated by considering each phase separately, with two sets of conservation equations for mass, momentum, species mass fractions, and energy. Each of these equations features interaction terms coupling the two phases through jump conditions such as the one discussed earlier. Eulerian–Eulerian models thus incorporate a two-way coupling, which is expected to be essential for situations featuring large values of the liquid volume fraction. However, the corresponding framework relies on rather complex closure relations, the extensive discussion of which is outside the scope of this chapter. The reader is therefore referred to the works of Ishii, Delhaye, and coworkers [18, 19] for further details about such two-fluid models.

The liquid phase function $\chi_l(\mathbf{x}, t)$ may also be used as a direct starting point of the two-phase flow description. It is, indeed, at the basis of one of the most well-known interface-tracking strategies, namely the volume-of-fluid (VOF) method. In the VOF algorithm, a color function or volume-of-fluid function is introduced. Its value is unity in the liquid phase, zero in the gas, and it is intermediate in cells where the interface is present. The VOF method is based on the consideration of a set of *volume averaged* conservation equations, with one of them, similar to Eq. (8.8), describing the advection of the volume-of-fluid function.

Among the other interface tracking methods that are available, the level-set interface tracking method, which makes direct use of Eq. (8.8), is rather easy to implement and allows for a flexible treatment of surface phenomena. The level-set method uses a continuous function G to describe the interface [20, 21]. This function is defined as the signed distance between any points of the domain and the interface. The zero-level curve ($G = 0$) therefore provides the interface location. The transport equation that describes the motion of the interface is given by Eq. (8.8). It is the hyperbolic type, and its discretization must combine robustness and high convergence order. High-order weighted essentially non-oscillatory (WENO). schemes [22] can be used to discretize the convective term in this equation. Unfortunately the level set G no longer remains a distance function when solving the G-equation numerically, and a renormalization algorithm is thus applied to keep it as the signed distance to the interface [23].

However, even if combined with such a redistancing algorithm, the numerical resolution may induce a loss of mass in underresolved regions. Therefore, many extensions of the method have been proposed in the literature, see for instance [24]. The coupled level-set/volume-of-fluid method (CLSVOF) has been shown well suited to capture atomization processes [25]. The main concept behind the CLSVOF method is to take benefit from the advantage of both level-set and VOF strategies: mass loss is limited through the VOF method, and a detailed description of interface properties is ensured with the level-set framework.

Marker particles can also be used to track the liquid–gas interface. The particles are located at the interface, move with the fluid (Lagrangian procedure), and are subsequently used to reconstruct the interface. The precision of this method depends directly on the spacing between the particles: marker particles must be added and deleted dynamically during the computation. The sharpness of the interface is maintained as it is advected, which enables an accurate computation of interface curvature and associated surface tension forces. However, the method raises some difficulties when breakup or coalescence processes take place. The particles tend to move apart or so close to each other that the reconstruction step becomes difficult.

It should be explicitly acknowledged that it is not possible to provide an extensive review of all the frameworks (LHF models, drift-flux models, DQMOM, etc.) that are available for numerical simulation of turbulent two-phase flows. The reader may refer to the book by Kuo and Acharya [26] for details on existing strategies. However, it is at the same time impossible to close this section without providing a few comments about the standard Lagrangian–Eulerian approach, which remains one of the most widely used to describe two-phase flow combustion in geometries of practical interest. For this purpose, we consider the spray joint probability density function (PDF) $\mathcal{F}(x,t;r,T,v)$, where x is the position, t the time, r the radius, T the temperature, and v the velocity of the droplet. Its transport equation, which is often referred to as the Williams equation [27, 28], is

$$\frac{\partial \mathcal{F}}{\partial t} + \nabla_x(v\mathcal{F}) + \nabla_r(\langle dr/dt \mid r,T,v \rangle \mathcal{F})$$
$$+ \nabla_T(\langle dT/dt \mid r,T,v \rangle \mathcal{F})$$
$$+ \nabla_v(\langle dv/dt \mid r,T,v \rangle \mathcal{F})$$
$$= \mathcal{Q}(\mathcal{F},\mathcal{F}). \tag{8.11}$$

The left-hand side (LHS) of Eq. (8.11) gathers different terms that correspond respectively to the temporal evolution, the droplet convection in the physical space, the rates of change of its characteristic size r and temperature T, and finally its acceleration. It is worth noting that a detailed consideration of the acceleration term reveals that diffusion-like contributions will also arise in the right-hand side (RHS) of Eq. (8.11). The term $\mathcal{Q}(\mathcal{F},\mathcal{F})$ appearing in Eq. (8.11) is a nonlinear quadratic term accounting for formation and destruction of particles through the nucleation, breakup, and coalescence processes. This term is often omitted for dilute sprays.

Since the PDF under consideration includes the velocity of the particle, the second term in the LHS appears in a one-point one-time closed form and therefore does not require any further consideration. The closure of the third term often relies on the classical droplet vaporization model associated with the conduction limit, which leads to the well-known d^2-law, i.e., $r_k^2(t) = r_k^2(t=0) - \mathcal{K}_{0,k} t$ where $\mathcal{K}_{0,k}$ is a positive constant. The vaporization rate may also be evaluated using the rapid mixing limit [29, 30]. Thus, the effects associated with variable properties and transient liquid heating, as well as the influence of the Stefan flow on the thickness of both thermal and mass boundary layers, may also be incorporated. These phenomena may be of significant importance if the liquid fuel (or oxidizer) is injected at a thermophysical condition that is far from boiling conditions [31]. The droplet acceleration is generally represented through a classical approach that incorporates the drag effects, i.e., the difference between the droplet velocity and the gas velocity that the droplet *sees*. In the corresponding model, it is often assumed that the gas velocity fluctuations are distributed according to a Gaussian distribution. More details on such representations can be found in reference [32].

From the numerical point of view, it is not possible to compute the evolution of all of the droplets involved in practical situations. Furthermore, we are not really interested in such a precise behavior of each individual droplets but rather in their average (statistical) behavior. In this respect, the usual procedure consists of considering N_p numerical (notional) droplets representing a group of N_k physical droplets having the same velocity, the same radius, and the same temperature at a given location in the physical space, and the joint PDF is obtained as $\mathcal{F}(\boldsymbol{x},t;r,T,\boldsymbol{v}) = \sum_{k=1}^{k=N_p} \mathcal{F}^k(\boldsymbol{x},t;r,T,\boldsymbol{v})$. Among the various methods available to perform the corresponding numerical integration, the stochastic particle method (SPM) of Dukowicz [32] is widely used for spray calculations; see also reference [33] for further details.

The evolution of each parcel is tracked using their Lagrangian equations,

$$\frac{d\boldsymbol{x}_k}{dt} = \boldsymbol{v}_k \; ; \; \frac{d\boldsymbol{v}_k}{dt} = \boldsymbol{\beta}_k \; ; \; \frac{dr_k}{dt} = R_k \; ; \; \frac{dT_k}{dt} = \mu_k, \qquad (8.12)$$

where $\boldsymbol{v}_k$ incorporates the processes of convection and dispersion, $\boldsymbol{\beta}_k$ accounts for the influence of drag, R_k represents the effects of vaporization on the droplets' radii, and μ_k is associated with the droplet heating, including the vaporization effects that have been discussed earlier. The corresponding terms can also describe atomization and coalescence for situations in which the spray can no longer be considered as dilute.

In this section, it has been established that the difficulties encountered in deriving the field and constitutive equations appropriate to multiphase flow systems stem from the presence of the interface and from the fact that both the steady and dynamic characteristics of multiphase flows heavily depend on the interfacial structure of the flow. For example, the steady-state and the dynamic characteristics of dispersed two-phase flow systems depend on the collective dynamics of droplets interacting with each other and with the surrounding continuous phase. In the case of separated flows, these characteristics depend on the structure and dynamics of the interface. To determine the collective interaction of particles and the dynamics of the interface, it is necessary to describe first the local properties of the flow and then to obtain a macroscopic description by means of appropriate averaging procedures. For dispersed flows, it is necessary to determine the rates of nucleation, evaporation (or condensation), motion, and disintegration of single droplets, as well as droplet collisions and coalescence processes. Some of these processes appear as source or sink terms in the evolution of the spray (or Williams) equation; others appears as flux of probability along the directions associated with each sample-space variable. This, is for instance, the case of the evaporation terms, the representation of which will be discussed in detail in the next section.

8.3 Modeling Issues Relevant to Evaporation and Combustion

The multiple interactions between the physical phenomena taking place in combustion chambers (i.e., atomization, dispersion, vaporization, and combustion) make the study of inert and reactive sprays in conditions that are typical of real applications very complex. For this reason, the study of an isolated single- or multicomponent droplet is a necessary step toward the modeling of more representative situations.

In reality, a rigorous modeling of droplet evolution would be required to take into account, inter alia, the deformation of droplet in the flow, the flow inside the droplet (Hill vortex), as well as the inhomogeneity of gaseous field properties (temperature and species) around the droplet. However, many single droplet model studies reported in the literature focus on spherical droplets in a quiescent homogeneous atmosphere. This allows to take advantage of the symmetry of the system, with all the quantities depending only on the distance from the droplet center, making it a one-dimensional problem.

Based on previous assumptions, we follow the presentation of Kazakov and coworkers [34] and consider two distinct sets of equations, written in spherical coordinates, so as to model the evaporation and combustion of multicomponent droplets. The first set of equations is used to resolve the gaseous phase and the second one concerns the liquid phase.

8.3.1 Description of the Gaseous Phase

The one-dimensional governing equations of mass, chemical species mass fractions, and energy conservation in the gaseous phase are given by

$$\frac{\partial \rho}{\partial t} + \frac{1}{r^2}\frac{\partial \rho u r^2}{\partial r} = 0, \tag{8.13}$$

$$\frac{\partial \rho Y_\alpha}{\partial t} + \frac{\partial \rho u Y_\alpha}{\partial r} = -\frac{1}{r^2}\frac{\partial r^2 \rho Y_\alpha V_{r,\alpha}}{\partial r} + \dot{\omega}_\alpha, \tag{8.14}$$

$$\frac{\partial \rho C_p T}{\partial t} + \frac{\partial \rho u C_p T}{\partial r} = \frac{1}{r^2}\frac{\partial}{\partial r}\left(r^2 \lambda \frac{\partial T}{\partial r}\right)$$

$$- \rho \sum_{i=1}^{N_{sp}} (Y_\alpha, V_{r,\alpha} C_{p,\alpha})\frac{\partial T}{\partial r} + \dot{\omega}_T, \quad (8.15)$$

where t denotes time and r is the radial coordinate, while ρ, T, C_p, and λ correspond to the density, temperature, constant-pressure specific heat capacity, and thermal conductivity, respectively. The radial fluid velocity is u, while Y_α, $V_{r,\alpha}$, $\dot{\omega}_\alpha$, and $C_{p,\alpha}$ denote, respectively, the mass fraction, diffusion velocity, chemical production rate, and constant-pressure specific heat capacity of the species α^{th}. Finally, $\dot{\omega}_T = -\sum_{\alpha=1}^{N_{sp}} \dot{\omega}_\alpha h_\alpha$ is the heat release rate, with h_α the enthalpy of species α, and N_{sp} the total number of chemical species. The absence of momentum equation is related to the nature of the flowfield that develops around the droplet, which is caused only by the *Stefan flux*, i.e., the mass flux at the droplet interface in the direction normal to the liquid–gas interface.

8.3.2 Description of the Liquid Phase

In the liquid phase, the one-dimensional governing equations of mass, chemical species, and energy conservation are

$$\frac{\partial \rho_l}{\partial t} + \frac{1}{r^2}\frac{\partial \rho_l u_l r^2}{\partial r} = 0, \quad (8.16)$$

$$\rho \frac{\partial Y_{l,\alpha}}{\partial t} + \rho_l u_l \frac{\partial Y_{l,\alpha}}{\partial r} = \frac{1}{r^2}\frac{\partial}{\partial r}\left(r^2 \rho_l D_{l,\alpha}\frac{\partial Y_{l,\alpha}}{\partial r}\right) \quad (8.17)$$

$$\rho C_l \frac{\partial T_l}{\partial t} + \rho_l C_l u_l \frac{\partial T_l}{\partial r} = \frac{1}{r^2}\frac{\partial}{\partial r}\left(r^2 \lambda_l \frac{\partial T_l}{\partial r}\right), \quad (8.18)$$

where $Y_{l,\alpha}$ and $D_{l,\alpha}$ denote the mass fraction and diffusion coefficient of the αth species in the liquid phase. The quantities v_l, ρ_l, C_l, T_l, and λ_l denote the flow velocity, density, specific heat capacity, temperature, and thermal conductivity, respectively, of the liquid. In the following, we will assume the same diffusion coefficient for all species, i.e., $D_{l,\alpha} = D_l$.

8.3.3 Boundary Conditions

The solution to this type of system relies on Dirichlet boundary conditions far from the droplet ($r \to \infty$), while Neumann boundary conditions are specified at the center of the droplet ($r = 0$)

$$\begin{cases} Y_\alpha = Y_\alpha^\infty & \text{as } r \to \infty \\ T = T^\infty & \text{as } r \to \infty \\ \dfrac{\partial Y_{l,\alpha}}{\partial r} = 0 & \text{at } r = 0 \\ \dfrac{\partial T_l}{\partial r} = 0 & \text{at } r = 0, \end{cases} \quad (8.19)$$

where Y_α^∞ denotes the mass fraction of the αth species in the gaseous environment, and T^∞ is the ambient temperature. In addition to the boundary conditions at $r = 0$ and $r \to \infty$, the temperature continuity and conservation of mass and energy across the liquid–gas interface ($r = \delta$), where $T = T_l$, are considered.

The evolution of the interface location δ is deduced, from the momentum equation for both phases at the liquid–gas interface, as

$$\frac{d\delta}{dt} = \left(u_l - (u - u_l) \frac{\rho}{\rho_l} \right)_{r=\delta}. \quad (8.20)$$

The balance equations for the mass and heat fluxes at the interface are given respectively by

$$\left(\rho_l D_l \frac{dY_{l,\alpha}}{dr} - (u - u_l) \rho_g \left(Y_{l,\alpha} - \varepsilon_\alpha \right) \right)_{r=\delta} = 0, \quad (8.21)$$

$$\left(\lambda_l \frac{dT_l}{dr} - \lambda \frac{dT}{dr} - \rho_g (u - u_l) L_v \right)_{r=\delta} = 0, \quad (8.22)$$

where L_v denotes the total latent heat of vaporization of the multicomponent mixture, which is given by

$$L_v = \sum_{\alpha=1}^{N_{s,l}} \varepsilon_\alpha L_{v,\alpha}, \quad (8.23)$$

where ε_α the fractional gasification rate of the species αth, which is defined as

$$\varepsilon_\alpha = \frac{\dot{m}_{l,\alpha}}{\sum_{\alpha=1}^{N_{s,l}} \dot{m}_{l,\alpha}}, \quad (8.24)$$

where $\dot{m}_{l,\alpha}$ is the mass flow rate of the αth species that crosses the liquid–gas interface. A positive fractional gasification rate indicates that there is an outward flux from the droplet surface to the gas environment (vaporization process). On the contrary, a negative fractional gasification rate indicates that there is an inward flux from the gas phase into the liquid phase over the interface (condensation process). Clearly, this quantity is zero for chemical species that do not change its phase (i.e., the corresponding species remains either at the liquid or at the gaseous state).

Due to its relatively large computational cost (fine grid resolution requirements), the one-dimensional model presented earlier is generally restricted to the simulation of single droplet case [34–36]. Simplified models are used to perform reactive spray simulations. In fact, in addition to the general assumptions that have been stated earlier, the gaseous phase is assumed to satisfy the quasi-steady condition, even in the presence of the evaporation process. This means that, as far as changes in the gas and vapor flow conditions are concerned, the response of the gaseous system is very fast. As a consequence, the partial derivatives with respect to time are zero for the gaseous phase variables.

In the gas, the set of conservation equations becomes

$$\frac{1}{r^2}\frac{d\rho u r^2}{dr} = 0, \tag{8.25}$$

$$\rho u \frac{dY_\alpha}{dr} = \frac{1}{r^2}\frac{d}{dr}\left(r^2 \rho D \frac{dY_\alpha}{dr}\right) + \dot{\omega}_\alpha, \tag{8.26}$$

$$\rho C_p u \frac{dT}{dr} = \frac{1}{r^2}\frac{d}{dr}\left(r^2 \lambda \frac{dT}{dr}\right) + \dot{\omega}_T, \tag{8.27}$$

where $\dot{\omega}_T = -\dot{\omega}_F \Delta_R h = -r_{st} \dot{\omega}_O \Delta_R h$, with r_{st} the fuel vapor to oxidizer mass stoichiometric ratio. The quantity $\Delta_R h$ is the heat of reaction per unit mass of fuel, $\dot{\omega}_F$ and $\dot{\omega}_O$ represent the consumption rates of fuel vapor and oxidizer, respectively. It is noteworthy that (1) the Fick's diffusion approximation has been used and (2) the unity Lewis number assumption has been retained. The molecular diffusion coefficient is denoted by D.

Considering the expression of the boundary conditions at infinity, at the interface location as well as at the flame position in reactive case, the mass flow rates of fuel can be expressed from the following two formulas:

$$\begin{cases} \dot{m} = 4\pi\delta\rho D \ln(1 + B_m) \\ \dot{m} = 4\pi\delta\frac{\lambda}{C_p} \ln(1 + B_T), \end{cases} \tag{8.28}$$

where B_m and B_T are, respectively, the mass and heat Spalding coefficients. The evaluation of these two coefficients depends on the physical process involved. Typically, in the case of evaporation without any chemical reactions, the oxidizer diffuses toward the liquid–gas interface, while fuel diffuses toward the gaseous environment, the Spalding coefficients may then be expressed as follows:

$$B_m = \frac{Y_v^s - Y_v^\infty}{1 - Y_v^s} \quad B_T = \frac{C_p(T^\infty - T^s)}{L_{eq}}, \tag{8.29}$$

where $L_{eq} = L_v + Q_l/\dot{m}$, with Q_l being the energy associated with droplet heating.

This is in contrast to the droplet burning case, where the formulations of B_m and B_T take into account the heat release that is transferred from the flame (located at flame radius r_f) to the liquid droplet and to the gaseous environment:

$$B_m = \frac{r_{st}Y_O^\infty + Y_v^s}{1 - Y_v^s} \quad B_T = \frac{C_p(T^\infty - T^s) - r_{st}Y_O^\infty \Delta_R h}{L_{eq}}, \tag{8.30}$$

where Y_O^∞ is the oxidizer mass fraction at a large radius from the droplet.

Considering a single-step infinitely fast chemistry, the adiabatic temperature T_f^{ad} and location r_f of the flame sheet can be estimated from the following two expressions:

$$r_f = \delta \frac{\ln(B_m + 1)}{\ln\left(1 + Y_{O,\infty} r_{\text{st}}\right)} \tag{8.31}$$

$$T_f^{\text{ad}} = T^s + \frac{r_{\text{st}}(-\Delta_R h - L_v) + C_p(T^\infty - T^s)}{C_p(1 + Y_{O,\infty} r_{\text{st}})}. \tag{8.32}$$

As shown in Eqs. (8.29) and (8.30), the evaluations of B_m and B_T require inter alia a satisfactory estimation of fuel mass fraction and temperature at the liquid–gas interface.

The mass fraction of fuel vapor at the droplet surface may be determined using

$$Y_v^s = \left[1 + \frac{W}{W_v}\left(\frac{P}{P_v^s} - 1\right)\right]^{-1}, \tag{8.33}$$

where W and W_v denote the molecular weights of the considered mixture and vapor, respectively. The quantities P and P_v^s are the total pressure of the system and the partial pressure of the vapor at the interface.

One of the most accurate models to describe the evaporation process at the interface is by considering the phase equilibrium there through the Clausius–Clapeyron relation

$$\frac{d\ln(P_v^s)}{dT} = \frac{L_v}{R_F T^2}, \tag{8.34}$$

where R_F is the specific gas constant for the gaseous fuel. It leads to the following expression for the partial pressure P_v^s of fuel vapor at the droplet surface:

$$P_v^s = P_{\text{ref}} \exp\left[-\frac{L_v}{R_F}\left(\frac{1}{T^s} - \frac{1}{T_{\text{ref}}}\right)\right], \tag{8.35}$$

where P_{ref} and T_{ref} are two reference parameters. The fuel boiling temperature T_{ref}, which corresponds to the reference pressure P_{ref}, has been introduced. The gas temperature at the droplet surface is T^s. In general, the simplified models assume that the liquid droplet temperature T_l is uniform and $T_l = T^s$.

The model described earlier assumes that the gas phase transport and thermo-physical properties around the droplet are constant in space, and therefore they are independent of chemical composition and temperature. The spatial variations of these properties can be included by evaluating them using a representative temperature, T^R, and vapor mass fraction, Y_v^R. These representative values are typically evaluated using the one-third rule [37, 38]:

$$T^R = T^s + \frac{1}{3}(T^\infty - T^s); \quad Y_v^R = Y_v^s + \frac{1}{3}(Y_v^\infty - Y_v^s) \tag{8.36}$$

After evaluating the single-component properties using T^R, the corresponding mixture values are computed using an appropriate mixture averaging by using the representative mass fraction. For example, the specific heat capacity of the mixture, appearing in Eqs. (8.29) and (8.30), is determined using

$$C_p = Y_v^R C_{pv} + (1 - Y_v^R) C_{pg}, \tag{8.37}$$

where C_{pv} and C_{pg} represent the constant pressure specific heat capacities of the vapor and mixture without the vapor respectively.

Using Eq. (8.28), the squared droplet diameter law can be obtained by taking Spalding number to be constant. This neglects the preliminary heating of the liquid droplet and therefore assumes that the droplet temperature does not change. This law is readily determined from Eq. (8.28) by noticing that $\dot{m}$ is also equal to $\rho_l \pi (a/4) da^2/dt$ with a as the droplet diameter:

$$a^2/a_0^2 = 1 - \theta t/a_0^2, \tag{8.38}$$

where $\theta = 8(\rho/\rho_l) D \ln(1 + B_m)$ denotes the magnitude of the negative slope of the straight line describing the decay rate of $a^2(t)$. It is noteworthy that the time required to completely vaporize a droplet of initial diameter a_0 may be readily deduced from Eq. 8.38, it is $T_a = a_0^2/\theta$. The squared droplet diameter law provides a satisfactory approximation of the decrease of the droplet diameter, especially when vaporization takes place in an environment without any large fluctuations of temperature and pressure. However, in situations featuring large variations of temperature, the change in the slope, θ, must be considered for the evolution of thermochemical properties at the droplet surface [39].

To characterize the evaporation process more accurately, the droplet temperature evolution equation is required and it is

$$\frac{dT_l}{dt} = \frac{\dot{m}}{m_l C_l} \left(\frac{C_p (T^\infty - T^s)}{B_T} - L_v \right), \tag{8.39}$$

where m_l represent the mass of liquid droplet. This equation states that the heating of droplet is through the difference between the energy received from the surrounding gas and the energy consumed by the vaporization process. At equilibrium, i.e., when the droplet temperature reaches the wet-bulb temperature, it is characterized by $C_p (T^\infty - T^s)/B_T = L_v$.

Equation (8.39) assumes that the liquid thermal conductivity is very large and the droplet temperature is uniform. This approximation can be avoided by considering finite conductivity models such as those discussed in the text that follows [40].

From a general viewpoint, the heat and mass transfer can be characterized by the Nusselt and Sherwood numbers. In the case of single-droplet evaporation in a quiescent environment, the expressions of the Nusselt and Sherwood numbers can be obtained using

$$\text{Nu} = \frac{h_c a}{\lambda} = -\frac{a}{(T^s - T^\infty)} \left(\frac{dT}{dr} \right)_s \tag{8.40}$$

$$\text{Sh} = \frac{K_c a}{D} = -\frac{a}{(Y_v^s - Y_v^\infty)} \left(\frac{dY}{dr} \right)_s, \tag{8.41}$$

where h_c and K_c are the convective heat and mass transfer coefficients, respectively. Based on the model in Eq (8.28), the transfer numbers can be expressed as

$$\text{Nu} = 2\frac{\ln(1+B_T)}{B_T} = \text{Nu}_c \frac{\ln(1+B_T)}{B_T} \tag{8.42}$$

$$\text{Sh} = 2\frac{\ln(1+B_m)}{B_m} = \text{Sh}_c \frac{\ln(1+B_m)}{B_m}, \tag{8.43}$$

where Nu_c and Sh_c represent the modified heat and mass transfer, respectively. In practice, they are equal to 2 in the case of droplet evaporation in a quiescent environment. At moderate temperature and evaporation rate, as B_T and B_m tend toward zero, $\text{Nu} = \text{Sh} = \text{Nu}_c = \text{Sh}_c = 2.0$. The values of Nu_c and Sh_c are used to take into account the effect of the convective flow on the evaporation rate. Hence, Eq. (8.28) now depends on Nu_c and Sh_c. According to theoretical and experimental studies, it has been shown that the effects of convection on heat and mass transfer rates could be accommodated by using a correction factor that is a function of the Reynolds and Prandtl (or Schmidt) numbers.

Based on a set of experiments conducted on monocomponent droplet vaporization at atmospheric pressure and moderate ambient temperatures (with low transfer rates), the following expression has been proposed [41]:

$$\text{Nu}_c = 2 + 0.6\,\text{Re}^{1/2}\text{Pr}^{1/3} \tag{8.44}$$

$$\text{Sh}_c = 2 + 0.6\,\text{Re}^{1/2}\text{Sc}^{1/3} \tag{8.45}$$

$$\text{where}\quad \text{Re} = \frac{\rho a\,|\mathbf{u}-\mathbf{v}_d|}{\mu};\quad \text{Pr} = \frac{\mu C_p}{\lambda};\quad \text{Sc} = \frac{\mu}{\rho D}, \tag{8.46}$$

with $|\mathbf{u}-\mathbf{v}_d|$ the relative gas-droplet velocity. At this level, it should be emphasized that the Ranz and Marshall correlations are particularly suited to high Reynolds number flows. However, they overestimate the transfer rate at low Reynolds numbers [42].

Faeth proposed [45] a correlation that gives the correct limiting values for both small and large Reynolds numbers by combining the correlations of Acrivos and Taylor [43] and Frossling [44]. The revised correlation is

$$\text{Nu}_c = 2 + \frac{0.555\text{Re}^{1/2}\,\text{Pr}^{1/3}}{\left(1+\dfrac{1.132}{\text{Re}\,\text{Pr}^{4/3}}\right)^{1/2}}. \tag{8.47}$$

Subsequently, Abramzon and Sirignano have proposed other correlations based on the film theory [46] $\text{Nu}_c = 2 + (\text{Nu}_0 - 2)/F_T$, with $F_T = (1+B_T)^{0.7}\ln(1+B_T)/B_T$ for $0 \leq B_T \leq 20$ and $1 \leq \text{Pr} \leq 3$, and where the value of Nu_0 is obtained from the correlation proposed by Clift et al. [47]: $\text{Nu}_0 = 1 + (1+\text{Re}\,\text{Pr})^{1/3}f(\text{Re})$, where $f(\text{Re}) = 1$ for $\text{Re} \leq 1$ and $f(\text{Re}) = \text{Re}^{0.077}$ for $\text{Re} \leq 400$. It should be noted that, in this context, the computation of the modified Sherwood number Sh_c is performed with the same correlations, by replacing Pr and B_T by Sc and B_m, respectively.

Another issue concerns the evolution of both position and velocity. According to the literature, it is possible to account for many forces to characterize the droplet dynamics. These are related to the drag force, Basset force, added mass effects, Magnus effects, Faxen forces, Saffman force, and jet propulsion effect. However, considering dilute sprays of small droplets featuring a high density ratio between the liquid and gas phases, only the drag force, which is the dominant contribution, will be considered.

The complexity associated with the analytical solution for temporal evolution of fully coupled two-phase flow system restricts its application to a few simplified configurations only. Therefore, many experimental studies have been conducted to evaluate these effects. These studies have led to empirical and semiempirical correlations for the drag coefficient, with most of them based on the solid-sphere assumption. Numerical simulations have been also performed to account for complex geometries resulting from the droplet deformation.

The equation of motion for a particle or droplet using the steady-state drag factor can be expressed as follows:

$$m\frac{d\mathbf{v}_d}{dt} = 3\pi\mu a f_D(\mathbf{u} - \mathbf{v}_d), \tag{8.48}$$

where f_D is the drag factor, i.e., the ratio between the drag coefficient and the Stokes drag coefficient. Assuming a spherical droplet,

$$f_D = \frac{C_D Re_d}{24} = \frac{C_D}{C_D^{st}}, \tag{8.49}$$

with $C_D^{st} = 24/Re_d$. Obviously, the value f_D tends to unity for the Stokes flow. Otherwise, Eq. (8.48) can be rewritten as

$$\frac{d\mathbf{v}_d}{dt} = \frac{\mathbf{u} - \mathbf{v}_d}{\mathcal{T}_v}, \tag{8.50}$$

where $\mathcal{T}_v = (\rho_l a^2)/(18\mu f_D) = \mathcal{T}_v^{st}/f_D$ denotes the kinetic characteristic (or velocity response) time, with $\mathcal{T}_v^{st}$ the Stokes characteristic time scale. As shown in Table 8.4, there are several correlations available for f_D in the literature.

From a practical viewpoint, the drag coefficient of an evaporating droplet has been found smaller than that of a nonvaporizing solid particle at the same Reynolds numbers and correlations $C_D^{vap} = C_D/(1 + B_T)^\beta$ have been proposed in the past. These correlations are based on the drag coefficient C_D for a solid sphere, the Spalding heat transfer number B_T, and the coefficient β, which can be set to unity [50], 0.32 [51],

Table 8.4 Standard expressions for the drag factor

	Validity	Expression
Schiller and Naumann [48]	Re $\leq$ 800	$f = 1 + 0.15 Re^{0.687}$
Putnam [49]	Re $\leq$ 1000	$f = 1 + Re^{2/3}/6$
Putnam [49]	$1000 \leq Re \leq 3.10^5$	$f = 0.0183 Re$

or 0.2 [52]. Subsequently, the following correlation has been recommended based on numerical simulations [53]:

$$C_D^{vap} = \frac{24.432}{Re^{0.721}} \frac{1}{(1+B_T)^{0.27}}. \tag{8.51}$$

As emphasized in the previous section, liquid sprays can be described following different mathematical approaches and/or strategies, i.e., full resolution, two-fluid or multicontinua framework, probabilistic, and discrete-particle methods. In this section, attention is focused on the last strategy, which is one of the most popular approaches to describe dispersed spray combustion, especially when a high level of resolution is required for the flowfield. This method allows following each droplet (or group of droplets having similar properties) and calculates the evolution of their properties within the surrounding gas. Such simulations generally rely on an Eulerian description of the gas coupled to a Lagrangian description of the liquid phase. The coupling of the continuous and dispersed phases is achieved by introducing source terms in the mass, momentum, and energy equations for the continuous gas phase, while the effects of the gas phase are introduced in the form of boundary conditions applied at the boundary layer (gas film) surrounding the droplets.

The conservation equations that describe the reactive multispecies two-phase flows may be written as

Continuity equation

$$\frac{\partial \rho}{\partial t} + \frac{\partial \rho u_i}{\partial x_i} = \dot{\mathcal{M}}, \tag{8.52}$$

Momentum equation

$$\frac{\partial \rho u_i}{\partial t} + \frac{\partial \rho u_i u_j}{\partial x_j} = -\frac{\partial P}{\partial x_i} + \frac{\partial \tau_{ij}}{\partial x_j} + \dot{\mathcal{V}}_i, \tag{8.53}$$

Species mass fraction transport equation

$$\frac{\partial \rho Y_\alpha}{\partial t} + \frac{\partial \rho u_i Y_\alpha}{\partial x_i} = \frac{\partial \rho Y_\alpha V_{\alpha,i}}{\partial x_i} + \dot{\omega}_\alpha + \dot{\mathcal{M}} \delta_{\alpha v}, \tag{8.54}$$

Energy equation

$$\frac{\partial \rho h_s}{\partial t} + \frac{\partial \rho u_i h_s}{\partial x_i} = \frac{\partial}{\partial x_i}\left[\lambda \frac{\partial T}{\partial x_i} - \rho \sum_{\alpha=1}^{N} h_{s,\alpha} Y_\alpha V_{\alpha,i}\right]$$

$$+ \frac{DP}{Dt} + \tau_{ij}\frac{\partial u_i}{\partial x_j} + \dot{\omega}_T + \dot{\mathcal{E}}, \tag{8.55}$$

Equation of state

$$\rho = \frac{PW}{RT}.$$

In Eqs. (8.52)–(8.55), ρ, u_i, and T are the density, velocity component in the x_i-direction, and temperature of the mixture respectively. The quantity τ_{ij} is the viscous stress tensor, $h_s = \sum_{\alpha=1}^{N_s} Y_\alpha h_{s,\alpha}$ represents the specific sensible enthalpy of the mixture, and $h_{s,\alpha}$ is the sensible enthalpy of species α. The mass fraction of chemical species α is Y_α, $\delta_{\alpha v}$ denotes the Kronecker tensor, the value of which is unity if the species α matches the vapor index v and zero otherwise. The quantities W_α and $W = (\sum_{\alpha=1}^{N_s} Y_\alpha/W_\alpha)^{-1}$ are the molecular weight of species α and the molecular weight of the reactive mixture, respectively. Finally, R is the universal gas constant.

Various source terms are present in the previous set of equations: $\dot{\omega}_\alpha$, and $\dot{\omega}_T$ are related to the chemical reaction processes whereas $\dot{\mathcal{M}}$, $\dot{\mathcal{V}}_i$, and $\dot{\mathcal{E}}$ result from the two-way coupling between the carrier phase and the spray and these terms will be discussed later.

For the liquid phase, the Lagrangian model is used for droplets. The evolution of droplet diameter and temperature is derived from the quasi-steady isolated droplet models, while the evolution of droplet velocity and position are derived from the drag model.

By denoting $\mathbf{v}_k$ and $\mathbf{x}_k$ as the velocity and position vectors of the kth droplet with diameter a_k, the following set of equations

$$\frac{d\mathbf{x}_k}{dt} = \mathbf{v}_k, \tag{8.56}$$

$$\frac{d\mathbf{v}_k}{dt} = \frac{1}{\mathcal{T}_{v,k}}[\mathbf{u}(\mathbf{x}_k,t) - \mathbf{v}_k], \tag{8.57}$$

$$\frac{da_k^2}{dt} = \frac{-a_k^2}{\mathcal{T}_{a,k}}, \tag{8.58}$$

$$\frac{dT_k}{dt} = \frac{1}{\mathcal{T}_{T,k}}\left[T(\mathbf{x}_k,t) - T_k - \frac{B_{T,k}L_v}{C_p}\right] \tag{8.59}$$

are used to track their evolution throughout the computational domain. The scalar $T(\mathbf{x}_k,t)$ and the vector $\mathbf{u}(\mathbf{x}_k,t)$ represent the gas temperature and velocity at the droplet position $\mathbf{x}_k$. The quantities $\mathcal{T}_{v,k}$, $\mathcal{T}_{a,k}$, and $\mathcal{T}_{T,k}$ represent the kinetic, evaporation, and heat relaxation time scales given, respectively, by

$$\mathcal{T}_{v,k} = \frac{\rho_l a_k^2}{18 C_{D,k} \mu}, \tag{8.60}$$

$$\mathcal{T}_{a,k} = \frac{Sc}{4Sh_c}\frac{\rho_l}{\mu}\frac{a_k^2}{\ln(1+B_{m,k})} \tag{8.61}$$

$$\mathcal{T}_{T,k} = \frac{Pr}{6Nu_c}\frac{C_l}{C_p}\frac{\rho_l a_k^2}{\mu}\frac{B_{T,k}}{\ln(1+B_{T,k})}. \tag{8.62}$$

8.3.4 Eulerian/Lagrangian Couplings

The terms $\dot{\mathcal{M}}$, $\dot{\mathcal{V}}_i$, $\dot{\mathcal{E}}$ are modifying the mass, momentum and temperature in the gaseous phase due to a distribution of the Lagrangian quantities on the Eulerian grid. Every Lagrangian source term has to be distributed over the Eulerian nodes on the basis of droplets contributions. A particle-source-in-cell (PSI-cell) method is generally used. The instantaneous distribution of mass, momentum, and energy source terms in the nodes surrounding the droplet of interest are inversely proportional to the distance between the cell and the droplet. As far as the DNS-DPS[1] framework is concerned, a weak numerical dispersion does appear. To address this specific issue, the droplets could be fully resolved but this would add an enormous computational burden, limiting the simulations. Another solution would be to introduce a diffusion delay between the droplet and the surrounding nodes. However, in the case of reacting flows, some of the evaporated fuel would not directly interact with the local flames. This procedure is unsatisfactory and yet to be addressed.

At every computational node, a control volume $V_c = \Delta^3$ is defined by the midpoint between the neighboring nodes. The mass source term applied to any Eulerian node n is

$$\dot{M}^{(n)} = -\frac{1}{V_c} \sum_k \alpha_k^{(n)} \frac{dm_k}{dt}, \qquad (8.63)$$

where $\sum_k$ is the sum performed over all droplets affecting the node n and the mass of the kth droplet is m_k. The symbol $\alpha_k^{(n)}$ denotes the distribution coefficient of the k droplet source term on the node n. Considering all the nodes affected by the droplet k, it is necessary to have $\sum_n \alpha_k^{(n)} = 1$ to ensure mass, momentum, and energy conservation during the Lagrangian–Eulerian coupling. In fact, $\alpha_k^{(n)}$ is the portion of the control volume for the node n intersecting the control volume of the droplet k.

The kth droplet mass is $m_k = \rho_l \pi a_k^3/6$ and, using Eqs. (8.58) and (8.63), one may write

$$\dot{M}^{(n)} = \rho_l \frac{\pi}{4 V_c} \sum_k \alpha_k^{(n)} \frac{a_k^3}{T_{a,k}} \qquad (8.64)$$

Similarly, the following relation:

$$\dot{V}_i^{(n)} = \frac{1}{V_c} \sum_k -\alpha_k^{(n)} \frac{dm_k v_{k,i}}{dt}, \qquad (8.65)$$

where $v_{k,i}$ is the kth droplet velocity in the direction i, leads to the expression of the momentum source term:

$$\dot{V}^{(n)} = \frac{-\pi \rho_l a_k^3}{4 V_c} \sum_k \alpha_k^{(n)} \left[\frac{2(u_i(\mathbf{x}_k, t) - v_{k,i})}{3 T_{v,k}} - \frac{v_{k,i}}{T_{a,k}} \right]. \qquad (8.66)$$

The droplet contribution to the gase phase energy equation is

$$\dot{\mathcal{E}}^{(n)} = \frac{1}{V} \sum_k -\alpha_k^{(n)} \frac{d(m_k C_l T_k)}{dt}, \qquad (8.67)$$

which gives, after using the appropriate droplet Lagrangian equations and some algebraic rearrangements,

---
[1] Discrete particle simulation.

$$\dot{\mathcal{E}}^{(n)} = \frac{-\pi \rho_l C_l}{4 V_c} \sum_k \alpha_k^{(n)} a_k^3 \left[\frac{2}{3} \frac{T(\mathbf{x}_k, t) - T_k - B_{T,k} L_v / C_p}{\mathcal{T}_{T,k}} - \frac{T_k}{\mathcal{T}_{a,k}} \right]. \quad (8.68)$$

Equation (8.68) tells how energy fluxes reaching the droplet surface are distributed (evaporation and liquid core heating) as well as the variation of energy due to the loss of liquid mass.

8.4 Two-Phase Flow Turbulent Combustion Regimes and Diagrams

Automotive engines, gas turbine engines, and industrial furnaces involve the formation of dense sprays issued from the atomization of the liquid fuel. These sprays immersed in a hot environment will evaporate. The resulting oxidizer–fuel vapor mixture can be ignited under appropriate conditions and the resulting combustion modes strongly depends on the attributes of the droplet cloud such as droplet number density, etc., as well as the liquid evaporation characteristics and combustion kinetics. All these parameters have been introduced in the development of combustion diagrams able to describe the properties of the flames that develop in a spray core. The development of these diagrams is described in the text that follows.

8.4.1 Genesis of Two-Phase Flow Combustion Diagrams

The research group hereafter denoted by Chiu et al. [54–58] was among the first to propose a spray combustion diagram that still remains as a working basis to date. The genesis of this work has been detailed in the book by Kuo and Acharya [26] and in the review paper by Sirignano [59]. First, Onuma and Ogasawara [60] hypothesized that the factor determining the combustion rate of a kerosene jet was the efficiency of the air and fuel-vapor mixing rather than the rate of evaporation of the liquid droplets. This marked, for the first time, the introduction of the relative impact of two physical phenomena, namely the mixing and vaporization processes, in comparison to combustion chemistry or flame propagation effects. A second important study that allowed for the refinement of spray combustion diagrams was performed by Chigier [61]. In his work, devoted to the atomization of pressurized jets, the existence of regions featuring low oxygen concentration and low temperature, which promote the flame to settle toward the outer limits of the atomized spray, has been emphasized. From this observation, Chiu and coworkers deduced the so-called *group behavior* of the droplets coming from the atomization process of the liquid injection, which generates a rich fuel mixture in the region surrounding the spray core. This mixture is nonflammable because of the lack of air penetration. Then, the radial transport of the fuel vapor leads to the formation of a flammable mixture at a certain distance from the jet axis, thus leading to the birth of a gaseous diffusion flame. As the droplets move past the dense core region of the jet, the separation distances between them increase

(a) Sheath combustion

(b) Group combustion

(c) Critical group combustion

(d) Partial group combustion

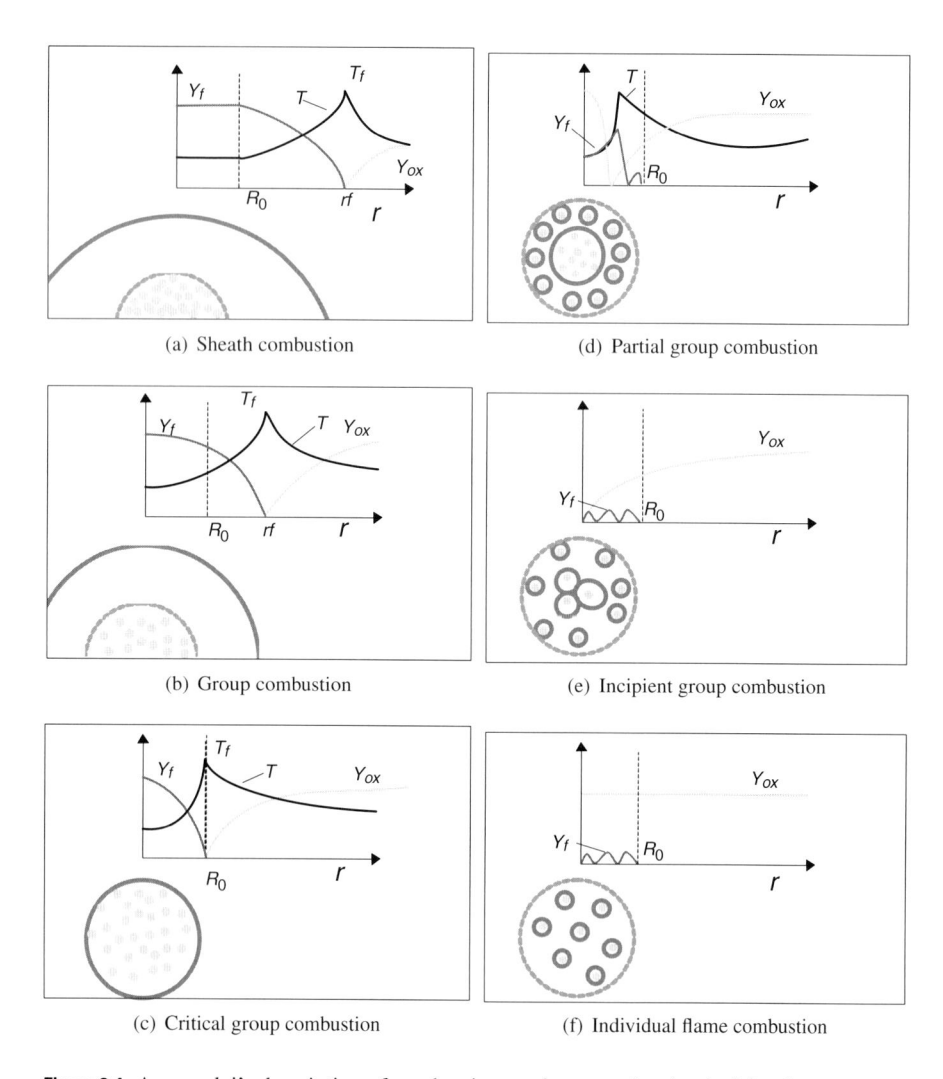

(e) Incipient group combustion

(f) Individual flame combustion

Figure 8.4 Annamalai's description of combustion modes around a cloud of droplets.

while the sizes of the droplets decrease. Then, the oxygen concentration increases and some of the droplets can burn individually with flames surrounding them and some can burn as groups. In general, the central (or inner) region consists of droplets vaporizing in an atmosphere with low oxygen concentration, while the outer region may contain droplets burning with multidrop flames. In group combustion models, the collective behavior of the droplets may be explained by the simultaneous analysis of an internal heterogeneous region and a homogeneous external gas phase region.

8.4.2 Spray Flame Structures

Prior to the description of the spray combustion diagrams, it may be useful to present the various flame structures that can be found. This has been previously documented

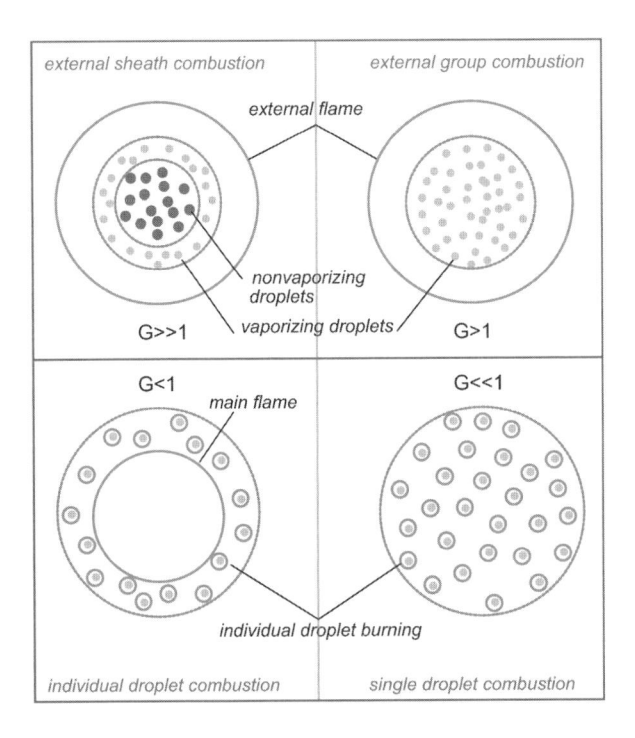

Figure 8.5 Spray combustion modes according to Chiu and coworkers.

by Annamalai [62], who also introduced some terminology relevant to their description. It is based on a cloud of droplets that is plunged into a hot oxidizer atmosphere. The droplet number density, i.e., the number of droplets per unit volume, is denoted by n. Let us start from high n values, i.e., very dense cloud of droplets. This corresponds to a situation that is referred to as the *sheath combustion* mode by Annamalai or to the *external sheath combustion* mode by Chiu et al.; see Figs. 8.4a and 8.5. In this case, the cloud behaves as an equivalent single droplet with a radius equal to the cloud radius. If the droplet density n in the cloud is diminished then the *group combustion* mode (according to Annamalai) or *external group combustion* mode (according to Chiu et al.) can be observed (Fig. 8.4b). As for the *sheath combustion* mode, the flame always remains outside the cloud although it is closer to the liquid. The vapor flux still prevents the oxygen from penetrating into the cloud. However, in contrast to the *sheath combustion* mode, slight variations of oxygen and temperature may reach its boundaries. A further decrease of n leads to a crossing of the so-called *critical combustion* regime, where the flame stands at the cloud frontier (Fig. 8.4c). The temperature at the cloud boundary is the flame temperature. It should be noted that only Annamalai used this description, since it is a short transient regime toward the *partial group combustion* regime (according to Annamalai) or *internal group combustion* regime (according to Chiu et al.). In this regime, the outer droplets consume all the oxygen that manages to penetrate the cloud boundary while the inner droplets simply evaporate. Then, the fuel vapor diffuses outward and establishes a group diffusion flame inside the cloud while the outer droplets burn individually (Fig. 8.4d). Again, if *n* decreases further, Annamalai introduced a specific transient regime: the *incipient group combustion* (Fig. 8.4e). In this case, outer droplets still burn individually and consume the oxygen. However, a weak oxidizer flux manages to reach the core of the cloud, which results in the breakup of the group flame mentioned earlier and leads to individual droplet burning in the cloud core but with a larger flame radius compared to the outer droplets. Finally, for very small values of *n*, the regime of *isolated droplet combustion* (Annamalai) or *single droplet combustion* (Chiu et al.) is reached. Since the droplets are far away from each other, an isolated combustion mode can be observed and the combustion intensity associated with each of them remains unaffected by the presence of the other neighbouring flames (Fig. 8.4f).

8.4.3 Combustion Diagrams

Labowsky and Rosner [63] were among the first to propose a spray combustion diagram (Fig. 8.6) with a strict dichotomy between two possible regimes: *group* and *individual* combustion. They developed this diagram assuming (1) either individual combustion of all droplets in the cloud or (2) simultaneous evaporation of all droplets in the cloud, while combustion takes place outside of it. This diagram is based on two coordinates: the total number of droplets in the cloud N and δ_s/a, which is the ratio between the averaged spacing between droplets, δ_s, and their mean radius a. It is worth noting that the averaged spacing can be directly determined from the droplet number density n as $\delta_s = n^{-1/3}$ provided the dilution level is sufficiently high (i.e., the volume of the liquid phase is much smaller than that of the gaseous phase).

This first step was, indeed, important because it sets the salient parameters defining the axes for the spray combustion diagram. Some further improvements were, however, required. Indeed, the combustion of a liquid fuel is also strongly altered by droplet interactions, which prevent mixing of the fuel vapor with the oxidizer and thereby forming a nonflammable mixture within the spray region. Moreover, it is important not to neglect thermal aspects; the presence of a large number of liquid droplets may lead to a significant decrease of the temperature during the evaporation process, which increases the difficulty to ignite the reactive mixture. At the very least, autoignition delays can be significantly lengthened.

A major step toward a global characterization of spray combustion was made by Chiu and coworkers when they described the combustion of a dense cloud of droplets and proposed a description of the corresponding combustion modes. The key element of their work was to disprove a previous classical theory that involved a unique combustion mode with individual flames around each droplet. In Chiu et al. diagram, spray combustion regimes are generally classified according to a *group combustion number G* [54]. Several other group parameters such as the Thiele modulus [63] or the sheath combustion parameter [64] were defined in later studies. In this respect, it seems worth noting that the relations that exist between these parameters have been analyzed by Sichel and Palanniswamy [65]. Sirignano [59] raised some concerns about these various works and identified two major shortcomings. First, those theories do not account for the fact that the vaporization of each droplet is dependent on its

Evolution of the spray combustion diagrams

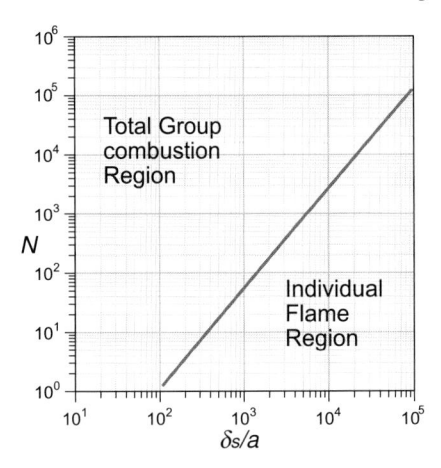

Figure 8.6 Labowsky and Rosner diagram.

Figure 8.7 Chiu et al. diagram.

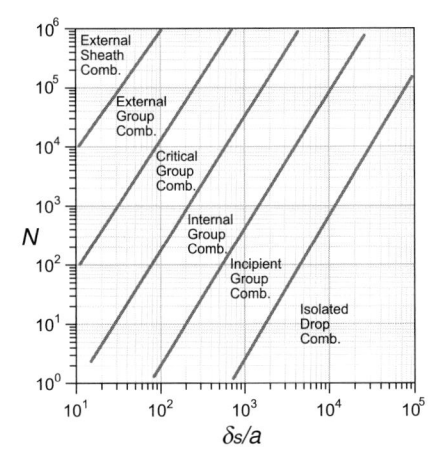

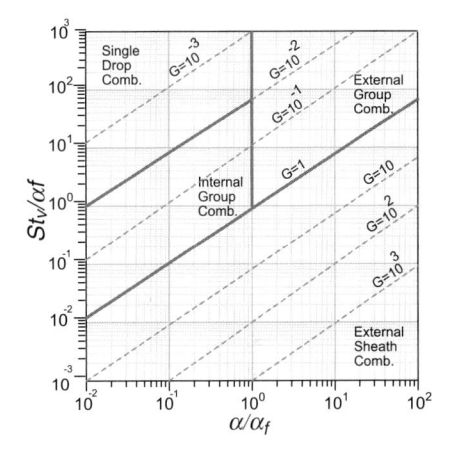

Figure 8.8 Annamalai diagram.

Figure 8.9 Urzay diagram.

environment, i.e., on the surrounding droplets. Second, they are based on the quasi-steady approximation and do not account for transient droplet heating or unsteady gas-phase conduction across the clouds. Nevertheless, the G parameter coming out of these past theories prevails to describe the spray flame structure.

Many expressions have been proposed for the group number G over the years [26]. Initially, Chiu and coworkers defined this dimensionless number as the ratio of the characteristic evaporation rate to the rate of temperature diffusion within the cloud. However, one of the most practical definition of G has been proposed by Urzay [66], who rewrote its expression as the ratio of two characteristic time scales describing the vaporization and diffusion processes in the fuel cloud.

Let us consider $t_c = R_c^2/\alpha_T$ as the characteristic time of heat diffusion into the cloud with radius R_c and thermal diffusivity α_T. The characteristic thermal penetration length L_t in the spray cloud can be approximated by $L_t = \sqrt{3/(4n\pi a)}$ by balancing the heat diffusion and vaporization terms in the gas phase energy budget; see reference [66]. This thermal penetration causes droplets to evaporate leading to variation of mass and thermal energy in the ambient gas. The characteristic time scale for this variation is defined as $t_d = L_t^2/\alpha_T$. It is also noteworthy that the vaporization characteristic time can be defined as $t_v = (\rho_l/\rho) \cdot (a^2/\alpha_T)$, where a is the mean droplet radius in the cloud.

Using these two characteristic timescales, t_c and t_d, the group combustion number may be defined as

$$G = \frac{t_c}{t_d} = \left(\frac{R_c}{L_t}\right)^2. \tag{8.69}$$

Large G values imply longer diffusion times in the cloud and large amounts of energy is required for droplet vaporization, thus delaying or preventing evaporation and combustion in certain regions of the cloud. On the contrary, small values of G are associated with short diffusion times and rapid cloud heating without any significant impact on the carrying gas, thus improving the combustion and vaporization of the droplets present in the cloud core.

As detailed in the work of Chiu and coworkers, it is possible to relate the group number G and a separation parameter S defined as $S = \delta_s/\delta_{rf}$, i.e., the ratio between the average droplet spacing δ_s and a characteristic diffusion flame radius δ_{rf} [67]. It denotes the radius of a diffusion flame surrounding a single vaporizing droplet in a quiescent oxidizer and having the mean properties of the spray cloud (in terms of radius and evaporation delay).

When the separation number S decreases, there is a point at which the flame topology evolves from individual droplet combustion to group combustion. For a given value of S, varying N, the total number of droplets in the spray cloud, two major modes (Fig. 8.7) of spray combustion may be identified with respect to the group number G. For the case $G \gg 1$, an external sheath-combustion regime, as described earlier, occurs. Thus, an external layer of droplets is evaporated and the resulting flame remains at a given standoff distance from the spray boundary. Heat diffusion balances the energy required for the vaporization process and the droplet temperature evolves from the saturation temperature at the cloud boundary down to the initial temperature of the liquid droplets in the cloud. Provided that the Damköhler number is sufficiently large, the fuel that is vaporized in the outer shell of the cloud is consumed within a diffusion flame that surrounds the cloud. This regime is encountered only in sufficiently large and dense droplet clouds. Under the opposite limiting condition, $G \ll 1$, droplets are sparse enough so that each droplet may burn individually with a surrounding flame. Those conditions delineate the so-called single droplet combustion regime. A smooth transition between these limiting regimes was anticipated by Chiu and coworkers [57], leading to intermediate submodes depending on the magnitude of G. As G decreases toward order-unity values by, for instance, decreasing the droplet-number density n, the vaporization zone tends to extend toward the core of the cloud. In this external group-combustion regime, the time needed for heat diffusion in the cloud is long enough to vaporize the droplets inside. Then, the liquid in the central region of the cloud is at the saturation temperature, while the fuel vapor burns in a diffusion flame that surrounds the cloud. If $G \approx 1$ internal group-combustion regime takes place as an intermediate limit between external sheath-combustion and single-droplet combustion. In this regime, droplets vaporize without combustion in the central region of the cloud, which is surrounded by (1) a diffusion flame and (2) an outer region of similar size within which individual droplet burning occurs. The flame is unable to penetrate into the cloud, which is too rich or/and too cold when the G value is low (i.e., below 10^{-2}). The flame locates at a given distance, while being fed by the evaporation of the internal droplets and the diffusion of the resulting vapor. However, on the boundary of the spray core droplets may burn individually because they are sufficiently far from each other to be surrounded by hot gases.

Annamalai [62] proposed an improved spray classification (see Fig. 8.8) in which two additional transient regimes are considered: the critical group combustion ($1 > G > 10$) and the incipient group combustion ($0.01 > G > 0.1$). All these regimes are summarized in Fig. 8.4.

Later on, Urzay [66] improved the diagram of Chiu et al. significantly by introducing (1) the flame radius of a burning droplet as an additional length scale and (2) a measure of air–fuel stoichiometry characteristics in the classification. Indeed, the single-droplet combustion regime cannot occur unless the droplet interspacing δ_s is larger than the flame diameter. The authors included a second parameter that relates the droplet radius to the droplet interspacing independently of the cloud dimension or the number of droplets within the cloud. To this effect, the parameter α is defined as

$$\alpha = \frac{t_v}{t_d} = \frac{\rho_l}{\rho}\left(\frac{a}{L_t}\right). \tag{8.70}$$

As the value α increases, the spray vaporization rate increases and the external flame expands in search of an equilibrium position where the value of the fuel-vapor mass fraction has decayed sufficiently for the fuel to burn with oxygen in stoichiometric proportions. The critical value of α, for which the droplet interspacing is equal to the flame diameter (i.e., $\delta_s = 2\delta_{rf}$), is denoted α_f. Its detailed definition may be found in reference [66]. The author added a second parameter, namely, the vaporization Stokes number:

$$St_v = \frac{t_v}{t_c} = \frac{\alpha}{G} = \frac{\rho_l}{\rho}\left(\frac{a}{R_c}\right), \tag{8.71}$$

which represents the ratio of the single-droplet vaporization time t_v to the diffusion time t_c of heat into the cloud. Urzay suggested using a (α, St_v) diagram instead of the classical (S, N) diagram because, depending on the value of the mass-loading parameter α, various combustion regimes may develop for the same value of the group-combustion number G. Thus, the group-combustion number G cannot be the sole parameter to describe the occurrence of one or another spray-combustion mode.

The spray-combustion regime diagram of Urzay is shown in Fig. 8.9. The axes are normalized by α_f to make them to be fuel independent. The external sheath combustion regime ($G \gg 1$) is present in this revised description. For $G \approx 1$, the vaporization layer thickness has grown inward the cloud and, past this front, a transition occurs from external sheath-combustion to either external group-combustion or internal group-combustion regimes depending on the liquid mass loading α/α_f. External group-combustion thus occurs for $\alpha/\alpha_f > 1$. In this regime, all droplets in the cloud are vaporizing, and, since the droplet interspacing is smaller than the flame radius of a single droplet, a single diffusion flame envelops the cloud. On the other hand, internal group combustion occurs for $\alpha/\alpha_f < 1$. This is an intermediate and transient regime toward incipient single-droplet combustion, in which all droplets present in the cloud undergo vaporization and combustion. Since the droplet interspacing is larger than the flame radius of a single droplet, diffusion flames surrounding each droplet can be found in the outer region of the cloud. A diffusion flame, which envelops the spray core where external group-combustion is taking place, progresses toward the center of the cloud in search of the stoichiometric condition. For $G \ll 1$, the time needed for the ambient thermal energy to diffuse inside the cloud is much faster than the spray-gas interaction time and all droplets undergo vaporization. However, unless $\alpha/\alpha_f < 1$, a diffusion flame around a single droplet cannot be sustained. Therefore, in the small-G limit, single-droplet combustion occurs for $\alpha/\alpha_f < 1$, and external group combustion takes place for $\alpha/\alpha_f > 1$.

All spray combustion diagrams that have been described above consider a quasi-stationary situation with a group of droplets with no oxidizer and surrounded by a diffusion flame [68]. Therefore, the combustion regimes are dependent on the ability of the oxidizer to penetrate into the cloud. However, the atomization of a liquid fuel is such that the droplets are initially surrounded by oxidizer (usually air). Thus, before reaching the stationary group combustion regimes, there is an ignition stage that will consume the oxidizer that is initially present in the spray. This ignition results in the propagation of a premixed flame in the mixture consisting of the oxidizer and the first fuel vapors. It should be noted, however, that in this case the characteristic timescales for evaporation and flame propagation will come into play.

Umemura [68], taken up by Mikami [69], classified the modes of flame propagation inside the group according to five distinct modes (Fig. 8.10). These modes are dependent on two main factors: the distance between the droplets and the temperature of the mixture (to be compared with the energy required for evaporation).

Mode 1 considers an expanding flame around an isolated droplet in an environment featuring a fairly low temperature and a droplet interspace smaller than the flame radius. Then, the reacting front incorporates directly the neighboring, droplets which are then heated up and evaporate, thus feeding the flame front, which goes on expanding thus leading to group combustion.

In *Mode 2*, the front of the diffusion flame reaches a flammable zone formed around the next droplet, which is being evaporated. Then, a premixed flame front, or more precisely a triple flame [70], propagates into the mixture before forming a diffusion flame around the droplet. This mode represents the case in which a new diffusion flame merges with an approaching one. However, if the spacing between the droplets increases, the flame does not come into contact with the flammable mixture and a third mode comes into play.

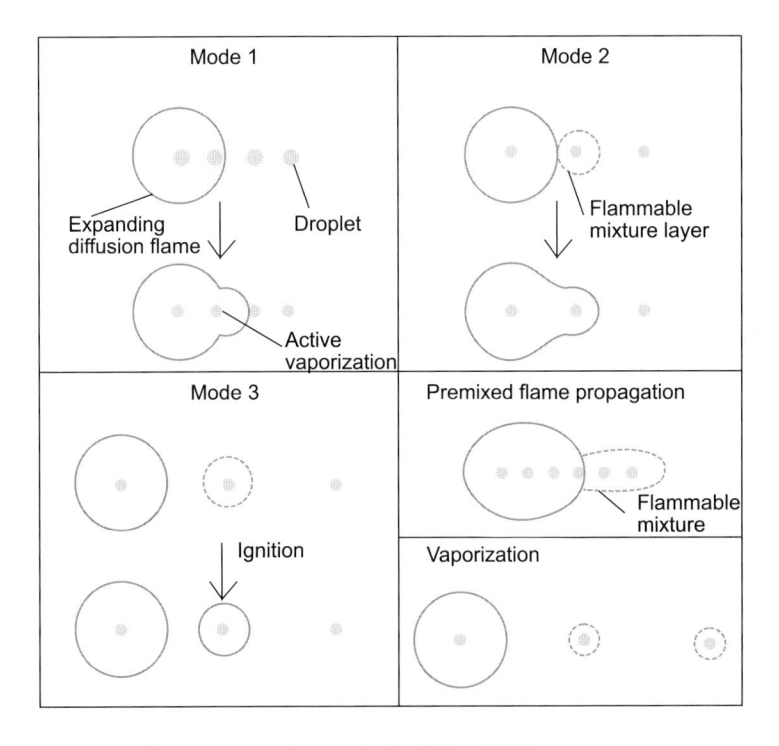

Figure 8.10 Flame propagation modes by Mikami [69].

In *Mode 3*, as for the second mode, the droplets close to a diffusion flame are surrounded by a flammable mixture due to their evaporation, which is in progress. Because they are more spaced from each other, the flame cannot reach the neighboring flammable area. Nevertheless, these areas will undergo self-ignition only by diffusion of heat generated by the flame.

It is possible to define two limiting, *premixed flame propagation* and *vaporization* modes. The premixed propagation mode appears when the set of droplets (which are very close to each other) has formed a continuous and rather homogeneous mixture of fuel vapor and oxidizer surrounding all droplets. In this case, a first premixed flame front crosses the core of the group before a diffusion flame is initiated subsequently. It can be visualized in the direct numerical simulation (DNS) study conducted by Reveillon [70]. In the second limiting mode, the vaporization mode, the droplets are too far apart to successful propagation of a premixed flame.

Borghi and his coworkers [71–73], hereafter referred to as Borghi et al., added physical parameters that characterize the reactive front itself, namely the characteristic flame time t_f and its thickness δ_f, for the analysis. Provided that the vaporization delay is sufficiently small ($t_v \ll t_f$), the mixture is locally premixed and a propagating premixed flame develops (Fig. 8.11a). This regime should be observed whatever the values of the mean droplets spacing, δ_s, and flame thickness, δ_f. In practice, the equivalence ratio of the mixture may not be fully uniform and a weakly varying partially premixed flame front propagates. If the evaporation time is large enough

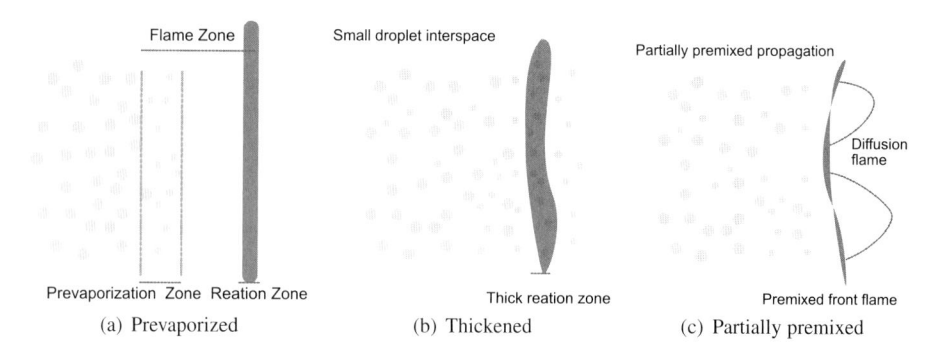

Figure 8.11 Spray flame structures by Borghi et al.

(a) Prevaporized — Prevaporization Zone, Reation Zone, Flame Zone
(b) Thickened — Small droplet interspace, Thick reation zone
(c) Partially premixed — Partially premixed propagation, Diffusion flame, Premixed front flame

for $\delta_f > \delta_s$, the collection of droplets penetrates the reactive-diffusive layers since the flame thickness is larger than the mean droplet spacing δ_s. This situation should rapidly promote the thickening of the flame (Fig. 8.11b). Aside from these extreme cases, the separation number, S, is the reference parameter. After the propagation of a primary partially premixed flame front, some remaining droplets may lead to a secondary reaction zone (Fig. 8.11c). The topology of this secondary combustion zone depends on the magnitude of S. For small values of S, droplets are likely to burn individually or as clustered in small groups surrounded by a flame. This is the so-called group combustion regime found here in a transient regime. In addition to this, Borghi et al. delineated a *percolation* combustion regime and a *pocket* combustion regime for increasing values of the separation number S (Fig. 8.12). The percolation regime consists of a combination of premixed group combustion that propagates into the clouds of droplets surrounded by oxidizer. Then, the remaining fuel burns in a diffusion flame located at a standoff distance from the spray. The pocket regime corresponds to a diffusion flame surrounding a pocket of oxidizer that itself is surrounded by vaporizing droplets feeding the flame with fuel.

Chiu et al. and Borghi et al. have defined these flame structures in the case of a quiescent spray without considering the global liquid fuel/air mass ratio. Within a real spray combustion system, this key ratio is known to modify flame stability along with the overall properties of the combustion chamber. Modifying this parameter would affect the distribution of the local equivalence ratio in the gaseous mixture. Specifically, the topology of the primary and secondary reaction zones may vary with this parameter. For instance, local extinctions may be observed if the local equivalence ratio is outside the flammability limits. Moreover, the droplet and mixture composition are also sensitive to advection, which plays a crucial role in spray combustion.

A new combustion diagram for transient configurations has thus been constructed using direct numerical simulations [70]. It appears that the flame structures may be organized into three main categories, which may be further divided into subgroups as shown in Fig. 8.13. (a) *External combustion*: this concerns combustion with a continuous flame interface. Two subregimes may be observed, depending on the location and topology of the reaction zone. (1) A *closed external* combustion regime

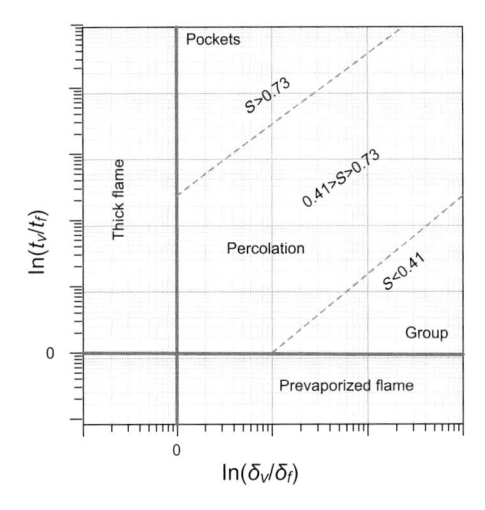

Figure 8.12 Spray combustion regimes by Borghi et al.

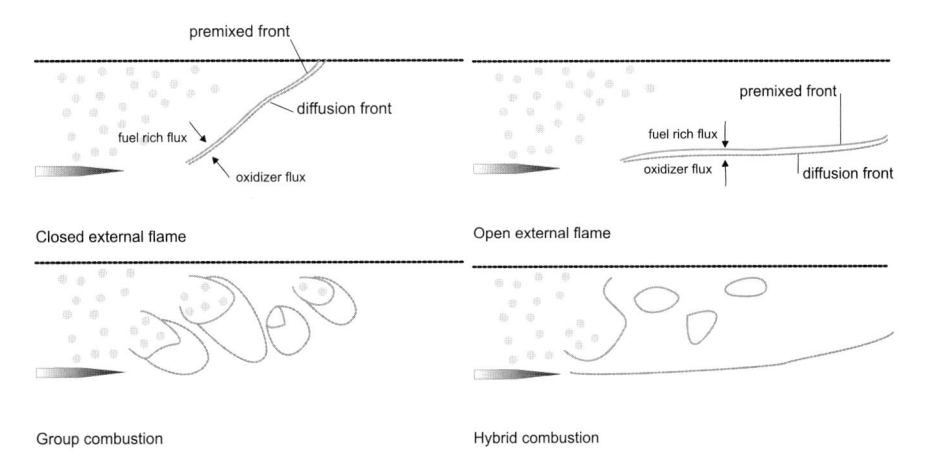

Figure 8.13 Spray flame structures by Reveillon and Vervisch.

ensues when a single flame front, mostly premixed, engulfs the droplets and their evaporated gaseous fuel converting them into combustion products. (2) An *open external* combustion regime occurs when two reaction zones develop on each side of the central jet. (b) *Group combustion*: the droplets are organized into several groups, with flames consuming each cluster independently. Both rich premixed and diffusion flames are observed for this. (c) *Hybrid combustion*: this regime is a combination of the previous two regimes. The premixed flames are burning in a group combustion mode, whereas the diffusion flames cannot percolate between the droplet clusters because of a fuel-rich environment there. Remaining fuel burns with the coflowing oxidizer in an external diffusion flame.

Again, these regimes may be organized in a combustion diagram, but at least three axes are necessary to classify them. The first is the ratio of characteristic evaporation to flame time scales, t_v/t_f. The second is the length scale ratio $\delta s/\delta f$. The third is the equivalence ratio, ϕ, within the spray jet required to characterize the amount of oxidizer entrained in the spray core.

If the evaporation time scale for every droplet is very small compared to the flame characteristic time then purely gaseous combustion occurs. Therefore, when $t_v/t_f \ll 1$, the droplets are vaporized much earlier than they reach the flame front and combustion develops in a fully gaseous phase. The jet is thus a mixture of fuel-vapor and oxidizer, the equivalence ratio of which depends on the initial droplet load. A diagram for this partially premixed jet is presented in Fig. 8.14, where various scenarios possible for a situation with short evaporation time scale. The main combustion modes discussed above are recovered by fixing δ_s, the mean droplets interspace distance, and traveling along a direction of increasing equivalence ratio.

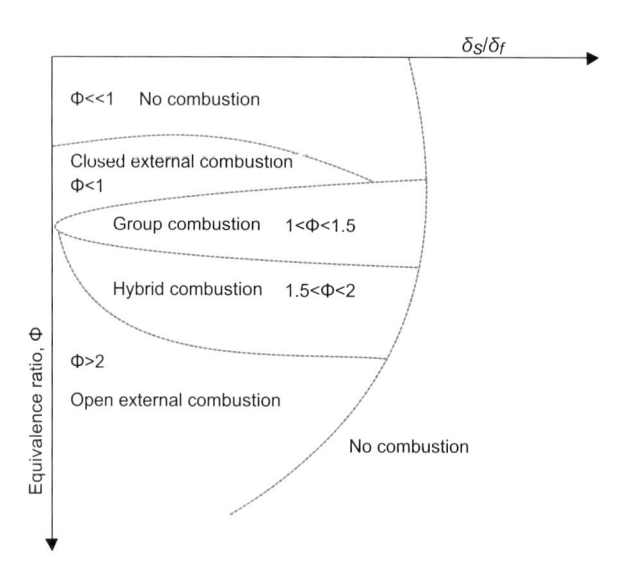

Figure 8.14 Two-phase flow combustion regimes by Reveillon and Vervisch.

In the leanest case, closed external combustion (Fig. 8.13) is observed. It is wrinkled and continuous around the fuel issuing from the spray evaporation. As the equivalence ratio increases, local fluctuations of the mixture fraction appear because of the fuel-vapor mixing with the surrounding oxidizer. For a slightly richer situation, vapor-rich areas (or pockets) burn first in a premixed regime that consumes the carrying oxidizer, followed by diffusion flame rings appearing between the remaining fuel and the external oxidizer. If the equivalence ratio continues to increase, there is still a premixed group combustion in the core of the jet to burn all the carrying oxidizer. However, enough fuel remains to maintain an external diffusion flame, leading to the hybrid combustion regime. If the equivalence ratio is further increased, the central vapor/oxidizer mixture becomes too rich to allow for the propagation of any premixed flame and both premixed and non-premixed fronts are pushed away from the core of the spray, to burn in the open external combustion regime, which is very similar to a gaseous non-premixed jet flame. It is worth noting that the droplets may reach the flame front when the evaporation timescale is comparable to the flame time scale, $t_v/t_f \approx 1$, but it is unlikely that they will cross the burning zone without being fully evaporated. However, they may interact with the turbulent structures while releasing their vapor, which will result in highly nonuniform fuel-vapor distribution.

8.5 Conclusion

The characterization of two-phase flow combustion regimes has been the subject of many studies over the past 40 years. Indeed, it is an essential element for developing and optimizing combustion systems using liquid fuels; thus, many efforts have been made toward a better understanding of the flow structures and topologies in spray combustion. However, it is noteworthy that the development of the actual simulation tools, which include more advanced physical and numerical models, offers the possibility to refine with great precision these spray combustion diagrams. This appears as a major objective to be considered for future work. In particular, impressive progresses have been recently made to handle the numerics of reactive two-phase flow simulations.

References

[1] D. A. Drew and S. L. Passman, *Theory of Multicomponent Fluids*. Berlin/Heidelberg: Springer, 1999.

[2] F.-X. Demoulin, P.-A. Beau, G. Blokkeel, A. Mura, and R. Borghi., "A new model for turbulent flows with large density fluctuations," *Atom. Sprays*, **17**, 315–345, 2007.

[3] Z. Bouali, B. Duret, F. X. Demoulin, and A. Mura, "DNS analysis of small-scale turbulence-scalar interactions in evaporating two-phase flows," *Int. J. Multiph. Flow*, **85**, 326–335, 2016.

[4] A. Neophytou, E. Mastorakos, and R. S. Cant, "The internal structure of igniting turbulent sprays as revealed by complex chemistry DNS," *Combust. Flame*, **159**, 641–664, 2012.

[5] E. Mastorakos, "Forced ignition of turbulent spray flames," *Proc. Combust. Inst.*, **36**, 2367–2383, 2017.

[6] A. M. Al Taweel and J. Landau, "Turbulence modulation in two-phase jets," *Int. J. Multiph. Flow*, **3**, 341–351, 1977.

[7] G. Hetsroni, "Particles-turbulence interaction," *Int. J. Multi. Flow*, **15**, 735–746, 1989.

[8] P. Marmottant and M. H. Villermaux, E. Freedman, "On spray formation," *J. Fluid Mech.*, **498**, 73–111, 2004.

[9] L.-P. Hsiang and G. M. Faeth, "Drop deformation and breakup due to shock wave and steady disturbances," *Int. J. Multiph. Flow*, **21**, 545–560, 1995.

[10] C. Arcoumanis, M. Gavaises, and B. French, "Effect of fuel injection processes on the structure of diesel sprays," *SAE Paper 970799*," 1997.

[11] G. M. Faeth, L.-P. Hsiang, and P.-K. Wu, "Structure and breakup properties of sprays," *Int. J. Multiph. Flow*, **21**, 99–127, 1995.

[12] C. A. Chryssakis, D. N. Assanis, and F. X. Tanner, "Atomization models," in *Handbook of Atomization and Sprays*. Berlin/Heidelberg: Springer, 2011, 215–231.

[13] M. Pilch and C. A. Erdman, "Use of breakup time data and velocity history data to predict the maximum size of stable fragments for acceleration-induced breakup of a liquid drop," *Int. J. Multiph. Flow*, **13**, 741–757, 1987.

[14] R. A. Mugele and H. D. Evans, "Droplet size distribution in sprays," *Indust. Eng. Chem.*, **43**, 1317–1324, 1951.

[15] R. W. Tate, "Some problems associated with the accurate representation of droplet size distributions," *Proc. 2nd Int. Conf. Liquid Atomization and Spray Systems (ICLASS),"* 1982.

[16] J. M. Delhaye, "Jump conditions and entropy sources in two-phase systems, local instant formulation," *Int. J. Multiph. Flow*, **1**, 359–409, 1974.

[17] I. Kataoka, "Local instant formulation of two-phase flows," *Int. J. Multiph. Flow*, **12**, 745–758, 1986.

[18] M. Ishii, *Thermo-Fluid Dynamic Theory of Two-Phase Flow*. Eyrolles (Paris), 1975.

[19] J. M. Delhaye, M. Giot, and M. L. Riethmuller, *Thermohydraulics of Two-Phase Systems for Industrial Design and Nuclear Engineering*. New York: Hemisphere and McGraw-Hill, 1981.

[20] J. A. Sethian, "A fast marching level set method for monotonically advancing fronts," *Proc. Natl. Acad. Sci. USA*, **93**, 1591–1594, 1996.

[21] S. Osher and R. P. Fedkiw, "Level set methods: an overview and some recent results," *J. Comput. Phys.*, **169**, 463–502, 2001.

[22] C.-W. Shu, "Essentially non-oscillatory and weighted essentially non-oscillatory schemes for hyperbolic conservation laws." Institute for Computer Applications in Science and Engineering, Tech. Rep. NASA CR-97-206253 ICASE Report No. 97-65, 1997.

[23] M. Sussman, E. Fatemi, P. Smereka, and S. Osher, "An improved level set method for incompressible two-phase flows," *Comput. Fluids*, **27**, 663–680, 1998.

[24] M. Sussman and E. G. Puckett, "A coupled level set and volume-of-fluid method for computing 3D and axisymmetric incompressible two-phase flows," *J. Comp. Phys.*, **162**, 301–337, 2000.

[25] T. Menard, S. Tanguy, and A. Berlemont, "Coupling level set/VOF/ghost fluid methods: validation and application to 3D simulation of the primary break-up of a liquid jet," *Int. J. Multiph. Flow*, **33**, 510–524, 2007.

[26] K. K. Kuo and R. Acharya, *Fundamentals of Turbulent and Multiphase Combustion*. Hoboken, NJ: John Wiley & Sons, 2012.

[27] F. A. Williams, "Spray combustion and atomization," *Phys. Fluids*, **1**, 541–545, 1958.

[28] F. A. Williams, *Combustion Theory*. San Francisco: Benjamin/Cummins, 1985.

[29] C. K. Law, "Recent advances in droplet vaporization and combustion," *Prog. Energy Combust. Sci.*, **8**, 171–201, 1982.

[30] W. A. Sirignano and C. K. Law, "Unsteady droplet combustion with droplet heating II: conduction limit," *Combust. Flame*, **28**, 175–186, 1977.

[31] L. Gomet, V. Robin, and A. Mura, "Lagrangian modelling of turbulent spray combustion under liquid rocket engine conditions," *Acta Astronautica*, **94**, 184–197, 2014.

[32] J. K. Dukowicz, "A particle-fluid numerical model for liquid sprays," *J. Comput. Phys.*, **35**, 229–253, 1980.

[33] P. J. O'Rourke, "Statistical properties and numerical implementation of a model for droplets dispersion in a turbulent gas," *J. Comput. Phys.*, **83**, 345–360, 1989.

[34] A. Kazakov, J. Conley, and F. L. Dryer, "Detailed modeling of an isolated, ethanol droplet combustion under microgravity conditions," *Combust. Flame*, **134**, 301–314, 2003.

[35] A. J. Marchese and F. L. Dryer, "The effect of liquid mass transport on the combustion and extinction of bicomponent droplets of methanol and water," *Combust. Flame*, **105**, 104–122, 1996.

[36] A. Cuoci, M. Mehl, G. Buzzi-Ferraris, T. Faravelli, D. Manca, and E. Ranzi, "Autoignition and burning rates of fuel droplets under microgravity," *Combust. Flame*, **143**, 211–226, 2005.

[37] G. L. Hubbard, V. E. Denny, and A. F. Mills, "Droplet evaporation: effects of transients and variable properties," *Int. J. Heat Mass Trans.*, **18**, 1003–1008, 1975.

[38] M. C. Yuen and L. W. Chen, "On drag of evaporating liquid droplets," *Combust. Sci. Technol.*, **14**, 147–154, 1976.

[39] Z. Bouali, C. Pera, and J. Reveillon, "Numerical analysis of the influence of two-phase flow mass and heat transfer on n-heptane autoignition," *Combust. Flame*, **159**, 2056–2068, 2012.

[40] W. A. Sirignano, *Fluid Dynamics and Transport of Droplets and Sprays*. Cambridge: Cambridge University Press, 1999.

[41] W. E. Ranz and W. R. Marshall, "Evaporation from drops," *Chem. Eng. Prog.*, **48**, 141–146, 1952.

[42] T. Yuge, "Experiments on heat transfer from spheres including combined natural and forced convection," *J. Heat Trans.*, **82**, 214–220, 1960.

[43] A. Acrivos and T. D. Taylor, "Heat and mass transfer from single spheres in stokes flow," *Phys. Fluids*, **5**, 387–394, 1962.

[44] N. Frössling, "Uber die verdunstung fallernder tropfen," *Gerlands Beitrage zur Geophysik*, **52**, 170–216, 1938.

[45] G. M. Faeth, "Current status of droplet and liquid combustion," *Prog. Energy Combust. Sci.*, **3**, 149–182, 1979.

[46] B. Abramzon and W. Sirignano, "Droplet vaporization model for spray combustion calculations," *Int. J. Heat Mass Trans.*, **32**, 1605–1618, 1989.

[47] R. Clift, J. R. Grace, and M. E. Weber, *Bubbles, Drops, and Particles*. New York: Academic Press, 1978.

[48] L. Schiller and A. Naumann, "Uber die grundlegenden berechnungen bei der schwerkraftaufbereitung," *Zeitschrift Verein Deutscher Ingenieure (VDI)*, **77**, 318–321, 1933.

[49] A. Putnam, "Integratable form of droplet drag coefficient," *ARS J.*, **31**, 1467–1468, 1961.

[50] P. Eisenklam, S. A. Arunachalam, and J. A. Weston, "Evaporation rates and drag resistance of burning drops," in *Symp. (International) Combust.*, **11**, no. 1, 1967, 715–728.

[51] C. H. Chiang, M. S. Raju, and W. A. Sirignano, "Numerical analysis of convecting, vaporizing fuel droplet with variable properties," *Int J. Heat Mass Trans.*, **35**, 1307–1324, 1992.

[52] M. Renksizbulut and M. C. Yuen, "Numerical study of droplet evaporation in a high-temperature stream," *J. Heat Trans.*, **105**, 389–397, 1983.

[53] C. H. Chiang and W. A. Sirignano, "Interacting, convecting, vaporizing fuel droplets with variable properties," *Int. J. Heat and Mass Trans.*, **36**, 875–886, 1993.

[54] H. H. Chiu and T. M. Liu, "Group combustion of liquid droplets," *Combust. Sci. Technol*, **17**, 127–142, 1977.

[55] H. H. Chiu, R. Ahluwalia, B. Koh, and E. J. Croke, "Spray group combustion," *AIAA Paper 7875*," 1978.

[56] H. H. Chiu and E. J. Croke, "Group combustion of liquid fuel sprays," University of Illinois at Chicago, Tech. Rep. 81-2, 1981.

[57] H. H. Chiu, H. Y. Kim, and E. J. Croke, "Internal group combustion of liquid droplets," in *Symp. (International) Combust.*, **19**, 1982, 971–980.

[58] H. H. Chiu and X. Q. Zhou, "Turbulent spray group vaporization and combustion," University of Illinois at Chicago, Tech. Rep., 1983.

[59] W. A. Sirignano, "Advances in droplet array combustion theory and modeling," *Prog. Energy Combust. Sci.*, **42**, 54–86, 2014.

[60] Y. Onuma and M. Ogasawara, "Studies of the structure of a spray combustion flame," in *Symp. (International) Combust.*, **15**, 1974, 453–465.

[61] N. A. Chigier, *Energy, Combustion and the Environment*. New York: McGraw-Hill, 1981.

[62] K. Annamalai and W. Ryan, "Interactive processes in gasification and combustion. Part I: Liquid drop arrays and clouds," *Prog. Energy Combust. Sci.*, **18**, 221–295, 1992.

[63] M. Labowsky and D. E. Rosner, ""Group" combustion of droplets in fuel clouds I: Quasi-steady predictions," in *Evaporation-Combustion of Fuels*. American Chemical Society, 1978, **166**, 63–79.

[64] S. M. Correa and M. Sichel, "The group combustion of a spherical cloud of monodisperse fuel droplets," in *Symp. (International) Combust.*, **19**, 1982, 981–991.

[65] M. Sichel and S. Palaniswamy, "Sheath combustion of sprays," in *Symp. (International) Combust.*, **20**, 1985, 1789–1798.

[66] J. Urzay, "A revised spray-combustion diagram of diffusion-controlled burning regimes in fuel-spray clouds," *Annual Research Briefs, Center for Turbulence Research*, 193–198, 2011.

[67] A. R. Kerstein and C. K. Law, "Percolation in combusting sprays I: Transition from cluster combustion to percolate combustion in non-premixed sprays," in *Symp. (International) Combust.*, **19**, no. 1, 1982, 961–969.

[68] A. Umemura and S. Takamori, "Percolation theory for flame propagation in non- or less-volatile fuel spray," *Combust. Flame*, **141**, 336–349, 2005.

[69] M. Mikami, H. Oyagi, N. Kojima, Y. Wakashima, M. Kikuchi, and S. Yoda, "Microgravity experiments on flame spread along fuel-droplet arrays at high temperatures," *Combust. Flame*, **146**, 391–406, 2006.

[70] J. Reveillon and L. Vervisch, "Analysis of weakly turbulent diluted-spray flames and spray combustion regimes," *J. Fluid Mech.*, **537**, 317–347, 2005.

[71] R. Borghi, "The links between turbulent combustion and spray combustion and their modelling," in *Transport Phenomena in Combustion*. London: Taylor and Francis, 1996.

[72] R. Borghi, "Background on droplets and sprays," in *Combustion and turbulence in two-Phase Flows, Lecture Series 1996-02*. Sint-Genesius-Rode, Belgium: Von Karman Institute for Fluid Dynamics, 1996.

[73] R. Borghi and M. Champion, *Modélisation et Théorie des Flammes*. Paris: Technip, 2000.

9 Solid Fuel Combustion

N. E. L. Haugen, K. Umeki, M. Liberman, I. Rogachevskii, and F. Picano

9.1 Introduction

Tens of thousands of years ago, when humans first learned how to make fire, their only source of fuel was solid biomass, such as, e.g., wood. The fire was primarily used for heating, cooking, light, and protection. This was the situation for millennia, but since the Industrial Revolution, the variety of combustion applications and corresponding fuels has increased dramatically. Even though the importance of solid fuels is still great, its relative importance has decreased during the last hundred years or so. This is due to the invention of new applications that utilize combustion, such as, e.g., the internal combustion engine (liquid fuel) and the gas turbine (gaseous fuel).

Solid fuel can be split into six main types: wood, other biomasses except wood, peat, coal, coke, and municipal solid waste. In addition to these main types, there are also more specialized types, such as rocket fuel. There is a large variety of different applications for solid fuels, ranging from domestic heating appliances to large pulverized coal power plants and fluidized bed boilers. In the following, we will mostly concentrate on pulverized solid fuel, where the pulverized particles are embedded in a turbulent gaseous fluid, such as, e.g., air. This is relevant for all kinds of pulverized burners and entrained flow gasifiers. In addition to the useful applications mentioned earlier, solid fuels are also central in various hazardous phenomena, such as dust explosions.

In 2014, 41% of the global electricity production came from solid fuels, which primarily means coal, biomass, or solid wastes. This fraction is even lower for North America (35%) and Europe (30%).

9.1.1 Governing Equations

In the following, we will define the equations describing particle transport within a fluid. The change in position x_p of a solid object, such as a solid fuel particle, is entirely determined by its velocity v_p, such that

$$\frac{dx_p}{dt} = v_p. \tag{9.1}$$

Furthermore, due to Newton's second law of motion, the velocity of a particle with mass m_p is controlled by the sum of forces $\boldsymbol{F}_p$ acting on the particle:

$$\frac{d\boldsymbol{v}_p}{dt} = \frac{\boldsymbol{F}_p}{m_p}. \qquad (9.2)$$

Please see Eq. (9.34) for a more detailed version of Eq. (9.2). The forces acting on a typical pulverized fuel particle are fluid drag and gravity. In addition, for dense particle suspensions, such as, e.g., fluidized beds, one must also account for the forces due to particle collisions. The drag force from the fluid on the particle is given by

$$\boldsymbol{F}_{p,\text{drag}} = \frac{1}{\tau_p}(\boldsymbol{u} - \boldsymbol{v}_p), \qquad (9.3)$$

where $\boldsymbol{u}$ is the velocity of the surrounding fluid. For a spherical particle, the particle response time is given by

$$\tau_p = \frac{\rho_p d_p^2}{18\mu(1 - f_c)} \qquad (9.4)$$

where d_p is the particle diameter, μ is the dynamic viscosity of the fluid and ρ_p and ρ are the particle and fluid densities, respectively. The correction factor f_c is introduced to account for the fact that the flow around the particle is not Stokesian for larger Reynolds numbers [1].

It is convenient to define two different particle Stokes numbers. The first one is based on the turbulence integral scale, such that

$$\text{St} = \frac{\tau_p}{\tau_i} \qquad (9.5)$$

where $\tau_i = L/u_{\text{rms}}$, L is the integral scale and u_{rms} is the root mean square velocity. The other Stokes number of interest,

$$\text{St}_\eta = \frac{\tau_p}{\tau_\eta} \qquad (9.6)$$

is based on the time scale of the Kolmogorov eddies, τ_η.

A fuel particle embedded in a hot and potentially reactive environment may exchange heat, mass, and momentum with the surrounding fluid. Three important phenomena in this respect are, drying, devolatilization, or char conversion. In the following, we will present some further details on the physical and numerical aspects of these three phenomena. (For a description of the corresponding influence on the fluid, see, e.g., [2].)

9.1.2 Thermal Conversion Reactions

When a solid fuel particle is heated to about 100°C, any remaining water in the particle will evaporate and leave the particle through the internal pore structure. The exact temperature where the water evaporates depends on the gas pressure and if the water is free or bound. Numerically, the drying process can be described by three different methods [3]. In the equilibrium method, the phase equilibrium is calculated to find the amount of evaporation. This method is accurate, and it even accounts for low-temperature drying, but it is more computationally expensive than the other methods. In addition, low-temperature drying is typically not relevant for combustion applications. For the second drying method, the so-called kinetic approach, the process of drying is considered a chemical process that can be described via Arrhenius kinetics. For this approach, the drying occurs over a range of temperatures around the boiling temperature. If this temperature distribution is narrowed down, to make it more accurate, the solver becomes more numerically stiff. Finally, for the thermal method, one assumes that for temperatures at or above the boiling temperature all the heat transferred to the particle is used to evaporate the water. With this method, all evaporation occurs exactly at the boiling temperature.

Particle devolatilization starts at above 200°C, depending on the fuel. During devolatilization, the volatile part of the dry particle escapes the particle, while leaving inorganic elements and carbonaceous char behind. A large number of different numerical methods for describing devolatilization exists in the literature. They vary largely in complexity, from the simple first-order reaction, competitive reactions, parallel reactions, and distributed activation energy model (DAEM) [4]. Meanwhile, some advanced models consider molecular structure of original fuels to obtain kinetic constants, such as FLASHCHAIN [5], chemical percolation and devolatilization (CPD) [6], and functional-group depolymerization vaporization cross-linking (FG-DVC) [7] models for coal, and independent competitive reactions of three major components for biomass [8].

After devolatilization, the remaining char can be converted either through oxidation or through gasification by steam or carbon dioxide. The conversion will proceed on char surfaces that are exposed to the reactant. Since char is a porous structure, this means exposed surfaces within the pores and the external surface of the particle. To identify the amount of pore conversion, the reactant concentration within the pore structure of the particle must therefore be known. This means that also information about the pore size evolution is required. If all this, together with an intrinsic heterogeneous kinetics, is known, an accurate char conversion rate can be determined. The intrinsic heterogeneous kinetics gives the Arrhenius parameters for a given char surface of a particular type of char. Such a kinetic will be applicable to all particle sizes and densities. The intrinsic chemical reaction mechanisms may be detailed, with larger sets of subreactions, involving adsorption and desorption processes. If all aforementioned information is known, and the potentially catalytic effects of any inorganic components can be neglected, the intrinsic approach is therefore very accurate, but it comes at the cost of having to know a significant amount of detail about the char. A popular but less general alternative is to use apparent kinetics. With apparent kinetics, which typically have just one subreaction for each main reaction (oxidation or gasification with steam or carbon dioxide), one does not explicitly account for the reactions within the pores of the char particle. For this approach, the Arrhenius parameters must therefore be obtained experimentally for char particles and conditions that are similar to the particles and conditions for which they will be used. In general, the rate of heterogeneous reactions is therefore expressed as a product of rate coefficient and structure function as:

$$r_{cg} = k_{cg}(p_A, T) \cdot f(X). \tag{9.7}$$

Rate coefficients are expressed as either nth order Arrhenius expressions of three overall reactions with O_2, CO_2, and H_2O, or considering elemental reaction steps including adsorption and desorption processes. The structure function $f(x)$ accounts for the change in density of the active sites. Most common structural functions are overall reaction, shrinking core, and random pore models. When considering the effect of annealing, catalytic effects, ash inhibition, and so on, more advanced models such as the CBK (char burnout kinetics) model are also used [9].

9.2 The Effect of Turbulence on the Heterogeneous Conversion of Powders

It has been known for a long time that turbulence has an effect on the transport coefficients of a fluid, such as the turbulent viscosity or diffusivity. A range of different models have been developed to account for this in Reynolds-averaged Navier–Stokes (RANS) simulations, for example the k-ϵ and k-ω models. Even for large-eddy simulations (LESs), models are required to account for the effect of turbulence on transport coefficients on the subgrid level.

When chemical reactions in the gas phase are of interest, such as, e.g., in the case of gas-phase combustion, also the effect of turbulence on the chemical reactions has to be modeled. Popular turbulent combustion models are the eddy dissipation model (EDM), the eddy dissipation concept (EDC) or the probability density function (PDF)–based models. These models handle the gas-phase combustion, which also includes combustion of the volatiles originating from a solid phase. None of these models do, however, account for the effect of turbulence on the conversion of the char fraction from pulverized coal or biomass combustion/gasification reactions. It is remarkable that so much effort has been put into the development of good gas-phase combustion models while, until the past years, nothing has been done to study the effect of turbulence on solid-phase (char) conversion.

Turbulence will affect both the mass and the heat transfer between the fluid and a reactive particle. Here, we will focus on the effect of turbulence on the mass transfer, but the heat transfer is expected to follow similar trends. Since either oxygen or a gasifying species (typically steam or carbon dioxide) is required for any kind of char conversion, the mass transfer from the fluid to the particle is one of the limiting factors on the char conversion rate. The other limiting factor is the reactivity of the char. For kinetically controlled conversion, it is the latter factor that is controlling the conversion rate, while for diffusion controlled conversion it is the former.

There are two different ways that turbulence may influence the mass transfer to a char particle. One is due to the turbulence-induced velocity difference between the fluid and the particle, while the other is due to turbulent particle clustering. In the following we will look more into both ways. It should be noted that the effect of the turbulence in both cases is to augment the mass transfer to the particle. This means that in the case of purely kinetically controlled char conversion, turbulence will not have an effect on the conversion rate.

9.2.1 Velocity-Induced Mass Transfer Increase

The relative velocity difference between a particle and its surrounding fluid, which is nonzero for finite Stokes numbers [10], is induced due to the rapid change in fluid velocity experienced by particles in a turbulent environment. For very large Stokes numbers, where the particle response time is much longer than the turnover time of the integral scale eddies, the relative velocity between the particles and the fluid can be approximated by the root-mean-square velocity of the turbulence. In the other extreme, when the Stokes number is very small, the particle will essentially follow the fluid, such that the relative velocity will be close to zero. In general, for $St < 1$ and $St_\eta > 1$, the particle-fluid velocity difference is given by [10]

$$u_{rel} = \beta u_{rms} \sqrt{\frac{St L^{2/3} - \eta^{2/3}}{L^{2/3} - \eta^{2/3}}}, \quad (9.8)$$

where β is 0.41 and L and η are the integral and Kolmogorov scales of the turbulence, respectively. The Sherwood number (Sh) can be obtained by using Eq. (9.8) in, e.g., the Ranz and Marshall [11] correlation:

$$Sh = 2 + 0.69 Re_p^{1/2} Sc^{1/3}, \quad (9.9)$$

where Sc is the Schmidt number and

$$Re_p = u_{rel} d_p / \nu \quad (9.10)$$

is the particle Reynolds number. The mass transfer coefficient is now easily found from

$$\kappa = \frac{Sh D}{d_p}, \quad (9.11)$$

when D is the mass diffusivity.

For large Reynolds numbers, such that $L \gg \eta$, it can be found by combining Eqs. (9.8)–(9.10) that

$$Sh = 2 + 0.69 \sqrt{St} (18 \beta^2 Re / S)^{1/4}, \quad (9.12)$$

where $S = \rho_p / \rho$ is the ratio of the material densities of the char and the fluid. In Fig. 9.1, the corresponding Sherwood number is plotted as a function of Stokes number for a fluid Reynolds number of $Re = u_{rms} L / \nu = 1000$. The different lines in the figure correspond to different values of S, and it is seen that the Sherwood number is largest for the lowest particle material density. This is because a low particle material density correspond to a large particle radius for a given Stokes number.

From Eq. (9.11), it is clear that an increase in Sherwood number yields a corresponding increase in mass transfer rate to the particle surface. For diffusion-controlled

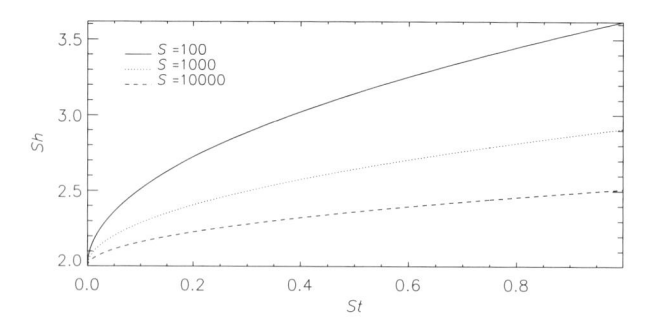

Figure 9.1 Sherwood number as a function of Stokes number for three different particle material densities. The relative particle material density is given as $S = \rho_p/\rho$ where ρ_p and ρ are the particle and fluid material densities, respectively.

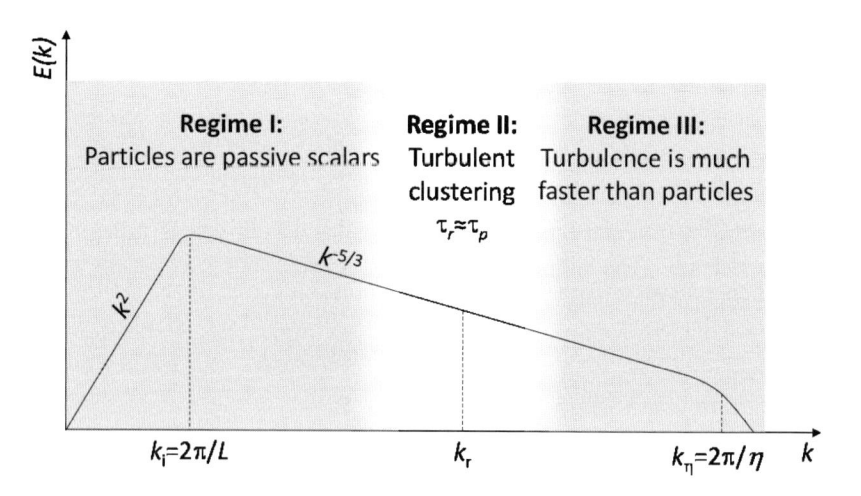

Figure 9.2 The turbulence energy spectrum. Wavenumber ranges for the different regimes are identified.

char reactions, this then results in an increase in char conversion rate. This effect is not very strong, though. For the parameters used here, we can see from Fig. 9.1 that the increase is less than a factor of 2.

9.2.2 Cluster-Induced Mass Transfer Decrease

Traditionally, turbulence is thought to disperse particles and to act as a source of increased particle diffusion that will smooth sharp gradients in the particle number density field. This is identified as Regime I in Fig. 9.2. In this regime, the motion of the particles is essentially like the motion of a passive scalar. Regime I is defined as all scales that are significantly larger than the turbulence-particle resonance (TPR) scale. Here, the TPR scale is defined as the turbulent scale that is associated with a time scale, τ_r, that is equal to the particle response time τ_p. The TPR time scale is given by $\tau_r \sim l_r/u_r$, where $l_r \sim k_r^{-1}$ is the size of the TPR eddies, and u_r and k_r

are the corresponding velocity and wave number, respectively. For a turbulent eddy within the intertial range, we know that $u_r \sim k_r^{-1/3}$ such that $\tau_r \sim k_r^{-2/3}$. Turbulence at scales that are significantly smaller than the TPR scale will not have any influence on the motion of the particles. This is because the turbulent eddies at such small scales are associated with time scales that are much shorter than the response time of the particles. Hence, the particles do not have time to respond to the influence of the eddies during the lifetime of the eddy. This corresponds to Regime III in Fig. 9.2. As discussed before, these eddies will, however, yield a relative velocity between the fluid and the particles. This effectively results in an increased heat and mass transfer between the two phases.

Finally, turbulent eddies at scales that are of the order of the TPR scale will tend to cluster the particles in the low vorticity regions between the turbulent eddies (Regime II in Fig. 9.1). One can think of this clustering as a result of particles being centrifuged out of TPR-scaled eddies, which makes them end up in the regions with low vorticity between the turbulent eddies. Based on the above, we can conclude that at least for monodispersed particle sizes, particles will tend to form dense particle clusters as long as the turbulent energy spectrum contain TPR-scale eddies, i.e., as long as St ≤ 1 and St$_\eta \leq 1$. However, it has been observed that clustering is maximum when the particle relaxation time τ_p is of the order of the Kolmogorov dissipative flow time scale τ_η, i.e., when St$_\eta \approx 1$. For this reason, it is quiet common to use the Kolmogorov-based Stokes number when studying particle clustering. For the remainder of this subsection we will nevertheless use the Stokes number based on the integral scale. From a practical point of view it has been observed that the local particle concentration may become even thousands times the bulk one for St$_\eta \approx 1$. The local particle consentrations will be lower for larger Stokes numbers, but they will be significant as long as the integral–scale based Stokes number, St, is not much larger than unity. Different measures of the turbulence-induced particle segregation have been proposed: attractor fractal dimension, density of particle pairs, and others [12, 13]. This phenomenon occurs in all kinds of turbulent flows and can be enhanced in reactive environments with variable-density conditions [14].

The existence of particle clusters may have a profound influence on the solid-phase reactions. This is due to the fact that particles that are concentrated in particle clusters will soon consume most of the reactant species (oxygen, carbon dioxide, or steam) within the cluster. In this way, a particle inside a cluster will have access to less reactant species than a particle outside the cluster [15]. Therefore, the conversion of these particles will be slower than for a similar fluid-particle realization that is not clustered.

It is important to realize that for particle clustering to have an effect on the conversion rate of the particles, the lifetime of a typical cluster cannot be much shorter than the time it takes for the particles to consume a significant fraction of the surrounding reactant. This can be nondimensionalized through the clustering Damköhler number:

$$\text{Da}_c = \tau_i / \tau_c, \tag{9.13}$$

where $\tau_c = d_p/(2n_p A_p D)$ is the chemical time scale, n_p is the overall particle number density, and A_p is the external particle surface area. Here it has been assumed that the chemical reactions at the particle surface are diffusion controlled, which means that the chemical time scale equals a diffusive time scale. For small Damköhler numbers, one will not see an effect of the particle clustering on the char conversion rate, while for large values ($Da_c > 0.1$) the effect may be dramatic.

Haugen et al. [10] show that the effect of the particle clustering can be accounted for by introducing a modified Sherwood number in Eq. (9.11). This highlights the fact that the effect of particle clustering is a mass-transfer effect, which is also the case for the previously discussed turbulence-induced relative velocity effect. When the effect of the turbulence-induced relative velocity is incorporated into the Sherwood number by combining Eqs. (9.8)–(9.10), to find an average Sherwood number $\overline{Sh}$, Haugen et al. [10] showed that the modified Sherwood number is given by

$$Sh_{mod} = \overline{Sh} \frac{A_1 A_2}{A_1 A_2 + Da_c St/2} \qquad (9.14)$$

where $A_1 A_2 = 0.08 + St/3$.

Everything leading up to Eq. (9.14) was developed in a simplified setup where surface reactions were assumed to be diffusion controlled and isothermal. This is clearly not the case in reality, but the main trends should be unaffected by this. This was recently confirmed by Kruger et al. [2] by running a series simulations with more realistic conditions, such as non-isothermal reactions and Arrhenius kinetics.

The conclusion is therefore that when the TPR scale eddies are not resolved in a numerical simulation, Eq. (9.14) should be used to account for the effect of turbulence on the mass transfer of reactants to the char surface. With that being said, it still remains to be shown what the effect of polydispersed particle size distributions is. One should also investigate the effect of particle clustering on the heat transfer to the particles.

9.3 Radiation-Induced Mechanism of Unconfined Dust Explosions

Dust explosions occur when an accidentally ignited flame propagates through a cloud of fine particles suspended in a combustible gas. They have been a significant hazard to humans and property for many centuries in the mining industry and in grain elevators [16, 17]. Fatalities in coal mine explosions range from dozens to hundreds annually, with significant damage to property, plant, and equipment. Currently, the danger of dust explosions constitutes a permanent threat in many industries in which fine particles are involved.

Understanding the mechanism of dust explosions is essential for minimizing the dust explosion hazard. Historically, large-scale chemical explosions that involve methane (propane, butane, etc.) mixed with air were considered as unconfined dust explosions. Admittedly, large-scale coal mine explosions share traits with large vapor cloud incidents and may be considered as either partly or almost fully unconfined dust explosions.

In the majority of large-scale vapor cloud explosions (VCEs) there was clear forensic evidence that a severe explosion had propagated into open uncongested areas. This was a feature of all of the large vapor cloud incidents for which detailed primary evidence was available, e.g., the 2005 Buncefield fuel storage depot explosion and several VCEs of severity, similar to that of the Buncefield [18]. However, despite considerable efforts over more than 100 years, the mechanism governing flame propagation in dust explosions remains a major unresolved issue.

Since detonation is the only established theory that allows sufficiently rapid burning while producing a high pressure that can be sustained in open areas, this led many researchers to conjecture that deflagration to detonation transition (DDT) following turbulent flame acceleration is the mechanism that explains the high rate of combustion and overpressures in dust explosions [19–21]. However, DDT has been validated only for highly reactive fuels, such as hydrogen and ethylene, in laboratory-scale experiments [22–27]. Applying it to industrial accidents is taking the models beyond its validation range, where they cannot be used as a predictive tool. In particular, in the case of the Buncefield and similar VCE incidents, the extent and density of congestion are substantially less than those required for DDT.

Forensic evidence and detailed reviews of many incidents similar to the Buncefield explosion have shown serious discrepancies between the presumption that the overpressures and damages were caused by detonation and what has been observed at most VCE incidents [18].

A detailed analysis of the physical damages and data available from CCTV cameras led to the conclusions [28] that the observed damages are not consistent with what would occur in a detonation. It is also shown that [18] "the combustion in Buncefield was unsteady (episodic), with periods of rapid flame advance (producing high local overpressures) being punctuated by pauses. Overall, the average rate of flame progress was subsonic ($\sim$ 150 m/s)." In most large-scale dust explosion accidents, a series of explosions consisting of a primary weak explosion, followed by a devastating secondary explosion, has been reported [16, 29]. While the hazardous effect of the primary explosion is relatively small, the secondary explosions propagate with a speed of up to 1000 m/s, producing overpressures in the range of 8–10 bar.

In the present study, it is shown that the effect of the strong increase of the radiation transparency caused by turbulent clustering of particles ahead of the primary ignited flame can be an alternative mechanism explaining the episodic flame propagation mode, featuring periods of high rates of combustion and overpressures punctuated by slow flame propagation in unconfined dust explosions.

9.3.1 Effect of Turbulent Clustering of Dust Particles on Radiative Heat Transfer

In turbulent flows ahead of the primary flame, dust particles with material density that is much larger than the fuel–air gas density assemble in small clusters with sizes about several Kolmogorov turbulent length scales. The turbulent eddies, acting as small centrifuges, push the particles to the regions between the eddies, where the pressure fluctuations are maximum and the vorticity intensity is minimum. Therefore, suspended small particles in a turbulent flow tend to assemble in clusters with much higher particle number densities than the mean particle number density. This effect, known as inertial clustering, has been investigated in a number of analytical, numerical, and experimental studies [30–35].

Analytical studies and laboratory experiments [36, 37] have shown that the particle clustering is significantly enhanced in the presence of a mean temperature gradient, so the turbulence is temperature stratified and the turbulent heat flux is not zero. This causes correlations between fluctuations of fluid temperature and velocity and nonzero correlations between fluctuations of pressure and fluid velocity. This enhances the particle clustering in the regions of maximum pressure fluctuations. As a result, the particle concentration in clusters rises by a few orders of magnitude as compared to the mean concentration of evenly dispersed particles [36–38].

We will show that the clustering of particles in the temperature-stratified turbulence ahead of the primary flame gives rise to a strong increase of the radiation penetration length. To investigate how clustering of particles affects the radiation penetration length, and how this effect depends on the turbulence parameters, we consider a turbulent flow with suspended particles exposed by a radiative flux. The radiative transfer equation for the intensity of radiation, $I(r,\hat{s})$, in a two-phase flow reads [39, 40]:

$$(\hat{s}\cdot\nabla)\,I(r,\hat{s}) = -\left[\kappa_g(r) + \kappa(r) + \kappa_s(r)\right] I(r,\hat{s})$$
$$+ \kappa_g(r)\,I_{b,g}(r) + \kappa(r)I_b(r) + \frac{\kappa_s(r)}{4\pi}\int \phi(r,\hat{s},\hat{s}')\,I(r,\hat{s}')\,d\Omega,$$

(9.15)

where $\kappa_g(r)$ and $\kappa(r)$ are the absorption coefficients of gas and particles, respectively, $\kappa_s(r)$ is the particle scattering coefficient, ϕ is the scattering phase function, $I_{b,g}(r)$ and $I_b(r)$ are the black-body radiation intensities for gas and particles, respectively, r is the position vector, $\hat{s} = k/k$ is the unit vector in the direction of radiation, k is the wave vector, (k,θ,φ) are the spherical coordinates in k space, and $d\Omega = \sin\theta\,d\theta\,d\varphi$. Taking into account that the scattering and absorption cross sections for gases at normal conditions are very small, the contribution from the gas phase is negligible in comparison with that of particles. We also take into account that for typical dust particles [41] the scattering coefficient is negligibly small compared with the particle absorption coefficient. Therefore, Eq. (9.15) is reduced to $(\hat{s}\cdot\nabla)\,I(r,\hat{s}) = -\kappa(r)(I(r) - I_b(r))$.

In the framework of a mean-field approach, all quantities are decomposed into the mean and fluctuating parts: $I = \overline{I} + I'$, $I_b = \overline{I}_b + I'_b$ and $\kappa = \overline{\kappa} + \kappa'$. The fluctuating parts, I', I'_b, κ', have zero mean values, and overbars denote averaging over an ensemble of fluctuations. The radiation absorption length for evenly dispersed particles is $L_a = 1/\overline{\kappa}$, where $\overline{\kappa} = \sigma_a \overline{N}$ is the mean particle absorption coefficient, $\sigma_a \approx \pi d_p^2/4$, and $\overline{N}$ is the mean number density of evenly dispersed particles. The instantaneous particle number density is $n = \overline{N} + n'$, with n' being fluctuations of the particle number density. The instantaneous particle absorption coefficient is $\kappa = n\,\sigma_a$, so that the fluctuations of the absorption coefficient are $\kappa' = n'\,\sigma_a = n'\,\overline{\kappa}/\overline{N}$.

Averaging the reduced radiative transfer equation for the intensity of radiation, $I(\mathbf{r},\hat{\mathbf{s}})$, over the ensemble of the particle number density fluctuations, we obtain the equation for the mean irradiation intensity: $(\hat{\mathbf{s}}\cdot\nabla)\,\overline{I}(\mathbf{r},\hat{\mathbf{s}}) = -\overline{\kappa}(\overline{I} - \overline{I}_b) - \langle \kappa'\,I'\rangle + \langle \kappa'\,I'_b\rangle$, where the angular brackets $\langle \cdots \rangle$ denote averaging over an ensemble of fluctuations. The expression for the one-point correlation function $\langle \kappa'\,I'\rangle$ has been derived by [41].

The small correlation between the particle number density fluctuations and the gas temperature fluctuations can be neglected, which implies that $\langle \kappa'\,I'_b\rangle = 0$. Substituting the correlation function $\langle \kappa'\,I'\rangle$ into the mean-field equation yields an equation for the mean radiation intensity $(\hat{\mathbf{s}}\cdot\nabla)\,\overline{I}(\mathbf{r},\hat{\mathbf{s}}) = -\kappa_{\text{eff}}\left(\overline{I} - \overline{I}_b\right)$, where the effective absorption coefficient κ_{eff} takes into account the particle clustering in a temperature-stratified turbulence.

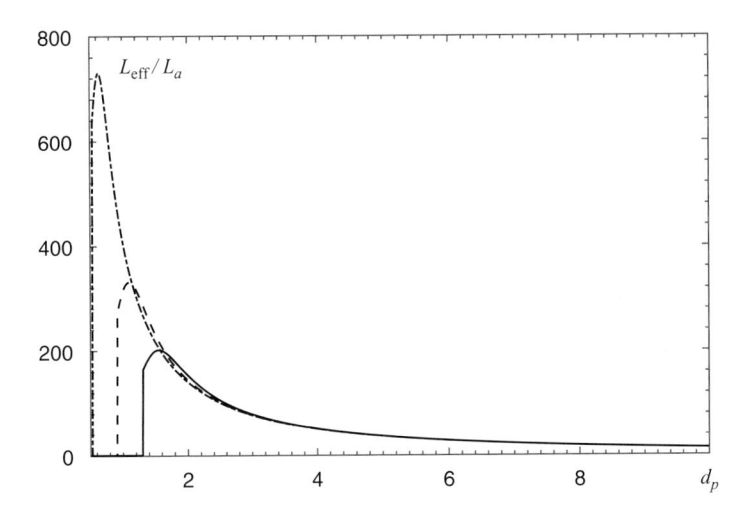

Figure 9.3 The ratio L_{eff}/L_a versus the particle diameter d_p for different mean temperature gradients $|\nabla \overline{T}|$: 0.5 K/m (solid), 1 K/m (dashed), 3 K/m (dashed-dotted).

Using the effective absorption coefficient κ_{eff} [41] with help of the normalized two-point correlation function of the particle number density fluctuations [36–38], one can obtain the following expression for the effective penetration length of radiation $L_{\text{eff}} \equiv 1/\kappa_{\text{eff}}$:

$$\frac{L_{\text{eff}}}{L_a} = 1 + \frac{2a}{\text{Sc}^{1/2}}\left(\frac{n_{\text{cl}}}{\overline{N}}\right)^2 \left(\frac{\ell_\eta}{L_a}\right)\left[1 + \frac{\mu - 1}{(\mu - 1)^2 + \alpha^2}\right]. \qquad (9.16)$$

Here μ and α are functions of the degree of compressibility, $\sigma_T = \left(\sigma_{T0}^2 + \sigma_v^2\right)^{1/2}$, of the turbulent diffusion tensor and the degree of compressibility, σ_v, of the particle velocity field [41].

Figure 9.3 shows the ratio L_{eff}/L_a versus the particle sizes determined for different values of the mean temperature gradients: 0.5, 1, and 3 K/m and for the following parameters: $\text{Re} = \ell_0 u_0/\nu = 5 \times 10^4$ with the integral turbulence scale $\ell_0 = 1$ m, the turbulent velocity $u_0 = 1$ m/s; $\nu = 2 \cdot 10^{-5}$ m^2/s, $c_s = 450$ m/s, $\sigma_{T0} = 1/2$ for methane–air at normal conditions, and $n_{cl}/\overline{N} = 500$.

9.3.2 Radiation-Induced Secondary Explosions

The role of radiation in dust explosions is an important issue that was not fully examined so far. In unconfined dust explosions, the radiative flux emitted from the advancing flame into the reactants is significantly enhanced by the increased emissivity of the large volume of burned products with hot particles. It is close to the black-body radiation at stoichiometric flame temperatures, 2200–2500 K [42, 43]. The experimentally measured thermal radiation in dust explosions is in the range 140–556 kW/m^2 [43, 44]. Recent advances in diagnostics methods have shown that radiation can make a decisive contribution to the overall energy transport, structure, and velocity of the flame [42–45]. Particles heated by thermal radiation ahead of the advancing flame transfer the heat to a flammable gas mixture, which can enhance the flame speed and cause ignition of the surrounding fuel–air mixture. The probability of ignition depends on the amount of thermal radiation absorbed by the particles, which must be heated to sufficiently high temperatures.

In the case of evenly dispersed particles, radiation is absorbed by particles in the nearest layers ahead of the flame front, and its intensity decreases exponentially. The situation is completely different if particles are nonuniformly dispersed, for example, in the form of an optically thick dust layer separated from the flame front by a gaseous gap with sufficiently small concentration of particles. It was shown [46] that such optically thick layer of particles can ignite either a deflagration or detonation ahead of the advancing flame.

It was hypothesized [47] that fibrous particles can be heated and ignited sufficiently quickly by the forward radiation, which in turn may lead to multipoint ignition of a gas mixture ahead of the flame front, resulting in a high rate of combustion. However, in order for a secondary explosion to occur, a multipoint ignition of a sufficiently large volume of fuel–air ahead of the advancing flame must be ignited by the radiatively heated particles. The fast multipoint ignition of a large volume of fuel–air ahead of the flame front ensures that the pressure from the ignited fuel–air would rise faster than it can be equalized by sound waves.

This requires the penetration length of radiation, L_{eff}, to be so large that $L_{\text{eff}} \gg c_s \tau_{\text{ign}}$, where c_s is the speed of sound in the flow ahead of the flame and τ_{ign} is the timescales of fuel–air ignition by the radiatively heated ignition kernels.

For example, the radiation absorption length for evenly dispersed micron size particles with the mass loading 0.03–0.05 kg/m^3 is in the range of a few centimeters, then, according to Fig. 9.3, turbulent clustering leads to an increase in the absorption length up to 10–20 m ahead of the advancing flame. The level of thermal radiation of $S \approx 200\text{--}400$ kW/m^2, emanating from hot combustion products in dust explosions, is sufficient to rise the temperature of particles by $\Delta T_p \approx 1000$ K during the time $\tau_{\text{ign}} = (\rho_p d_p c_{p,p} \Delta T_p)/2S < 10$ ms, where $c_{p,p}$ is the particle specific heat.

The particle–gas energy exchange time is $\tau_{pg} = \rho d_p^2 c_p / 6\lambda \mathrm{Nu} < 1$ ms, where $\lambda = \chi \rho c_p$ is the thermal conducitivity, χ is the thermal diffusivity, c_p is the heat capacity of the gas, and $\mathrm{Nu} \approx 2$ is the Nusselt number.

For ignition to occur, a gas volume with a size that is of the order of the typical flame thickness, r_f, has to be heated to the ignition temperature. This means that the time it takes for an isolated particle to heat the surrounding gas to a level where it can ignite is given by $t_\chi \sim r_f^2/\chi \approx 500$ ms. A cluster of radiatively heated particles ignites the fuel–air mixture in a much shorter time, since in this case the mixture is heated by thermal conduction within a small distances, $\Delta r = 1/n_{cl}^{1/3}$, corresponding to the particle separation in the cluster. For the typical values of ignition time of the fuel–air mixture by radiatively heated clusters of particles of the order of 10 ms, the effective propagating rate of combustion caused by the secondary explosion can be estimated as $L_{\mathrm{eff}}/\tau_{\mathrm{ign}} \sim 10^3$ m/s. Such a rate of the secondary explosion corresponds to the intensity of a shock wave with Mach number, $\mathrm{Ma} = 2{-}3$, producing overpressures of 8–10 atm. This explains the mechanism of the secondary explosion and the level of damages observed in unconfined dust explosions. This mechanism of the multipoint radiation-induced ignitions due to the turbulent clustering of particles explains the occurrence of the secondary explosion regardless of whether the particles are combustible or not.

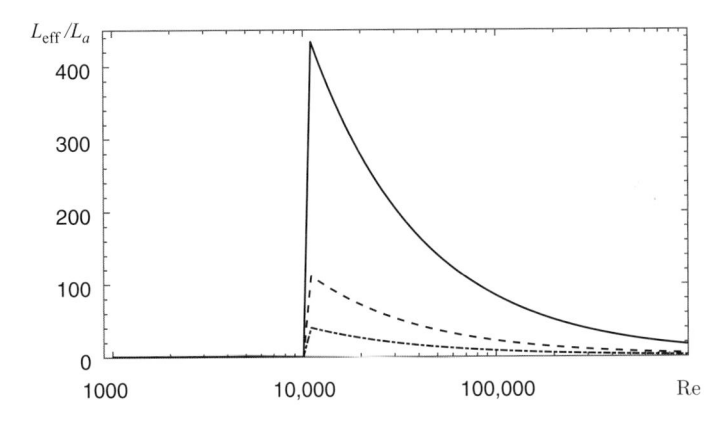

Figure 9.4 The ratio L_{eff}/L_a versus the Reynolds number Re for different particle diameter d_p: 3 μm (solid), 6 μm (dashed), 9 μm (dashed-dotted), and for the temperature gradients $|\nabla T| = 1$ K/m.

Figure 9.4 shows the values of L_{eff}/L_a versus Reynolds numbers calculated for particles of different sizes. It is seen that a significant increase of the radiation penetration length caused by the particle clustering occurs within a rather narrow interval of Reynolds numbers. The effect is much weaker if the flow parameters ahead of the flame front are out of the range of "radiation transparency." This explains the episodic nature of the rate of combustion in dust explosions. According to the analysis of the Buncefield explosion [28], "The high overpressures in the cloud and low average rate of flame advance can be reconciled if the rate of flame advance was episodic, with periods of very rapid combustion being punctuated by pauses when the flame advanced very slowly." Since the primary flame is a deflagration, propagating with velocity of the order of meters per second, the duration of this stage is the longest time scale in the problem. During this time the particle clusters ahead of the flame are heated by the forward radiation for a sufficiently long time to become ignition kernels in a large volume ahead of the flame. This initiates the secondary explosion.

The parameters of turbulence change after the secondary explosion, so that the rapid phase of combustion is interrupted until the shock waves produced by the secondary explosion dissipate. The next phase continues until the parameters of turbulence in the flow ahead of the combustion wave fall within the interval corresponding to the "transparent window," which may cause the particle clustering and a proper conditions for the next secondary explosion.

In summary, it is shown that the clustering of dust particles in a turbulent flow gives rise to a significant increase of the thermal radiation absorption length ahead of the advancing flame front. This ensures that clusters of dust particles are heated by the radiation from hot combustion products of large gaseous explosions to become multipoint ignition kernels in a large volume ahead of the flame front. The turbulent clustering of particles is an alternative mechanism explaining the origin of the secondary explosion that produce the high speeds of combustion and high overpressures in unconfined dust explosions instead of the deflagration-to-detonation transition suggested in earlier studies.

9.4 Intraparticle Transport Phenomena in Solid Fuel Combustion

When considering particle scale, each thermal conversion reaction involves subprocesses: (1) transfers of heat and/or mass to external surface of particles through the gas boundary, (2) intraparticle thermal conduction or diffusion of reactants, (3) thermal conversion reaction including adsorption and desorption, (4) transport of products to external surface, and (5) transport of products through a gas boundary. Chemical reactions also involve heat release and physical changes such as swelling, shrinking, deformation, and fragmentation. Dependent on the type of fuels and the design of combustion devices, the relative importance of intraparticle phenomena varies mainly according to temperature and particle diameters. This section discusses the role of intra-particle transport phenomena mostly for biomass combustion and gasification, albeit physical principles are also applicable for coal.

9.4.1 Time-scale Analyses

The relative importance of physical and chemical processes in overall kinetics of solid combustion can be estimated by dimensionless numbers. The most relevant dimensionless numbers during pyrolysis are the Biot number, Bi, and internal and external (thermal) Damköhler numbers, Da_i and Da_e:

$$\text{Bi} = \frac{\text{Conduction time scale}}{\text{External heat transfer time scale}} = \frac{h d_p}{\lambda_{\text{eff}}} \quad (9.17)$$

$$\mathrm{Da}_i = \frac{\text{Conduction time scale}}{\text{Pyrolysis reaction time scale}} = \frac{\rho_p c_{p,s} d_p^2}{\lambda_{\text{eff}} k_{\text{dev}}} \qquad (9.18)$$

$$\mathrm{Da}_e = \frac{\text{External heat transfer time scale}}{\text{Pyrolysis reaction time scale}} = \frac{\rho_p c_{p,s} d_p}{h k_{\text{dev}}}. \qquad (9.19)$$

Here, h is the external heat transfer coefficient including the effect of radiation, d_p is the particle diameter, λ_{eff} is the effective thermal conductivity of particles, ρ_p is the density of the particle, $c_{p,s}$ is the specific heat of the particle, and k_{dev} is the rate coefficient of devolatilization.

Figure 9.5 shows the effect of reactor temperature and particle diameter on dimensionless numbers using kinetic and thermophysical parameters of woody biomass. Both Damköhler numbers are above 10 and the Biot number is around unity in major combustion devices. Therefore, it is important for combustion of wood to consider both thermal conduction and external heat transfer during devolatilization. In contrast, most coal combustion plants use pulverized burners with particle size less than 100 μm, raising the importance of chemical kinetics while reducing the effect of thermal conduction. Note that time-scale analyses in solid fuel combustion should be used only for the order of magnitude due to its inherent heterogeneity in chemical and thermophysical parameters.

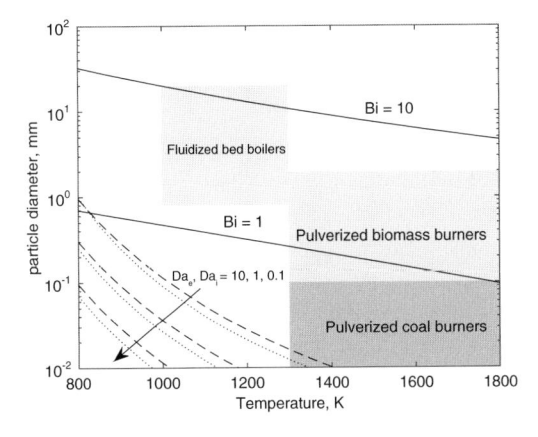

Figure 9.5 Time-scale analyses for devolatilization of wood particles. Solid lines: Biot number; dash lines: internal Damköhler number; dotted lines: external Damköhler number.

Char conversion is affected mostly by mass transfer and chemical kinetics. Since both physical and chemical time scales vary by a few orders of magnitude dependent on the type of fuels and oxidizing agent, detailed analyses are omitted here. Under the reaction conditions of practical applications, char conversion is usually affected by intra-particle mass diffusion through pores. This is often called zone II, in comparison with zone I (chemically controlled) and zone III (under the influence of external mass transfer).

9.4.2 Resolved Particle Models

Resolved particle models consider intraparticle transports of heat, mass and momentum with varying physical parameters [48, 49]. Most resolved particle models are 1D, but anisotropic fuels such as wood logs require 2D/3D solution for accurate modeling results. The Eulerian–Eulerian approach is commonly applied for transport equations. The change in the density of solid species, $\rho_{i,p}$, can be directly described with the rate of formation/consumption, $\dot{\omega}_{i,p}$, as

$$\frac{\partial \rho_{i,p}}{\partial t} = \dot{\omega}_{i,p}. \tag{9.20}$$

Transport equations for the gas mixture and gas specie in the particle pores are given by

$$\frac{\partial (\varepsilon \rho)}{\partial t} + \nabla \cdot (\rho \mathbf{u}) = \dot{\omega}_{p \to g}, \tag{9.21}$$

and

$$\frac{\partial (\varepsilon \rho Y_i)}{\partial t} + \nabla \cdot (\rho \mathbf{u} Y_i) = \nabla \cdot (\rho D_{\text{eff}} \nabla Y_i) + \dot{\omega}_{i,g}, \tag{9.22}$$

when ε is the porosity of the particle, Y_i is the mass fraction of gas specie i, D_{eff} is the effective diffusivity, $\dot{\omega}_{p \to g}$ is the rate of formation of the gas mixture, and $\dot{\omega}_{i,g}$ is the rate of formation of gas specie i.

Momentum transport is usually described by Darcy's law as

$$\mathbf{u} = -\frac{\mathbf{K}}{\mu} \nabla p, \tag{9.23}$$

while the state equation is given by

$$p = \frac{\rho R_g T}{M_g}. \tag{9.24}$$

Here, $\mathbf{K}$ is the permeability, μ is the dynamic viscosity, p is the total gas pressure, R_g is the ideal gas constant, M_g is the molecular mass of the gas mixture, and T is temperature. Local thermal equilibrium between solid and gas is a reasonable assumption. Hence, overall enthalpy transport can be written as

$$\left(\sum \rho_{i,p} c_{pi,p} + \varepsilon \rho \sum Y_i c_{pi,g} \right) \frac{\partial T}{\partial t} + \left(\rho \sum Y_i c_{pi,g} \mathbf{u} \right) \cdot \nabla T = \nabla \cdot (\lambda_{\text{eff}} \nabla T) + \sum r_j \Delta h_j, \tag{9.25}$$

where λ_{eff} is the effective conductivity and Δh_j is the heat of reaction j. The source terms in the transport equations originate from drying, devolatilization, and heterogeneous reactions (char combustion and gasification).

Since the pore size of typical char particles are of the same order as the mean free path of the gases, surface and configurational diffusion can be ignored. Hence, the effective diffusivity can be calculated from molecular diffusivity, $D_{A,m}$, and Knudsen diffusion, $D_{A,k}$, with tortuosity, τ, as

$$D_{\text{eff}} = \frac{\varepsilon}{\tau} \frac{1}{\frac{1}{D_{A,m}} + \frac{1}{D_{A,k}}}. \qquad (9.26)$$

The effective conductivity contain contributions from both solid and gas, in addition to radiation across pores:

$$\lambda_{\text{eff}} = \frac{\sum \rho_{i,p} \lambda_{i,p}}{\sum \rho_{i,p}} + \varepsilon \sum Y_{i,g} \lambda_{i,g} + f(\varepsilon) \cdot \epsilon \sigma d_{\text{pore}} T^3, \qquad (9.27)$$

where the radiative term varies based on the pore size distribution. For example, wood pellets and coal have isotropic pore structures, while wood has pores that are aligned parallel with the direction of growth, giving different values in the longitudinal and lateral directions. Classic calculation methods and values for thermophysical parameters can be found in [3] while recent studies address the effect of pore structure by pore resolved particle simulation [50].

For the boundary conditions at the particle surfaces, it is common to use classic correlations, such as the Ranz–Marshall correlation. Correlations for various shapes of particles can be found in [51]. However, some recent studies resolve the quiescent boundary layer [52] or gas flow surrounding the particles [53–55], showing that Stefan flow caused by chemical reactions affects the gas flow in the gas boundary and modifies the transfer coefficients. For a thorough discussion of resolved particle simulation, readers are directed to the recent review by Haberle et al. [3].

9.4.3 Effect of Thermal Conduction on the Devolatilization of Biomass

As discussed previously, pyrolysis of biomass particles is affected by thermal conduction while char burnout is controlled by mass diffusion, albeit to various degrees. Figure 9.6 shows the evolution in local temperature and conversion, $X = (\rho_0 - \rho)/(\rho_0 - \rho_c)$, for typical reaction conditions in grate or fluidized bed boilers (a) and pulverized burners (b), calculated by the resolved particle model described in [56].

Reaction zone of pyrolysis appears to be narrow and progresses from the surface to the center in case (a). This is because the local temperature has a steep gradient, and the narrow region with the temperature relevant to pyrolysis, ca. 600–800 K, moves from the surface to the center relatively slowly. Fuel particles that behave similar to this case are often called thermally thick particles. If the pore distribution is isotropic, the convective cooling from outgoing gas flow from devolatilization cannot be neglected. In addition, secondary char formation from the reactions of large molecule gases often affects the final char yield. From a modeling perspective, it is reasonable to divide the particle into two zones, i.e., the char layer and the core of unreacted fuel. Such a treatment simplifies the calculation of thermophysical parameters in resolved particle models.

Case (b) shows little difference in local temperature across the particle, resulting in a rather uniform progress of pyrolysis throughout the particle. Such fuel particles

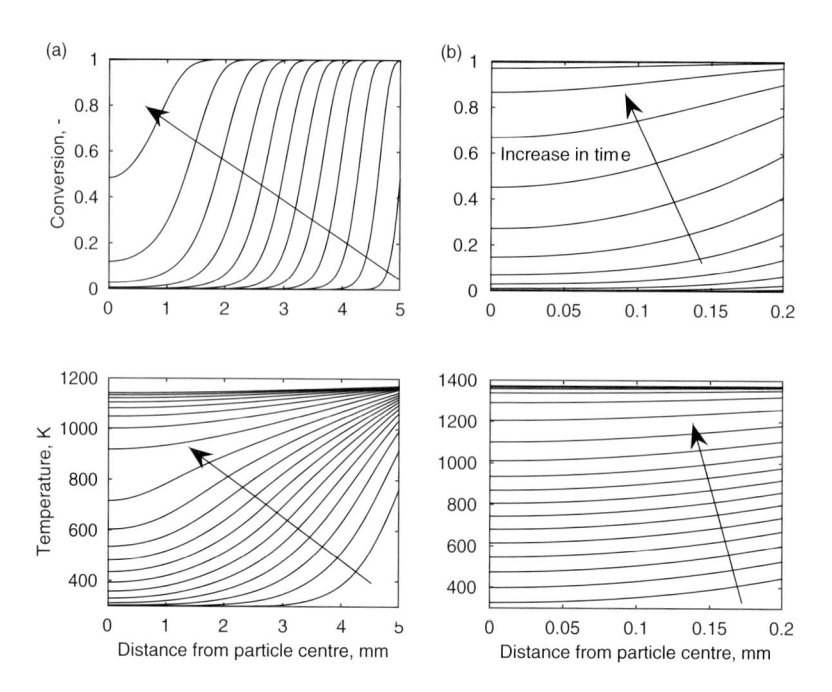

Figure 9.6 Local conversion and temperature during biomass pyrolysis. (a) $T_g = 1173$ K and $d_p = 10$ mm (Bi = 7.7). Line is plotted every 2.5 s. (b) $T_g = 1373$ K and $d_p = 0.4$ mm (Bi = 1.2). Line is plotted every 0.04 s.

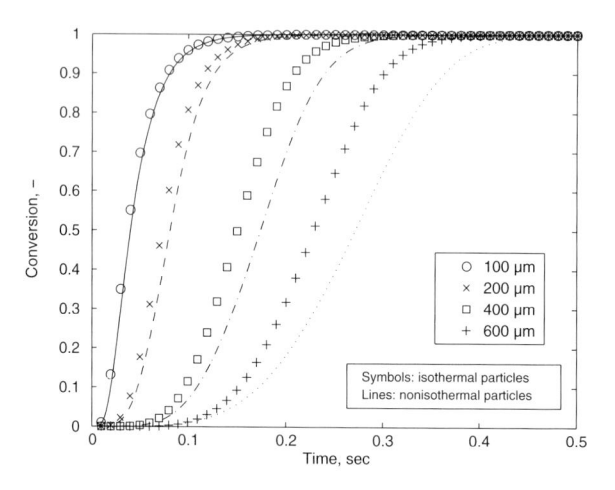

Figure 9.7 Comparison of simulation results with and without thermal conduction during pyrolysis of free-falling biomass at $T = 1273$ K.

are called thermally thin. For thermally thin particles, it is important to clarify if the conversion rate of the overall particle can be described without considering intraparticle transport phenomena. In Fig. 9.7, the overall conversion history of a particle from the resolved particle model is compared with simulations using the isothermal particle model, which solves the heat balance equation,

$$m_p c_p \frac{dT_p}{dt} = h A_p (T_g - T_p) + \epsilon \sigma (T_w^4 - T_p^4) + \sum r_j \Delta h_j \qquad (9.28)$$

where A_p is the surface area of the particle. This has been done for conditions that correspond to a pulverized burner. As is apparent from the definition of the Biot number (Eq. 9.17), a change in particle size will affect the relevance of thermal conduction on the overall conversion rate. It is clear from the figure that thermal conduction cannot be neglected for particles larger than ca. 200 μm. Similar analyses for char burnout can be found in, e.g., [57].

9.4.4 Simplified Models and Application in Burner Simulation

Although some studies implemented resolved particle models in multiparticle simulation platforms [58], intraparticle models require simplifications when applied to large-scale simulations. After consideration of physical processes, the simplification is usually carried out by ignoring certain physical processes, and then by converting partial differential equations to ordinary differential equations or by deriving analytical solutions.

The most common approach for high Biot number particles during pyrolysis is (1) ignoring the effect of intraparticle gas flow, (2) assuming very thin reaction zone, and (3) using constant thermophysical properties. Then, fuel particles are divided into nonreacting char layer and raw fuel core [4]. This approach is called the shrinking core model (SCM), and various analytical solutions have been derived [56, 59, 60].

Char burnout in practical combustion applications often progresses in Regime II, where both intrinsic reactivity and pore diffusion affect the overall reaction rate. A common approach is to use an analytical solution of the effectiveness factor, which is a ratio between observed reaction rate and the reaction rate one would experience if mass diffusion within the particle was infinitely fast. The effectiveness factor for an nth-order reaction with respect to oxidizing agent is

$$\eta = \frac{1}{\phi} \left[\frac{1}{\tanh(3\phi)} - \frac{1}{3\phi} \right], \qquad (9.29)$$

where

$$\phi = \frac{V_p}{S_p} \sqrt{\frac{n+1}{2} \frac{\rho_c k_c C_{As}^{n-1}}{D_{eA}}}. \qquad (9.30)$$

The effectiveness factor for Langmuir–Hinshelwood type kinetic expression can be found in [61]. One challenge is the changes of radius and density due to the consumption of carbon. In zone II, consumption of carbon inside the particle is nonuniform, and it is faster at the external surface. Therefore, char conversion without the consideration of ash has two stages: (1) consumption of carbon near the particle surface while no change in diameter, and (2) decrease in diameter due to the complete consumption of carbon at the particle surface. A few studies have addressed such changes in particle density and diameter during char conversion [62, 63].

9.5 Preferential Transport of Solid Dispersed Phase in Turbulent Flows with Implications for Combustion Phenomena

Solid fuel combustion is often constituted by a particulate phase that burns in a turbulent flow. Particles need to be gasified via a pyrolysis process before reacting with the oxidant in hot regions. Transport and mixing of the solid and fluid phases determine the local mixture fraction and impact the level of pollutant emissions and combustion efficiency. Hence, the main transport properties of a solid dispersed phase are here discussed.

9.5.1 Dispersed Solid Phase in Turbulent Flows: Coupling Mechanisms

The interaction between dispersed solid and fluid phases is first discussed by analyzing the fundamental laws of mechanics. To reduce the vast number of controlling parameters, a fluid with constant density ρ and dynamic viscosity $\mu = \rho \nu$ laden with rigid solid spheres is considered. Actually, the essence of the coupling mechanism is not different from reactive cases. The fluid flow velocity $\boldsymbol{u}$ is governed by the incompressible Navier–Stokes equations:

$$\nabla \cdot \boldsymbol{u} = 0, \qquad \rho \frac{D\boldsymbol{u}}{Dt} = \nabla p + \mu \nabla^2 \boldsymbol{u} + \rho \boldsymbol{g}, \tag{9.31}$$

where p is the pressure and $\boldsymbol{g}$ is the gravitational acceleration. Navier–Stokes equations should be completed with initial and boundary conditions. In particular, dealing with particle-laden flows, boundary conditions have to be imposed on all surfaces of the moving sphere by assigning the local value of the solid phase velocity $\boldsymbol{v}_s = \boldsymbol{v} + \boldsymbol{\omega} \times \boldsymbol{r}$, where $\boldsymbol{v}$ and $\boldsymbol{\omega}$ are the linear and angular velocities at the sphere center. Particle dynamics follow the Newton–Euler equations:

$$\rho_p \frac{\pi d_p^3}{6} \dot{\boldsymbol{v}} = \int_S \boldsymbol{\tau} \cdot \boldsymbol{n} \, dA + (\rho_p - \rho) \frac{\pi d_p^3}{6} \boldsymbol{g} + \boldsymbol{F}_c \tag{9.32}$$

$$\rho_p \frac{\pi d_p^5}{60} \dot{\boldsymbol{\omega}} = \int_S \boldsymbol{r} \times (\boldsymbol{\tau} \cdot \boldsymbol{n}) dA + \boldsymbol{T}_c, \tag{9.33}$$

with $\boldsymbol{\tau} = -p\boldsymbol{I} + \mu(\nabla \boldsymbol{u} + \nabla \boldsymbol{u}^T)$ the stress tensor and $\boldsymbol{F}_c$ and $\boldsymbol{T}_c$ the force and torque induced by collisions among the particles with diameter d_p and density ρ_p. Hence, the fluid pressure and velocity gradient at particle surfaces determine forces and torques on the particle from the fluid. In turn, the fluid behavior around the particle strongly depends on the particle velocity "imposed" as boundary condition. This complex mutual coupling localized on the moving particle boundaries makes the problem very difficult to tackle using a direct fundamental approach; see [64] for a recent discussion on the topic. Although the coupling mechanism to solid fuel combustion is similar to the situation described earlier for isothermal nonreacting systems, additional complexities are given by the usual turbulent flow regime with strong density, temperature, and energy fluctuations.

Given the extreme difficulties to directly tackle the problem, and in order to proceed towards simplified models, it is crucial to analyze the effect of the main controlling parameters on the global coupling mechanisms between the phases. Considering a turbulent flow (in-compressible) laden with rigid spheres, three nondimensional numbers control the coupling between the two phases: the volume fraction $\Phi = V_s/V_t$, the mass fraction $\Psi = M_s/M_t = (\rho_p/\rho_t)\Phi$ or equivalently the density ratio $R = \rho_p/\rho$, and the scale ratio $S_\eta = d_p/\eta$ with $\eta = (\nu^3/\varepsilon)^{1/4}$ the Kolmogorov dissipative length (ε is the energy dissipation rate).

The volume fraction gives the ratio of the distance between the particles and their size, $l/d_p \simeq (\Phi_M/\Phi)^{1/3}$, with Φ_M being the maximum packing fraction, which is around 0.6 for random distributions. The closer the particles are, the more important is their mutual interaction via collisions and short-range hydrodynamic forces. In particular, since these mechanisms concern particle pairs, the collisional intensity is proportional to the square of the volume fraction (leading order), $\propto \Phi^2$. In other words, at low volume fractions, particles are too far from each other to have direct interaction. The mass fraction Ψ determines the repartition of the overall momentum between the phases. Actually, since the bulk velocities of the two phases are usually similar, the global momentum ratio is proportional to the mass fraction Ψ. Hence, at low mass fraction ($\Psi \ll 1$) the fluid-phase momentum is unaffected by the presence of the solid phase and the opposite limit holds for very high Ψ (granular media). A sketch of this is given in Fig. 9.8.

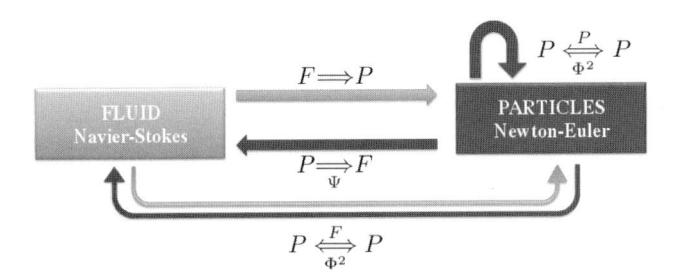

Figure 9.8 Global scheme of the coupling mechanisms between solid and fluid phases. Each arrow represents a different way of coupling: light gray (rectangle at left and arrows directed to the right) induced by the fluid phase (F) and dark gray (rectangle at right and arrows directed to the left, as well as curved arrow on top) the particle phase (P). Four coupling mechanisms are represented: Fluid action to Particles ($F \longrightarrow P$); Particle feedback on the Fluid ($P \longrightarrow F$); Particle–Particle interaction via collisions (direct $P \longleftrightarrow P$) and Particle–Particle interaction via hydrodynamic short-range interaction ($P \longleftrightarrow P$ via fluid). Below each arrow, the leading order term as a function of volume/mass fraction is reported.

From this global analysis, different couplings emerge. In very dilute conditions, at small mass and volume fractions, i.e., $\Psi < 10^{-3}$–10^{-2} and $\Phi < 10^{-3}$–10^{-2}, the so-called one-way coupling regime takes place where the fluid phase forces the solid phase, but not vice versa. Collisions and hydrodynamic coupling between particles are also negligible. Particles are forced and transported only by the fluid phase, whose dynamics can be approximated as a single phase flow. Despite the crude simplification, this regime frequently occurs in several applications, such as very dilute aerosols, particulate pollutants, clouds, dilute spray combustion, and soot transport [65, 66]. At higher mass fraction $\Psi > 10^{-3}$–10^{-2}, but still keeping the volume fraction small $\Phi < 10^{-3}$–10^{-2}, the solid phase back-reaction on the carrier fluid cannot be neglected anymore because of the momentum exchange between the phases. This is the so-called two-way coupling regime where fluid affects particles and vice versa, while short-range particle pair interactions, such as collisions or lubrication forces, can still be neglected. This regime may occur at high density ratio R since $\Psi \sim R\Phi$, which is the case of solid particles transported in a gas, as in solid combustion where $R \sim 10^3$–10^4. Increasing the volume fraction $\Phi > 10^{-3}$–10^{-2}, the probability to have collisions and short-range hydrodynamic interactions suddenly increases. This regime corresponds to the so-called four-way coupling mechanism. This is the most complex case, since the mutual and diverse coupling mechanisms among the phases spans a wide range of scales and they do not allow any simplification of the problem. This is the typical regime of fluidized bed and of the initial dense regime of liquid/solid fuel injectors.

Some additional remarks need to be given on the scale ratio S_η. The presented picture holds from a global point of view, while locally around a single particle strong differences emerge when the particle size is larger than the smallest scale of turbulence $S_\eta > 1$, the so-called finite-size particle regime. Even if the global picture is unchanged, locally the particle always alters the flow even in very diluted regimes. On the other hand, when $S_\eta < 1$ the flow around the particles can be considered smooth, locally homogeneous, and unaltered by the presence of particles in the dilute regimes. These assumptions are crucial in order to simplify the problem. Another important point is the effect of gravity/buoyancy on the solid-phase dynamics. Dealing with sufficiently high-speed flows, these effects can often be neglected. Otherwise, another parameter needs to be included to account for the importance of the buoyancy/settling of the dispersed phase [67].

9.5.2 Preferential Transport of Solid Dispersed Phases

9.5.2.1 Dilute Conditions

The dynamics of the dispersed phase in turbulent flows is first discussed for the dilute one-way coupling limit and with particle size smaller than the Kolmogorov length, i.e., $\Phi \sim \Psi \simeq 0$ and $S_\eta \ll 1$. Each particle is assumed as a material point that is forced by an unperturbed time-varying flow. In this limit, each particle feels the surrounding fluid flow sampled at its position and Eq. (9.32) reduces to [68]

$$\frac{d\mathbf{v}_p}{dt} = \frac{\mathbf{u} - \mathbf{v}_p}{\tau_p} + \frac{\rho}{\rho_p} \frac{D\mathbf{u}}{Dt} + \left(1 - \frac{\rho}{\rho_p}\right)\mathbf{g} + \frac{1}{2}\frac{\rho}{\rho_p}\left(\frac{D\mathbf{u}}{Dt} - \frac{d\mathbf{v}_p}{dt}\right)$$
$$+ \sqrt{\frac{9}{2\pi}\frac{\rho}{\rho_p}\frac{1}{\tau_p}} \int_0^t \frac{1}{\sqrt{t-\tau}} \frac{d}{d\tau}(\mathbf{u} - \mathbf{v}_p)\,d\tau, \qquad (9.34)$$

with $\boldsymbol{u}|_p$ as the fluid velocity at the particle position and $\tau_p = \rho_s d^2/(18\mu)$ the particle relaxation time (a small particle Reynolds number is assumed). All terms related to fluid velocity are calculated using the (unperturbed) flow velocity at the position of the particle. The terms on the RHS of Eq. (9.34) are the drag, pressure gradient, gravity, added mass, and basset history terms, respectively. It has been shown that in the limit of high density ratio $\rho_p/\rho \gg 1$, as for solid in gas, the only surviving terms are the drag and gravity terms [69], with the latter being negligible in high-speed flows (high Froude number). In this situation Eq. (9.34) reduces to

$$\frac{d\boldsymbol{v}_p}{dt} = \frac{\boldsymbol{u} - \boldsymbol{v}_p}{\tau_p}. \tag{9.35}$$

The particle relaxation time measures the lag for a particle to adapt its velocity $\boldsymbol{v}_p$ to the local fluid velocity $\boldsymbol{u}$. It represents the ratio between the particle inertia and viscous drag effects. Once fixed a turbulent flow configuration, the only parameter controlling the coupling between particles and fluid is the so-called Stokes number, $St = \tau_p/\tau_f$, which compares the particle relaxation time with a typical flow time scale. Different limits emerge:

- Small Stokes number $St \ll 1$, Eq. (9.35), reduces to $\boldsymbol{v}_p = \boldsymbol{u}$ because particles are very fast in adapting to flow velocity. These particles behave as tracers and follow fluid trajectories.
- Large Stokes number $St \gg 1$, Eq. (9.35) reduces to $d\boldsymbol{v}_p/dt = 0$ so particles are in the ballistic limit and travel without feeling the flow. Solid phase velocity becomes uncorrelated with respect to local fluid velocity.
- Intermediate Stokes number $St \sim 1$, Eq. (9.35), cannot be simplified. In this limit particles are strongly coupled with the flow dynamics. In particular, since the particle relaxation time, τ_p, is of the order of the flow time scale τ_f, particles tend to follow flow trajectories, but with a time lag that results in continuous *outward* drift. Hence particles appear to be centrifuged by flow structures with characteristic time τ_f. It should also be remarked that particle/fluid velocities and trajectories differ.

In the former two limits the solid phase tends to occupy the whole space without showing segregation. In the intermediate Stokes number regime, several peculiar transport phenomena have been observed as a local or a mean drift of particles showing spatial segregation. In the present section, two common phenomena will be discussed: small-scale clustering and turbophoresis. Because of these effects, the local particle concentration may grow by orders of magnitude with respect to the bulk one, while the small-scale clustering occurs in all turbulent flows and appears as an inhomogeneous local particle distribution that is arranged in clusters. Turbophoresis is characteristic only of turbulent nonhomogeneous flows and amounts to a mean drift of particles induced by the nonhomogeneous turbulence that promotes a mean accumulation in areas of low turbulent intensity. Turbophoretic effects are typically seen in wall-bounded flows.

Turbophoresis

Turbophoresis in wall bounded flows is a mean drift of particles toward the wall and is induced by the inhomogeneous turbulent flow near the wall [70]. From a phenomenological point of view, particles showing strong turbophoresis have a sufficiently small inertia to follow the turbulent flow when traveling from the bulk region toward the wall. However, when they slow down near the wall in the viscous sublayer, their inertia prevents them from being reentrained by ejection flow motions and remain trapped at the wall [70, 71]. This unbalance creates the mean drift toward the wall. Turbophoresis peaks when the particle relaxation time is of the order of the time scale of the *buffer* layer structures. In particular, this near wall region is populated by important flow structures, such as low- and high-velocity streaks and hairpin vortices, which can sustain the turbulence dynamics injecting energy from mean to fluctuating flow. In wall flows the typical turbulent time scale is usually expressed by the wall viscous time $\tau_* = \mu/\tau_w$, with τ_w the wall friction. Hence the commonly used (viscous) Stokes number is defined as $St^+ = \tau_p/\tau_*$. Turbophoresis peaks when the viscous Stokes number is around 20 ($St^+ \sim 20$), i.e., when the particle relaxation time is of the order of the time scale of the vortical structures populating the buffer layer. From a statistical point of view, the drift velocity can be related to the gradient of the turbulent kinetic energy (or better to the Reynolds stress components), so the drift is controlled from the inhomogeneous behavior of the turbulent wall flow [72]. Turbophoresis may increase the mean particle concentration at the wall even up to thousands times the bulk value.

9.5.2.2 Dense Conditions

When the mass fraction is high enough, but not the volume fraction, a two-way coupling regime takes place so the dispersed solid phase affects the turbulent flow behavior. In these conditions, clustering and turbophoresis are still important. The back reaction slightly reduces the intensity of the segregation even though the main phenomenology remain similar. However, the important aspect that cannot be neglected is the modification of the flow. Turbulence is modulated; the forcing applied by the particle clusters enhances the energy content of the small scales of turbulence while altering the classical energy cascade [73]. These motions may also increase the small-scale mixing between clusters and fluid. This is an important aspect that can mitigate the clustering effect in combustion problems. Similar arguments have been reported also for turbophoresis. In this context it should be noted that also the mean flow behavior may be strongly altered by the solid phase, e.g., jet spreading rate. Increasing the volume fraction and entering in the four-way coupling regime makes the analysis quite complex. Here the dynamics is controlled by particle induced stress, collisions, and hydrodynamic interactions. Clustering is usually strongly reduced compared to dilute cases and may also be induced by viscous interaction between the particles. The back reaction is so important that the mean flow behavior is usually extremely altered. It is difficult to draw general consideration in this regime, because of its strong dependence on the flow configuration and particle features [74]. In addition, both experimental and numerical approaches are difficult because of the flow opacity and of the complexity of the interactions.

9.5.2.3 Implications for Combustion

Both segregation phenomena mentioned earlier may be important in turbulent combustion of solid powders (and also of small droplets) because they strongly affect the local fuel concentration [65]. Pollutants formation (soot) and incomplete combustion may occur in clusters. For example, the soot is essentially composed by carbon agglomerates that originate in regions characterized by a rich mixture and high temperatures [66]. If the turbulent transport conveys such agglomerates in low temperature or regions with low oxygen concentration, their oxidation is abruptly interrupted, leading to soot formation, which may be successively dispersed in the atmosphere. Similar arguments hold for the production of unburned fuel emissions. It should be noted that also the growth rate of soot can be affected by the small-scale clustering, since it can strongly increase the collision rate. Another important aspect, which can counteract the particle segregation effect, is the difference of particle/fluid velocities and trajectories. It should be remarked that the fuel gas produced by particles behaves as the fluid. First, the different trajectories followed by the particles and the gas result in an additional source of mixing [75]. Second, the different particle to gas velocity may speed up the gasification process via local convection. Focusing on the global mixing, these aspects, together with the micro-mixing at high mass fraction, may counteract the negative segregation effect. It has been observed that, depending on the typical reaction time scale, the global effect may result in an increase or decrease of the reaction rate [15].

9.6 Final Remarks and Perspectives

Combustion of solid powders requires a fast mixing of particles, oxidant and high-temperature regions. The discussed peculiar particle transport phenomena in turbulent flows, such as segregation or small-scale clustering, may cause either an increase or a decrease of the reaction rate depending on the local conditions. These aspects also impact the overall combustion efficiency and the pollutant emissions. From an engineering point of view, standard commercial CFD models do not often consider the coupling between turbulence and particle transport, which is at the base of these transport phenomena. To avoid problem-specific parameters, a better understanding and improved modeling of these phenomena is mandatory. The other perspective for future research is to analyze and understand the dense regime behavior in turbulent conditions, especially when density variations are considered. Given the numerical and experimental difficulties to tackle this problem, the investigation of this regime will be a challenge for the next years.

References

[1] C. T. Crowe, J. D. Schwarzkopf, M. Sommerfeld, and Y. Tsuji, *Multiphase flows with droplets and particles*. New York: CRC Press, 2011.

[2] J. Kruger, N. E. L. Haugen, and T. Løvås, "Correlation effects between turbulence and the conversion rate of pulverized char particles," *Combust. and Flame*, **185**, 160–172, 2017.

[3] I. Haberle, Ø. Skreiberg, J. Lazar, and N. E. L. Haugen, "Correlation effects between turbulence and the conversion rate of pulverized char particlesnumerical models for thermochemical degradation of thermally thick woody biomass, and their application in domestic wood heating appliances and grate furnaces," *Prog. Energy Combust. Sci.*, **63**, 204–252, 2017.

[4] C. Di Blasi, "Modeling chemical and physical processes of wood and biomass pyrolysis," *Prog. Energy Combust. Sci.*, **34**, 47–90, 2008.

[5] S. Niksa and A. R. Kerstein, "FLASHCHAIN Theory for Rapid Coal Devolatilization Kinetics. 1. Formulation," *Energy & Fuels*, **5**, 647–665, 1991.

[6] D. M. Grant, R. J. Pugmire, T. H. Fletcher, and A. R. Kerstein, "Chemical Model of Coal Devolatilization Using Percolation Lattice Statistics," *Energy & Fuels*, **3**, 175–186, 1989.

[7] P. R. Solomon, G. Hamblen, and A. Serio, "A characterization method and model for predicting coal conversion behaviour," *Fuel*, **72**, 469–488, 1993.

[8] E. Ranzi, A. Cuoci, T. Faravelli, A. Frassoldati, G. Migliavacca, S. Pierucci, and S. Sommariva, "Chemical kinetics of biomass pyrolysis," *Energy & Fuels*, **22**, 4292–4300, 2008.

[9] R. H. Hurt, M. M. Lunden, E. G. Brehob, and D. J. Maloney, "Statistical kinetics for pulverized coal combustion," *Symp. (Int.) Combust.*, **26**, 3169–3177, 1996.

[10] N. E. L. Haugen, J. Kruger, D. Mitra, and T. Løvås, "The effect of turbulence on mass and heat transfer rates of small inertial particles," *J. Fluid Mech.*, **836**, 932–951, 2018.

[11] W. E. Ranz and W. R. Marshall, Jr., "Evaporation from drops Part I," *Chem. Eng. Prog.*, **48**, 141–146, 1952.

[12] J. Bec, L. Biferale, M. Cencini, A. Lanotte, S. Musacchio, and F. Toschi, "Heavy particle concentration in turbulence at dissipative and inertial scales," *Phys. Rev. Lett.*, **98**, p. 084502, 2007.

[13] R. Monchaux, M. Bourgoin, and A. Cartellier, "Analyzing preferential concentration and clustering of inertial particles in turbulence," *Int. J. Multiph. Flow*, **40**, 1–18, 2012.

[14] F. Battista, F. Picano, G. Troiani, and C. Casciola, "Intermittent features of inertial particle distributions in turbulent premixed flames," *Phys. of Fluids*, **23**, p. 123304, 2011.

[15] J. Kruger, N. E. L. Haugen, D. Mitra, and T. Løvås, "The effect of turbulence on the reaction rate of particles with heterogeneous surface reactions," *Proceedings of the Combustion Institute*, **36**, 2333–2340, 2017.

[16] R. K. Eckhoff, *Dust Explosions in the Process Industries: Identification, Assessment and Control of Dust Hazards*. Oxford: Gulf Professional Publishing, 2003.

[17] T. Abbasi and S. Abbasi, "Dust explosions–cases, causes, consequences, and control," *J. Hazard. Mater.*, **140**, 7–44, 2007.

[18] G. Atkinson and J. Hall, "A review of large vapour cloud incidents," in HSL Report MH15/80 http://primis.phmsa.dot.gov/meetings/MtgHome.mtg, 2015.

[19] S. Kundu, J. Zanganeh, and B. Moghtaderi, "A review on understanding explosions from methane–air mixture," *J. Loss Prevent. Proc. Indust.*, **40**, 507–523, 2016.

[20] D. Bradley, G. Chamberlain, and D. Drysdale, "Large vapour cloud explosions, with particular reference to that at buncefield," *Phil. Trans. R. Soc. A*, **370**, 544–566, 2012.

[21] A. Pekalski, J. Puttock, and S. Chynoweth, "Deflagration to detonation transition in a vapour cloud explosion in open but congested space: Large scale test," *J. Loss Prevent. Proc. Indust.*, **36**, 365–370, 2015.

[22] K. Shchelkin, "Influence of the wall roughness on initiation and propagation of detonation in gases," *Zh. Eksp. Teor. Fiz*, **10**, 823–827, 1940.

[23] A. Oppenheim and R. Soloukhin, "Experiments in gasdynamics of explosions," *Annu. Rev. Fluid Mech.*, **5**, 31–58, 1973.

[24] G. Ciccarelli and S. Dorofeev, "Flame acceleration and transition to detonation in ducts," *Prog. Energy Combust. Sci.*, **34**, 499–550, 2008.

[25] A. Teodorczyk, P. Drobniak, and A. Dabkowski, "Fast turbulent deflagration and ddt of hydrogen–air mixtures in small obstructed channel," *Int. J. Hydrog. Energy*, **34**, 5887–5893, 2009.

[26] C. Wang, F. Huang, E. K. Addai, and X. Dong, "Effect of concentration and obstacles on flame velocity and overpressure of methane-air mixture," *J. Loss Prevent. Proc. Indust.*, **43**, 302–310, 2016.

[27] C. Wang, X. Dong, J. Cao, and J. Ning, "Experimental investigation of flame acceleration and deflagration-to-detonation transition characteristics using coal gas and air mixture," *Combust.*, **187**, 1805–1820, 2015.

[28] G. Atkinson and L. Cusco, "Buncefield: A violent, episodic vapour cloud explosion," *Proc. Saf. Environ. Protect.*, **89**, 360–370, 2011.

[29] G. Atkinson, E. Cowpe, J. Halliday, and D. Painter, "A review of very large vapour cloud explosions: Cloud formation and explosion severity," *J. Loss Prevent. Proc. Indus.*, **48**, 367–375, 2017.

[30] F. Toschi and E. Bodenschatz, "Lagrangian Properties of Particles in Turbulence," *Annu. Rev. Fluid Mech.*, **41**, 375–404, 2009.

[31] Z. Warhaft, "Why we need experiments at high Reynolds numbers," *Fluid Dyn. Res.*, **41**, p. 021401, 2009.

[32] S. Balachandar and J. K. Eaton, "Turbulent Dispersed Multiphase Flow," *Annu. Rev. Fluid Mech.*, **42**, 111–133, 2010.

[33] T. Elperin, N. Kleeorin, and I. Rogachevskii, "Self-Excitation of Fluctuations of Inertial Particle Concentration in Turbulent Fluid Flow," *Phys. Rev. Lett.*, **77**, 5373–5376, 1996.

[34] T. Elperin, N. Kleeorin, V. S. L'vov, I. Rogachevskii, and D. Sokoloff, "Clustering instability of the spatial distribution of inertial particles in turbulent flows," *Phys. Rev. E*, **66**, p. 036302, 2002.

[35] E.-W. Saw, G. P. Bewley, E. Bodenschatz, S. Sankar Ray, and J. Bec, "Extreme fluctuations of the relative velocities between droplets in turbulent airflow," *Phys. Fluids*, **26**, p. 111702, 2014.

[36] T. Elperin, N. Kleeorin, M. Liberman, and I. Rogachevskii, "Tangling clustering instability for small particles in temperature stratified turbulence," *Phys. Fluids*, **25**, 085 104–085 104, 2013.

[37] A. Eidelman, T. Elperin, N. Kleeorin, B. Melnik, and I. Rogachevskii, "Tangling clustering of inertial particles in stably stratified turbulence," *Phys. Rev. E*, **81**, p. 056313, 2010.

[38] T. Elperin, N. Kleeorin, B. Krasovitov, M. Kulmala, M. Liberman, I. Rogachevskii, and S. Zilitinkevich, "Acceleration of raindrop formation due to the tangling-clustering instability in a turbulent stratified atmosphere," *Phys. Rev. E*, **92**, p. 013012, 2015.

[39] Y. B. Zeldovich and Y. P. Raizer, *Physics of Shock Waves and High-Temperature Phenomena*. New York: Academic Press, 1966.

[40] J. R. Howell, M. P. Menguc, and R. Siegel, *Thermal Radiation Heat Transfer*. Boca Raton, FL: CRC Press, 2010.

[41] M. Liberman, N. Kleeorin, I. Rogachevskii, and N. E. L. Haugen, "Mechanism of unconfined dust explosions: Turbulent clustering and radiation-induced ignition," *Phys. Rev. E*, **95**, p. 051101, 2017.

[42] G. Nathan, P. Kalt, Z. Alwahabi, B. Dally, P. Medwell, and Q. Chan, "Recent advances in the measurement of strongly radiating, turbulent reacting flows," *Prog. Energy Combust. Sci.*, **38**, 41–61, 2012.

[43] M. Bidabadi, S. Zadsirjan, and S. A. Mostafavi, "Radiation heat transfer in transient dust cloud flame propagation," *J. Loss Prevent. Proc. Indust.*, **26**, 862–868, 2013.

[44] M. Hadjipanayis, F. Beyrau, R. Lindstedt, G. Atkinson, and L. Cusco, "Thermal radiation from vapour cloud explosions," *Proc. Saf. Environ. Protect.*, **94**, 517–527, 2015.

[45] M. Liberman, M. Ivanov, and A. Kiverin, "Radiation heat transfer in particle-laden gaseous flame: Flame acceleration and triggering detonation," *Acta Astron.*, **115**, 82–93, 2015.

[46] M. Ivanov, A. Kiverin, and M. Liberman, "Ignition of deflagration and detonation ahead of the flame due to radiative preheating of suspended micro particles," *Combust. Flame*, **16**, 3612–3621, 2015.

[47] S. Moore and F. Weinberg, "High propagation rates of explosions in large volumes of gaseous mixtures," *Nature*, **290**, 39–40, 1981.

[48] H. Lu, E. Ip, J. Scott, P. Foster, M. Vickers, and L. L. Baxter, "Effects of particle shape and size on devolatilization of biomass particle," *Fuel*, **89**, 1156–1168, 2010.

[49] X. Shi, F. Ronsse, and J. G. Pieters, "Finite element modeling of intraparticle heterogeneous tar conversion during pyrolysis of woody biomass particles," *Fuel Process. Technol.*, **148**, 302–316, 2016.

[50] P. A. Nikrityuk and B. Meyer, *Gasification Processes: Modeling and Simulation*. Weinheim, Germany: Wiley-VCH Verlag GmbH & Co. KGaA, 2014.

[51] R. B. Bird, W. E. Stewart, and E. N. Lightfoot, *Transport Phenomena*, 2nd ed. Hoboken, NJ: John Wiley & Sons, 2007.

[52] O. Karlström, M. Costa, A. Brink, and M. Hupa, "CO_2 gasification rates of char particles from torrefied pine shell, olive stones and straw," *Fuel*, **158**, 753–763, 2015.

[53] A. Galgano, C. Di Blasi, A. Horvat, and Y. Sinai, "Experimental validation of a coupled solid- and gas-phase model for combustion and gasification of wood logs," *Energy & Fuels*, **20**, 2223–2232, 2006.

[54] S. Farazi, M. Sadr, S. Kang, M. Schiemann, N. Vorobiev, V. Scherer, and H. Pitsch, "Resolved simulations of single char particle combustion in a laminar flow field," *Fuel*, **201**, 15–28, 2017.

[55] G. L. Tufano, O. T. Stein, A. Kronenburg, A. Frassoldati, T. Faravelli, L. Deng, A. M. Kempf, M. Vascellari, and C. Hasse, "Resolved flow simulation of pulverized coal particle devolatilization and ignition in air- and O_2/CO_2-atmospheres," *Fuel*, **186**, 285–292, 2016.

[56] A. K. Biswas and K. Umeki, "Simplification of devolatilization models for thermally-thick particles: Differences between wood logs and pellets," *Chem. Eng. J.*, **274**, 181–191, 2015.

[57] H. Umetsu, H. Watanabe, S. Kajitani, and S. Umemoto, "Analysis and modeling of char particle combustion with heat and multicomponent mass transfer," *Combust. Flame*, **161**, 2177–2191, 2014.

[58] A. H. Mahmoudi, F. Hoffmann, M. Markovic, B. Peters, and G. Brem, "Numerical modeling of self-heating and self-ignition in a packed-bed of biomass using XDEM," *Combust. Flame*, **163**, 358–369, 2016.

[59] Y. Haseli, J. A. Van Oijen, and L. P. H. De Goey, "A simplified pyrolysis model of a biomass particle based on infinitesimally thin reaction front approximation," *Energy & Fuels*, **26**, 3230–3243, 2012.

[60] H. Ström and H. Thunman, "CFD simulations of biofuel bed conversion: A submodel for the drying and devolatilization of thermally thick wood particles," *Combust. Flame*, **160**, 417–431, 2013.

[61] G. W. Roberts and C. N. Satterfield, "Effectiveness factor for porous catalysts. Langmuir-Hinshelwood Kinetic Expressions," *Indust. Eng. Chem. Fund.*, **4**, 288–293, 1965.

[62] N. E. L. Haugen, M. B. Tilghman, and R. E. Mitchell, "The conversion mode of a porous carbon particle during oxidation and gasification," *Combust. Flame*, **161**, 612–619, 2014.

[63] K. Umeki, S.-a. Roh, T.-j. Min, T. Namioka, and K. Yoshikawa, "A simple expression for the apparent reaction rate of large wood char gasification with steam," *Biores. Technol.*, **101**, 4187–4192, 2010.

[64] A. Prosperetti, "Life and death by boundary conditions," *J. Fluid. Mech.*, **768**, 1–4, 2015.

[65] P. Jenny, D. Roekaerts, and N. Beishuizen, "Modeling of turbulent dilute spray combustion," *Prog. Energy Combust. Sci.*, **38**, 846–887, 2012.

[66] A. Attili, F. Bisetti, M. E. Mueller, and H. Pitsch, "Formation, growth, and transport of soot in a three-dimensional turbulent non-premixed jet flame," *Combust. Flame*, **161**, 1849–1865, 2014.

[67] L.-P. Wang and M. R. Maxey, "Settling velocity and concentration distribution of heavy particles in homogeneous isotropic turbulence," *J. Fluid Mech.*, **256**, 27–68, 1993.

[68] M. R. Maxey and J. J. Riley, "Equation of motion for a small rigid sphere in a nonuniform flow," *Phys. Fluids*, **26**, 883–889, 1983.

[69] S. Olivieri, F. Picano, G. Sardina, D. Iudicone, and L. Brandt, "The effect of the basset history force on particle clustering in homogeneous and isotropic turbulence," *Phys. Fluids*, **26**, p. 041704, 2014.

[70] A. Soldati and C. Marchioli, "Physics and modelling of turbulent particle deposition and entrainment: Review of a systematic study," *Int. J. Multiph. Flow*, **35**, 827–839, 2009.

[71] F. Picano, G. Sardina, and C. M. Casciola, "Spatial development of particle-laden turbulent pipe flow," *Phys. Fluids*, **21**, p. 093305, 2009.

[72] M. Reeks, "The transport of discrete particles in inhomogeneous turbulence," *J. Aeros. Sci.*, **14**, 729–739, 1983.

[73] P. Gualtieri, F. Picano, G. Sardina, and C. M. Casciola, "Clustering and turbulence modulation in particle-laden shear flows," *J. Fluid Mech.*, **715**, 134–162, 2013.

[74] W. Fornari, A. Formenti, F. Picano, and L. Brandt, "The effect of particle density in turbulent channel flow laden with finite size particles in semi-dilute conditions," *Phys. Fluids*, **28**, p. 033301, 2016.

[75] J. Reveillon and F.-X. Demoulin, "Effects of the preferential segregation of droplets on evaporation and turbulent mixing," *J. Fluid Mech.*, **583**, 273–302, 2007.

10 Challenges in Practical Combustion

The previous chapters discussed the recent advances in various topics related to turbulent reacting flows, specifically in the views of fundamental challenges that have been overcome to pave ways to build "greener" combustion systems for practical use. Despite these advances, practical implementation brings its own challenges, as has been discussed in the books edited by Swaminathan and Bray [1] and Dunn-Rankin [2]. The discussion in this chapter highlights the combustion concepts of and challenges involved in using the computational methodologies and tools for designing gas turbine combustors and internal combustion (IC) engines. The first section focuses on the stationary gas turbines employed for power generation and aero-gas turbine combustors are discussed in the second section. The final section discusses IC engines, specifically diesel engines, and related topics.

10.1 Stationary Gas Turbine Combustion Challenges

D. Lörstad

Stationary gas turbines are typically used for power generation and mechanical drive applications. A main benefit of a gas turbine is that it has very high power density compared to size and weight, which explains why it is commonly used in the aero industry. However, this is also a benefit for many stationary gas turbines, since it greatly simplifies quick erection and may be used where weight and size is an issue, such as on oil platforms. In addition, it is an advantage in locations of limited access that may be the situation at isolated areas. Another advantage is the short time needed to start and stop gas turbines, which makes them suitable as reserve power commonly used for communities, hospitals, industry, etc. A gas turbine on its own (single cycle applications) usually cannot reach the same efficiency as a piston engine or diesel engine; however, a gas turbine has the advantage that most losses are contained within the exhaust gases. Therefore it is suitable to be connected to a heat recovery boiler (steam turbine) to further increase efficiency (combined cycle applications). For larger gas turbines combined cycle efficiency of larger than 60% electric efficiency may be obtained and a total efficiency of higher than 95% is feasible if a co-generation such as district heating or process water heating is included. Because of this possibility of excellent efficiency, gas turbines are common for power generation.

A gas turbine consists of a compressor, a combustion chamber, and a turbine arranged sequentially one after another. The turbine drives the compressor through a shaft and its excess power is used for other purposes (either for electricity generation or for mechanical drive applications). Compressed air is continuously delivered from the compressor to the combustor, where fuel is injected at constant pressure into the air stream and the combusted mixture with high velocity enters the turbine, where the energy is transformed to rotational mechanical energy. High efficiency of a gas turbine depends on high efficiency of the compressor and turbine, and minimum pressure losses in the combustor, which are strongly linked to the optimum pressure level, high turbine inlet temperature (TIT), and minimum air flow for turbine cooling. The turbine inlet conditions are defined by the combustor exit flow and temperature field. Detailed knowledge of the flow and heat distribution is required to optimize the efficiency, cooling, and life of both the turbine and the combustor. However, the need for high TIT to attain high efficiency may be in conflict with the requirement for nitric oxides (NOx) emission limit, since this emission increases with flame temperature [3–5]. Conventional or non-premixed combustion systems burn fuel close to stoichiometric conditions (locally), resulting in local high temperature, and therefore generates high NOx independently of TIT but this NOx may be reduced by injecting steam/water into the combustion zone. However, the associated cost is significant since purified water is required. Therefore, dry low emissions (DLE) combustion systems were introduced at the end of the twentieth century, which uses lean premixed combustion where the excess oxidizer cools the flame to reduce NOx. To minimize NOx emission for a given flame zone temperature, the mixing of the fuel, air, and burnt gases is essential to avoid local hot spots and the combustor flow residence time must be low. Perfectly premixed fuel–air mixture to minimize NOx can be obtained using a predefined fuel distribution to balance the air distribution. However, this limits the freedom to control the commonly occurring high level of combustion dynamics (self-excited thermoacoustic oscillations), since the most common strategy for this is to adjust the fuel distribution [6, 7], which in this case would cause a NOx penalty. Combustion oscillations are inevitable in turbulent combustion and these oscillations can grow to a sufficiently large level, causing damage to the structure when the heat release fluctuations interact with the acoustics and flow dynamics of the combustor system. The damage to the structure is caused by high-frequency fatigue or fretting. Especially when the frequencies of the thermoacoustic oscillations (combustion dynamics) are close to the Eigen frequencies of the structure (resonant condition), damage may occur rapidly and this has to be avoided fully. Therefore the structure has to be designed to have sufficiently different Eigen frequency compared to the potential range for the thermoacoustic oscillation, and this may be achieved by adding stiffeners. Devices for passive acoustic damping are typically also added to limit the combustion dynamics [8]. A second drawback of a close to perfect premixing of fuel–air mixture is increased risk of flame instability leading to potential flame flashback.

Gas turbines must be able to operate over a broad range of load with low emissions. TIT is typically reduced with load due to compressor and turbine constraints leading to an upper TIT limit at high loads for NOx reasons and a lower TIT limit at part load conditions due to potential flameout conditions. As flame temperature is reduced, increased quenching and local flameout leads to increased amount of unburnt hydrocarbons and carbon monoxide (CO) emissions and risk of severe combustion dynamics. Therefore a separate fuel stage, often called the pilot stage, is usually added that operates at or close to non-premixed combustion mode at low TIT and hence supports burnout of the fuel from the main fuel stage(s). The pilot flames cause a NOx penalty and therefore the pilot fuel fraction has to be sufficiently low at high loads, but it may be dominant at ignition and start-up of the engine. The pilot fraction is then gradually reduced toward full load conditions to balance flame stability and CO and NOx emissions. Since NOx formation rate increases with pressure there is a wider flexibility to reach NOx emission targets by stepping away from the close to perfectly premixed operation at part load conditions. Furthermore, the combustor flow residence time requirement for NOx emission is in direct conflict with the time required for reburn of CO at part load conditions. The dilution and cooling air entering the combustion chamber downstream of the flame zone may increase CO emissions by quenching reactions locally, while using a hot thermal barrier coating to protect combustor walls may help to reduce CO emissions and thereby improve the burnout limit. Figure 10.1 shows an example of the SGT-800 combustor that is optimized for both low NOx at full load and low CO at part load conditions [9, 10].

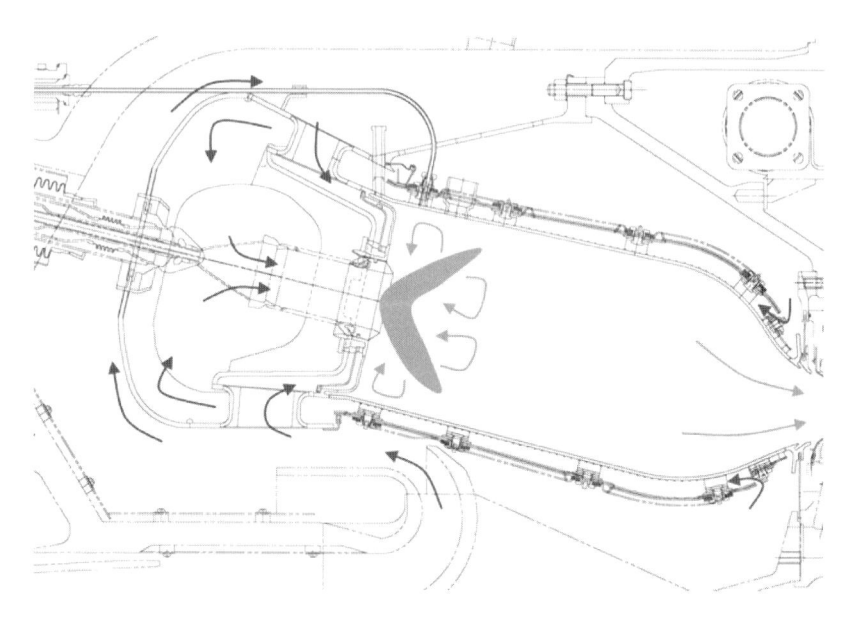

Figure 10.1 SGT-800 combustor flow paths, flame position, and recirculation zones [10].

The ability to use wide range of gaseous and/or liquid fuels is an important technical aspect that has to be considered when designing combustion systems for stationary gas turbines. The fuel type and energy content addressing broad fuel flexibility influences the fuel system acoustics, fuel–air mixing, combustion stability, emission levels etc., strongly. The difference in fuel pressure required for running high and low calorific fuels using the same fuel nozzles may be significant. There is a large density difference between gaseous and liquid fuels and hence the total nozzle area for a liquid fuel has to be smaller than for a gasesous fuel. However, the minimum acceptable hole size is limited by manufacturing capabilities to obtain sufficient tolerances and due to blocking risk, and therefore it is difficult to obtain as good level of liquid fuel–air mixing. Liquid fuels introduce additional complexities such as carbon deposits on hot walls (coking), primary and secondary breakup phenomena in fuel spray, atomization, spatial and temporal distribution of droplets, rate of evaporation, combustion kinetics, etc. Also, liquid fuels in general consists of different high hydrocarbons, which strongly affects the aforementioned topics.

10.1.1 Combustor Design Process

The development of a combustor involves several disciplines as shown in Fig. 10.2, where the arrows indicate some (not all) of the interactions among the disciplines. The "Testing" discipline is marked separately due to its importance for the final verification stage that a certain design has met the project goals on engine system performance including combustion specific objectives. As shown, the "Testing" interacts with most or all of the other disciplines. Combustor performance is usually validated with measurements of the following quantities, static, total, and differential pressures in the air and fuel paths; metal temperatures using thermal paint or thermocouples; strain and structure dynamics; exhaust gas composition including emissions; combustion dynamics through dynamic pressures; flame visualization using cameras; flame flashback; blow out, etc. [9, 11]. Usually the optical access in the combustor operating at high pressure and temperature conditions is limited. Even though the optical access offers a valuable opportunity to explain unexpected phenomena in combustion behavior, only visual or measurable quantities may be confirmed and there is a risk of flame images to be misinterpreted. This limits the possibilities to explain certain observations deviating from the expected behavior, and hence identifying the required design improvements based on testing becomes an open question. Although the testing is expensive (hardware manufacturing and test setup, preparation, and operation) it is always essential for the final validation of a new or revised combustor concept.

The disciplines aero design, CFD and kinetics, heat transfer, and thermoacoustics are directly related to combustion predictions using computational fluid dynamics (CFD), while mechanical design, structural analysis, and lifing interact strongly with the combustion through geometry and aerothermal boundary conditions. In the past decade, the CFD has grown to be important in supporting design improvements, to guide scope and locations for measurements and to some extent to reduce the amount of test iterations required to fulfill the final project (product development) goals and objectives. The CFD has the following advantages: (1) non-manufactured designs may be simulated and studied, (2) the additional cost to investigate high-pressure conditions is marginal compared to the atmospheric conditions [12], (3) a wealth of data is available (not only for the specific locations and parameters predefined in

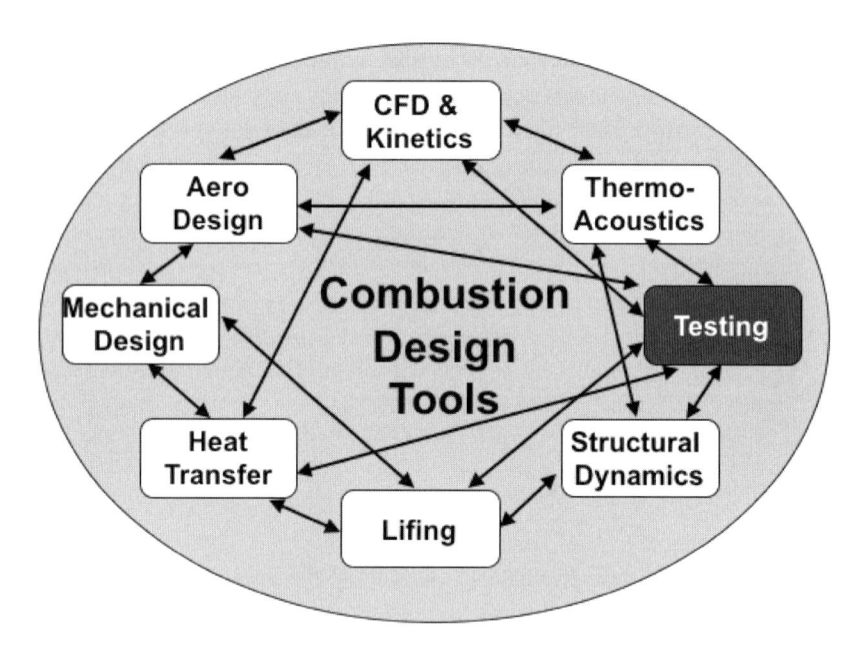

Figure 10.2 Combustion design multidiscipline interaction overview.

measurement campaigns), and (4) any off-design condition may also be simulated, etc. The CFD simulations employing relatively inexpensive models such as Reynolds-averaged Navier–Stokes simulations (RANS), flamelet-based combustion models, and simplified spray models for liquid fuels are usually standard tool for design iterations and more expensive high-fidelity methods such as large-eddy simulation (LES) with more advanced combustion and spray models are growing in importance to support the combustor design iterations and to be part of the verification stages. However, there is still a need for further accuracy improvement, reduction in simulation cost, and more experience with high-fidelity simulations in order to be a fully accepted part of the whole design process. High-fidelity tools have potentials to (1) reduce development cost and lead time due to reduced need for middle-step testing, (2) reduce risk of failed tests leading to further test iterations, and (3) reach difficult goals and objectives for product development. It should also be kept in mind that a test (physical experiment) includes a tenth or hundreds of test conditions of interest and each test point is usually run for several minutes or hours, since this is often required to make sure the test condition is stabilized and combustion has reached a steady state or a limit cycle. However, each high-fidelity CFD calculation is typically run for a certain amount of flow-through time, which usually corresponds to a few hundred milliseconds or a second and this duration is far from the simulation period required according to practical testing experience.

An overview of a generic combustor development process is shown in Fig. 10.3 describing the steps involved from concept phase to basic and detailed design until the validation phase. In the beginning of a project, first the goals are defined based on input from market analysis and engine performance evaluations that specify the requirements for the combustor. Then the combustor development may be initiated with a concept analysis followed by basic CFD calculations and atmospheric cold flow and combustion testing of preliminary designs. For a cost-effective development process, the analysis cost should typically be in proportion to the risk of not meeting the intermediate goals set for the combustor, as it would require further design and testing iterations. Since the cost for the atmospheric combustion testing (noted as AC in Fig. 10.3) is relatively low the corresponding analysis cost is also expected to be low. A rough cost estimation is marked in red in Fig. 10.3, which is typical for industrial gas turbine projects depending on the project scope. When the project matures and reaches the detailed design stage, usually the cost for a high-pressure (HP) test required increases by an order of magnitude. This justifies certain budget to perform detailed analysis to reduce the risk of missing the intermediate project goals that would lead to additional HP test iterations. Finally, the validation phase including prototype design and engine testing is reached, which for the same reason may require detailed analysis to reduce risks and optimize the outcome.

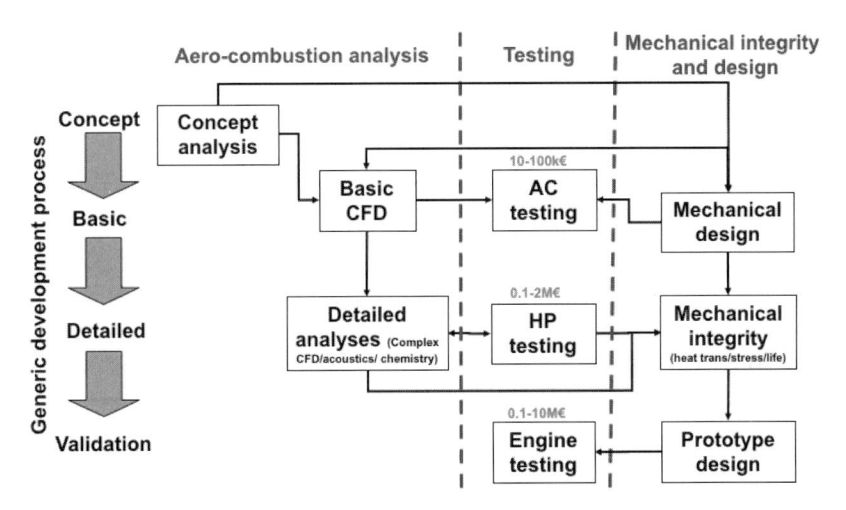

Figure 10.3 Overview of generic combustor development process.

To fulfill the development process requirements and since computationally inexpensive methods typically do not have high accuracy for all cases, there is a need to have different tools for different stages of the development process. Figure 10.4 shows an overview of typical combustion tools used according to the cost or effort spent versus accuracy or design resolution. The cheapest tools refer to low-order models, which includes aerothermal flow and thermal acoustic network 1D tools, empirical correlations etc. For some simplified cases, they may be as accurate as or better than currently available high-fidelity simulations but in general the accuracy is insufficient. CFD, such as steady and unsteady RANS or LES, with flamelet or transported probability density function (PDF)- based combustion models, are orders of magnitude costlier than the low-order tools. It is expected that higher computational cost is justified by increased accuracy or applicability, but it has to be proven and validated that the added value for the expensive tool is worth the additional cost. If not, the cheaper method should be used, which has to be evaluated through validation using measurement [13–15]. It is important to reduce development cost and lead time for commercial efficiency and hence the cheapest tool giving sufficient accuracy to perform the design improvement with acceptable risk is usually preferred. For these reasons the left part of Fig. 10.4 is typically used in the concept and basic design phases while the right part may be considered during the detailed design and validation phases.

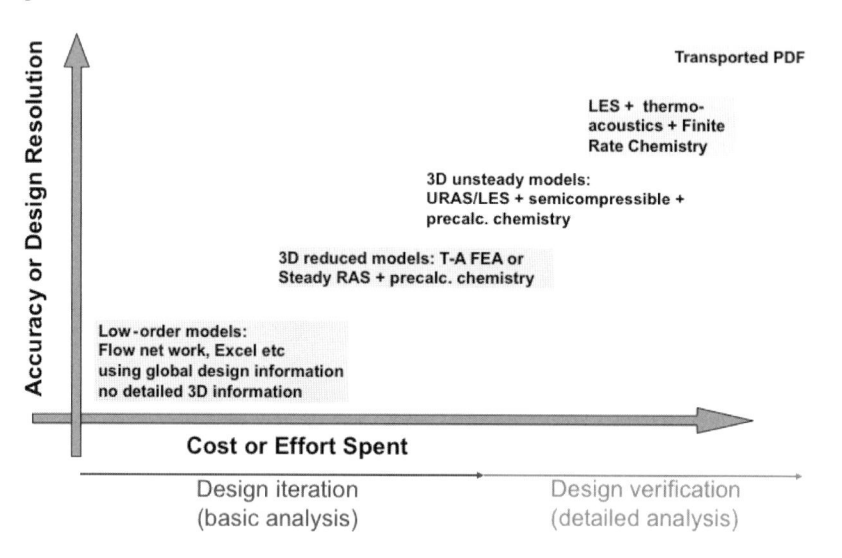

Figure 10.4 Combustion tools overview.

10.1.2 Future Simulation Efficiency

The detailed analysis is expected to be used more and more in the future as the cost for computational resources declines; however, there will always be a need for cost-effective computational tools and therefore the basic analysis will most likely remain as a part of the design process. Simulation times, less than a second for example, allowing direct interaction with a designer is currently possible only for simplified low-order tools. CFD is far from this, and RANS, which is still commonly used for design iterations, may require minutes for steady simplified two-dimensional (2D) models to at least overnight simulations for large three-dimensional (3D) models. RANS will likely remain an option in the future, leading to even quicker simulations for small computational models. RANS may also be used for very large models where a larger portion of the gas turbine is included, for example, to include a full annular combustor instead of one-burner sectors or to include compressor and turbine parts, etc. However, the RANS approach has its limitations for its applicability and therefore it has to be used for tasks where it is suitable. Such tasks may include getting a quick overview of flow distribution, to compare the basic effect of design changes, etc., but the user must be aware of the limitations. For example, the use of two equation models such as k-ω SST turbulence model [16, 17] has been shown to be unable to predict dynamic behavior of the flame and changes in flame position due to changes in the mean reaction rate. This drastically limits the confidence of this model when applied to GT combustors, at least for burners where the averaged flow field has large gradients so that the mean reaction rate has very limited influence on the flame position compared to the mean flow velocity gradient [18].

High-fidelity simulations such as LES and advanced combustion models are used more and more for validation purposes and sometimes during design iterations but are often limited to a smaller geometry of the combustor such as one-burner or can-sector, or limited to mixing only (no combustion) or to part of the cooling system. As computer power becomes more affordable, high-fidelity simulations will become more common. In addition, to predict phenomena such as transversal (azimuthal) thermoacoustic modes, tangential periodic boundary conditions cannot be used and hence all cans or the full-annular combustor must be simulated, which leads to very large computational models [19]. Therefore, there is a need to develop computationally cheaper, yet accurate, methods. Approaches such as the scale adaptive simulation (SAS) model [20, 21] or Detached Eddy Simulation models and reduced combustion models [18, 22] lie between the RANS and LES in terms of accuracy and computational cost. The main idea behind SAS is to resolve the flow dynamics, similar to those captured in LES, in highly turbulent regions of the flow domain and similar to RANS in the rest of the flow domain, which could be a way to save grid size in the near-wall region and also to avoid small time steps in case the highest Courant–Friedrichs–Lewy (CFL) number is located in regions such as fuel nozzles where RANS solution may be acceptable. This has been proven to be of practical use for combustor calculations to identify flame location, low-frequency combustion dynamics, pressure drop, temperature field, effect of high fuel reactivity etc. [10, 18]. However, the fuel–air mixing computed using the SAS model has been shown to deviate from measurements and to drastically damp high frequency oscillations [23]. Since correct prediction of fuel–air mixing is a basic requirement for any flame calculation, the SAS model needs improvement to have confidence in its performance for GT combustor calculations.

Another topic of key importance for the gas turbine industry is the combustor exit or turbine inlet temperature (TIT) because of its major influence on the gas turbine load and efficiency. Therefore the prediction of mass-flow-averaged temperature field at the combustor exit plane is expected to be accurate within a few degrees Kelvin. This demand is tougher by an order of magnitude compared to other flow features, where a few percentage error margin is usually acceptable. Otherwise, the CFD results have to be adjusted to fulfill this requirement, which introduces undesirable uncertainties into the computational model. Hence, energy-conserving combustion models are important and also sufficient simulation duration is required to flush out the initial transient effects before meaningful statistics can be collected for analysis. In addition, the reduced or global chemical kinetic schemes used in the simulations should be very accurate on flame temperatures over the potential range of equivalence ratio expected in the combustor flow. It is important that the flamelet-based approaches handle fresh air streams (cooling and dilution air) downstream of the flame zone without giving spurious reaction zones and their effects [10], and in case some calibration (tuning) is required to capture the correct flame position there may also be a need for further tuning to achieve sufficient burnout at the combustor exit. The importance of TIT to represent similar engine load conditions also emphasizes that studies involving parametric variation of combustion model parameter(s) should be performed for a given TIT rather than for a given equivalence ratio.

Another topic that requires further attention is the prediction of combustion or thermo-acoustic instabilities. The duration for combustion instabilities to develop may vary from a fraction of a second, when there is a clear coupling between flow, acoustics, and combustion under conditions conducive for instabilities to develop – given by the Rayleigh criterion to a month or many months of engine operation. Figure 10.5 shows an example of engine operation for a month with a daily load variation as shown by the red line. The level of thermoacoustic oscillation, as shown by the purple and green lines, is relatively low for about 4 weeks followed by a rapid increase of combustion instability with high amplitude of pressure fluctuations. Practical experiences suggest that this can happen due to changes in ambient conditions, for example, seasonal changes, a change in fuel composition, a slight drift in design points because of changes in tolerances after a certain period of operation, etc. Also, the level of combustion dynamics can be different between two engines of the same rating placed side by side and running on the same fuel. This reveals that the pressure fluctuation amplitude may depend on very small differences such as manufacturing tolerances, which makes it extremely difficult to reliably predict thermoacoustics using CFD. Furthermore, it is quite possible to assume that very small errors due to geometry, boundary conditions, periodic boundary condition assumptions, sizes of numerical grid and time step, modeling of turbulence and combustion, simulation time period of the order of seconds, neglecting structural vibrations, etc. may have similar influences on the predicted combustion dynamics amplitude.

Also, model simplifications such as using a limited number of tangential burner sectors for an annular combustor or a limited number of cans for a can-annular system cannot be used for prediction of most azimuthal modes. However, using the full annular geometry or all cans would make the computational model very large [19].

All these factors cause a large number of uncertainties for predicting thermoacoustic oscillations using CFD and makes it difficult to verify if the expected pressure fluctuation levels would be below the acceptable level or not. Nevertheless, CFD is useful to investigate if self-excited modes initiate and grow in the simulations. This helps to understand if certain frequencies will occur at all, what frequencies are likely to grow, what triggers the growing modes, and how to avoid those growing modes. However, there is always a risk (or a surprise) for thermoacoustic oscillations to show up in the final testing phase of an engine development program or after a sufficiently long period of operation. Therefore there is a need to combine CFD simulations of self-excited modes with other approaches; for example, machine learning to predict these risks would be a very useful step forward. Another example could be to combine CFD with acoustic solvers, where the CFD simulations are forced by single or multiple acoustic modes and the response from the CFD simulation could be used to evaluate the risk of certain modes to develop.

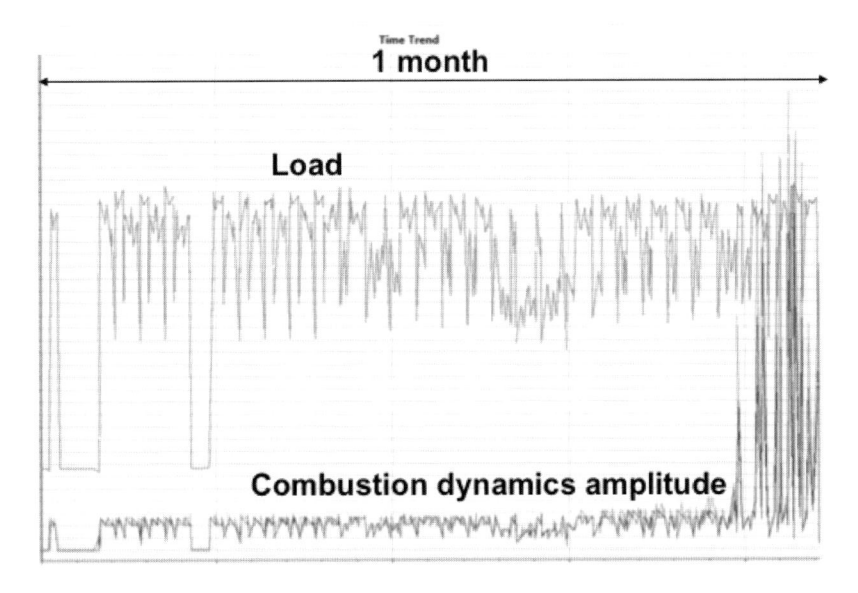

Figure 10.5 Example of a real engine operation with thermoacoustic instabilities.

10.1.3 Challenges for Stationary Gas Turbines

The traditional challenges for the gas turbine industry are related to increased efficiency, increased reliability (the machine should work as planned) and availability (few and short stops for planned services), small capital cost (cost for new engine), and reduced service cost (extended component life). These challenges affect the total life cycle cost, which is the main driver of competiveness.

The further challenges are imposed by emissions, fuel flexibility, and load flexibility. The emission legislation in most countries has become increasingly stringent in the past few decades. This is an example of a situation in which the interaction between political decisions and industrial development has been successful in the way that certain markets are made available if reduced emission levels are achieved. This motivated industries to develop engines meeting the emission targets, leading to technological innovations.

The fuel flexibility requirement has increased drastically in the past decade due to increased fuel cost for the traditional fuels (natural gas consisting mainly of methane and diesel consisting of various large hydrocarbons). Therefore the profitability of using alternative gases has increased, for example, from chemical or petroleum industries, process or syngas industries etc. Common gaseous constituents in these nontraditional fuels are methane, hydrogen, ethane, propane, higher hydrocarbons, and carbon monoxide. In addition, the fuel may contain high amounts of inert gases such as carbon dioxide and nitrogen. Previously those nontraditional gases might have been used in less efficient power plants or in the worst case wasted (through chimneys and stacks) but they can be used as fuel for gas turbines to reduce fuel cost. Depending on the amount and availability of the alternative gases and their composition, they may be mixed with natural gas. Furthermore, the oil and gas industries flare or vent enormous amounts of gases, which could be used as fuels, that are estimated to be equivalent to a large portion of the EU's gas consumption [24]. Both due to greenhouse effects and for profitability reasons there is a push to make use of those gases, and gas turbines are an excellent fit for that purpose. Also, the intermittent production of power from the growing share of renewables leads to increased demand for energy storage and there is a need for converting wind or solar energy during periods of excess energy into a storable energy carrier. For example, hydrogen could be such a carrier, as it can be stored in metal powder (iron, aluminium) for practical and safety reasons [25] and it may be used in gas turbines during periods of less renewable energy production. This market and also the technology is still in the early stages but gas turbines may be a key component in such a transition to a green society. However, already the increasing amount of renewable power production from sun and wind has increased the demand for gas turbine load flexibility, due to its ability to respond to quick load variations, to match the variation of renewable energy production leading to increased demand for low emissions at low load, quicker starts and stops, etc.

10.1.4 Challenges for Combustion Prediction

For combustion prediction, the challenges may be listed as follows:

- Development cost: Even though an acceptable level of accuracy is often obtained for flow distribution, pressure drop, fuel–air mixing, flame position, heat load distribution, turbine inlet flow distribution, etc., there is a need to further improve the accuracy and to make simulations more cost effective, i.e., optimizing accuracy to simulation cost.
- Design optimization: To reach more complex goals and to include CFD fully into design iterations, prediction methods need improved accuracy for topics (problems), such as NOx/CO emissions, thermoacoustics oscillations and their damping, flashback, blowout, liquid fuel spray/combustion, and effects of gas fuel flexibility on combustor performance and operation, for which the current CFD models often fail to perform satisfactorily.
- Combustion stability: To reduce risk of high level of combustion dynamics that may occur in engine operation and since CFD simulations of self-excited modes are not sufficient, simulation methods and approaches to predict and down-select the potential unstable modes are needed.

10.2 Aero-engine Combustor Design Methods: Approach and Challenges

M. Zedda

Aero-engine combustors have to meet a wide range of requirements. A short list is provided below [4].

1. *High combustion efficiency*: Fuel has to burn completely within the combustor for high cycle efficiency at high power. At low power, efficiency has to be high enough to deliver fast acceleration. At mid-power, efficiency is one of the main criteria defining the point of transition from pilot-only to pilot and main operation for lean burn systems. In high-temperature rise cycles that are typical of military aero-engines, combustion efficiency has to be maximized during transient maneuvers to minimize the chances of heat release occurring in the high-pressure turbine, which in turn can lead to durability issues. In general, high efficiency means low emission of CO and unburnt hydrocarbons (UHCs).
2. *Reliable ignition, both on the ground and at altitude*: Ignition is strongly dependent on primary zone aerodynamics, stoichiometry, and fuel breakup, as well as igniter location and discharge power capability. In aero-engines, the combustor has to be able to relight reliably even at the low inlet pressure and temperature levels delivered by the compression system at windmilling conditions after a flameout at altitude.
3. *Wide stability limits*: The flame has to stay alight over wide ranges of pressure, velocity, and air-to-fuel ratios. Wide rich and weak stability limits are required across the engine operating envelope to avoid extinction.
4. *Quiet thermoacoustics*: The combustor has to be free from coupling between pressure oscillations induced by acoustics and heat release fluctuations. Any axial, circumferential, and mixed thermoacoustic modes have to be of sufficient low amplitude not to impact the engine operability and durability.
5. *Low pressure loss*: This has to be minimized for low specific fuel consumption. On the other hand, a minimum pressure loss across the combustor walls and the fuel injector is required for good mixing, satisfactory atomization, and turbine cooling.
6. *Acceptable gas temperature outlet profile*: The exit temperature traverse has to comply with the constraints imposed by the turbine design. In particular, maximum hub and tip temperatures have to be within prescribed limit to maximize turbine life. Furthermore, given a turbine design the combustor exit temperature traverse can have an impact on the turbine performance parameters (e.g., turbine efficiency). Both 1D and 2D profiles have to be within control for life and performance reasons.
7. *Low emission of pollutant species*: Across the operating envelope, NOx, CO, UHC, and smoke have to be minimized. While NOx and smoke are a concern at high-power conditions, CO and UHC increase at mid- to low-power conditions. Requirements for low emissions are becoming increasingly stringent, which has led to development of lean burn design solutions for aviation applications.

8. *Minimum cost*: The focus on cost is strongly dependent on the application. However, there is a general trend toward a reduction of the unit cost, which can have a significant impact on the design based on the materials and manufacturing techniques used.
9. *Maximum maintainability*: The combustor design has to make it easy to change components during the periodic overhaul. Emphasis on total care contracts has increased focus on design solutions allowing easy swap of components during overhauls.
10. *Durability*: The drive toward higher cycle efficiency leads to high engine core temperatures, which in turn poses significant life challenges to the hot components. The combustor has to be designed to withstand the severe temperature levels and the corresponding gradients for the ever more demanding target cycle lives. Combustor durability is key to maximize profits especially for civil large and military engines, for which a large proportion of the revenues are related to the aftermarket.
11. *Size and shape compatibility with the engine envelope*: In aero-engine applications, space is at a premium. The inlet and outlet of the combustion system have to match compressor and turbine requirements, which can have a large impact on the combustor layout and overall aerodynamics.
12. *Low weight*: As for the other engine components, the combustor has to be designed so as to minimize weight to avoid triggering the snowball weight increase effect affecting aircraft systems.
13. *Multifuel capability*: For aviation applications, kerosene (Jet A1) is most widely used. While a relatively wide range of compositions and corresponding properties are allowed within the standard fuel specification, which can have a non-negligible impact on emissions and coke formation, new types of fuels are being considered, either synthetically produced or resulting from blends with biofuels. The fuel system, fuel injector, and combustor have to be capable of running on a relatively wide range of kerosene or kerosene-like fuels without compromising performance.

Design of a combustion system meeting all the requirements listed above is bound to be a complex trade-off exercise. In fact, introduction of a design change improving one performance parameter often has a detrimental impact on another one. Development of a modern aero-engine combustor relies on a cascade of experimental and numerical assessments, which allows going from preliminary design to entry into service within ever shorter time scales. Figure 10.6 shows the process of development and validation that is commonly used in industry for a combustion system.

Experimental tests of increasing technology readiness levels (TRLs) allow filtering promising design solutions that will be ultimately tried as part of engine runs. Increasing TRLs mean more engine representative geometries and operating conditions.

For development of combustion systems that can be regarded as derivatives of well-established designs, some of the steps shown in Fig. 10.6 can be skipped. On the contrary, development of novel combustion system concepts is likely to go through

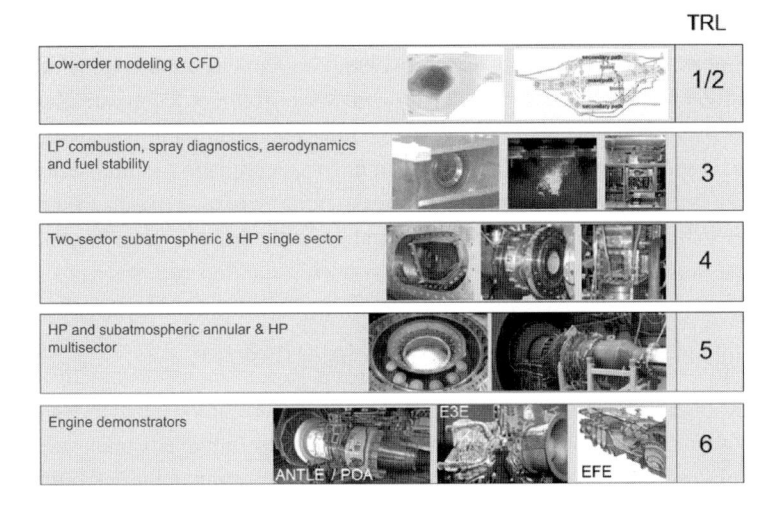

	TRL
Low-order modeling & CFD	1/2
LP combustion, spray diagnostics, aerodynamics and fuel stability	3
Two-sector subatmospheric & HP single sector	4
HP and subatmospheric annular & HP multisector	5
Engine demonstrators	6

Figure 10.6 Technology development and verification cascade for aero-engine combustion systems.

all the steps in order to deliver a comprehensive characterization of the new design features and the corresponding design trade-offs.

Currently combustor development is heavily reliant on testing. While other engine components can be developed using mainly numerical methods and their performance verified experimentally at high TRL at the end of the process, the complexity of the physicochemical phenomena involved in a combustor makes it particularly challenging to devise accurate and reliable numerical models that could entirely replace experimental tests. Modeling of free shear, turbulent two-phase flows with chemical reactions generating pollutant species and convective and radiation loads is bound to pose formidable challenges. As a result, testing is expected to dominate aero-engine combustor development for decades to come. However, the significant advances in computing power achieved in the past 10 years or so have produced remarkable improvements in the level of resolution and accuracy of the models that can be used to simulate various combustion phenomena. In particular, more and more details of combustor and fuel injector geometry can now be modeled. Similarly, models simulating unsteady effects, finite-rate chemistry, chemistry–turbulence, and two-phase flows have improved markedly and can now be applied to a number of combustion design challenges. Because of this, the combustor development process is nowadays supported significantly by numerical simulations. While not all combustion phenomena can be accurately tackled with numerical methods at the moment, some are amenable to simulation. By using a combination of tests and simulations, the duration and cost of the combustion system development process can be reduced.

Due to the inherent challenge of simulating the complex phenomena taking place in a combustion system, predictive methods have to be constantly validated against experimental data. As a matter of fact, the various TRL tests shown in Fig. 10.6 generate a large amount of valuable data that can be used to validate and anchor numerical predictions. While high-TRL tests provide more definitive proof of the actual performance of a combustion system, they tend to produce less validation data than low-TRL tests, where a wider range of measurement techniques can be applied due to the less harsh operating environment. The significant advances achieved in the development and application of diagnostic techniques to combustion tests have actually been a key enabler for the evolution of numerical methods used in combustor design. When properly validated, such predictive methods can then be used throughout the design process as well as during the service life of the engine.

10.2.1 The Role of Low-Order Methods in Combustor Design

A combination of low- and high-order methods is currently used in industry to support combustor design. Low-order methods have the outstanding advantage of low computational cost as well as ease of incorporation of experimentally derived correlations. While by their very nature low-order models cannot capture complex 3D phenomena, they can be more easily tuned to match experimental data. As low-order methods rely on compliance to governing equations (conservation of mass and energy and balance of momentum) in usually 1D frameworks, their usefulness mainly resides in the embedded correlations, which are built on experimental data and are usually characterized by a limited range of applicability.

A wide range of low-order methods are used to support combustor design. They go from hand-calculation-like spreadsheets to network models. Network models are particularly effective in the calculation of pressure drops and flow distributions [26]. The inherent advantage of flow networks is the flexibility they offer to model arbitrary flow features. Such networks can be used in predictive mode to calculate pressure losses and flow distributions before any test is run. The accuracy of such predictive calculations depends on the applicability of the underlying correlations to the combustor modeled, i.e., on the similarity of the new combustor with previous combustor designs. After tests are carried out to characterize the combustion system's aerodynamics, the flow network model can be tuned to reproduce the measured pressure losses and flow distributions. Given the impact of these on combustor performance, the usefulness of a matched flow network model cannot be underestimated. In subsequent phases of the design, such a network can then be used to define changes to cooling or in general combustor porosity.

A network approach is often relied on to model the thermoacoustics of combustion systems as well [27]. This allows to quickly identify acoustic modes that could couple with unsteady heat release effects. A 2D network approach can be used for aeroengine combustors due to their thin annular shape, for which radial modes occur at too high frequencies to couple with the flame. As a result, only circumferential and axial modes can be considered as relevant for thermoacoustics. Accurate prediction of the acoustics relies on availability of damping/impedance characteristics for the combustion system components (e.g., fuel injector, cooling) as well as appropriate acoustic boundary conditions. Even if the acoustic components of the network model are accurately derived either experimentally or computationally, prediction of thermoacoustic instabilities requires use of a flame transfer function (FTF or alternatively a flame transfer matrix), which expresses the flame response to pressure fluctuations as a ratio of unsteady heat release to velocity fluctuations in the flame front region. Once such FTF is available, the model can be used to identify the combustion system's susceptibility to thermoacoustic instabilities across regions of the operating envelope for which the FTF is valid. Approaches for deriving the FTF are discussed later on.

10.2.2 The Role of High-Order Methods in Combustor Design

In conjunction with low-order methods, high-order methods based on computational fluid dynamics (CFD) are widely used to simulate a range of combustion phenomena in support of combustor design. CFD is routinely used in industry to calculate (1) system aerodynamics, (2) temperature traverse, (3) injector aerodynamics, (4) emissions trends, and (5) metal temperature.

CFD is also sometimes used to assess (1) relight/extinction, (2) thermoacoustics, and (3) fuel coking. Such simulations are used in support of both rich and lean burn combustion design concepts. Similar requirements apply to the CFD used in support of afterburners. However, the simulation methods specific to afterburner design support will not be covered here.

10.2.3 CFD for System Aerodynamics

Successful development of a combustion system hinges on achievement of a number of aerodynamic functionalities. The flow exiting from the compressor has to be slowed down to allow flame stabilization. Pressure loss has to be reduced down to the level required for good mixing, atomization, and turbine cooling. Fuel injector, mixing ports, and cooling features have to be fed based on the requirements defined in the preliminary design phase. The impact of unsteady aerodynamic phenomena on various performance parameters has to be assessed. Given the importance of the system aerodynamics, for a new combustion system a detailed characterization of the aerodynamics is usually carried out through extensive isothermal testing. Before the testing is done, an estimation of pressure losses and flow distributions is obtained using a flow network model to start with. After that, a detailed CFD model of the entire system is built and run [28]. While accurate prediction of pressure losses is a challenge for the CFD, running a model of the entire system to capture the 3D flow field around and inside the combustor can provide valuable insight on the proposed design. At this stage, the simulation does not account for spray and combustion. For a typical combustion system, the model starts at prediffuser exit and ends at combustor exit, as shown in Fig. 10.7.

All the gas washed features of the fuel injector are resolved in this model, while the cooling flows are modeled as boundary conditions. These cooling flows are calculated by the flow network model. The CFD model is typically run as a steady state Reynolds-averaged Navier–Stokes (RANS) simulation, unless the aerodynamics of the system is

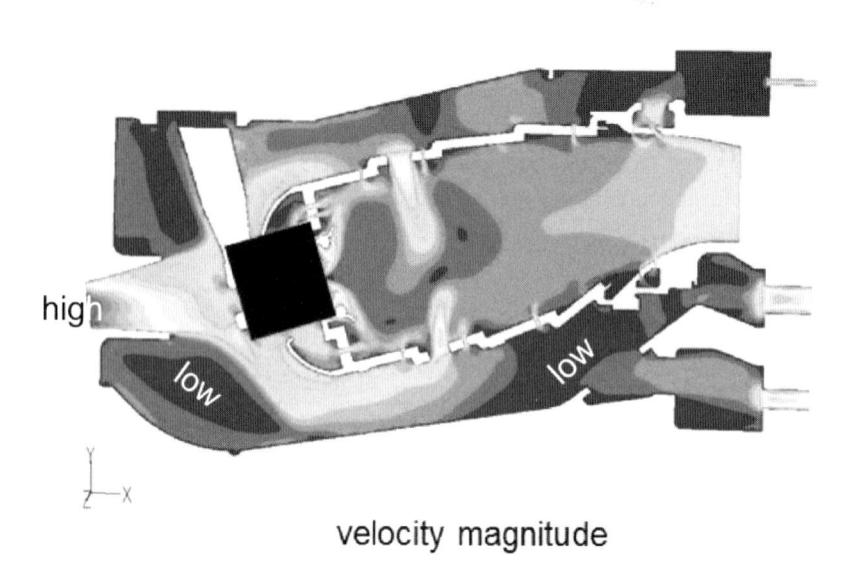

Figure 10.7 Simulation of full system aerodynamics.

characterized by large-scale unsteadiness requiring adoption of an unsteady approach. While the RANS approach does not allow capturing unsteady effects properly, it can provide a first-pass assessment of 3D aerodynamic effects, some of which will be measured in the subsequent aerothermal testing. Turbulence is usually modeled via a two-equation realizable k-ε approach with all variables discretized using second-order schemes. Running this full-system model enables investigation of the aerodynamic feed to various combustor features and calculation of flow distribution as well as pressure losses. Some consideration can also be drawn about the mixing inside the flametube (e.g., fuel injector flow cone angle or port jet penetration), although the heat release from combustion can have a dramatic impact on the combustor aerodynamics.

If at first sight accurate assessment of the 3D aerodynamics through CFD looks straightforward, a number of uncertainties at play have to be taken into account. To start with, in an engine the combustor will be subject to thermally generated distortion that may have an impact on the aerodynamic feed quality. Displacement effects can be taken into account in the generation of the CFD model if some estimation of the thermal expansion of the various components can be obtained from a thermomechanical model, which during the initial stages of the design is bound to be rather crude. Furthermore, the representativeness of the inlet boundary conditions located at the prediffuser inlet is key for an accurate prediction of the 3D aerodynamics in the combustion system. These boundary conditions usually come from a CFD simulation of the compression system, which will be affected by its own model errors. To predict the mixing out losses of the compressor outlet guide vane wakes it is important to use an accurate estimation of the prediffuser inlet turbulent quantities (e.g., turbulence kinetic energy and its dissipation rate). For this reason, the compressor CFD simulation has to use a two-equation approach for the modeling of turbulence (e.g. k-ω SST) [17]. An inaccurate calculation of the mixing out effects in the prediffuser can have an impact on the downstream flow distribution and correlated pressure losses. Another source of uncertainty is given by the need to account for cooling flows through boundary conditions stemming from an, as yet, uncalibrated flow network model. Cooling flows usually cannot be backed out as part of this CFD calculation because of the small scales characterizing the cooling features. Proper resolution of such small scales would increase the mesh size to an unmanageable level. On the other hand, underresolution of the cooling features would lead to large errors in the calculation of the cooling flows. Mitigations to the uncertainty due to cooling flows can be introduced by flow testing of the features of interest or by detailed submodeling, whereby given the smaller extent of the computational domain the small features can be captured properly. Similar considerations can be made for the turbine bleeds, for which the flow characteristic will be a function of the 2D inlet profiles. Eventually, errors in the CFD prediction can be due to separation phenomena, which are notoriously challenging for a steady-state CFD to simulate properly. For cowled combustors, the flow moving along the cowl wall toward the annuli can separate from the wall with potentially dramatic impact on the annulus feed. These effects are unlikely to be captured properly through simulation and may be cause for discrepancy between prediction and measurements.

Once the experimental survey of the aerodynamics is completed, measurements of velocity and pressure can be compared against predictions. In particular, detailed 2D mapping of velocity and turbulence quantities at prediffuser inlet can be used to revise the model boundary conditions. Comparison with 2D maps of velocity and pressure on downstream planes allows validating the model set up in terms of the mesh and turbulence model chosen. After this phase, the full-system CFD model can be used to quickly assess any changes to the geometry that may be of interest for system performance improvement.

10.2.4 CFD for Fuel Injector Design

The main functions of the fuel injector are to atomize the fuel, mix it with air, and establish a downstream aerodynamic pattern conducive for flame stabilization. The fuel injector is a key component of the combustion system, and its design is usually carried out through a combination of experiments and simulations.

CFD is widely used to characterize the aerodynamics of fuel injectors [29]. At an early stage of the design, the effective area is the main parameter of interest. This can be usually predicted by a RANS CFD model; adoption of a uRANS (unsteady RANS) model is required only if the injector generates a flow field characterized by large-scale unsteadiness that prevents achieving a steady-state solution. When a (u)RANS approach is used, care is required for the choice of turbulence model. As the majority of injector concepts create a swirling aerodynamic pattern, two-equation turbulence models are expected to be outperformed by second-order closure methods (i.e., Reynolds stress models - [RSMs]). However, the RSM can have difficulties to converge. So, it is not uncommon to use two-equation turbulence models, as long as the approach has been validated against measurements for the same class of injectors. Particle image velocimetry (PIV) techniques are often used to measure 2D and 3D maps of time-averaged velocity and corresponding fluctuations, which are used for the validation.

After the first-pass RANS analysis, LES is often used to understand aerodynamics in greater detail [30, 31]. Grids used for LES have to be constructed carefully, with a view to avoid sudden jumps in cell size so as not to compromise the mesh-based filtering the method relies on. Usually a number of iterations are required for the meshing, whereby the density and cell size distribution are assessed against a criterion based on estimating the percentage of turbulence kinetic energy directly resolved (i.e., Pope's criterion [30]). At the same time, a velocity subgrid scale (SGS) model has to be selected. For simulation of fuel injector aerodynamics, the typical choices are constant Smagorinsky, dynamic Smagorinsky, and Vreman models and they are chosen based on the kind of mesh used, especially in terms of the near-wall cell size. For complex geometries that are typical of aero-engine fuel injectors, it is not uncommon to have to accept some compromises in terms of cell size at the wall, expressed through the y^+ value. Resolution of at least 80% of the turbulence kinetic energy near the walls would call for a prohibitively large mesh count and small time steps so as to make the simulation impractical. In a way, the process of setting up an accurate LES model for a class of injectors is similar to the one used for RANS. However, if attention is paid to mesh quality the choice of the SGS model is bound to have less of an effect than the choice of the turbulence model in RANS. The inherent advantage of using LES is its superior ability, in comparison to RANS, to predict unsteadiness and many important features of the time-averaged flow field. For example, the airflow cone angle, recirculation zone sizes and velocity fluctuations obtained using LES are likely to be more accurate than those from RANS simulations in general.

To start with, the fuel injector aerodynamics is studied in an airbox configuration, whereby the injector is fed by plenum and discharges into a single-sector box. The downstream box depth has to be chosen so as to be close to the depth of the combustor the injector is meant to be fitted to. This ensures that any potential wall effect on the aerodynamic cone angle is captured properly. For very wide cone angle fuel injectors, heatshield/baseplate cooling flows should be modeled as well, because of the potential interaction with the fuel injector flow. At a later stage, velocity profiles more representative of the ones feeding the injector in the combustor will have to be used to assess the impact on effective area and overall aerodynamic features. Usually, the investigation on the feed profile effects is carried out in a geometry that is more representative of the actual combustor. Similarly, sector-to-sector interaction may need to be studied by imposing cyclic boundary conditions on an annular sector model.

Adoption of relatively highly resolved CFD models allows assessing the fuel injector swirlers and passage design in details. In particular, it will highlight any problem areas, for instance any separation that may occur on the suction side of swirler vanes. In general, it will enable optimizing the fuel injector design with a view to minimizing the parasitic losses and maximizing the air-to-fuel momentum transfer to achieve satisfactory atomization. At the same time, studies can be performed to characterize any trade-off between aerodynamic performance and manufacturing costs.

If LES is used, the unsteady aerodynamics can be analyzed using proper orthogonal decomposition (POD), which allows isolating the higher energy-containing modes. A close look at POD modes can help clarify the nature of the unsteady velocity fluctuations for the injector at hand. Moreover, the full spectral content of the velocity fluctuations can be extracted from LES for a number of monitoring points placed inside the computational domain. The derived spectra can then shed light on the velocity fluctuations across a wide range of frequencies; in particular, frequencies of dominant modes can be spotted (e.g., due to a vortex shedding occurring at a fixed frequency). In other words, LES can be used as a computational equivalent to hot wire anemometry.

Although CFD simulation of fuel injector aerodynamics is a relatively mature discipline, validation against PIV measurements is of paramount importance. First of all, it is the only means of choosing the most appropriate setting, which is most crucial for RANS but still required for LES. Second, measurements can help spot whether the injector aerodynamics is characterized by bimodal behaviors and the associated bifurcation points. For instance, there may be conditions at which either a wide or a narrow cone angle can be generated by the fuel injector. While these bimodal behaviors can be easily spotted experimentally, they are usually hard for the CFD to predict correctly. On the other hand, if airflow testing is performed using rapid prototyped geometries, attention has to be paid to surface roughness, which may be considerably higher than in production parts. Depending on the swirl levels imposed on the flow by the swirlers, large surface roughness can have a significant impact on the bulk fuel injector aerodynamics. In this case, surface roughness wall models have to be used in CFD, but their application requires validation against experimental data.

10.2.5 CFD for Temperature Traverse

Gas temperature at the exit of the combustor is one of the most important parameters in a Brayton (or Joule) cycle. The trend toward low specific fuel consumption leads to smaller and hotter cores. However, hot combustor exit temperatures make the design of durable high-pressure turbines a challenge. So, knowledge of the detailed temperature map at the combustor exit is key to a successful engine development. Before the key full annular traverse test is performed, CFD is used to predict combustor exit gas temperature profiles [32]. This can give advance warning about potential discrepancies between the actual profiles and the ones assumed for the turbine design. Figure 10.8 shows an example of the comparison between the mean sector traverse predicted by CFD and profiles measured for a number of sectors in a full annular rig test at pressure.

The single-sector CFD prediction of traverse can only predict the profiles of an average sector. Sector-to-sector variations, which can be due to a number of factors such as nonuniformity of the aerodynamic feed to the combustor and uneven fuel injector flow number, cannot be capture by such a simplified CFD approach. While 2D maps of temperature are key to drive the high-pressure turbine nozzle guide vane

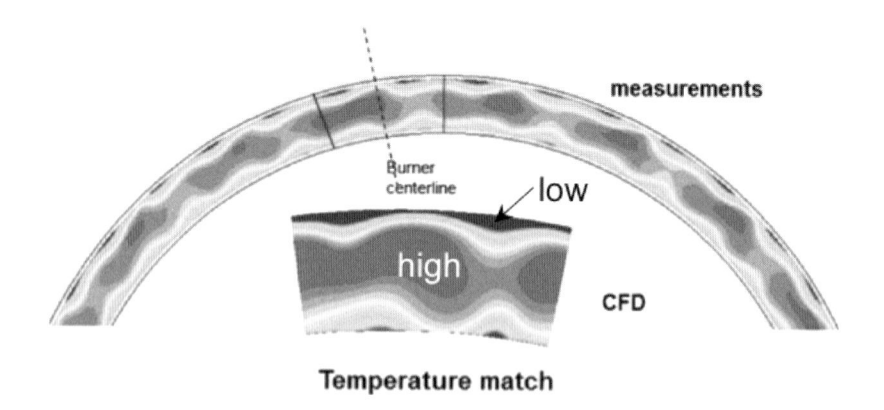

Figure 10.8 Predicted versus measured temperature distribution.

(NGV) cooling design, rotor design is mostly driven by the circumferentially averaged radiation temperature distribution function (RTDF). Figure 10.9 shows a typical comparison between predicted and measured RTDFs. The plot also shows how the match can be influenced by the cooling flow boundary conditions, which are derived from the flow network model.

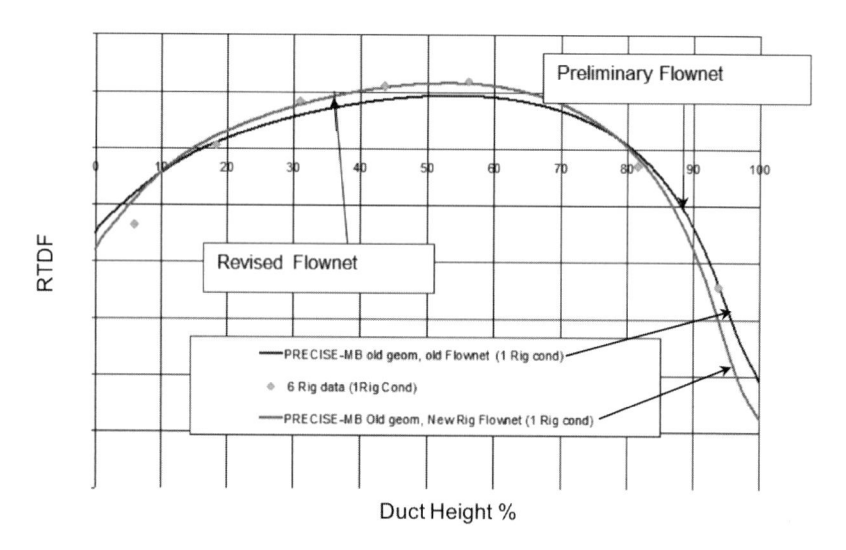

Figure 10.9 Comparison of computed and measured RTDF.

To produce such a prediction of temperature, the full system model described in the previous section can be rerun in reactive mode. Alternatively, a submodel can be extracted from it, which contains only the flametube and annuli. For the two simulations to yield similar results it is necessary that the submodel uses velocity and turbulence profiles extracted from the full system.

Various combustion modeling approaches can be used for prediction of combustor exit traverse. Choice of the combustion model is a function of a number of factors:

the combustion regime prevailing within the flametube (i.e., diffusion vs. premixed) as well as the time scale for delivery of such a prediction. For aviation applications, the combustion mode in the flametube will be diffusion dominated, with some level of premixing depending on the specific design. In general, the combustion mode for aero-engine combustors can be assumed to be partially premixed. A suitable approach for prediction of temperature profiles for such combustors is based on flamelets [33–36]. In this framework, detailed kinetic mechanisms are used upfront of the CFD simulation to calculate tables that can be looked up by the solver at run time (see tabulated chemistry approach described in Section 3.4.4). The principle underpinning flamelets is that a complex flame front can be approximated by laminar flames as long as chemistry–turbulence interaction and their effects are included using a probability density function (PDF) approach. To account for the fact that the flame will not relax instantaneously to its equilibrium behavior, a progress variable concept can be adopted. In this way, the lookup tables become functions of mixture fraction, progress variables, as well as the corresponding variances. The advantage of the flamelet approach is that detailed chemistry effects can be efficiently parameterized through a small number of variables that can be calculated by the CFD solver locally. While flamelet-based prediction of chemical species linked to slow time scales can be affected by nonnegligible errors, the approach is good enough to predict heat release and consequently exit traverse.

Prediction of traverse is subject to the uncertainties affecting the characterization of the full-system aerodynamics, as briefly mentioned in the previous section. Furthermore, the predicted combustor exit traverse is a function of the spray boundary conditions used. Different assumptions about this initial spray placement will yield different air-to-fuel ratio distributions close to the injector, which in turn can have an impact on primary zone mixing and ultimately exit traverse. The sensitivity of exit traverse on spray boundary conditions is actually dependent on the combustor being simulated, as for different mixing mechanisms the relationship between front end stoichiometry and back end temperature profiles will be potentially different.

In general, spray boundary conditions can be regarded as the weakest point in reactive CFD models aimed to calculate exit traverse and potentially pollutant species emissions. The standard approach used in industry consists of specifying a spatial distribution of Sauter mean diameters (SMDs) together with an assumed droplet size PDF. Moreover, initial fuel velocity distributions have to be assumed, together with the initial fuel temperature. With these boundary conditions, Lagrangian tracking of the spray parcels can be performed. Such Lagrangian models will allow for secondary breakup and evaporation. Due to the high density of the spray in the region close to injection, two-way coupling is required.

Definition of appropriate spray boundary conditions can be informed by two different sources: spray testing or detailed modeling of the primary breakup. While the experimental approach is preferred in that accurate characterization of the spray can in principle be obtained [37], application of diagnostic techniques to the very high-pressure conditions that are typical of modern turbofan cycles (e.g., 50 bar) is rather challenging. More often, phase Doppler anemometry (PDA) [38] or shadowgraph [39] measurements can be taken at moderate pressures. These data can then be used to derive correlations expressing the quantities of interest for definition of the spray boundary conditions for the injector at hand across part of the operating envelope. However, the actual state of the spray at very high-pressure conditions is seldom known in detail due to the challenge of performing measurements at such high pressure, at which the conditions are nominally supercritical (i.e., surface tension of the liquid phase goes to zero). An alternative approach to gain insight into the primary breakup of jets and/or films so as to have enough information to define spray boundary condition accurately is to set up and run a detailed model for simulation of the fuel breakup. Prediction of the spray statistics at a range of operating condition of interest can in principle be obtained by Eulerian volume of fluid (VoF) type of calculations [40], whereby the interface between liquid and gaseous phases is calculated. However, accurate prediction of all the relevant length scales leads to unmanageable computational demands, which makes this kind of simulation a formidable challenge. In practice, VoF-like simulations can be performed from time to time to better understand the fuel–air mixing mechanisms associated with a specific design. These assessments are bound to provide only a qualitative assessment of the spray statistics rather than a proper quantification of the spray size PDF.

Further uncertainty can affect prediction of the traverse at part power for certain types of designs. Lean burn combustors, for instance, will suffer from a deterioration of efficiency with decreasing power. This will have an impact on the temperature distribution and ultimately exit traverse. In this case, attention has to be paid to the combustion model's ability to predict CO and UHC accurately. Targeted validation exercises have to be carried out to verify if reasonable levels of temperature can be calculated in cases of incomplete burning. If the flamelet approach turns out to be incapable of predicting efficiency trends reliably enough, more advanced and computationally expensive combustion models have to be adopted. In particular, the stochastic fields method [41] allows taking into account finite rate chemistry effects directly. The modeling challenge in this case is to identify a small enough reaction mechanism producing results in short time scales but still capturing the relevant finite-rate chemistry effects.

A challenge specific to very hot cycles is the significant amount of CO_2 dissociation taking place inside the flametube, leading to high concentrations of CO at combustor exit. While calculation of dissociation CO does not pose particular problems to the combustion CFD model, care has to be paid to the methods used for calculation of any heat release that may happen in the downstream NGV resulting from reassociation of CO with oxygen. Either a finite rate chemistry approach or a flamelet method augmented with an additional independent parameter related to CO have to be used.

The vast majority of full annular rig tests are carried out with an open end at combustor exit. In this case the combustion CFD simulation can be performed by using an incompressible solver and prediction at the sampling plane can then be compared against the gas temperature measurements. However, in the engine configuration the high-pressure turbine NGV will create blockage effects and locally modify the static pressure field. Data exchange between the combustor and high-pressure turbine CFD has to happen at a sufficient axial distance from the NGV leading edge to ensure that the pressure bow wave effects are negligible. While this practice is good enough for combustors whose aerodynamics is dominated by large-scale mixing due to port jets, lean burn combustors may require the CFD to include the NGVs to account for the impact they have on the long recirculation zones spanning the entire length of the combustor. In this instance, a compressible solver has to be utilized.

10.2.6 CFD for Emissions Ranking

As stated earlier, NOx, CO, UHC, and soot are the pollutant species of interest in combustor design. A significant proportion of the design effort is related to attempts to minimize these emissions. In aero-engine applications, combustors can be broadly categorized as rich–quench–lean (RQL) or lean burn. In the former design concept, a rich primary zone is subject to quick dilution via large port jets. In this case admission of mixing air is distributed among the front end of the combustor (fuel injection and heatshield) and primary and dilution jets, on top of the cooling flow that has to be budgeted to keep walls cool enough to enable achievement of the agreed on life targets. In the latter design concept, the majority of the mixing air is introduced at the front (through the fuel injector as well as the heatshield), so that the remainder of the air injected into the flametube is for cooling. In this case, the mixing features are pretty much entirely dominated by the fuel injector. Absence of a rich primary zone, which on the contrary is an inherent feature of RQL systems, calls for staging of the fuel injector through splitting of the fuel into main and pilot. The main objective of the pilot is to provide flame stability at operating conditions at which extinction could otherwise occur.

Introduction of lean burn has allowed a significant reduction of NOx and soot emissions, due to the uniform gas temperature distribution that can be achieved within the flametube. At the same time, at part power conditions relatively high values of CO and UHC concentrations lead to inefficiencies that have to minimized at design stage.

Combustion CFD can be used effectively to drive combustor design toward low NOx [33]. The bulk of the NOx emitted by aero-engine combustors is thermal; i.e., it is generated by oxidation of the N contained in the mixing air through high-temperature–dependent chemical reactions. A secondary contribution to NOx formation is the so-called prompt NOx, which is created in low-temperature fuel-rich regions. While thermal NOx is associated with long residence times (time at high temperature), prompt NOx formation time scales are much shorter.

The standard approach to calculate NOx is to solve a transport equation for the species of interest whereby the source terms are derived from the pretabulated flamelet results. In practice, given the dominance of thermal NOx, an accurate prediction of mixing and gas temperature distribution inside the flametube is likely to lead to reasonable predictions of NOx trends. This capability can be used to rank different design concepts with respect to NOx. An example is provided in Figs. 10.10 and 10.11, whereby prediction of traverse and NOx emission indices (EIs) for two different two-sector periodic combustors are compared with experimental data.

As can be seen from Fig. 10.11, the errors in the absolute values of NOx predicted still allows calculating trends between different combustor designs and against operating conditions.

More accurate predictions of traverse and NOx can usually be achieved using a large-eddy simulation (LES) approach for simulation of the unsteady effects,

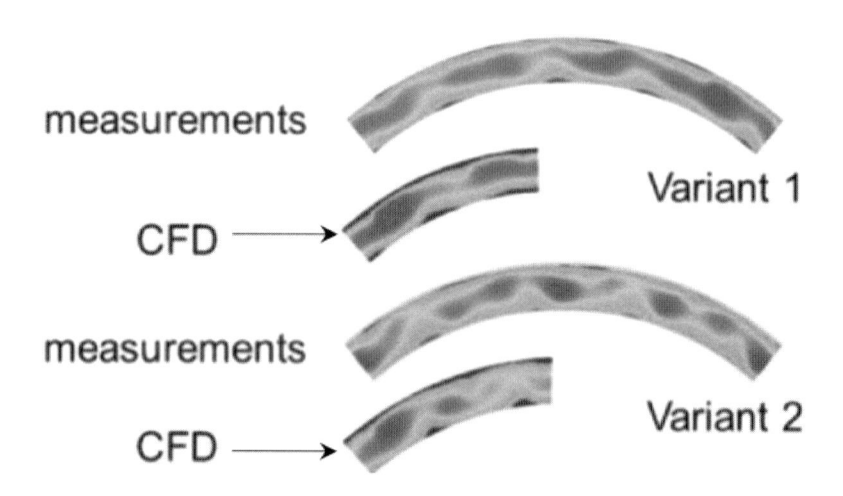

Figure 10.10 Measured and computed temperature traverse for two combustor designs.

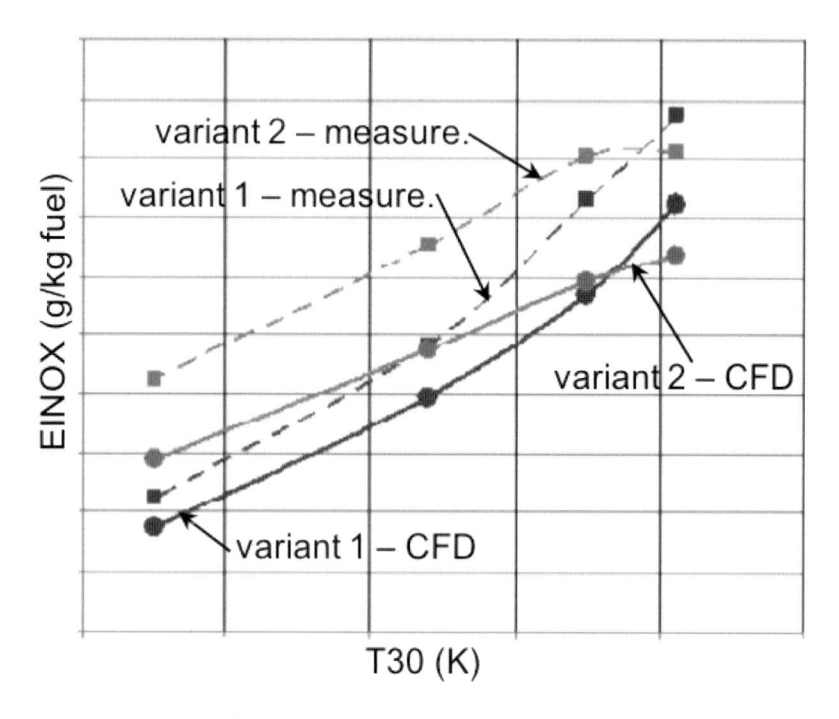

Figure 10.11 Trends of measured and computed NOx emission index for two combustor designs. The relative change is more important than the absolute values.

which will impact the mixing. LES is currently being used as one of the methods in the toolkit of gas turbine industries for assessment of combustor aerothermal performance. Uptake of LES methods in the simulation of unsteady effects has been more widespread for combustors than for turbomachinery components due to the free shear nature of the mixing taking place inside the flametube, as opposed to the wall-dominated aerodynamics that is typical of bladed systems. However, in industry utilization of computationally expensive LES methods has to be weighed against the usually tight design time scales. Another route for improvement of NOx emissions is more accurate reaction mechanisms, which can lead to a better quantification of the absolute levels of the pollutant species.

Even if combustion CFD is routinely used to rank different combustor designs in terms of traverse and NOx, predictions are subject to the same uncertainties mentioned earlier. In simple terms, if the prediction of mixing, fuel, and air placement is affected by significant errors, both temperature traverse and NOx calculations can be misleading.

Predictive techniques can also be used for calculation of CO and UHC trends. This is of particular interest for lean burn combustors that can suffer from relatively low efficiency when operating in pilot and main mode at mid-power. In this case CO and UHC predictions can again help drive the design in the right direction, even if the discrepancy between measured and predicted absolute values can be large.

The main challenge in the prediction of CO and UHC is that finite-rate chemistry effects can be captured only to a limited extent by the computationally cheap tabulated chemistry approaches. Therefore, switching to an approach solving the reaction mechanism on the fly may offer a more accurate means of calculating inefficiencies. However, the computational cost increases considerably. Adoption of a stochastic fields method in this case is likely to yield improved accuracy, also because it significantly reduces chemistry–turbulence model errors introduced by the use of a presumed PDF approach, which is typical of flamelet methods. However, the modeling of micro-mixing for the stochastic fields method can introduce errors (see Sections 3.4 and 3.6).

Another challenge in the prediction of CO and UHC species in lean burn is the need to resolve near wall mixing quite accurately, which in turn calls for the requirement to capture the small length scale associated with the mixing of cooling flow and main flow. A fine-grained LES approach with good resolution near wall is likely to deliver improvements in the prediction of CO and UHC trends.

The most challenging pollutant species to predict is by far soot. In this case, the complexity of the physics and chemistry involved means that only relatively crude models can be used for routine calculations. The phenomena of soot formation and consumption can be regarded as composed of four steps: nucleation, surface growth, coagulation/agglomeration, and oxidation. Therefore, for a soot model to have a chance of providing some useful predictive capability these four phases have to be modeled, even if in a simplified fashion.

The standard approach used for prediction of aero-engine combustor soot trends is based on a two-equation model, whereby soot mass concentration and number density are transported across the computational domain [42]. The two equations have chemical source terms to model the four steps mentioned earlier. In a flamelet approach, the source terms can be expressed as functions of chemical reaction rates that are pre-tabulated. Although heavily simplified, the method has proven to have some predictive capability. Due to the high nonlinearities involved in the various steps of soot formation and consumption, it is paramount to start from an accurate prediction of the air-to-fuel ratio distribution. Because of the intermittent nature of soot formation, LES is a preferred option. In this case, soot is calculated at each time step and averaged at the end of the simulation. Figure 10.12 shows an example of experimental verus numerical ranking of four different injectors. In the CFD simulations, LES was used together with a two-equation soot model.

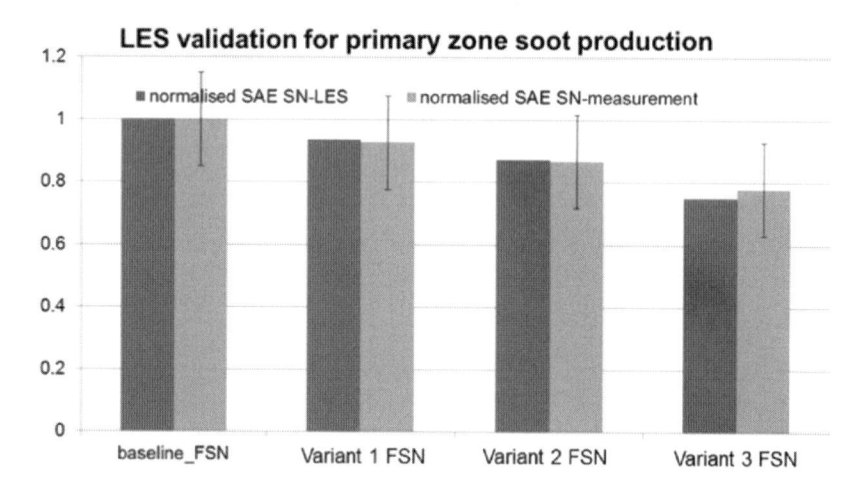

Figure 10.12 Ranking of four fuel injectors using measurements and computations. Green bars (on the right) represent relative smoke number measured at exit of full annular rig, red bars (on the left) represent relative smoke numbers at exit of primary zone as predicted by LES. All values are normalized against baseline.

One of the challenges making reliable prediction of soot trends hard is the fact that soot at combustor exit is the result of the difference between two usually large numbers, one for production and the other for oxidation. Errors in the calculation of either of them can lead to a completely wrong trend. This is particularly true for RQL systems designed to rely heavily of the high oxidation rates that can be achieved in elevated temperature regions. Furthermore, soot oxidation chemistry for kerosene fuels at high pressure is hard to obtain. As a result, it is safe to state that soot modeling is likely to be more useful to tackle production rather than oxidation differences.

In general, a number of approaches can be pursued to try to improve soot predictions. First, accurate predictions of time and space resolved air-to-fuel ratio is key to derive reasonable production and consumption rates. This can be achieved using LES together with experimentally derived spray boundary condition [43]. Second, finite-rate chemistry approaches have the potential to improve the simulation of chemistry effects. This can be achieved using a sectional method, whereby soot and its precursors are binned based on their mass with chemistry calculated on the fly [44]. To capture chemical effects better, improved oxidation chemistry at high pressure is required. Third, accurate modeling of chemistry–turbulence effects is bound to lead to improvements as well. This can be achieved using a transported PDF approach, for instance in a Eulerian framework through the stochastic fields method [41]. The issue is that the resultant computational cost will increase so much as to make application of the method impractical in an industrial environment.

This points to the need to carry out substantial research in the field of soot modeling so as to arrive at an approach offering improved accuracy with respect to current methods whilst incurring an acceptable computational cost. Key to achieving this objective is the ability to account for chemical effects by reaction mechanisms reduced in terms of number of species up to a point where they can be used as part of the design process.

Another fundamental challenge for accurate prediction of soot trends is the wide range of length scales at play. Soot particle sizes range up to two orders of magnitudes and the widely used assumption of particle sphericity is a gross approximation. As surface growth and oxidation are strongly dominated by surface effects, models that do not directly capture the particle size distribution (PSD) are unlikely to provide the level of accuracy and reliability required for the various soot problems encountered during combustor design.

To add to the challenge, while current regulations impose soot limits just in terms of smoke number, which is mainly a function of soot mass concentration, future regulations will impose limits in terms of both mass concentration and particle sizes. This change in requirements suggests that models able to calculate the full PSD will have to be used for prediction of soot trends.

10.2.7 CFD and Finite-Element Analyses of Metal Temperature

Safety is a primary requirement for aero-engines. Hot components are particularly exposed to thermally induced durability issues, so the combustor design has to include effective cooling of its various components to keep metal temperatures and the corresponding gradients low enough to avoid mechanical failures. Moreover, increasingly demanding life targets put further emphasis on development of efficient cooling solutions.

Combustion system components exposed to the flame are heated by both convection and radiation. While convection from hot gases is present in all kinds of combustors, RQL liners are subject to a significant radiation load, which adds to the convective load, as radiation is caused not just by combustion products (e.g., CO_2 and H_2O) but also soot. In fact, the radiation load from RQL liners at high power conditions can locally be larger than the convective load.

During the combustion design process, allowance has to be made for sufficient cooling flow to keep metal temperature levels and gradients low. A significant effort is made to devise and implement efficient cooling schemes in order to minimize the cooling flow budget. In general, the greater flow is used for cooling and the lower flow can be used to control emissions. So, cooling schemes utilizing a combination of film and impingement coolings for maximum efficiency are widely used.

Even for cooling design, a mix of experimental and numerical assessments are required. Databases of total efficiency (i.e., nondimensional metal temperature) expressed as a function of nondimensional cooling mass flow can be used at preliminary design stages. Once the overall cooling flow distribution has been defined, the detailed design can be worked out. The challenge of defining an appropriate cooling solution is linked to the often strong dependence of component life on metal temperature, which in turn calls for very accurate prediction of thermal levels. For instance, a change in metal temperature of just 20°C can make a substantial difference in terms of component life. The sensitivity of component life to metal temperature actually depends on the life-limiting failure mechanism, but in general very accurate estimations of metal temperature are required.

Owing to the 2D nature of the metal temperature peaks (the so-called "hot spots"), 3D numerical models have to be used in support of detailed cooling design. Either detailed CFD models capturing fine details of the cooling schemes or finite-element (FE) thermal models can be used. In the first instance, aerodynamic and heat transfer effects can be assessed through calculation of film cooling effectiveness, heat transfer coefficients, and radiation loads. On the other hand, FE thermal models can derive metal temperature distributions by solving the conduction equation in the solid domain. To be able to do so, the FE model has to be provided with thermal boundary conditions, usually expressed as 2D maps of heat transfer coefficients, gas temperatures, and radiation heat loads. Thus, predictions of these quantities have to be produced for the FE model to be run. Adoption of an FE approach carries advantages and disadvantages. Given the high level of accuracy required in the derivation of metal temperatures, "out of the box" FE predictions may need the CFD-derived thermal boundary conditions to be changed. This can be done if metal temperature measurements are available for at least a baseline design. Changes to the thermal boundary conditions allow tuning the FE model for it to match thermocouple or thermal paint measurements. On the other hand, the number of parameters that can be used as part of this matching exercise is usually larger than the number of measurements. So, measurements can be matched using different combination of correction factors for the thermal boundary conditions. This introduces errors in the scaling of metal temperatures up to the high power conditions that are most important for combustor component life.

An alternative approach is to use conjugate heat transfer (CHT) models in a CFD framework, whereby a single computer code is used for both solid and fluid domains. In this case all physics is embedded in one model. Although a CHT model can be modified to make its calculated metal temperatures more consistent with a set of measurements through changes to meshes and turbulence models, for example, this approach offers much less flexibility than the one based on FE. The CHT approach is more suitable than FE for assessment of design changes based on significant modification to the hot side heat load.

Whether a CHT or an FE approach is used, prediction of heat transfer effects on both cold and hot sides of the liner is challenging. For simplicity, four different maps have to be predicted by the CFD solver: heat transfer coefficient, film temperature, and radiation load on the hot side and heat transfer coefficient on the cold side.

Calculation of accurate film temperature requires resolution of the small scales driving near wall mixing, which is usually of unsteady nature. So, the CFD has to be based on fine near wall meshes and an unsteady solution approach (either scale-adaptive simulation [SAS] [21] or LES). If LES is used, the computational demand may be too large for practical application because of the need to combine very small cells to very small time steps. In general, even if the small scale near wall mixing is captured adequately, accuracy in the calculation of film cooling effectiveness strongly depends on the velocity, turbulence, and air-to-fuel ratio predictions for the main stream flow. Large-scale unsteady effects may need to be included as well, which makes derivation of a zoomed-in submodel for detailed simulation of near wall effects difficult.

Derivation of hot side heat transfer coefficients is characterized by similar issues as the ones outlined for film cooling effectiveness, although the temperature field is bound to have less of an impact. For accurate prediction of heat transfer coefficients, small cells have to be used in the CFD to capture the boundary layer. The k-ω SST turbulence model can provide improved predictions with respect to other two-equation turbulence models in that the boundary layer can be resolved as long as a sufficiently fine grid is used at the wall.

Derivation of cold side heat transfer coefficients is mainly a matter of resolution. Even here, adoption of the k-ω SST approach used in conjunction with fine grids on the wall allows deriving heat transfer coefficient maps of adequate accuracy.

For cooling schemes relying on small effusion holes, resolution of the relevant scales poses a challenge in terms of computational affordability, especially if the component modelled includes a large number of holes. In this case, use of correlations is likely to be a more reasonable choice.

One of the most uncertain boundary conditions required for a thermal analysis is the radiation load. As stated earlier, radiation is given by the products of combustion (H_2O and CO_2) as well as soot. For an RQL combustor, the prevailing contribution is from soot. Due to the thick optical nature of the soot cloud, accurate prediction of the radiation load depends mainly on the near wall gas temperature and soot distribution. The uncertainty affecting prediction of both parameters, and in particular of soot, is bound to lead to significant uncertainties in the calculation of soot-induced radiation load. A discrete ordinate method is usually adequate to capture the main physical effects considering the uncertainties at play. It is common to use gray models and therefore neglect the dependence on wavelength.

Due to the high heat flux impacting on liner walls, it is common to cover the surfaces exposed to the flame with a ceramic insulator, known as thermal barrier coating (TBC). Its impact on metal temperature can be accounted for if the conductive and radiative properties of the TBC material are known.

As stated previously, metal temperatures have to be calculated with a high degree of accuracy to be of some use in the component life assessment. However, the uncertainties at play are significant, as already explained. As a result, the reliability of blind pretest predictions is somewhat limited, which implies that measurements have to be taken to anchor the models.

At the same time, a number of research activities are to be undertaken to improve the accuracy of thermal models for combustors. Improvements in the predictions of film cooling effectiveness are bound to stem from adoption of unsteady methods, with the potential of embedding some of the physics in hot side submodels to reduce computational costs in routine calculations. Improvements in the prediction of radiative loads call for more accurate predictions of soot concentration. Moreover, there is a need for development of techniques to measure the spatial distribution of radiation load on the liner walls.

The wide range of length scales to be captured means that availability of more computing power will in itself deliver more accurate predictions of metal temperatures.

Last but not least, geometric nonconformances due to manufacturing tolerances or thermally induced distortions have to be accounted for in the simulation to be a closer representation of reality. To account for manufacturing tolerances, it is mandatory to build statistical databases for proper quantification of the dimensional PDFs. On the other hand, thermomechanical models have to be used to assess the impact of displacements.

10.2.8 Low- and High-Order Methods for Thermoacoustics

Combustors can be affected by thermoacoustic instabilities. In the most general sense, these instabilities are due to coupling between combustor acoustics and flame. Modern aero-engine combustors are of annular type. As aero-engine combustor flametubes can be considered thin geometries, radial acoustic modes are bound to be higher frequency than axial and circumferential ones. As a result, the combustors are exposed to thermoacoustic modes that can be axial, circumferential, or axi-circumferential (i.e., mixed). As such, they do not couple with the flame to create self-excited oscillations.

Axial mode instabilities, sometimes referred to as low-frequency rumble (LFR), are mainly due to entropy fluctuations generated in the flame front region and propagating downstream due to convective effects. During the advection process, dissipation effects take place as a result of the unsteady mixing of cold and hot gases. When the fluctuation of entropy, to which temperature and density fluctuations are associated, hits the choked nozzle at combustor exit, a pressure wave is generated that moves backwards toward the combustor at the local speed of sound. When the pressure and associated velocity perturbation hits the flame front, a new entropy fluctuation is created. In this way, a self-excited instability can occur, which will quickly grow in amplitude up to a level at which damping effects balance off instability effects and a so-called limit cycle, characterized by a pressure fluctuation amplitude, behavior results. Because of the convective contribution to this kind of instability, the resulting frequency is low, i.e., between 100 and 200 Hz – hence the LFR denomination.

Circumferential modes instabilities, sometimes referred to as high-frequency rumble (HFR), are due to a direct coupling between the acoustic pressure fluctuations and heat release fluctuations. The acoustic modes will induce a local fluctuation in airflow velocity, which results in a fluctuation in air-to-fuel (AFR) ratio. The AFR fluctuation will lead to heat release fluctuations, which can be in phase with the acoustic mode. If that happens, small perturbations will grow into an instability that will quickly reach its limit cycle amplitude. The frequency at which these instabilities occur depend on the wavelengths at play; typically, the frequency range is from 300 up to 1000 Hz – hence the HFR denomination. Circumferential (or azimuthal) modes can sometimes couple with axial modes to give mixed modes.

The challenge posed by thermoacoustic instabilities to combustor development is significant, as rumble can be experimentally observed only when the combustion system is tested at high TRL, i.e., in a full annular rig with exit restrictor or in engine. If significant rumble fluctuations are found, a fix to the problem has to be found when the product development is at a relatively advanced stage. Rumble-induced fluctuations can lead to mechanical integrity problems that are unacceptable in service. Mitigations can take the form of changes to the operating lines or the design. Especially in the latter case, the cost and delay associated with rumble can be severe.

Low-order modeling can be used to mitigate rumble issues [27]. The approach relies on application of the linear Euler equation in a 2D framework where axial and azimuthal acoustic modes can be derived given the combustor geometry and acoustic boundary conditions. The combustor geometry leads to the definition of acoustic paths as well as relevant flow parameters such as AFR distributions. Impedance boundary conditions have to be defined at inlets and outlets. The equation is usually cast in the frequency domain. A complex frequency can be introduced whose real component is the mode frequency and the imaginary part is the growth rate. Positive growth rates lead to instability as an initially small perturbation will grow up to the limit cycle level. Stable modes have negative growth rates, so that small perturbations will be damped and will quickly die out. For practical applications, it is common to implement the solver in an acoustic network, which offers the flexibility of modeling different combustor layouts or acoustic features.

For HFR, a flame transfer function (FTF) has to be fed to the model to define the flame response to the pressure fluctuations induced by acoustics. The FTF is usually defined as the ratio of the unsteady heat release to the unsteady fuel injector mass flow, whereby both quantities are normalized by their time-averaged value. The FTF can be regarded as a black box providing the link between acoustics and flame.

For LFR, a temperature transfer function (TTF) has to be fed to the model to define the relationship between the pressure fluctuation at the flame front and the temperature fluctuation at the choke.

The FTF can come from experiments or simulations. The experimental route is most widely used and consists of subjecting the flame to perturbations generated by forcing devices (referred to as speakers or sirens) and measuring the FTF or parameters that can be related to it [45]. Such flame forcing is usually introduced in a single sector rig and multi-microphone techniques or OH/CH chemiluminescence measurements are taken to back out the FTF. While HFR modes are usually azimuthal, the forcing is often obtained through axial perturbations. Single-sector rigs can be used to measure HFR TFTs because the azimuthal modes wavelength is usually larger than the pitch between fuel injectors that is typical of annular combustors. For the measured FTF to provide accurate predictions of instabilities, the single-sector combustor has to produce flame shape and aerodynamic features consistent with the ones expected in the aero-engine combustor. Once the FTF is obtained by sweeping through the relevant frequency range, that can be used by the low-order model to predict rumble behavior of the annular combustor. The low-order model will predict a number of modes and their stability (i.e., frequencies and growth rates).

The most widely used approach aims to investigate linear stability. For this purpose, the flame will be forced with small-amplitude perturbations in the rig. The derived FTF will then be used in the linear Euler framework of the low-order model. This method allows calculating whether any of the acoustic modes will couple with the flame to generate instabilities. However, if unstable modes are found the amplitudes of the fluctuation of various parameters are undefined.

An alternative approach is to derive a nonlinear FTF, for which the amplitude of the forcing has to be large enough to respond nonlinearly. Nonlinear FTFs are sometimes referred to as flame describing functions (FDFs). The saturation level of the FDF derived from the forced rig testing will be a function of the damping due to the wall cooling. Because of the different surface-to-volume ratio of the single-sector rig with respect to the full annular combustor, FDF saturation levels can be different in the two configurations.

For LFR, a similar experimental approach can be pursued, but different diagnostic techniques have to be employed for measuring the evolution of the temperature fluctuations along the can.

It is worth highlighting that the design of a forced single-sector rig is quite challenging. In particular, attention has to be paid to avoid rig resonances that could mask the flame response. Furthermore, designing sirens able to force a high-energy flame reliably can be regarded as an engineering challenge in itself.

CFD simulations can be used to derive FTFs as well [46, 47]. In this case, the flame is forced by a sinusoidally varying input (e.g., mass flow) and the FTF can be backed out by monitoring the volume integral of the heat release as well as the fuel injector mass flow rate. The exercise can be repeated for the range of frequencies of interest, which will ultimately deliver a frequency-dependent FTF. Alternatively, the model can be forced by a forcing obtained by summing up a wider range of sinusoidal signals (the so-called "sum of sines" method). As in the case of forced experiments, the simulation will be limited to one sector, with an amplitude small enough to avoid the nonlinear regime but large enough to introduce a signal-to-noise ratio enabling the extraction of the FTF.

If designing a forced rig for derivation of FTFs requires careful consideration, CFD methods for calculation of FTFs are not mature yet, especially for spray flames. However, FTFs have been successfully derived by CFD for gaseous flames. The approach is based on use of a compressible solver and impedance boundary conditions. The model has to include the whole combustion system, including a fully resolved fuel injector, which allows calculating the impedance of the various ari passages. As explained in previous sections, an LES approach is favored because of its superior ability to predict unsteady phenomena accurately. The most significant source of uncertainty affecting the derived FTF is the spray response. Pressure fluctuations induced by acoustic waves will lead to local airflow velocity fluctuations. In turn, these will impact the atomization and the related AFR distribution. While in theory it would be possible to perform forced spray (e.g., volume of fluids) calculations relating the pressure fluctuations hitting the fuel injector to the oscillating AFR, the computational cost of such simulation would be prohibitive, as even small length scales would have to be captured. Moreover, the extent of the computational domain would have to be large enough to allow capturing the relevant details of the fuel injector and corresponding aerodynamics. In practice, at the present time this kind of approach is only capable of yielding a coarse approximation of the spray response to acoustic pressure fluctuations. A more promising route to deriving the spray response is through experiments [48]. This calls for application of the phase Doppler anemometry (PDA) techniques to a forced fuel injector, preferably at reacting conditions, for derivation of spray transfer functions (i.e., particle size distributions and the three velocity components) for the range of frequencies of interest. Once obtained through experiments, the spray transfer functions can be embedded in the CFD model to provide the frequency dependent spray response to pressure fluctuations. A quasi-steady approach, reliant on use of steady-state primary breakup correlations fed with time- and space-dependent air velocities related to the pressure forcing, is unable to provide acceptable accuracy over the full range of frequencies. Owing to the need to run injector-specific (and challenging) experiments to back out the spray transfer functions, the approach based on spray transfer functions cannot be regarded as completely predictive.

Another challenge for the running of these compressible forced simulations is the need to account for the damping effects due to cooling. While these can be reliably predicted using CFD methods, the level of resolution required for the small cooling features like effusion holes makes them impractical for use in large combustion system level simulations. Therefore, impedance-based boundary conditions have to be used for the damping due to cooling flows.

CFD methods can also be used to derive TTFs for LFR analysis. In this case, the dominant time scale is related to the propagation and dissipation of entropy waves from flame front to combustor exit. In this respect, LFR is bound to be easier to capture by CFD than HFR. However, the spray response is still necessary to account for the impact of acoustic waves on atomization.

The CFD approaches described so far aim to derive transfer functions to be used in low-order models. An alternative approach is to use CFD to simulate the self-excited instabilities occurring in the combustion system. In this case, the objective is to calculate the limit cycles directly. As far as HFR is concerned, the approach is of prohibitive computational cost, as the simulation domain would have to extend to the full annular geometry. On the contrary, direct CFD calculations of self-excited LFR instabilities are bound to be easier to carry out, due to the axial nature of the modes involved [49].

Thermoacoustic instabilities can be hard to investigate due to the wide range of possible causes. One of the most difficult effects to account for is interaction between fuel system vibrations and rumble. Vibrations of parts of the fuel system can lead to unsteady fuel feed to the combustor, which in turn can interact with thermoacoustic modes. Simulating these effects required adoption of a fluid–structure interaction approach to the fuel system, to provide input to the forced CFD model outline earlier. Given the complexity of the fuel system operating in aero-engines, simulation of these effects is particularly challenging.

10.2.9 Relight and Extinction Methods

Aero-engine combustors have to be designed to relight after a flameout at altitude, even if that is an unlikely event. After relight is achieved, sufficient temperature rise is required in the combustor to spool up the engine to idle (the so-called "pullaway") in the short time detailed in the engine specification. Furthermore, during operation the flame has to stay lit for the wide range of pressures, temperatures, and airflows dictated by the engine cycle.

Design rules have been developed by manufacturers to make sure that a new combustor will achieve the target relight and pullaway capability. These rules, derived from experimental databases, express the combustor relight capability usually in terms of a loading parameter, which is a function of combustor massflow, temperature, pressure, and volume. In general, a larger combustor will lead to higher efficiency and better stability. However, a larger volume will lead to higher thermal NOx emissions. As the combustor volume has to be defined at an early stage of the design, combustor relight, pullaway, and NOx are to some degree locked in early, although at later stages of the design there may be some margin to influence these attributes given a fixed volume.

To date, such design rules have been based just on experimental testing. The complexity of the physical and chemical phenomena involved in relight makes application of predictive methods particularly hard [50]. Key challenges are listed here. To start with, relight after a flameout at altitude has to happen when the engine is windmilling. In such event, the compression system operates at an extremely off-design condition, where significant flow separation occurs. In turn, this leads to combustor aerodynamic patterns and flow distributions that can be very different from the ones present at high power conditions. Moreover, the low-pressure drop across the fuel injector, combined with low inlet air pressures and temperatures, degrades the quality of atomization, up to a point where big lumps of fuel get released from the fuel injector rather than a fine mist of droplets. Eventually, for a meaningful simulation of the relight event, the location and energy content of the spark generated by the ignitor has to be known. On top of all that, the combustion model used has to be capable of accounting for local strain effects and ultimately of predicting extinction. Unsteady CFD models incorporating all these effects are being developed. Although results are encouraging [51–53], more physics and validation are required before they can be used in design practice.

An alternative approach to the full-blown LES method briefly outlined earlier is based on a semiempirical method that needs the inert two-phase flow field prediction as an input to calculate ignition probability maps once a critical Karlovitz number (Ka) is specified [54]. The advantage of the semiempirical approach is its low computational cost, the disadvantage is the need to calibrate it against experimental data, especially in terms of the value of the critical Ka.

While the ultimate goal of these methods is the accurate calculation of ignition loops and temperature rises, their relatively low maturity entails that much more research is required before they become reliable predictive tools. Nonetheless, even now such methods can be used to guide the design process by predicting how much change in relight and extinction capability is associated to a proposed design change.

The research areas that need to be worked on to improve methods for prediction of relight and extinction are briefly reviewed here.

Due to the predominance of separation effects on the systems aerodynamics at windmill relight conditions, representative experiments have to be run to characterize the flow field and understand any unsteady phenomena (e.g., separations) and their impact on flow distributions inside the combustion system. Such tests have to be run on rigs including some representation of the rotors.

Experiments also have to be carried out to measure fuel preparation at these extremely off-design conditions. These tests should be run at relevant (i.e., low-pressure and low-temperature) conditions to support definition of spray boundary conditions in the CFD models. Such experiments could also be used to validate high-fidelity two-phase flow models attempting to simulate the details of the air–fuel mixing. Diagnostic techniques able to assess the fuel distribution both quantitatively and qualitatively (e.g., shadowgraph) would have to be used. The numerical challenge would be due to the need to capture fuel ligaments with nonspherical shapes across the entire combustor, which would lead to a high computational cost mitigated only by the large size of the fuel lumps, thereby not requiring micron-sized levels of numerical resolution. Ultimate aim of the experimental and numerical research on spray would be to derive correlations that could be used to define spray boundary conditions.

Work is also required to estimate the amount of energy actually discharged into the combustor by the ignitor, which would have to take into account all the system losses at play. Moreover, introduction of the spark leads to generation of plasma. While a proper, physics-based characterization of the energy input and the very first phases of the relight would call for plasma modeling, the level of complexity associated with development and application of such models to a real combustor suggest that an experimental characterization starting from the formation of the first flame kernel, which would ignore the plasma physics, is a more sensible proposition.

10.2.10 Fuel Coking Methods

The striving for ever more efficient engine cycles has led to a significant amount of heat transferred to the fuel. While this is thermodynamically advantageous, fuel temperature levels have to be limited to values below which the thermooxidative breakdown of fuel resulting in formation of carbon deposit on the fuel passage surfaces can be avoided. Therefore, attention has to be paid to the fuel injector thermal management to prevent the fuel from getting too hot. This can be accomplished by use of FE (finite element) thermal models of the fuel injector to back out wetted wall temperatures. While a conservative approach for coking avoidance is to set low values of threshold wetted wall temperature (e.g., 150°C), residence time as well as fuel composition effects are important as well. For instance, if the fuel flow inside the passages is going through recirculation zones, there is increased likelihood of coke deposits forming there, especially if wetted wall temperatures are high. Moreover, the thermal stressing history of the fuel in the run up to the fuel injector as well as the constituents of the fuel can have a dramatic impact on the propensity to coke formation. For example, a high concentration of dissolved oxygen and especially heteroatomic species (e.g., reactive metals) can lead to increase coke deposition rates significantly.

To date, semiempirical conservative approaches have been used to design fuel passages to minimize the chance to coke formation. These are in the form of correlations relating wetted wall temperatures to deposition rates for nominal fuels. Safety margins are embedded in these correlations. At the same time, attention is paid at design stage to minimize the occurrence of slow moving regions and recirculations in the fuel passages. This can be achieved by performing single-phase CFD simulations of the fuel circuits.

Introduction of alternative aero-engine architecture (e.g., gear turbofan) has led to a requirement for increased amounts of heat to be transferred to the fuel. In turn, this calls for more accurate, physics and chemistry based methods for the prediction of fuel coking deposition rates and deposit thickness.

While the semiempirical approach mentioned earlier is used at the moment, methods are being developed to simulate the formation of coke precursors as well as their deposition rates and deposit thickness based on coupling a reduced reaction mechanism for precursors to a simple yet physics based deposition submechanism to work out deposition rates [55]. At the same time, deposit growth models can be used to estimate the increase in thickness of the carbonaceous layer on the walls over time for different inlet fuel compositions to be expected given the wide range of Jet A1 compositions allowed for in-spec fuels [56].

Availability of a computationally affordable yet accurate pseudo-mechanism for prediction of deposition rates coupled with a sticking probability model for calculation of the deposit growth paves the way to account for the evolution over time of the heat transfer inside the fuel injector due to the insulating effect of the layer of deposit. This can be taken into account by running CHT simulations associated to different levels of deposit thickness to work out the change in fuel passage flow number (i.e., effective area).

To make such reacting models applicable to real fuel injector coking problems, research is needed to improve understanding of the sensitivity against inlet fuel composition, which can be done via tests carried out on relatively simple geometries. Another aspect to be investigated is the impact of the fluid dynamic regime on the deposition steps of the pseudo-detailed mechanism to make sure that deposition rates are calculated correctly for both turbulent and laminar flows. Eventually, the impact of the increased surface roughness derived from use of additive layer manufacturing (ALM) techniques on coking has to be studied, again experimentally. A similar approach can be taken to quantify the impact of surface treatment on the fuel passages. Eventually, there is need to validate fuel injector thermal models, in particular for their ability to predict radiation loads for a range of different operating conditions. For this purpose, advanced techniques for local measurement of radiation load are required.

10.2.11   The Role of Spray Modeling

The challenges associated to application of predictive methods in support of the design of aero-engine combustors have been mentioned in the previous sections. For each combustor attribute, the uncertainties tend to be specific to the problem at hand. However, fuel placement has a key role, which is worth spending a few more words on.

Atomization of the fuel and corresponding mixing of fuel and air can be accomplished by different means. For instance, an airblast fuel injector relies on momentum transfer between fast moving air and a slow-moving film to obtain generation of fuel ligaments and then droplets, whereas a pressure jet fuel injector achieves atomization by breakup of a liquid jet at high pressure followed by formation of droplets. However, the process of atomization can be divided to two steps: primary breakup, leading to the formation of large droplets, and secondary breakup, leading to the formation of smaller droplets as a result of large droplets breaking up. Both primary and secondary breakups are complex time-dependent phenomena influenced by a number of factors (e.g., turbulence intensity and lengthscales) and usually concurrent with significant evaporation effects.

Fuel preparation can have an impact on

- Emissions: Soot, NOx, CO, and UHC are a function of the fuel-to-air ratio at injector exit.
- Fuel injector aerodynamics: The blockage and momentum transfer between air and fuel impact the fuel injector effective area and local aerodynamics.
- Temperature traverse: The combustor exit profiles are a function of the fuel-to-air ratio established close to the injector, although this sensitivity is more or less strong depending on the combustor design style.
- Relight and extinction: At relight conditions, atomization quality is usually poor, with big lumps of fuel being flung around inside the combustor. In general, it is expected that the relight is easier if the droplets are smaller. The extinction performance will be impacted by spray quality as well.
- Thermoacoustics: Both HFR and LFR are a function of the spray response to pressure fluctuations, which will change the air-to-fuel ratio in space and time.
- Metal temperature: Droplet size and initial velocities can have an impact on near wall gas temperature as well as radiation levels through changes in gas temperature and soot concentration.

It is worth pointing out that while fuel placement can significantly impact a number of combustion attributes, sensitivities can be more or less strong depending on the design style considered (e.g., rich burn vs. lean burn).

Models attempting to predict the attributes listed above require as accurate a fuel placement prediction as possible, often in both space and time. The standard approach for simulation of spray is to impose boundary conditions at a number of injection points close to the actual location of injection of the fuel. Such boundary conditions are defined through a particle size distribution function as well as the three initial velocity components. Given these initial conditions, a Lagragian model is then used to track the turbulent transport of the spray and account for evaporation effects. The challenge is to define a set of boundary conditions that are representative of the actual initial fuel placement. This can be done by using primary breakup correlations based on measurements taken on the fuel injector of interest or by interrogating the results of a high-fidelity two-phase flow simulation, again run for the injector of interest.

If a correlation is used, it has to provide information about both droplet size distribution and velocities for the injector at the operating condition of interest. This usually requires application of the phase Doppler anemometry (PDA) measurement technique, which is very time consuming and moreover cannot be used to probe the spray very close to the fuel injection location. For standard fuel injector designs, the measurement plane is located more than 5 mm downstream of the plane of injection. So, a matching exercise has to be carried out aimed at guessing the spray boundary conditions that can deliver the spray distribution measured downstream. Furthermore, the resulting method is not predictive, as for a new fuel injector geometry a new set of measurements may be required. Eventually, the correlational approach does not lend itself easily to accounting for unsteady effects.

If a high-fidelity simulation approach is pursued, the main challenge is the large computational cost to be incurred if adequate resolution for the spray boundary conditions has to be obtained. Eulerian methods (e.g., volume of fluids, level set) can be used in an LES framework to predict the evolution in space and time of the interface between the liquid and gaseous phases [57]. Due to the wide range of length scales that are physically relevant, extra fine grids and subsequently small time steps have to be used. In particular, it is important to calculate the curvature of the interface between liquid and gaseous phases accurately. Moreover, while the very fine resolution required (e.g., a few microns for a high-pressure condition) calls for using submodels, the extent of the computational domain considered has to be enough to allow capturing the large-scale aerodynamics that may have a crucial impact on the details mode of breakup. As a result of the computational cost of such simulations, they are run not routinely but on an ad hoc basis. More often, a compromise is accepted in terms of level of resolution and results are used only for qualitative assessments, for example, to compare two different injectors.

Whether an experimental or a numerical approach is pursued for characterizing fuel placement, significant challenges have to be overcome. This is particularly true for investigations of primary breakup at high-pressure and high-temperature conditions, which are referred to as supercritical. At supercritical conditions, the surface tension effects vanish, so that fuel and air mix as two different gases. However, locally the fuel conditions may not be supercritical, so that surface tension effects still apply. To better understand how to model fuel at high-pressure and -temperature conditions, experiments have to be run in high-pressure rigs to allow observing and measuring fuel behavior. Also, fundamental flame-related quantities such as S_L can show unexpected behavior at high-pressure and -temperature conditions; see page 253 in [58].

10.2.12 Trends in Aero-engine Combustor CFD

A number of trends can be observed in the CFD simulations used in industry to support design of aero-engine combustors.

- Massively parallel computations: While physics and chemistry based models have improved in accuracy in the past couple of decades, the most remarkable advance has been in the availability of relatively cheap large-scale computational power. This trend will accelerate in the coming years, as more of the combustor geometry and the important unsteady effects will be captured. Research in optimization of computational performance and parallel scalability of CFD codes will be key to make best use of the increasing computational power.
- LES as a routine tool: Academia has already switched to LES methods for most of the combustion CFD research. Industry is now following the same path. From an industrial perspective, it is important to understand the trade offs between accuracy and computational cost, so that educated choices can be made at different stages of the design process.
- Multicomponent simulations: By and large, the performance of aero-engine components is simulated in isolation. However, component–component interactions are being investigated and will be simulated routinely as part of the design process, especially to account for unsteady effects.
- Improvements on front and back end of CFD packages: Significant improvements in pre- and post-processing have introduced and will carry on delivering benefits to the CFD users (e.g., adaptive mesh refinements, slick post-processors). In the future, augmented reality facilities will enable full exploitation of large CFD solutions.
- Integration and automation: Various combustion methods will be integrated in a software framework enabling further speed up of the design process, starting from simple 1D calculation to full-blown 3D unsteady reacting two-phase flow CFD with parameterization of the geometry and automatic meshing generation included [59].
- Thermoacoustic CFD: A significant research effort is being made to develop the building blocks for future application of CFD methods to the prediction of thermoacoustic instabilities. This effort is likely to ramp up further as a result of introduction of lean burn combustors in service
- Conjugate heat transfer modeling: While FE methods are today's workhorse for calculation of metal temperature, conjugate methods will gradually replace FE methods. This will go hand in hand with the increase in computing power as

well as improvements in the physics-based approaches for the simulation of hot side effects.
- Soot chemistry: More detailed chemistry models will be used to simulate soot production and consumption, including coupling between soot and combustion chemistry and calculation of particle size distributions.
- Primary breakup modeling: The remarkable increase in computing power will lead to a gradual uptake of high-fidelity two-phase flow models for prediction of primary breakup and fuel placement.

It is the author's belief that the improvements to be expected from these trends can be achieved only if academia and industry work closely together. Furthermore, although this article focuses on predictive methods, these objectives are deemed to be achievable only if continuous investment is made in the development and application of advanced measurement techniques, which are crucial to provide validation data for such predictive methods.

Acknowledgments
The author would like to thank Emmanuel Aurifeille and Koulis Resvanis for providing results from the CFD simulations.

10.3 Internal Combustion Engines

D. Norling and X.-S. Bai

The internal combustion engine (ICE) is the main power source for the propulsion of road vehicles and ships. Spark-ignition (SI) engines and compression-ignition (CI) engines are two commonly used types of ICE. Combustion in an SI engine occurs in premixed flame mode and the mixture of fuel and air is fully premixed, typically at stoichiometric fuel/air ratio in order to use the three-way catalysis for high-efficiency after-treatment of the exhaust gas. The CI engine has a higher efficiency than the SI engine, and thus a lower CO_2 emission, while it suffers from high emissions of both NOx and soot. The combustion mode in CI engines is more complex: it is often at multiple modes, including spontaneous ignition, premixed flame propagation, and diffusion flames. One typical CI engine is known as a diesel engine, which runs with diesel fuel and combustion occurs when diesel is injected into the engine. The combustion mode of a diesel engine is referred to as conventional diesel combustion (CDC), which starts with spontaneous ignition of the charge, typically in the fuel-rich mixture, followed by the establishment of a diffusion flame. The flame is lifted off the fuel nozzle. The leading front of the lifted flame contains multiple combustion modes, for example a fuel-rich premixed flame in the center, surrounded by a diffusion flame [60, 61]. The structure of the leading flame front depends on the ambient condition, the liftoff length and the fuel type [62]. In this section we will focus our discussion on the challenges in diesel engine development; however, it is appropriate to start with a brief survey of other recently developed engine concepts.

10.3.1 Combustion Concepts in ICE

Figure 10.13 shows a schematic diagram of several different ICE concepts that have been developed up to date. These different concepts are developed to cope with the requirements for ICE: high engine efficiency and low emissions of pollutants and green-house gas CO_2. The homogeneous charge compression ignition (HCCI) engine makes use of the advantages of both the SI engine (by using a lean premixed mixture to avoid the formation of soot and NOx) and the CI engine (by using a high compression ratio to achieve high engine efficiency) [63–66]. HCCI engine combustion is mainly governed by spontaneous ignition wave propagation with a fraction of the reaction fronts in premixed flame mode, depending on the thermal stratification of the mixture [67–69]. Thus, the combustion process is highly sensitive to the temperature of the charge and it leads to high cycle-to-cycle variation and difficulty in engine control. At low load there is a tendency of misfire and at high load the pressure-rise rate is high, giving rise to high engine noise. Spark-assisted HCCI (SACI) has been tested to mitigate the problem of misfire at low load. However, the operation window for SACI can be narrow due to the switch of combustion modes between premixed flame propagation and spontaneous ignition wave propagation [70].

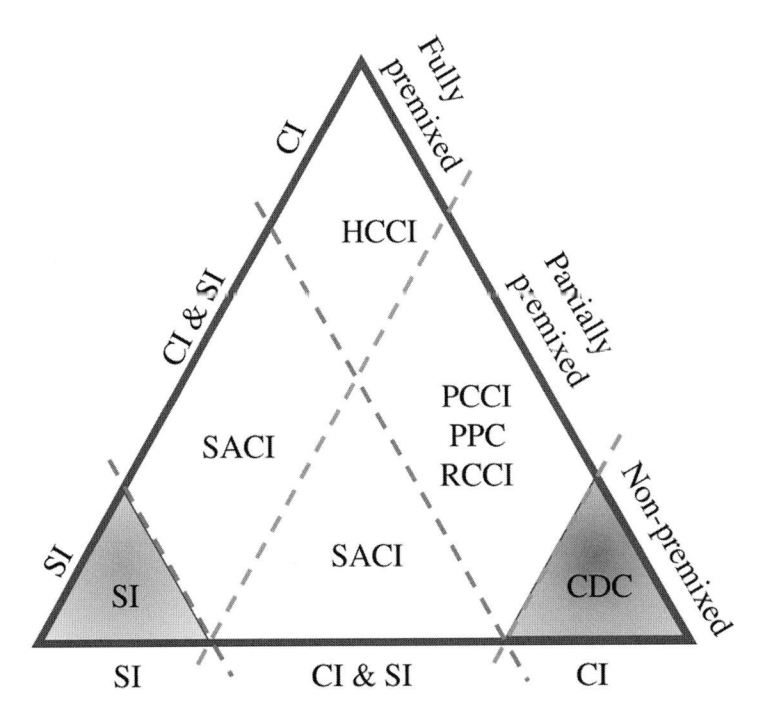

Figure 10.13 Fuel–air mixing, method of charge ignition, and modes of combustion in ICE. SI: spark-ignition; CI: compression ignition; HCCI: homogeneous charge compression ignition; PCCI: partially premixed charge compression ignition; PPC: partially premixed combustion; RCCI: reactivity-controlled compression ignition; SACI: spark-assisted compression ignition (to assist the ignition of HCCI, PCCI, or PPC); CDC: conventional diesel combustion.

To achieve better control of the combustion process in ICE it is advantageous to have certain stratification of the fuel in the charge. The partially premixed charge compression ignition (PCCI) engine is one example of such a concept. In PCCI the charge is in general fuel lean; thus low soot and NOx emissions can be achieved simultaneously. The stratification of the composition is beneficial for reducing the pressure-rise rate at high-load operation. Recent direct numerical simulation (DNS) studies of PCCI combustion have indicated that the combustion process is similar to that of HCCI combustion, where the spontaneous ignition wave is the dominant mode of combustion. Thus, the combustion process is highly sensitive to the stratification of both composition and temperature [71, 72]. Spark assistance may be needed at low load to avoid misfire.

Two different engine concepts have been developed to achieve reliable ignition at low load, to avoid too high pressure-rise rate at high load, and to make use of the advantages of the HCCI concept. The first is the reactivity-controlled compression ignition (RCCI) engine [73, 74] and the second is the partially premixed combustion (PPC) engine [75, 76]. In the RCCI engine, a low-reactivity fuel (such as gasoline) is injected first, to form a fuel-lean homogeneous mixture (as in the HCCI concept). The ignition process is initiated later in the compression stroke by injection of a high-reactivity fuel (such as diesel). Combustion of the high-reactivity fuel triggers the combustion of the low-reactivity fuel. The fuel-lean low-reactivity fuel mixture is advantageous for low NOx and soot emission, while the high-reactivity fuel injected later is used for the controlled ignition. An example of an RCCI engine is the methanol–diesel RCCI engine that shows benefits in engine efficiency, pollutant emissions, and combustion control [77–79]. A DNS study of methanol–n-heptane RCCI combustion indicated that interaction of the high-reactivity fuel (n-heptane) and the low-reactivity fuel (methanol) can be due to different mechanisms [80]: enhancement of ignition of the low-reactivity fuel–air mixture by the bulk flow compression due to the ignition of the high-reactivity fuel, and back-support of the hot reaction products from combustion of the high-reactivity fuel to the deflagration wave propagation in the low-reactivity fuel–air mixture.

In contrast to the RCCI engine, the PPC engine uses only a single fuel. The fuel is injected during the late compression stroke but earlier than in a conventional diesel engine; thus, a highly stratified spatial distribution of composition and temperature of the fuel–air mixture is formed. The fuel should not have too short ignition delay time (e.g., low cetane number) to avoid too early ignition. Thus, gasoline fuel has been tested and shown to have good performance [81]. If diesel fuel is used instead, certain dilution by exhaust gas recirculation (EGR) is needed to slow down the ignition process [75]. To optimize the composition and temperature stratification in a PPC engine the fuel injection can be split into multiple injections: an early injection to form a homogeneous mixture to retain the advantages of HCCI combustion and one or more late injections to slow down the heat release rate and thus the pressure-rise rate. PPC is different from PCCI in that, in PPC the fuel–air mixture varies from fuel-lean to fuel rich, while in PCCI the mixture is in general fuel-lean (with equivalence ratio less than 1). DNS studies of the PPC engine process have shown the existence of multiple combustion modes: ignition of the mixture in both the fuel-lean and fuel-rich mixtures, and diffusion-controlled combustion of the fuel-rich mixture by mixing with the fuel-lean mixture [82–84]. The former controls the emissions of NOx and the latter controls the CO and soot emission. A tradeoff of soot and NOx has been observed by optimizing the mass-split between the different injections [76, 85].

Among the different engine concepts shown in Fig. 10.13, the SI and diesel engines have been in commercialization for a long time; yet great challenges exist for these engines. In particular, diesel engines are known to be environmentally unfriendly due to the emission of particulate matter and NOx, whereas SI engines often have a lower efficiency and a higher CO_2 emission compared to diesel engines. Conventional SI engines operating at stoichiometric air/fuel ratio have the disadvantage of causing high emissions of CO, unburned hydrocarbons (UHC) and NOx. Several different strategies have been developed in the past 20 years to improve the efficiency and reduce the emission of SI engines. This includes lean-burn combustion and direct-injection of gasoline fuel to the engine cylinder (known as GDI). Urata and Taylor [86] outlined the benefits of lean-burn SI combustion: (a) reduced pump-loss and large work output, (b) high values of the ratio of specific heat that gives rise to high engine thermodynamic efficiency, and (c) reduced emissions of NOx and CO. Shuai et al. [87] reviewed the recent progress of GDI engine technologies, including high tumble flow design, cooled EGR, high compression ratios, overexpansion cycles with variable valve timing (VVT), boosting, high-pressure injection, and other design features. Mazda SKYACTIV-G is an example of a GDI engine with high compression ratio. A significant improvement in the engine efficiency is achieved due to the high compression ratio; however, a high compression ratio can also give rise to engine-knock that deteriorates the engine performance. SKYACTIV-G solved the knocking issue by reducing the residual hot exhaust gas in the cylinder. The most recent SKYACTIV-X engine makes use of the SACI concept, i.e., compression ignition controlled by spark discharge. Based on seamless switching between spark and compression ignition, the benefits of HCCI combustion with high engine efficiency and low emissions of pollutants were realized [87].

In the remainder of this chapter we will focus on conventional diesel engines, in particular truck diesel engines. Further discussion on other combustion concepts is found in the review papers of Dec [61], Reitz and Ganesh [73], and Musculus et al. [88].

10.3.2 Diesel Engine Combustion Chamber Design

Modern truck diesel engines typically have displacement volumes of 8 to 16 liters. The engines are turbocharged and have charge-air coolers. Maximum engine speeds are about 2500 rpm but 1000–1500 rpm is typical for highway cruising speed. The cylinder head is of four-valve design with two inlet and two exhaust valves. This design permits a large valve area together with a centrally mounted injector. The cylinder head is usually flat on the side toward the combustion chamber. This surface is known as the *firedeck*. The valves, when closed, are part of this flat surface. Note, however, that a completely flat firedeck when the valves are closed would severely restrict the gas flow in and out of the valves at low valve lifts. For this reason, a compromise is struck with small crevices surrounding the valves when closed. The inlet ports are designed so that the air entering the cylinder will get a swirling motion around the cylinder axis. This motion is called swirl and the strength of the swirl motion is an important parameter for the performance of a diesel combustion system, both for emission formation and fuel efficiency.

The centrally mounted fuel injector is usually fed with a high-pressure fuel rail. Maximum injection pressures usually range from 800 bar (typically for idle) to 2500 bar (for full load operation). The injection timing in relation to the piston position and the fuel rail pressure are the two most important control parameters available for combustion and emission control after the hardware is set. In engines where NOx emissions are mitigated by EGR, the ratio of EGR gas to air is also an important parameter that can be controlled through valves in the gas exchange system, or through the variation of turbine area for a variable geometry turbines. The fuel injector has a number of small nozzle holes. Typical numbers are between 6 and 10. Hole diameters for heavy duty engines are typically 0.15 to 0.2 mm. Each hole and its respective sector of the combustion chamber are made to be as similar to the other sectors as possible.

One important limitation of a diesel engine is the maximum allowed peak cylinder pressure (PCP). The limiting factor is the structural integrity of the engine. As many critical parts work under high temperatures, a combination of thermal and mechanical loads must be taken into account when the PCP limit is decided for a new engine design. The PCP limit of a given engine sets an envelope of possible combinations of geometrical compression ratio, boost pressures, and combustion phasing.

The piston is made out of steel and is cooled from the bottom side by one or more jets of oil. The oil jets enter a dedicated cooling gallery through which the oil is shaken forward toward an exit from which it falls back into the engine crankcase. On the top side of the piston, toward the combustion chamber, there is a piston bowl. The bowl has a smaller diameter than the cylinder and is usually rotationally symmetric. Along the periphery of the piston top there is a flat surface which is referred to as the *squish land*. The height of the squish land over the piston pin center is the *compression height*.

The piston is made to come as close as possible to the cylinder head at top dead center (TDC). A small clearance between the piston and the firedeck cannot be avoided as the piston should never hit the firedeck and the influence of thermal expansion, inertial forces, and production tolerances prescribe some margin of safety. The volume between the first compression ring and the top of the piston and between the cylinder liner and the piston mantle is known as the *topland volume*. The radial clearance between the piston and the liner are set by thermal expansion considerations and control over soot and coking residues. From a performance perspective the topland volume should be kept as small as possible.

It is desirable to minimize the clearance and topland volumes, since air trapped in these volumes is not available for combustion. The ratio of air mass in these crevices to the total air mass in the cylinder per engine cycle is larger than the volume ratio between the crevices and the total volume would suggest. The reason for this is the lower temperature due to the large area-to-volume ratios of these crevices. The temperatures of the combustion chamber walls are significantly lower than those of the bulk gas during the high-pressure phase of the cycle, and the large area-to-volume ratio of the crevices gives rise to intense heat loss in the crevices and as a consequence the gas density is much higher than for the bulk gas.

Minimizing the topland volume by decreasing the topland height puts more heat load on the upper compression piston ring, and the survival of this piston ring is the main obstacle in minimizing the topland volume. Minimization of the volume between the piston and the firedeck at TDC is limited by production tolerances.

10.3.3 The Combustion Process

Liquid diesel fuel is injected into the combustion chamber, where it quickly atomizes and vaporizes. The fuel vapor mixes with ambient air and the mixture then autoignites when the pressure and temperature conditions in the combustion chamber are suitable. Following the ignition of the spray jet, a lifted diffusion flame is established, where the leading front is a partially premixed flame that is affected by the local flow, the rate of mixing, and low-temperature chemistry upstream [62]. The main body of the flame is a mixing-controlled diffusion flame where the fuel vapor from the diesel jet is burnt at a rate limited by turbulent mixing of the fuel vapor and the surrounding air. Ultimately the fuel cannot be mixed with air faster than it is injected, this being the reason of the different vapor penetration lengths with different injection pressures and nozzle diameters. This stage of combustion is essentially spray driven. After the fuel injection ends the residual fuel vapor is burned in a diffusion flame mode that is governed by mixing with the surrounding air. This stage of combustion is referred to as *post-injection oxidation*, or *late-cycle oxidation*. Turbulence in the surrounding air governs the mixing process in this stage, since the fuel-jet induced turbulence is at small-scale and it is dissipated quickly. Thus, it is important to maintain the level of turbulence in the surrounding air after the injection is ended. To this end, the swirl motion of the gas charge is of importance, since turbulent kinetic energy generated by the spray jet can be partially maintained in the post-injection oxidation stage in the presence of swirl motion in the surrounding air [89].

The combustion chamber dimensions, together with the time required for the combustion event, make flame-to-wall interaction unavoidable. As an example, Fig. 10.14 shows the ignition sites in a CI engine firing with a primary reference fuel (PRF87), which is a mixture of 13% *n*-heptane and 87% iso-octane. The fuel distribution was measured using planar laser-induced fluorescence (PLIF) of fuel tracer acetone, while the ignition region was identified using chemiluminescence imaging. The PLIF was recorded from 2.4 to 3.4 CAD after TDC, whereas the boundaries of the ignition regions were recorded around the instance of time that the PLIF signal was recorded. It is clear that high concentration of fuel vapor is found in the vicinity of the piston wall, and the onset of ignition is in fact in the vicinity of the wall as well; cf. Fig. 10.14a. The heat transfer between the flame and the combustion chamber wall, usually the piston, is high during the spray-driven stage of the combustion. The heat loss from the gas charge to the walls results in a loss of potential work extraction from the gas. The flame-to-wall interaction makes the understanding and optimization of a diesel combustion system geometry challenging. Before the flame comes into the vicinity of the wall during the combustion stages, the burn rate is dominated mainly by parameters set by the injector, the fuel type, and the pressure and temperature of the compressed air or a mixture of air and recirculated exhaust gas. The mixing and burning of the fuel in the flame after the wall interaction is highly dependent on the combustion chamber geometry. A high mixing rate of fuel and air is desirable to achieve high burning rate, while it is less favorable to have the combustion product mixed with the combustion product itself. It is important to avoid the fuel being trapped in a dead end, or in a stable vortex, as this will decrease the burning rate and can possibly be sources of UHC and engine-out soot. The gas to wall heat transfer through this process should be kept as low as possible.

From a combustion point of view the following challenging research questions related to combustion in diesel engines require in-depth studies:

1. Conditions for the onset of ignition. An important question is at what composition (equivalence ratio), temperature/pressure, and flow/turbulence conditions the onset of ignition occurs.
2. Mechanisms of liftoff and stabilization of the lifted spray flame, and structures of the lifted flame front. Mechanisms for the stabilization of the lifted gas flames have been understood fairly in-depth [91]. However, mechanisms of flame liftoff in diesel engines are far more complex, owing to the challenge of the high-temperature and high-pressure environment.
3. Impact of liftoff on the formation of soot and NOx. There is evidence that soot formation is closely correlated with the liftoff height [92, 93]. However, it is unclear, from a fundamental point of view, how the soot/NOx formation and destruction are correlated with the flow and mixture properties at the liftoff position in diesel engines.
4. Post-injection oxidation of soot and unburned fuel. From the standpoint of emissions, post-injection soot oxidation is critically important, since this process determines the actual level soot in the engine-out exhaust gas.
5. Spray/wall interaction, flame/wall interaction and their impact on gas to wall heat loss.

To illustrate the impact of ambient air conditions on the liftoff of the spray flame and on the soot formation under diesel engine conditions, Fig. 10.15 shows the structure of an n-dodecane spray flame in a constant volume combustion chamber under different ambient temperature conditions [62]. The fuel injector has a 90 μm nozzle diameter and injection pressure of 140 MPa. The ambient air temperature varies from 900 to 1000 K with the corresponding ambient pressure of 5.94 and 6.62 MPa, respectively. The liftoff height and the formation of soot are shown to be sensitive to the ambient temperature. The lower ambient temperature (900 K) case has a higher liftoff height; the leading front of the flame is an autoignition-induced partially premixed

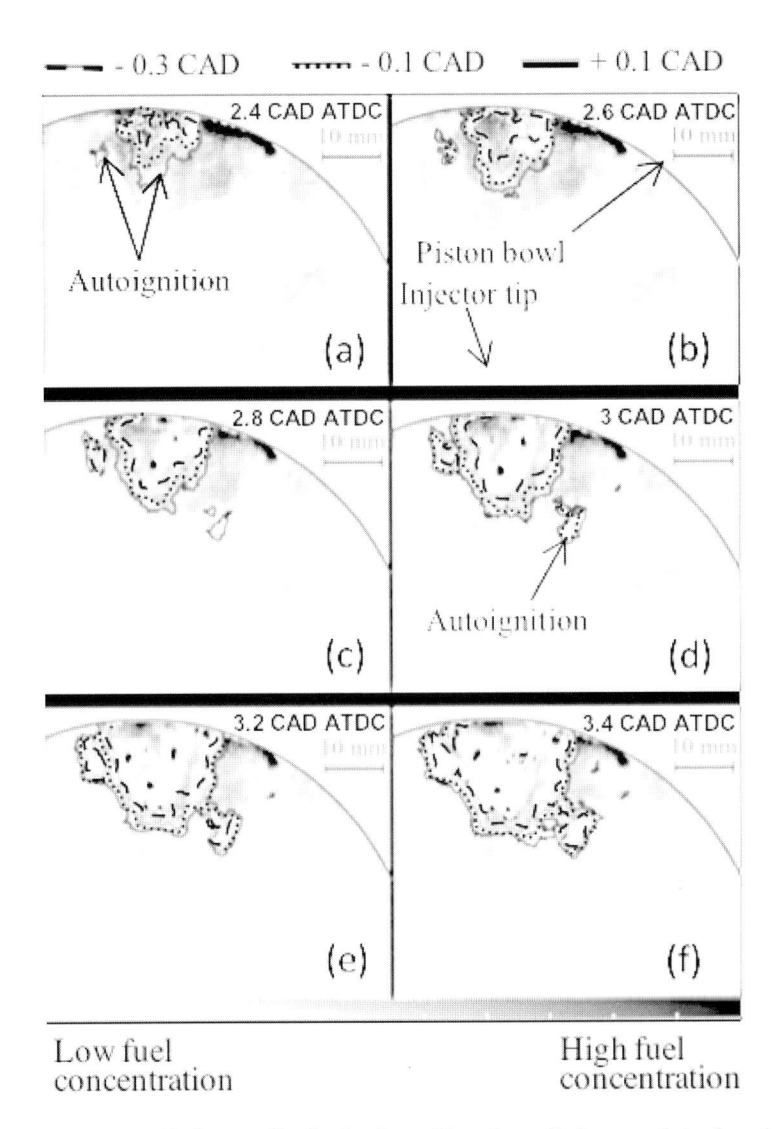

Figure 10.14 Fuel vapor distribution in an CI engine cylinder recorded using planar laser induced fluorescence (PLIF) of a fuel tracer (acetone), and ignition sites characterized using chemiluminescence imaging. The dashed, dotted, and solid lines indicate the boundary of the ignition region at respectively 0.3 CAD, 0.1 CAD before the instance of time at which the PLIF image was recorded, and 0.1 CAD after the PLIF image was recorded. (a–f): PLIF images taken at 2.4, 2.6, 2.8, 3, 3.2, and 3.4 CAD ATDC. For interpretation of the references to color in this figure legend, the reader is referred to the web version of Wang et al. [90]. Reprinted with permission from Elsevier.

flame: in the central fuel-rich region the fuel is converted to combustion intermediates (mainly CO and H_2), resulting in low fuel concentration in the fuel-rich center flame region and thereby lower level of soot formation. The higher ambient temperature case (1000 K), on the contrary, has a lower liftoff height; subsequently, the mixture

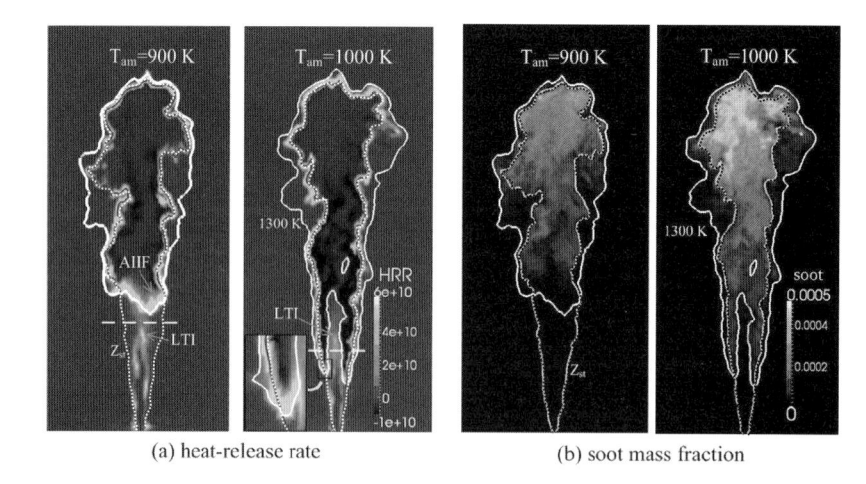

Figure 10.15 Structures of diesel spray flames from an LES investigation [62]. (a) Heat release rate. (b) Soot mass fraction. AIIF: autoignition induced flame front; LTI: low-temperature ignition; horizontal dashed line: lift-off position from experiment; solid white lines: isoline of temperature of 1300 K; dotted lines: stoichiometric mixture fraction. For intepretation of the references to color in this figure legend, the reader is referred to the web version of Cheng et al. [62]. Reprinted with permission from Elsevier.

in the center of the jet is fuel richer and there is no premixed flame (the mixture is beyond the rich flammability limit) in the center of the jet to oxidize the fuel. The fuel in the jet center is oxidized in the downstream diffusion flame, where more soot is formed. The leading front of the flame in the high ambient temperature case is similar to the triple-flame structure of a partially premixed flame [91]. In both cases the low-temperature ignition chemistry is important, as shown in the spatial distribution of heat release rate in front of the liftoff position. It is expected that, in a real diesel engine combustion chamber, the spray/wall interaction will further complicate the liftoff and soot formation process; this is not yet well explored.

10.3.4 Gas Motion and Turbulence in the Combustion Chamber

The turbulent gas motion in the combustion chamber has a significant impact on the fuel/air mixing, combustion, heat transfer to walls, and the formation/oxidization of pollutants. When the piston approaches the firedeck before TDC, the gas charge situated between the squish land and the firedeck is squished into the piston bowl, this indeed being the reason for the name *squish land*. After TDC, when the piston descends, there is a reverse gas motion when gases are sucked into the volume opening up between the squish land and the firedeck. This gas motion is referred to as the reversed squish motion. The squish motion affects the swirl angular velocity as the gas charge is pushed into a piston bowl with smaller diameter than the cylinder bore. As angular momentum is conserved the angular velocity is increased. During the reverse squish after TDC the opposite happens.

During the injection process the injector exerts work against the centripetal force field set up by the swirl motion. This process involves a competition between radial momentum induced by the injected fuel spray and the centrifugal force induced by the swirl that forces the fluid to move inward toward the center axis of the combustion chamber, through two mechanisms [89]. First, the high-speed spray jets entrain the ambient gas in the proximity of the nozzle to flow downstream toward the periphery of the piston bowl, resulting in the deflection of the flow stream by the piston bowl and a formation of a large-scale recirculation zones in the bowl. Second, the fuel spray jet flow is a source of radial flow momentum, which is transferred to the surrounding gas via entrainment and turbulent diffusion. Both mechanisms act to redistribute the swirl angular momentum within the bowl, displacing the high-momentum fluid inward.

Spray/swirl interaction generates not only significant change of the large-scale flow motion, but also significant enhancement of turbulence production. Figure 10.16 shows the turbulent kinetic energy measured in a direct-injection, swirl-supported diesel engine [89]. In the motored run case, turbulent kinetic energy shows a monotonic decrease in the expansion stroke, whereas with fuel spray injection a rapid increase of turbulent kinetic energy is noted after TDC (0 CA). It is shown that increase in both swirl ratio (Rs) and injection pressure (thus spray momentum) enhance the production of turbulence. The turbulent kinetic energy is dissipated late in the cycle, long after the spray is terminated. However, compared with the noninjection case (motored), turbulent kinetic energy generated due to swirl/spray interaction retains in the expansion stroke. How to use this stored kinetic energy for enhanced late cycle mixing and oxidation of soot, UHC and CO is a field of active research.

10.3.4.1 Work Extraction

A fast heat release is beneficial for maximizing fuel efficiency. As the piston moves after TDC while the heat release is taking place, the available expansion work decreases with increasing time of heat release. As a result, the potential work extraction for a slow heat release is less than for a fast one.

The above reasoning leads to the conclusion that heat release should be centered at the TDC. For an adiabatic engine this is true, but for a real engine the heat released before TDC leads to overly high temperatures, which gives rise to high heat loss to the wall. Also, the high temperature promotes more NOx formation. Furthermore, fast heat release rate centered at TDC gives rise to high peak cylinder pressure and high pressure-rise rate, and therefore high noise level. The injection timing is a compromise between NOx formation, expansion ratio, peak cylinder pressure, and heat losses.

Since diesel engines operate with excess air, not all air must be consumed to realize complete combustion of the fuel. The air excess is different for different operating conditions depending on the turbocharger specifications. A higher air excess ratio usually decreases the combustion duration and reduces the heat loss as increased gas charge mass decreases bulk temperature. On the other hand, increased air excess typically increases NOx formation. This seems to be contradictory as NOx formation is strongly increased at higher temperatures. However, the formation of NOx is a local process in the stoichiometric layer surrounding the spray jet. The temperature in this

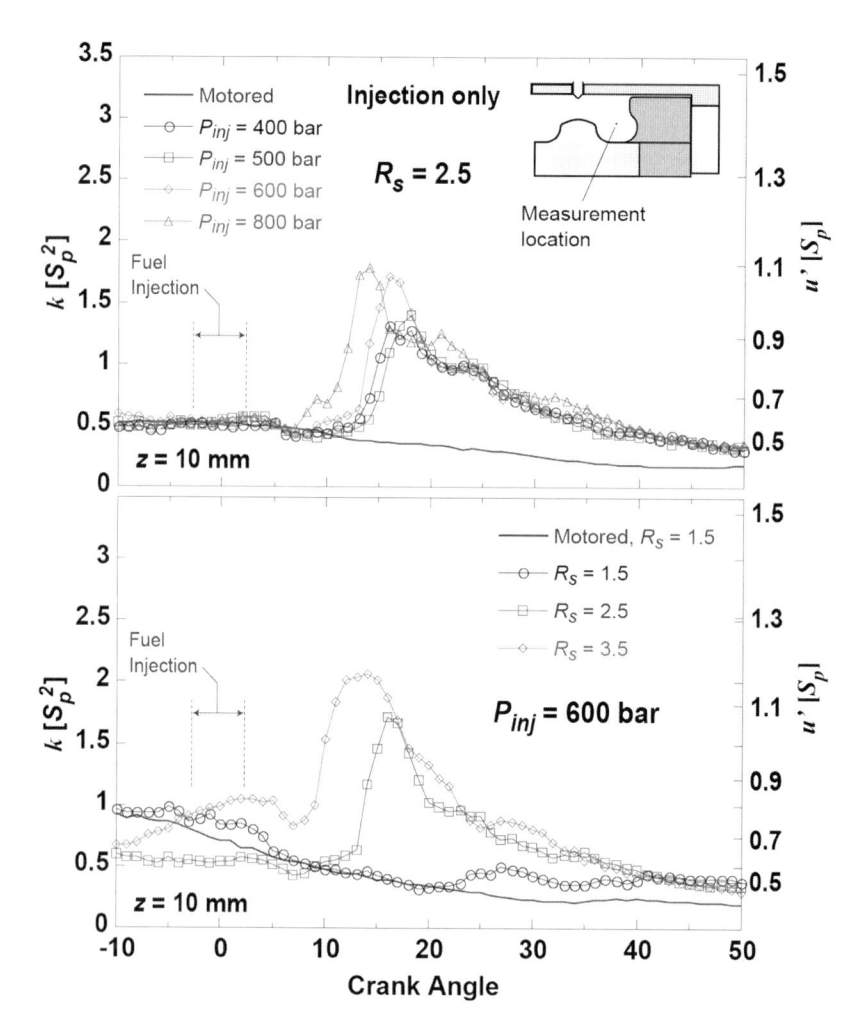

Figure 10.16 Measured evolution of turbulent kinetic energy for motored (without fuel injection) and injection (without combustion) engine operation, from Miles [89]. Reprinted with permission from Springer-Verlag.

zone is mainly controlled by the mixing rate, while the bulk gas temperature is strongly dependent on the total mass of the air.

Another minor factor affected by excess air through bulk temperature is the ratio of specific heats, γ ($\gamma = c_p/c_v$). A low γ value through the expansion stroke is beneficial as more work can be extracted from the gas. As the γ value decreases with increasing temperature, this effect mitigates some of the heat loss benefits with higher air excess.

10.3.4.2 Challenges in Diesel Combustion Modeling

CFD modeling is today heavily used in diesel engine development [94–96]. Engine design simulations are commonly carried out using the Reynolds-averaged Navier–Stokes (RANS) approach, while large-eddy simulation (LES) is also used in fundamental studies of flow and combustion dynamics, for example in studies of the dynamic behavior of the spray process and cycle-to-cycle variation. In both RANS and LES, several model challenges exist, e.g., spray modeling, and coupling of the flow simulation with detailed chemistry within reasonably short computational time.

10.3.4.3 Cycle-to-Cycle and Spray-to-Spray Variations

In diesel engine development there is often a preconception about the physical process having very little variation. This thinking comes from the fact that the variation in pressure traces from one engine cycle to another is very small. It is close at hand to compare to SI engines, for example, where the cycle-to-cycle variation is significant, and closely monitored, as it is an important characteristic of the engine. However, the observed low variation of the pressure trace does not mean a low variation from flame to flame of the different nozzle holes. It has been observed in spray bombs that the fuel jets from a multi-hole nozzle can have different start of injection and different penetration lengths. Also, the spray penetration can vary over one injection event. The connection between variation in spray penetration within one nozzle and injection/needle valve motion inside the injector is unclear. A possible reason for spray penetration variation could be unsteady fluid flow inside the nozzle. The nozzle holes are not exactly equally shaped: how large an impact does this have? Connected to the topic of injection variation, if the variation and distribution of variation are important for CFD simulations, will a RANS model have any chance of making useful predictions? It could be argued that soot formation and oxidation would be difficult to predict well with a RANS simulation, as coherent flow structures are not represented in the averaged flow.

10.3.4.4 Thermal Barrier Coatings

Another observed difficulty is the prediction of the combustion behavior in a diesel combustion chamber with ceramic thermal barrier coating (TBC). Experimental investigations have shown a decrease in apparent heat release rates with ceramic TBC in the combustion chamber. The physical mechanism responsible for this behavior is unknown. A wide variety of candidate explanations are the following:

1. Increased wall heat transfer due to thinner quenching layers, *convective vive*
2. Increased radiative heat transfer due to the property of ceramic TBC
3. Trapping of air and fuel in the pores of the TBC
4. Change in the interaction between the wall and the flame due to (a) increased surface roughness, (b) increased temperature, (c) catalytic activity of TBC, and (d) loss of swirl angular momentum due to increased wall friction

Since this problem seems to arise from some process not modeled in today's typical CFD models, the way forward must be from detailed experiments aiming to observe in detail the unusual behavior in a TBC combustion chamber that is not present in a metal chamber. After an experimental observation the modeling efforts can start. In the context of TBC, and maybe especially ceramic TBC, the influence from wall roughness on combustion and boundary layer behavior should be considered. It is desirable to properly model the combustion chamber wall with significant surface roughness or porosity.

10.3.4.5 Radiation Heat Transfer

Heat transfer is a key process in diesel engines that affects the efficiency and emissions. Radiative heat transfer is particularly important in diesel engines due to the presence of soot. Most current modeling strategies do not involve any radiation model. The effect of radiation on wall heat transfer, spray evaporation, and soot oxidation is not well explored.

10.3.4.6 Liquid/Wall Interaction

In diesel engines the fuel is injected close to TDC and the distance from the injector to the piston bowl is short. The spray can reach the relatively cold wall. The spray impinged on the wall may condensate and form a liquid film on the wall. This is especially true when alcohol fuels, such as methanol, are used in diesel engines. How should a liquid wall film be modeled? How does a wall film affect heat transfer? Modeling of a liquid wall film is a challenge in diesel engine combustion simulations.

10.3.4.7 Liquid Breakup

The breakup of the liquid fuel jet into droplets close to the nozzle exit is a complex physical process. It is difficult to observe the breakup experimentally due to the small time and length scales and the problem of achieving optical access into this very dense spray region [97]. It is not well known which parameters in the breakup region are important for the far field, from the standpoint of nozzle exits. Many jet-flame characteristics can be captured well with models not even considering the liquid phase at all [98, 99]. Answers to these questions could help the scientific community and industry to put research efforts into the most relevant tasks.

Cavitation of the fuel inside the injector sac and nozzle holes can change the flow characteristics of the exiting fluid jet significantly. The vapor in the fluid jet is not randomly or evenly distributed. Will this initial vapor/fluid distribution affect breakup? Will it affect heat release or emissions formation significantly? These questions need answers.

From an engineering standpoint two important requirements for a model are that it is fast and reasonably simple to use. Also, the possibility to be as predictive as possible is of great importance. It is a severe drawback if spray characteristics must be measured for every design change on the injector nozzle, for example, in order to be able to predict a given output. In such a case, it is important to know which outputs, if any, can be predicted without measurements, using a simpler model instead. It is important to investigate the possibility to model the injector behavior to a certain necessary degree. If the model is too expensive to use, the spray characteristics may be measured for a group of injectors with wide parameter design space, and the database can be used as a lookup table for engineering simulations.

10.3.4.8 Combustion modeling

In diesel engine development, improved efficiency and low emissions of pollutants are mandatory. Combustion models used in the engine design should take into account the soot, UHC, CO, and NOx chemistry in order to capture the emissions accurately. The confidence of emission predictions vary with the type of pollutants. With the current models, NOx prediction has the highest accuracy, followed by soot prediction, CO prediction, and UHC prediction, in descending order. CO and UHC are typically not considered in diesel combustion simulations. To capture this level of detail, it is important to employ finite-rate chemistry in the CFD modeling. To this end, two important questions are: (1) What are the penalties for using reduced chemical kinetic models for diesel combustion instead of a detailed chemistry? (2) How can we run full detailed chemistry mechanisms in 3D CFD calculations with reasonable run times?

Turbulence–chemistry interaction (TCI) is known to play an important role in the RANS and LES modeling of turbulent combustion, yet it is one of the major challenging issues that are not well resolved. Flamelet models are frequently used in diesel engine design simulations. For example, in the development of the Volvo 2.4-liter diesel engine [94], a representative interactive flamelet model [100] was used. However, in many diesel engine combustion simulations direct coupling of the finite-rate chemistry is needed. In most of the modeling strategies directly coupling finite-rate chemistry, TCI is bypassed assuming the combustion is mainly driven by mixing. State-of-the-art TCI models coupling finite-rate chemistry exist, for example, the transported probability density function models; however, the computational time required for comprehensive TCI models is still unfavorably long. Development of cost-effective TCI models for diesel engine combustion simulations is needed.

10.3.5 Challenges in Diesel Combustion Measurements

Diesel engine measurements can be divided into two types: (1) measurement of overall engine performance including in-cylinder pressure and engine-out emissions and (2) measurements of spatial distribution and temporal evolution of charge composition, temperature, and velocity in the cylinder. While the former has been developed to certain degree of maturity, the latter is still a challenge and therefore will be discussed here. Detailed diesel engine measurement of scalar and vector distributions in the cylinder is challenging due to its high temperature and high pressure environment, limited optical access and the existence of reflective surfaces. Unlike open flames it is difficult to probe the gas in the engine or apply laser techniques. The following quantities are of special interest to measure: (1) wall temperature; (2) fuel distribution; (3) distribution of important species such as radicals, soot, and NOx; and (d) the velocity field. In particular, the distribution of soot in the cylinder and in the engine-out exhaust gas is of great interest to measure for the development of clean diesel engines.

10.3.5.1 Challenges in Soot Measurements

An engine-out soot measurement includes soot mass, particle number, and particle size data. A simplified and indirect way to measure soot is through the measurement of the filter smoke number (FSN) that may be further related to soot mass fraction with empirical correlations [101]. However, what exactly are we counting on particle counter? Do we have enough measurement data to develop predictive models? The soot modeling accuracy has been continuously improved using the detailed soot models; however, there are still some soot trends which the models could not capture. Understanding soot and UHC interaction will shed light on these trends.

Optical techniques have been used to detect where soot is formed and oxidized in the engine [102]. Examples of laser diagnostics methods for in-cylinder soot measurements are laser extinction method (LEM), planar laser-induced incandescence (PLII), and natural luminosity imaging (NL). LEM can provide quantitative soot extinction data with high temporal resolution; however, LEM is in general limited to measurements along a single line of sight [103]. LEM can also suffer from beam-steering effects [104]. Beam steering occurs due to gradients in refractive index orthogonal to the beam path, which are caused by in-cylinder density and species gradients. PLII can provide two-dimensional (2D) images of in-cylinder soot intersected by the plane of a laser sheet, but it is in general not quantitative [105]. Although PLII may be calibrated to provide quantitative measurements, calibration typically requires the use of another optical technique, such as LEM, simultaneously [106], which increases the complexity of the measurements. Furthermore, PLII has a slow repetition rate of the required high-pulse-energy lasers. Thus, development of high-pulse-energy lasers with high repetition rate (higher than of the current state-of-the-art 10 Hz) is needed to allow for capturing multiple images during one engine cycle. Multiple pulse laser technique may be an intermediate solution; however, with multiple pulses, soot distributions might be affected by previous PLII laser pulses, such that the technique cannot be considered nonintrusive. NL is a line-of-sight technique that images the natural luminosity of all the soot within the camera field of view. Modern cameras have rather high frame rate to enable a high temporal resolution. However, the natural luminosity of soot is a function of many variables, including both soot density and soot temperature. Because of variations of soot temperature along the line of sight, even with spectroscopic techniques such as the two-color method, these quantities are difficult, if not impossible, to separate, making NL a technique with large uncertainties, especially when used to determine the soot volume fraction [107].

Recently, a new 2D imaging technique, the so-called "diffused back-illumination" (DBI) imaging techniques of soot extinction, has been developed to provide quantitative images of soot along the line of sight [108]. The main benefit of the DBI technique is its ability to compensate for beam steering by using sufficiently diffuse illumination. DBI measures soot optical density with high temporal and spatial resolution. The optical density (KL) is the product of the mean-effective extinction coefficient, K, and the effective optical length L, along the line-of-sight soot volume. Lind et al. [102] applied the DBI technique together with a high-frame-rate NL imaging technique to study the mechanisms of soot oxidation in an optical engine modified based on the heavy-duty Cummins N-series engine. Figure 10.17 shows two images taken simultaneously using the NL and DBI techniques. The images were normalized with respect to the maximum KL and NL value during the whole cycle. As shown, one can notice the poor correlation between the local NL and KL intensities in the images. The poor correlation between KL and NL is because NL is a function of both soot density and soot temperature, whereas KL is mainly a function of soot volume fraction. The simultaneous imaging of NL and KL can thus characterize the variation of the soot temperature. Although the DBI technique offers a temporal resolution of soot distribution in the cylinder within one engine cycle, the technique cannot differentiate the spatial distribution of soot along the line of sight. Neither can the technique tell the soot particle size or the mass fraction distribution.

10.3.5.2 Challenges in Temperature and Heat Loss Measurements

In the field of measuring heat losses, local measurements are difficult but not impossible to do. Fast response thermocouples and phosphor thermometry are examples of techniques available for temperature measurements on the combustion chamber surface. These temperatures are the base for heat flux calculations. The local measurements should be made in many different points in order to get the complete picture for the system heat losses. Another approach for heat transfer measurement would be to measure overall heat flow in the engine coolants. This is hard to do, as heat travels well through the engine structure and might escape measurements. Also, it is impossible to understand where and when the heat transfer occurred with such an approach.

10.3.5.3 Challenges in Composition and Velocity Measurements

It is desirable to have simultaneous measurements of multiple scalars and velocity to understand the interaction between in-cylinder flow/turbulence and chemical reactions. Challenges in measurements of these quantities have been discussed in Section 2.3, where state-of-the-art techniques are discussed, for example, the tomographic particle image velocimetry (TPIV) technique and the simultaneous OH, CH$_2$O, and CH PLIF technique. These techniques have been mostly demonstrated for laboratory flames under atmospheric conditions. Further development for diesel engine conditions requires the techniques to be able to capture short time scales within an engine cycle, which is 0.17 ms per CAD for an engine speed of 1000 rpm, and to be able to cope with the complex geometry of the engine combustion chamber, e.g., complex shaped bowls and cylinder head. Optical piston bowls with real diesel engine bowl geometry have so far seldom been studied due to the image distortion by the complex curved bowl surface.

10.3.6 Concluding Remarks

Modern ICEs are obliged to address the public concern on pollutant and greenhouse gas emissions. Novel in-cylinder solutions are needed to achieve high engine efficiency and low pollutant emissions. This calls for deeper understanding of the physics of in-cylinder turbulent combustion processes. The design of modern diesel engines

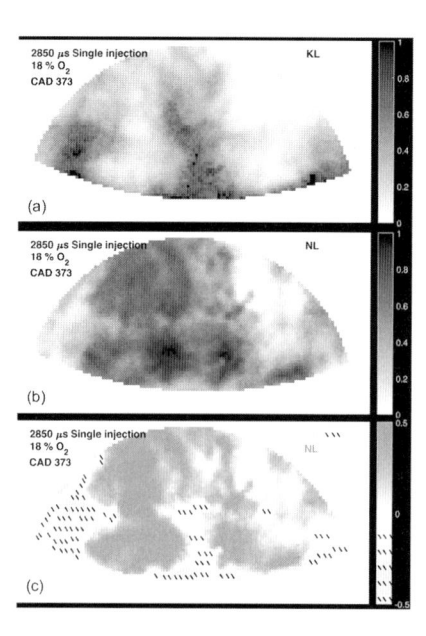

Figure 10.17 Normalized NL image (a), normalized KL image (b), and (c) composite image of the difference between images (a) and (b). Image timing at 373 CAD, courtesy of Ted Lind.

requires sophisticated CFD models that can simulate/predict the in-cylinder combustion process at an affordable cost. The combustion in ICE is a multiscale and multiphysics process, involving high-speed liquid fuel injection, three-dimensional turbulent swirl/tumble flow motion, liquid spray breakup, vaporization, fuel vapor mixing with ambient air, heat transfer, and turbulence–chemistry interaction. The physics of turbulent combustion in ICE is far from being understood. The high temperature and high pressure in-cylinder environment in the ICE gives rise to multiple combustion modes occurring simultaneously: spontaneous ignition, premixed, and diffusion flames. To understand the multiple combustion modes and their impact on engine performance, the sophisticated experimental methods discussed in Section 2.3 should be further developed for in-cylinder diagnostics, by addressing the challenges discussed in Section 2.3. The high-fidelity numerical methods (LES and DNS) discussed in Section 2.2 and Chapters 3 and 4 should be employed to study generic combustion processes that are directly relevant to ICEs, in order to gain detailed flow and scalar information to improve the understanding of the physics of turbulent combustion in ICEs. Based on the understanding of various combustion modes and reaction zone structures, cost-effective CFD models addressing the challenges discussed in Section 10.3.4 should be developed to help the design and development of next-generation IC engines and their optimization.

References

[1] N. Swaminathan and K. N. C. Bray, Eds., *Turbulent Premixed Flames*. Cambridge: Cambridge University Press. 2011.

[2] D. Dunn-Rankin (ed.), *Lean Combustion Technology and Control*. San Diego: Academic Press Academic/Elsevier, Elsevier. 2008.

[3] Y. A. Zeldovich, D. Frank-Kamenetskii, and P. Sadovnikov, *Oxidation of Nitrogen in Combustion*. Moscow: Academy of Sciences of USSR, 1947.

[4] A. H. Lefebvre, *Gas Turbine Combustion*, 2nd ed. London: Taylor & Francis.

[5] P. Glarborg, J. A. Miller, B. Ruscic, and S. J. Klippenstein, "Modeling nitrogen chemistry in combustion," *Prog. Energy Combust. Sci.*, **67**, 31–68, 2018.

[6] M. Andersson, A. A. Larsson, and A., A. M. Carrera, "Pentane rich fuels for standard Siemens DLE gas turbines," *ASME GT2011-46099*, 2011.

[7] G. Bulat, K. Liu, G. Brickwood, V. Sanderson, and B. Igoe, "Intelligent operation of Siemens (SGT-300) DLE gas turbine combustion system over an extended fuel range with low emissions," *ASME GT2011-46103*, 2011.

[8] D. Lörstad, J. Pettersson, and A. Lindholm, "Emission reduction and cooling improvements due to the introduction of passive acoustic damping in an existing SGT-800 combustor," *ASME GT2009-59313*, 2009.

[9] D. Lörstad, A. Lindholm, J. Pettersson, M. Björkman, and I. Hultmark, "Siemens SGT-800 industrial gas turbines enhanced to 50 MW: Combustor design modifications, validation and operation experience," *ASME GT2013-95478*, 2013.

[10] D. Lörstad, A. Ljung, and A. Abou-Taouk, "Investigation of Siemens SGT-800 industrial gas turbine combustor using different combustion and turbulence models," *ASME GT2016-57694*, 2016.

[11] M. Andersson, A. Larsson, A. Lindholm, and J. Larfeldt, "Extended fuel flexibility testing of Siemens industrial gas turbines: A novel approach," *ASME GT2012-69027*, 2012.

[12] D. Moëll, D. Lörstad, and X. S. Bai, "LES of hydrogen enriched methane/air combustion in the SGT-800 burner at real engine conditions," *ASME GT2018-76434*, 2018.

[13] A. Lantz, R. Collin, M. Aldén, A. Lindholm, J. Larfeldt, and D. Lörstad, "Investigation of hydrogen enriched natural gas flames in a SGT-700/800 burner using OH PLIF and chemiluminescence imaging," *J. Eng. Gas Turbines Power*, **137**, 031505-1–8, 2015.

[14] D. Lörstad, A. Lindholm, D. G. Barhaghi, A. Bonaldo, F. Fedina, C. Fureby, A. Lantz, R. Collin, and M. Aldén, "Measurements and LES of a SGT-800 burner in a combustion rig," *ASME GT2012-69936*, 2012.

[15] D. Moëll, D. Lörstad, and X. S. Bai, "Numerical investigation of hydrogen enriched natural gas in the SGT-800 burner," *ASME GT2015-44040*, 2015.

[16] F. R. Menter, "Two-equation eddy viscosity turbulence models for engineering applications," *AIAA J.*, **32**, 1598–1605, 1994.

[17] F. R. Menter, "Zonal two-equation k-omega turbulence models for aerodynamis flows," *AIAA Paper. 93-2906*, 1993.

[18] D. Moëll, D. Lörstad, and X. S. Bai, "Numerical investigation of methane/hydrogen/air partially premixed flames in the SGT-800 burner fitted to an atmospheric rig," *Flow Turbul. Combust.*, **96**, 987–1003, 2016.

[19] D. Flin, "Controlling combustion in gas turbines," *Power Eng. Int.* September 10–14, 2016.

[20] F. R. Menter, M. Kuntz, and R. Bender, "A scale adaptive simulation model for turbulent flow prediction," *AIAA Paper 2003-0767*, 2003.

[21] F. R. Menter and Y. Egorov, "A scale-adaptive simulation model using two-equation models," *AIAA Paper 2005-1095*, 2005.

[22] A. Abou-Taouk, D. Lörstad, N. Andersson, and L.-E. Eriksson, "CFD analysis and acoustic-mode identification of a SGT-800 burner in a combustion rig," *ASME GT2016-45853*, 2016.

[23] D. Moëll, D. Lörstad, A. Lindholm, D. Christensen, and X. S. Bai, "Numerical and experimental investigation of the Siemens SGT-800 burner fitted to a water rig," *ASME GT2017-64129*, 2017.

[24] C. D. Elvidge, D. Ziskin, K. E. Baugh, B. T. Tuttle, T. Ghosh, D. W. Pack, E. H. Erwin, and M. Zhizhin, "A fifteen year record of global natural gas flaring derived from satellite data," *Energies*, **2**, 595–622, 2009.

[25] J. M. Bergthorson, Y. Yavor, J. Palecka. W. Georges, M. Soo, J. Vickery, S. Goroshin, D. L. Frost, and A. J. Higgins, "Metal-water combustion for clean propulsion and power generation," *Appl. Energy*, **186**, 13–27, 2017.

[26] J. M. Rogero and P. A. Rubini, "Optimisation of Combustor Wall Heat Transfer and Pollutant Emissions for Preliminary Design Using Evolutionary Techniques," *Proceedings of ISOABE, 15th International Symposium on Airbreathing Engines*, September 2–7, Bangalore, India, 2001.

[27] A. P. Dowling and S. R. Stow, "Acoustic analysis of gas turbine combustors," *J. Prop. Power*, **19**, 751–764, 2003

[28] R. Adoua and G. J. Page, "Influence of turbulence modelling on swirl at a combustor exit," *ASME GT2016-57639*, 2016.

[29] D. Dunham, A. Spencer, J. J. McGuirk, and M. Dianat, "Comparison of URANS and LES CFD Methodologies for Air Swirl Fuel Injectors," *ASME GT2008-50278*, 2008.

[30] S. B. Pope, *Turbulent Flows*. New York: Cambridge University Press, 2000.

[31] M. Dianat, J. J. McGuirk, S. Fokeer, and A. Spencer, "LES of unsteady vortex aerodynamics in complex geometry gas-turbine fuel injectors," *ETMM10: 10th International ERCOFTAC Symposium on Engineering Turbulence Modelling and Measurements*, Marbella, Spain, September 17–19, 2014.

[32] M. S. Anand, R. Eggels, M. Staufer, M. Zedda, and J. Zhu, "An advanced unstructured grid finite volume design system for gas turbine combustion analysis," *ASME GTINDIA2013-3537*, 2013.

[33] W. J. S. Ramaekers, "Development of flamelet generated manifolds for partially-premixed flame simulations," Technische Universiteit Eindhoven, 2011.

[34] I. Langella, Z. X. Chen, N. Swaminathan, and S. K. Sadasivuni, "Large-eddy simulation of reacting flows in industrial gas turbine combustor," *J. Propul. Power*, **34**, 1269–1284, 2018.

[35] Z. X. Chen, N. Swaminathan, M. Stöhr, and W. Meier, "Interaction between self-excited oscillations and fuel-air mixing in a dual swirl combustor," *Proc. Combust. Inst.*, **37**, 2325–2333, 2019.

[36] I. Langella, J. Heinze, N. Swaminathan, T. Behrendt, L. Voigt and M. Zedda, "Turbulent flame shape switching at conditions relevant for gas turbines," *ASME Paper GT2019-91879*, 2019.

[37] A. H. Lefebvre and V. G. McDonnel, *Atomisation and Sprays*. Boca Raton, FL: CRC Press, 2017.

[38] Y. Hardalupas and S. Horender, "Phase Doppler anemometer for measurements of deterministic spray unsteadiness," *Particle and Particle Syst. Charact.*, **18**, 205–215, 2002

[39] S. Gepperth, D. Guildenbecher, R. Kock, and H. J. Bauer, "Pre-filming primary atomization: experiments and modelling," *ILASS – Europe 23rd Annual Conference on Liquid Atomization and Spray Systems*, Brno, Czech Republic, September 2010.

[40] E. De Villiers, A. D. Gosman, and H. G. Weller, "Large Eddy Simulation of primary diesel spray atomization," *SAE Technical Paper 2004-01-0100*, 2004.

[41] W. P. Jones, A. J. Marquis, and K. Vogiatzaki, "Large-eddy simulation of spray combustion in a gas turbine combustor," *Combust. Flame*, **161**, 222–239, 2013.

[42] K. M. Leung, R. P. Lindstedt, and W. P. Jones, "A simplified reaction mechanism for soot formation," *Combust Flame*, **87**, 289–305, 1991.

[43] A. Giusti. E. Mastorakos, H. Hassa, J. Heinze, E. Magens, and M. Zedda, "Investigation of flame structure and soot formation in a single sector model combustor using experiments and numerical simulations based on the large eddy simulation/conditional moment closure approach," *ASME GTP-17-1246*, also *J. Eng. Gas Turbines Power* **140**, 061506, 2018.

[44] C. Eberle, P. Gerlinger, and M. Aigner, "A sectional PAH model with reversible PAH chemistry for CFD soot simulations," *Combust. Flame*, **179**, 63–73, 2017.

[45] M. A. Macquisten, M. Whiteman, S. R. Stow, and J. A. Moran, "Exploitation of measured flame transfer functions for a two-phase lean fuel injector to predict thermoacoustic modes in full annular combustors," *ASME GT2014-25036*, 2014.

[46] A. Giusti, H. Zhang, A. M. Kypraiou, P. M. Allison, and E. Mastorakos, "Numerical investigation of the response of turbulent swirl non-premixed flames to air flow oscillations," *Proc. Combust. Inst.*, 2017.

[47] S. Ruan, T. Dunstan, N. Swaminathan, and R. Balachandran, "Computation of forced premixed flames dynamics," *Combust. Sci. Technol.*, **188**, 1115–1135, 2016.

[48] J. Su, A. Barker, A. Garmory, and J. Carrotte, "Spray response to acoustic forcing of a multi-passage lean-burn aero-engine fuel injector", *Proceedings of the ASME Turbo Expo 2018: Turbomachinery Technical Conference and Exposition. Volume 4A: Combustion, Fuels, and Emissions*. Oslo, Norway. June 11–15, 2018. V04AT04A040. ASME. https://doi.org/10.1115/GT2018-75554.

[49] S. Hochgreb, D. Dennis, I. Ayranci, W. Bainbridge, and R. S. Cant, "Forced and self-excited instabilities from lean premixed, liquid-fuelled aeroengine injectors at high pressures and temperatures," *ASME GT2013-95311*, 2013.

[50] S. R. Stow, M. Zedda, A. Triantafyllidis, A. Garmory, E. Mastorakos, and T. Mosbach, "Conditional moment closure LES modelling of an aero-engine combustor at relight conditions," *ASME GT2011-45100*, 2011.

[51] H. Zhang and E. Mastorakos, "Prediction of the global extinction conditions and dynamics in swirling non-premixed flames using LES/CMC modelling," *Flow Turbul. Combust*, **96**, 863–889, 2016.

[52] H. Zhang and E. Mastorakos, "LES/CMC modelling of a gas turbine model combustor with quick fuel mixing," *Flow Turbul. Combust*, **102**, 909–930, 2019.

[53] J. C. Massey, Z. X. Chen and N. Swaminathan, "Lean Flame Root Dynamics in a Gas Turbine Model Combustor," *Combust. Sci. Technol.*, **191**, 1019–1042, 2019.

[54] A. Neophytou, E. Mastorakos, E. S. Richardson, S. Stow, and M. Zedda, "A practical model for the high-altitude relight of a gas turbine combustor," *Seventh Mediterranean Combustion Symposium (MCS-7)*, 2011.

[55] S. Zabarnick, "Pseudo-detailed chemical kinetic modeling of antioxidant chemistry for jet fuel applications," *Energy Fuels*, **12**, (3), 547–553, 1998.

[56] E. Alborzi, S. Blakey, H. Ghadbeigi, and C. Pinna, "Prediction of growth of jet fuel autoxidative deposits at inner surface of a replicated jet engine burner feed arm," *Fuel*, **214**, 528–537, 2018.

[57] T. Pringuey and R. S. Cant, "Robust conservative level set method for 3D mixed-element meshes – application to LES of primary liquid-sheet breakup," *Com. Comput. Phys.*, **16**, 403–439, 2014.

[58] G. Ghiasi, I. Ahmed, and N. Swaminathan, "Gasoline flame behavior at elevated temperature and pressure," *Fuel*, **238**, 248–256, 2019.

[59] X. Zhang, X., D. J. J. Toal, N. W. Bressloff, A. J. Keane, F. Witham, J. Gregory, S. R. Stow, C. Goddard, M. Zedda, and M. Rogers, "Prometheus: a geometry-centric optimization system for combustor design," *ASME GT2014-25886*, 2014.

[60] J. E. Dec, "A conceptual model of DI diesel combustion based on laser-sheet imaging," in *SAE Technical Paper Series*. SAE International, February 1997.

[61] J. E. Dec, "Advanced compression-ignition engines—understanding the in-cylinder processes," *Proc. Combust. Inst.*, **32**, 2727–2742, 2009.

[62] C. Gong, M. Jangi, and X.-S. Bai, "Large eddy simulation of n-dodecane spray combustion in a high pressure combustion vessel," *Appl. Energy*, **136**, 373–381, 2014.

[63] S. Onishi, S. H. Jo, K. Shoda, P. D. Jo, and S. Kato, "Active thermo-atmosphere combustion (ATAC) - a new combustion process for internal combustion engines," in *SAE Technical Paper Series*. SAE International, February 1979.

[64] P. M. Najt and D. E. Foster, "Compression-ignited homogeneous charge combustion," in *SAE Technical Paper Series*. SAE International, feb 1983.

[65] R. H. Thring, "Homogeneous-charge compression-ignition (HCCI) engines," in *SAE Technical Paper Series*. SAE International, September 1989.

[66] M. Yao, Z. Zheng, and H. Liu, "Progress and recent trends in homogeneous charge compression ignition (HCCI) engines," *Prog. Energy Combust. Sci.*, **35**, 398–437, 2009.

[67] R. Sankaran, H. G. Im, E. R. Hawkes, and J. H. Chen, "The effects of non-uniform temperature distribution on the ignition of a lean homogeneous hydrogen–air mixture," *Proc. Combust. Inst.*, **30**, 875–882, 2005.

[68] J. H. Chen, E. R. Hawkes, R. Sankaran, S. D. Mason, and H. G. Im, "Direct numerical simulation of ignition front propagation in a constant volume with temperature inhomogeneities," *Combust. Flame*, **145**, 128–144, 2006.

[69] R. Yu and X.-S. Bai, "Direct numerical simulation of lean hydrogen/air auto-ignition in a constant volume enclosure," *Combust. Flame*, **160**, 1706–1716, 2013.

[70] T. Joelsson, R. Yu, and X. S. Bai, "Large eddy simulation of turbulent combustion in a spark-assisted homogenous charge compression ignition engine," *Combust. Sci. Technol.*, **184**, 1051–1065, 2012.

[71] C. S. Yoo, T. Lu, J. H. Chen, and C. K. Law, "Direct numerical simulations of ignition of a lean n-heptane/air mixture with temperature inhomogeneities at constant volume: Parametric study," *Combust. Flame*, **158**, 1727–1741, 2011.

[72] F. Zhang, H. Liu, R. Yu, M. Yao, and X. S. Bai, "Direct numerical simulation of h2/air combustion with composition stratification in a constant volume enclosure relevant to HCCI engines," *Int. J. Hydrogen Energy*, **41**, 13758–13770, 2016.

[73] R. D. Reitz and G. Duraisamy, "Review of high efficiency and clean reactivity controlled compression ignition (RCCI) combustion in internal combustion engines," *Prog. Energy Combust. Sci.*, **46**, 12–71, 2015.

[74] S. L. Kokjohn, R. M. Hanson, D. A. Splitter, and R. D. Reitz, "Fuel reactivity controlled compression ignition (RCCI): a pathway to controlled high-efficiency clean combustion," *International J. Engine Res.*, **12**, 209–226, 2011.

[75] G. T. Kalghatgi, P. Risberg, and H.-E. Angstrom, "Partially pre-mixed auto-ignition of gasoline to attain low smoke and low NOx at high load in a compression ignition engine and comparison with a diesel fuel," in *SAE Technical Paper Series*. SAE International, January 2007.

[76] V. Manente, B. Johansson, and P. Tunestal, "Partially premixed combustion at high load using gasoline and ethanol, a comparison with diesel," in *SAE Technical Paper Series*. SAE International, April 2009.

[77] C. Yao, C. Cheung, C. Cheng, Y. Wang, T. Chan, and S. Lee, "Effect of diesel/methanol compound combustion on diesel engine combustion and emissions," *Energy Conv. Manage.*, **49**, 1696–1704, 2008.

[78] A. B. Dempsey, N. R. Walker, and R. D. Reitz, "Effect of piston bowl geometry on dual fuel reactivity controlled compression ignition (RCCI) in a light-duty engine operated with gasoline/diesel and methanol/diesel," *SAE Int. J. Engines*, **6**, 78–100, 2013.

[79] Y. Li, M. Jia, Y. Chang, Y. Liu, M. Xie, T. Wang, and L. Zhou, "Parametric study and optimization of a RCCI (reactivity controlled compression ignition) engine fueled with methanol and diesel," *Energy*, **65**, 319–332, 2014.

[80] S. Hu, C. Gong, and X.-S. Bai, "Dual fuel combustion of n-heptane/methanol-air-EGR mixtures," *Energy Procedia*, **105**, 1913–1918, 2017.

[81] R. Hanson, D. Splitter, and R. D. Reitz, "Operating a heavy-duty direct-injection compression-ignition engine with gasoline for low emissions," in *SAE Technical Paper Series*. SAE International, April 2009.

[82] F. Zhang, R. Yu, and X. S. Bai, "Detailed numerical simulation of syngas combustion under partially premixed combustion engine conditions," *International Journal of Hydrogen Energy*, **37**, 17 285–17 293, 2012.

[83] F. Zhang, R. Yu, and X. S. Bai, "Direct numerical simulation of PRF70/air partially premixed combustion under IC engine conditions," *Proc. Combust. Inst.*, **35**, 2975–2982, 2015.

[84] F. Zhang, R. Yu, and X. S. Bai, "Effect of split fuel injection on heat release and pollutant emissions in partially premixed combustion of PRF70/air/EGR mixtures," *Appl. Energy*, **149**, 283–296, 2015.

[85] V. Manente, B. Johansson, and P. Tunestal, "Characterization of partially premixed combustion with ethanol: EGR sweeps, low and maximum loads," *J. Eng. Gas Turbines and Power*, **132**, p. 082802, 2010.

[86] N. Swaminathan and K. Bray, Eds., *Turbulent Premixed Flames*. Cambridge: Cambridge University Press, 2009.

[87] S. Shuai, X. Ma, Y. Li, Y. Qi, and H. Xu, "Recent progress in automotive gasoline direct injection engine technology," *Autom. Innov.*, **1**, 95–113, 2018.

[88] M. P. Musculus, P. C. Miles, and L. M. Pickett, "Conceptual models for partially premixed low-temperature diesel combustion," *Prog. Energy Combust. Sci.*, **39**, 246–283, 2013.

[89] P. C. Miles, "Turbulent flow structure in direct-injection, swirl-supported diesel engines," in *Flow and Combustion in Reciprocating Engines*, 173–256. Berlin and Heidelberg: Springer, 2008.

[90] Z. Wang, P. Stamatoglou, M. Lundgren, L. Luise, B. M. Vaglieco, A. Andersson, M. Aldén, Öivind Andersson, and M. Richter, "Simultaneous 36kHz PLIF/chemiluminescence imaging of fuel, CH_2O and combustion in a PPC engine," *Proc. Combust. Inst.*, **37**, 4751–4758, 2019.

[91] K. M. Lyons, "Toward an understanding of the stabilization mechanisms of lifted turbulent jet flames: Experiments," *Prog. Energy Combust. Sci.*, **33**, 211–231, 2007.

[92] G. Lequien, Öivind Andersson, P. Tunestal, and M. Lewander, "A correlation analysis of the roles of soot formation and oxidation in a heavy-duty diesel engine," in *SAE Technical Paper Series*. SAE International, October 2013.

[93] M. Jangi, T. Lucchini, G. D'Errico, and X.-S. Bai, "Effects of EGR on the structure and emissions of diesel combustion," *Proc. Combust. Inst.*, **34**, 3091–3098, 2013.

[94] O. Andersson, J. Somhorst, R. Lindgren, R. Blom, and M. Ljungqvist, "Development of the euro 5 combustion system for volvo cars 2.4.i diesel engine," in *SAE Technical Paper Series*. SAE International, April 2009.

[95] H. Persson, A. Babajimopoulos, A. Helmantel, F. Holst, and E. Stenmark, "Development of the combustion system for volvo cars euro6d VEA diesel engine," in *SAE Technical Paper Series*. SAE International, March 2017.

[96] H.-W. Ge, Y. Shi, R. D. Reitz, D. Wickman, and W. Willems, "Engine development using multi-dimensional CFD and computer optimization," in *SAE Technical Paper Series*. SAE International, April 2010.

[97] M. Linne, "Imaging in the optically dense regions of a spray: A review of developing techniques," *Prog. Energy Combust. Sci.*, **39**, 403–440, 2013.

[98] L. M. Pickett, J. Manin, C. L. Genzale, D. L. Siebers, M. P. B. Musculus, and C. A. Idicheria, "Relationship between diesel fuel spray vapor penetration/dispersion and local fuel mixture fraction," *SAE Int. J. Engines*, **4**, 764–799, 2011.

[99] Y. Pei, E. R. Hawkes, and S. Kook, "Transported probability density function modelling of the vapour phase of an n-heptane jet at diesel engine conditions," *Proc. Combust. Inst.*, **34**, 3039–3047, 2013.

[100] H. Pitsch, H. Barths, and N. Peters, "Three-dimensional modeling of NOx and soot formation in DI-diesel engines using detailed chemistry based on the interactive flamelet approach," in *SAE Technical Paper Series*. SAE International, October 1996.

[101] A. Dodd and Z. Holubecki, *The measurement of diesel exhaust smoke*. Motor Industry Research Association, 1965.

[102] T. Lind, G. Roberts, W. Eagle, C. Rousselle, Öivind Andersson, and M. P. Musculus, "Mechanisms of post-injection soot-reduction revealed by visible and diffused back-illumination soot extinction imaging," in *SAE Technical Paper Series*. SAE International, April 2018.

[103] Z. Li, Y. Gallo, T. Lind, O. Andersson, M. Alden, and M. Richter, "Comparison of laser-extinction and natural luminosity measurements for soot probing in diesel optical engines," in *SAE Technical Paper Series*. SAE International, October 2016.

[104] M. Musculus and L. Pickett, "Diagnostic considerations for optical laser-extinction measurements of soot in high-pressure transient combustion environments," *Combust. Flame*, **141**, 371–391, 2005.

[105] T. Lind, Z. Li, C. Micó, N.-E. Olofsson, P.-E. Bengtsson, M. Richter, and O. Andersson, "Simultaneous PLIF imaging of OH and PLII imaging of soot for studying the late-cycle soot oxidation in an optical heavy-duty diesel engine," *SAE Int. J. Engines*, **9**, 849–858, 2016.

[106] L. de Francqueville, G. Bruneaux, and B. Thirouard, "Soot volume fraction measurements in a gasoline direct injection engine by combined laser induced incandescence and laser extinction method," *SAE Int. J. Engines*, **3**, 163–182, 2010.

[107] M. P. Musculus, S. Singh, and R. D. Reitz, "Gradient effects on two-color soot optical pyrometry in a heavy-duty DI diesel engine," *Combust. Flame*, **153**, 216–227, 2008.

[108] J. Manin, L. M. Pickett, and S. A. Skeen, "Two-color diffused back-illumination imaging as a diagnostic for time-resolved soot measurements in reacting sprays," *SAE Int. J. Engines*, **6**, 1908–1921, 2013.

11 Closing Remarks

X.-S. Bai, N. E. L. Haugen, C. Fureby, G. Brethouwer, and N. Swaminathan

This book is an outcome of the workshop on Physics of Turbulent Combustion that took place in September 2016 at the Nordic Institute for Theoretical Physics (Nordita) in Stockholm, jointly sponsored by Nordita and the Combustion Institute. During the month-long workshop, experts in turbulent combustion from academia and industry worldwide gathered together in an attempt to identify the most significant scientific and engineering issues toward understanding and predictive modeling of turbulent combustion phenomena. This was done with a particular focus on the increasing demand in fuel diversification and combustion at extreme conditions in favor of clean and efficient energy production. The workshop involved leaders within industrial research and development (R&D) and academic experts in combustion theory, direct numerical simulation (DNS), large-eddy simulation (LES), mathematical modeling, combustion kinetics, multi-phase combustion, and laser diagnostics. On identification of the targeted grand challenge problems, the enabling key scientific issues, such as structures and propagation of turbulent premixed and non-premixed flames, ignition/extinction, supersonic combustion, solid and liquid fuel combustion, chemical kinetic mechanism, and diagnostic developments were discussed. This forms the basis of the present book, *Recent Advances in Turbulent Combustion*.

Turbulent combustion is the fundamental process in natural disastrous fire, in combustion engines that power road vehicles, airplanes, space transportation vehicles, and ships, as well as in power plants and numerous industrial sectors. It is also the key process in supernovae explosions and thermonuclear fusion. As of today, about 90% of the worlds primary energy supply comes from combustion of a wide range of fuels, the majority of them being fossil based. Despite the recent promotion of electric road vehicles and airplanes the situation is expected to remain similar in future decades. The concern on greenhouse gas (CO_2) emission and emission of pollutants (such as NOx and soot) from combustion calls for the development of novel clean and efficient combustion devices. The success of the development of novel combustion devices largely relies on the progress of fundamental knowledge on turbulent combustion and on the development of accurate, reliable, and predictive computational fluid dynamics (CFD) modeling tools that can be efficiently used in different stages of the design process. As of today, CFD predictions can achieve an acceptable accuracy for the flow distribution, fuel–air mixing, flame-front positions, and heat load distribution. However, CFD predictions of key details such as NOx/CO, soot and unburnt hydrocarbon emissions, burnout of solid fuels, thermoacoustic oscillations, and flame dynamics, including forced and autoignition, as well as lean blowout, are still not reliable enough to be fully included in the design iterations.

The challenge in predictive CFD design is not only in the accuracy of the prediction but also with respect to efficiency and robustness. The physics of turbulent combustion is essentially a manifestation of the interaction between turbulence, chemical reactions, heat transfer, and acoustics, when appropriate. For combustion of liquid or solid fuels, the interaction between turbulence and the fuel (particles or droplets) is an additional complication that must be understood. Turbulence itself is a challenging research area and over the past few decades significant developments in experiments and high-fidelity modeling and simulations have been made. In combustion, different elementary reaction steps occur over a range of time scales, and when coupled with turbulence they occur also in different spatial locations and a range of length scales. Depending on the mixture composition, temperature, pressure, and flow conditions, a number of different combustion phenomena, such as ignition, quenching, deflagration and detonation wave propagation, formation of pollutants, etc., ensue. In principle, all these details, regardless of combustion modes and length/time scales, can be described mathematically using second-order partial differential equations that are derived from mass conservation, momentum conservation, and energy conservation, with constitutive relations for the thermodynamic quantities, as well as data for the transport properties, and chemical kinetic mechanisms describing the paths and rates of chemical reactions (discussed in Chapter 1). These equations can be solved using DNS methods with supercomputers. However, the formidably long computational time for DNS is far beyond the design time allowed for industrial product development, and as a result, industrial design CFD tools are based on Reynolds-averaged Navier–Stokes (RANS) and LES models. The latter is becoming more popular as a compromise between the accuracy of the prediction and the efficiency of the simulation and is being used when the dynamics of the flame and flow is of concern.

The desire of having efficient and predictive CFD modeling tools has resulted in the development of various models for turbulent combustion, which often target specific combustion process, e.g., flamelet combustion, diffusion flames, premixed flames, MILD combustion, solid or liquid fuel conversion, and supersonic combustion. Recent progress in these model development and validation has been the focus of this book. Significant progress in model development and validation has recently been made due to the availability of high-fidelity experimental and DNS data. With DNS data it is now possible to evaluate the individual terms in the models a priori for careful assessment and validation of the models. When validating different turbulent combustion models it is important to compare different models and CFD codes for a single target flame. This is practiced very successfully in many national and international research programs.

Perhaps the most impressive recent progress in turbulent combustion has been made in the area of DNS. Due to the recent progress of supercomputers, DNS is currently being used in a number of research groups to study many challenging combustion process, e.g., high Karlovitz number premixed flames, moderate, intense, or low dilution (MILD) combustion, lifted jet flames, near-wall combustion, solid fuel combustion, and combustion processes relevant to new conceptual internal combustion engines, such as homogeneous charge compression ignition (HCCI) and partially premixed combustion (PPC) engines. The spatially and temporally fully resolved DNS data allow for the understanding and analysis of the fine detailed species distribution, which includes complex structures existing in the fine scales that are often not resolved in today's experiments. As an example, DNS has revealed the existence of a penta-brachial flame structure in lifted DME/air non-premixed jet flames at elevated temperatures and pressures, including a low-temperature combustion branch, a triple-flame branch, and a high-temperature branch upstream of the triple flames. It is expected that DNS will play a more important role in understanding the physics of turbulent combustion and in providing a complete dataset for model development and validation. The limitation of DNS is the long computational time required for high Reynolds number flames, which requires continued effort in the development of computer software and hardware, and in the optimization of computer codes, e.g., adaptive mesh refinement methods optimized for the underlying hardware. A DNS is also limited by the accuracy and efficiency of the chemical kinetic mechanisms used in the simulations. Continued development of chemical kinetic mechanisms under challenging conditions, e.g., at high pressures and temperatures, are needed (Chapter 5). In particular, there is a need for highly accurate and robust skeletal reaction mechanisms that can be used in DNS and LES simulations.

DNS provides a big data source that contains detailed information of the flame physics. However, analysis of the data is a challenging task. Methods developed in the machine learning (ML) area offer a powerful way to tackle complex problems. The deep learning method based on deep neural networks (DNNs), which is a subset of ML methods, has recently attracted the attention for their superior performance across a wide range of real-life challenging tasks, including voice recognition, video classification, and playing Go. In the past few years, there has been a surge of interest in applying ML methods for (nonreacting) turbulence modeling, both in the RANS framework and LES framework. Comparatively, there is less activity in applying ML methods to turbulent combustion problems. It is expected that there will be more effort in applying ML methods to the development and optimization of turbulent combustion modeling in the near future.

Recent development in experimental diagnostic methods has been truly impressive. Multiple-scalar simultaneous laser-induced fluorescence (LIF) imaging of OH, CH, HCO, CH_2O, temperature, and fuel tracer (four of these can be measured simultaneously) allows for the identification of the flame structures in experiments. Three-dimensional measurements of the velocity and scalars, and high-speed (100 kHz burst laser system synchronized with high-speed CMOS cameras) are other examples of new development that can capture unprecedented details of the evolution of individual eddies in turbulent flames. In the future, it is desirable that further development of quantitative measurements of species and high spatial resolution of flame structures shall be carried out, which will give insights into further details of the combustion and emission processes in turbulent flames, and provide data for model development and validation.

It is the objective of the contributors to this book that the material presented here will provide an overall broader picture of the field, from challenges in industrial design considerations, challenges in experimental measurements and numerical simulations, to the state-of-the-art CFD modeling of turbulent combustion. It is our wish that this will motivate further studies in this rapidly evolving socioeconomically important field of research.

Index

absorption coefficient, 376
alignment
  flame-strain, 70, 110
  scalar-strain, 72
  vorticity-strain, 72
analysis
  cave-to-case, 6
  cradle-to-grave, 4
  life cycle (LCA), 4
Arrhenius relation, 13, 204, 289
atomization, 407, 429, 433
  primary breakup, 331, 433
  secondary breakup, 331, 417, 433
atomization regime, 331
autocorrelation, 139

beam-steering, 77
Beta-PDF, 125
biofuel, 408
black-body radiation, 376, 378
BML, 118
  bimodal PDF, 120
  burning mode PDF, f, 120
Brayton cycle, 415

cavitation, 330, 448
CDC (conventional diesel combustion), 436
CDF (cummulative distribution function), 125
CFD, 411
Chapman–Jouguet, 281, 291
chemical kinetics, 200
chemical mechanism
  alcohols, 220
  complete, 213
  detailed, 213
  global, 217
  liquid fuel, 220
  natural gas, 218
  reduced, 215
  skeletal, 215
  stiffness, 218
closed external combustion, 361
CMC, 121, 177, 268
  e_Q term, 122

combustion
  chemistry, 202
  flameless, 240
  HiTAC, 242
  micro-volume, 32
  MILD, 8, 38–40, 44, 50, 240, 244
  non-premixed, 9, 25, 162, 167
  partially premixed, 9, 25, 129, 162, 169
  premixed, 9, 25, 100, 166
  stratified, 169, 185
  supersonic, 281
combustion regime, 28, 46, 102
  broken reaction zones, 32, 56
  corrugated flamelets, 29, 48, 102
  distributed combustion, 32, 50, 102
  distributed flamelets, 30, 32, 103
  distributed reaction zones, 32, 51, 56, 61
  non-premixed, 163
  partially premixed, 163
  thin reaction zones, 31, 51, 56, 102
  wrinkled flamelets, 29, 48, 102
combustion waves, 284
combustor
  aero engine, 407, 408
  stationary GT, 396
combustor development, 400
compressibility, 309
conditional diffusion, 116
conditional mean, 122
conduction, 385
conductivity, 379, 383
cooling flow, 411
counter-gradient flux, 18
Counterflow Flame, 248
counterflow flame, 53
critical combustion, 353
critical group combustion, 357
cutoff scale, 108
cycle-to-cyle variation, 447

d-square law, 339
Darcy's law, 382
DDT, 282, 297
deflagration, 282, 380

Index

deflagration to detonation transition, DDT, 375
design method
   CHT, conjugate heat transfer, 424
   combustor, 399
   conjugate heat transfer (CHT), 424
   fuel injector, 413
   high-order method, 411
   LES, 400, 401, 414, 421
   low-order model, 401, 410
   network model, 401, 410, 411, 413, 416
   RANS, 400, 401, 412
   transported PDF, 401
   uRANS, 401, 413
detonation, 26, 302, 375
diffusion velocity, 10
diffusivity, 379, 382
   thermal, 100
dilute regime, 331
direct numerical simulation (DNS), 14, 44, 257
   canonical configuration, 45
   complex flames, 51
   exascale, 52
   forcing, 46
   numerical experiment, 44
   petascale, 52
   small-eddy simulation, 47
disperse phase, 329
displacement speed, 107, 112
   LES, 113
   PDF, 146
drag
   particle, 368, 389
drag coefficient, 347
drag factor, 347
droplet number density, 353
droplet relaxation time scale, 349
droplet size distribution, 332
dry low emission (DLE), 397

eddy breakup model, 103
eddy diffusivity, 18
eddy dissipation concept (EDC), 370
eddy dissipation model (EDM), 104, 174, 267, 269, 370
effectiveness factor, 385
efficiency, combustion, 407
electric vehicle, 5
emission, 407, 419, 433
   CO, 201, 398
   GHG, 2, 8, 201
   greenhouse gases (GHGs), 1
   NOx, 9, 201, 240, 271, 397, 439
   prompt NO_x, 419
   soot, 439
   thermal NOx, 419
emission index, EI, 419

emissivity, 378
energy dissipation rate, 387
engine
   GDI, 439
   HCCI, 25, 437, 439
   IC, 436
   PCCI, 438
   PPC, 438
   RCCI, 438
   SACI, 437
   SI, 437
engine envelope, 408
enthalpy, 382
equation
   aerothermodynamic, 11
   calorific state, 12
   constitutive, 11
   G, 111
   governing, 10, 284
   ideal gas, 12
   state, 12, 282
Eulerian–Eulerian approach, 336, 382
exhaust gas recirculation (EGR), 242, 260, 271, 438, 440
explosion
   dust, 374, 375, 378, 379
      primary, 375
      secondary, 375, 378–380
   vapor cloud, 375
explosions
   dust, 378
external group combustion, 358
extinction, 212, 430

fall-off region, 205
fast chemistry, 104
Favre-filterting, 15
FDF, 428
filtering, 15
firedeck, 439, 441
flame
   Burke–Schumann sheet, 26
   lifted jet, 129
   liftoff height, 436, 442
   marker, 265
flame curvature, 107
flame front, 126
flame thickness, 45, 379
flamebrush, 126
flamelets, 9, 56, 165, 171, 201, 267, 404
flameout, 407
flue gas recirculation (FGR), 242
fluidized bed, 383
Fourier transform, 139
Froude number, 389

Index

FSD, 32, 56, 145
  algebraic model, 108
  approach, 106
  generalized, 109
  realizability, 109
  transport equation, 109
FSD approach, 106
FTF, 411, 427
fuel coking, 431
fuel injector, 411, 413

G-equation, 111, 149, 167
  distance function, 111, 114
  filtered, 113
  level-set, 112, 145
  PDF, 114
  reinitialisation, 114
gas turbine engines, 271
geometrical model, 106
global warming, 1
granulometry, 328
green energy, 406
group behavior, 351
group combustion, 351, 353, 354, 358, 360
  external, 356
  internal, 357
group number, 354–356

high-speed flows, 282
hybrid combustion, 362

ignition, 378, 407
  chemistry, 207
  kernel, 378, 380
  multi-point, 378–380
ignition delay, 25, 208, 210
incipient droplet combustion, 358
incipient group combustion, 354, 357
internal combustion engine, 271
internal group combustion, 353, 358
isolated droplet combustion, 354

jump condition, 337

kerosene, jet A1, 408
Klimov–Williams criterion, 29
Knudsen number, 383

Lagrangian equation, 339
Large Eddy Simulations (LES), 370
Laser Doppler Velocimetry, 254
late-cycle oxidation, 441
LCOE, 2
length scale
  Batchelor, 142
  Corrsin–Obukhov, 137

  integral, 26, 45
  Kolmogorov, 26
LES, 15, 270, 447
level-set, 166, 337
linear eddy model, 105, 179
  triplet map, 105
liquid combustion, 328
liquid fuel, 328, 448

Mach number, 281
machine learning, 405
mass loading, 358
mean-field approach, 376
metal temperature, 423
micromixing, 34, 116
  frequency, 117
micromixing model
  AMC, 117
  binomial Langevin, 116
  CD, 116
  EMST, 116
  LMSE, 116
Minkowski functionals, 261
mixture fraction, 26, 168
multienvironment PDF, 118

Navier–Stokes equations, 386
NGV, 416, 418
NTC region, 208
number
  CFL, 403
  Damköhler, 26, 102–104, 106, 111, 126, 140, 147, 164
  Damköhler number, particle clustering, 373
  Damköhler, ignition-delay, 33
  Damköhler, micromixing, 34
  Damköhler, thermal, 380
  Karlovitz, 27, 29, 47, 55, 102, 124, 126, 137, 140, 147, 431
  Lewis, 49, 101, 111, 122, 128, 246
  Mach, 379
  Nusselt, 345, 379
  Ohnesorge, 331
  Reynolds, 328
  Schmidt, 371
  Sherwood, 345, 346, 371
  smoke, 423
  Stokes, 371, 390
    particle, 389
    vaporization, 357
  Weber, 328, 331

OH* chemiluminescent, 256
OH-PLIF, 254
open external combustion, 361, 363

Index

partial group combustion, 353
partially stirred reactor (PaSR), 268
particle
   1-way coupling, 387
   2-way coupling, 388, 390
   4-way coupling, 388, 390
   Biot number, 380, 385
   boundary layer, 383
   cluster, 373, 391
   clustering, 374–376, 378–380, 389–391
   conductivity, 381
   heat transfer, 370, 380, 381
   mass transfer, 370, 374, 380
   Nusselt number, 379
   relaxation time, 389
   response time, 368
   Sherwood number, 371
      modified, 374
   specific heat, 381
   Stefan flow, 383
   Stokes number, 368, 371, 373
      viscous, 390
   transport, 391
particle scattering coefficient, 376
particle-source-in-cell, 350
particulate matter, 439
PaSR, 40, 104, 175
payback period, 3
percolation combustion regime, 360
perfectly stirred reactor (PSR), 40, 267
permeability, 382
photovoltaic, 2
PIV, 414
   stereo, 65
   tomographic, 65, 151
PLIF, 57
plug flow reactor (PFR), 250, 268
pocket combustion regime, 360
POD, 415
prediffuser, 412
preferential diffusion, 270
premixed flame
   Mallard and Le Chattlier analysis, 100
   preheat zone, 101
   reaction zone, 101, 102
   speed, S_L, 8, 100, 101, 164
   thickness, 100, 102
   time scale, τ_c, *see also* time scale
   turbulent mixing, 133
   wave propagation speed, 100
premixed flame mode, 359
pressure-dilatation work, 138
probability density function, 370
progress variable, 36, 101, 128, 130
   FDF, 115
   PDF, 115

propagation
   deflagration, 25
   detonation, 26
   flame, 25, 210
   ignition wave, 25
pseudo-shock, 302
PSI, 350
PSR, 104, 269, 316
pullaway, 430
pulverized burner, 383

radiation, 375, 380, 381, 383
   thermal, 378
radiation absorption length, 376
radiative flux, 378
radiative heat transfer, 375
radiative transfer equation, 376
Rayleigh criterion, 404
Rayleigh-Taylor instability, 331
reaction
   bimolecular, 204
   chain branching, 58, 203, 210
   chain carrying, 211
   chain terminating, *see* recombination
   elementary, 203
   order, 204
   recombination, 58, 211
   three-body, 204
   unimolecular, 204
reaction rate
   averaged, 118
   filtered, 118, 127
relight, 430, 433
Reynolds-averaged Navier–Stokes (RANS), 19, 267, 370, 447
RQL, 419, 422
RTDF, 416

Sauter mean diameter, 334, 417
scalar spectrum, 139
   inertial-convective range, 140, 141
   inertial-convective-reactive range, 141
   inertial-diffusive range, 140, 141
   stationary, 139
   viscous-diffusive range, 142
scramjet, 282
SDF, 106, 107
SDR (scalar dissipation rate), 108, 137
   algebraic model, 124
   bridging model, 117
   conditional, 117, 123, 178
   mixture fraction, 27
   resolved, 108
   SGS, linear relaxation model, 108, 123, 129, 130
   unconditional, 123

separation parameter, 356
SGS closures, 17
sheath combustion, 353, 356, 357
shrinking core model, 385
single droplet combustion, 354, 358
SKYACTIV, 439
solar, 2
solid combustion, 367
solid fuel, 367
   biomass, 380, 383
      wood, 367, 381, 383
         wood logs, 382
         wood pellets, 383
   char, 369, 385
   char conversion, 368, 371, 381–383
   char oxidation, 9
   coal, 367, 380, 383
   coke, 367
   deformation, 380
   devolatilization, 368, 369, 381
   drying, 368
   fragmentation, 380
   gasification, 369, 380
   mass transfer, 381
   municipal solid waste, 367
   peat, 367
   porosity, 382
   pyrolysis, 383, 385, 386
   secondary char formation, 383
   shrinking, 380
   swelling, 380
   toruosity, 383
soot, 391, 421, 445, 450
Spalding coefficient, 343
spectral analysis, 138
spectral flux, 139
spectral transfer, 139
speed of sound, 281
spray, 433
spray boundary condition, 417
spray combustion regime, 354
squish, 440
squish land, 444
stability limits, 407
statistical model, 115
Stefan flux, 341
strain rate
   normal additional, a_N^a, 135
   normal, a_N, 134, 135
   tangential additional, a_T^a, 135
   compressive, 110
   curvature induced, 109
   extensive, 111
   principal, 110
   tangential, 107, 109, 110
stress tensor, 386
stretch factor

flame, φ, 29
   Karlovitz, K, 29
stretch rate, 107
   flame, 29
   negative, 110
   surface-averaged, 109
subject, 372
surface
   isoscalar, displacement speed, 135
   isoscalar, evolution, 134
   isoscalar, flow strain, 135
   material, 110
   passive, 110
surface average, 107
swirl, 441
swirl motion, 440
system aerodynamics, 411

tabulated chemistry approach, 121, 125, 269, 417
   FGM, 125, 270
   FlaRe, 126, 150
   FPV, 270
   strained flamelet, 126
   unstrained flamelet, 126
technology readiness level (TRL), 408
temperature traverse, 415
thermal barrier coating, TBC, 425, 447
thermal choking, 303, 306
thermoacoustics, 404, 407, 426
   HFR, 429
   LFR, 429
   limit cycle, 426
   resonance, 397
   Rumble, HFR, 427
   Rumble, LFR, 426
   self-excited, 397
thickened flame model, 104, 132, 173
Thiele modulus, 354
time scale
   chemical, 137, 139, 147
   chemical or flame, 102, 126
   flame, τ_c, 100
   ignition, 35
   Kolmogorov, 27, 34, 102, 137, 142, 267
   mixing, 34, 116, 117, 137, 144
   SGS, 104
   TKE transfer, 148
   turbulence integral, 102, 104
TKE (turbulent kinetic energy), 138
topland volume, 440
transported PDF, 115, 176, 268
   Eulerian stochastic fields, 117, 133, 176, 418, 421, 423
   Monte Carlo method, 116, 117
TTF, 427
turbine inlet temperature (TIT), 397, 403

turbophoresis, 389, 390
turbulence model
    DES, detached eddy simulation, 403
    k-ω SST, 403, 413, 425
    k-ε, realizable, 412
    RSM, Reynolds stress model, 413
    SAS, scale adaptive simulation, 403, 425
turbulence modulation, 329
turbulence–chemistry interaction, TCI, 222, 314, 449
turbulence–particle resonance scale, 372, 374
turbulence–scalar interaction, 110
two-fluid framework, 336
two-phase flow, 409, 431
two-point correlation function, 377

vaporization mode, 359
variable valve timing, 439
variance
    averaged (RANS), 128
    resolved, 128
    SGS, 103, 127, 128
    SGS, algebraic model, 128
    SGS, OMA, 128
    SGS, transport equation, 124, 128
volume-of-fluid, 337

wavenumber
    Batchelor, 142
    Corrsin–Obukhov, 140
    Kolmogorov, 140
well stirred reactor (WSR), 247
Wiener process, 118
wrinkling factor, 108
    Ka dependence, 109
    fractal dimension, 108
    fractal model, 108
    inner cutoff, 108
    SGS, 109

ZND, 281